Illustrated Dictionary of Metalworking and Manufacturing Technology

Steve F. Krar

Editor in Chief

Consultant, Kelmar Associates
Welland, Ontario

McGraw-Hill

New York San Francisco Washington, D.C. Auckland Bogotá Caracas Lisbon London
Madrid Mexico City Milan Montreal New Delhi San Juan Singapore Sydney Tokyo Toronto

Library of Congress Cataloging-in-Publication Data

Illustrated dictionary of metalworking and manufacturing
technology / Steve F. Krar, editor in chief ; Consultant, Kelmar
Associates, Welland, Ontario
p. cm.
Includes index.
ISBN 0-07-038302-2
1. Metal-work—Dictionaries. 2. Manufacturing processes—
Dictionaries, I. Krar, Stephen F. II. Kelmar Associates.
TS204.I36 1999
671'.03—dc21 98-37884
CIP

McGraw-Hill
A Division of The ***McGraw-Hill*** *Companies*

1 2 3 4 5 6 7 8 9 0 DOC/DOC 9 0 3 2 1 0 9 8

ISBN 0-07-038302-2

The sponsoring editor for this book was Linda Ludewig, the editing supervisor was Caroline Levine, and the production supervisor was Sherri Souffrance. Interior design by Jaclyn J. Boone. It was set in CON2 using Garamond Light by Kim Sheran of McGraw-Hill's Professional Book Group Composition Unit, Hightstown, N.J.

Printed and bound by R. R. Donnelley & Sons Company.

CONTENTS

Preface *v*

Acknowledgments *vii*

About the Authors *ix*

SECTION 1 **Computers and the Internet** *Steve C. Yuen* ***1***

Computer Terms *1*

Internet Introduction *27*

Internet Terms *27*

SECTION 2 **CNC Machines and Programming** *Ken (Kun) Li* ***38***

SECTION 3 **Conventional Machine Tools and Accessories** *J. W. Oswald* ***57***

SECTION 4 **Cutting Tools and Holders** *Arthur R. Gill and Steve F. Krar* ***77***

Superabrasives *87*

SECTION 5 **Dies, Jigs and Fixtures, and Molds** *Steve F. Krar* ***99***

Dies *99*

Jigs and Fixtures *107*

Molds *111*

SECTION 6 **Hand Tools and Operations** *J. W. Oswald* ***114***

SECTION 7 **Fluid Power: Hydraulics and Pneumatics** *George C. Ku* ***126***

SECTION 8 **Manufacturing Processes** *Steve F. Krar* ***139***

SECTION 9 **Measurement and Inspection and Quality Control** *Paul J. Warner and Ken (Kun) Li* ***157***

Measurement and Inspection *157*

Quality Control *171*

SECTION 10 **Metallurgy, Heat Treating, and Testing** *J. W. Oswald* ***178***

SECTION 11 **Metalworking Fluids and Coolants** *Steve F. Krar* ***198***

SECTION 12 **Nontraditional Processes** *Paul J. Wanner* ***205***

SECTION 13 Plastics *Steve F. Krar* ***215***

SECTION 14 Lasers and Robotics *Ken (Kun) Li and Steve F. Krar* ***245***

Lasers in Manufacturing *245*

Industrial Robotics *250*

SECTION 15 Welding Processes *Steve F. Krar* ***258***

Common Abbreviations ***279***

Computers *279*

The Internet *280*

Metalworking and Manufacturing *280*

Plastics *282*

Welding *283*

APPENDIX A Geometric Symbols and Definition ***285***

APPENDIX B Tables ***291***

Preface

The tremendous development of new technologies over the past 40 years has a major effect on the way products are manufactured throughout the world. It is now possible to produce higher-quality goods faster, and at lower costs than was possible in the 1950s. The force driving new manufacturing technologies was the development of the computer and the use of numerical control to operate machine tools more accurately and efficiently.

Each new technology brings with it a set of unique terms, some coined specially for a specific technology. A few of the technologies that have a major effect on the rapid advancements, the ability to control machine tools accurately, and to remain competitive throughout the world are listed here.

- *Computers and the Internet.* The development of the computer has quickly spawned technologies such as computer numerical control (CNC), computer-aided design (CAD), computer-assisted manufacturing (CAM), and many others. This has resulted in better machine tools capable of producing quality products at reasonable costs. The Internet has provided the opportunity to communicate with anyone in the world and exchange all types of information quickly.
- *Electrical Machining Processes.* Electrical discharge machining (EDM), electrochemical machining (ECM), and electrolytic grinding have made it possible to accurately cut complex forms in all types of materials. These processes use some form of electric-spark erosion to remove materials.
- *Metrology.* Advancements in the fields of measurement and quality control have produced measuring tools capable of measuring to millionths of an inch. Noncontact in-processing measuring tools, such as the laser, make it possible to measure parts at any stage of production and make adjustments to machine tools as soon as parts near their tolerance levels.
- *Superabrasive technology.* The development of manufactured diamond and cubic boron nitride (CBN) created grinding and machining tools capable of cutting very hard, abrasive materials easily with a minimum of breakdown.
- *Stereolithography.* This technology combines laser, photochemistry, optical scanning, and computer software to produce a solid three-dimensional prototype (model) from a CAD design. It reduces the time to market for new products, reduces development costs, and allows companies to respond quickly to market changes.

To cover the various technologies such as computers, measurement, metal-removal processes, hydraulics, lasers and robotics, plastics, and welding, an author team was selected on the basis of each person's expertise. The authors have researched the material for this handbook with leading manufacturers in the world to include the most up-to-date terms. Naturally, one volume cannot cover all aspects or details of a subject; therefore, the most common terms in each technology were selected. The rapid development of new technologies will continually create a need for new terms and definitions.

It is important that everyone associated with the skilled trades, metalworking, and manufacturing have a ready source of information about the terms that affect their industry and their careers. The *Illustrated Dictionary of Metalworking and Manufacturing Technology* covers terms relating to these areas, from the basic to the most recent developments. Its aim is to present metalworking and related manufacturing technical terms in a clear, concise, and easy-to-understand language. While some technical terms never change and are included in this book, it is important that a source providing explanations of new terms be available. The following format is followed:

- Each term is followed by a brief definition, usually consisting of a sentence or two.
- In many cases, this is followed in point form by a concise explanation of the main points relating to the term or process.
- Selected illustrations are included to further clarify some terms.
- Many terms are cross-referenced to other sections in this handbook where they apply or are related.

Acknowledgments

The authors wish to thank Alice H. Krar for the countless hours spent in typing, proofreading, and checking the many terms that make up this work. Her attention to detail has enhanced a book that is a valuable reference for the student and layperson. A special note of thanks must go to all contributors, whose efforts have resulted in a reference book that covers many areas of communication and manufacturing.

Many thanks to the McGraw-Hill team: Linda Ludewig, Carol Levine, and Sherri Souffrance, and the Hightstown Desktop Publishing Department, including Jackie Boone, Pat Caruso, Kim Sheran, Joanne Morbit, Michele Pridmore, Michele Zito, Paul Scozzari, Paul Cattanach, Charles Burkhour, Steven J. Gellert, and Charles Napa.

The *Illustrated Dictionary of Metalworking and Manufacturing Technology* would not have been possible without the assistance of leading industries and professional organizations who provided valuable assistance, technical resources, and photographs. Most sections of this book were reviewed by industrial and professional experts, many of whose suggestions were incorporated to ensure that the book is as up to date and accurate as possible.

Listed under each section is a list of organizations whose assistance and cooperation are greatly appreciated.

Computers and the Internet

Corel Corporation; Deckel-Maho Inc.; Giddings & Lewis Inc.; Greco Systems; Hewlett Packard Co.; Internet for Everyone (The McGraw-Hill Companies); Modern Machine Shop; Numerical Control Computer Sciences; Prentice Hall Inc.; Rockwell International; Society of Manufacturing Engineers.

CNC Machines/Programming

Allen Bradley; Cincinnati Milacron, Inc.; DMG America; Fadal Engineering Div., Giddings & Lewis Inc.; Hewlett Packard Co.; Kelmar Associates; KPT Kaiser Precision Tooling Inc.; Modern Machine Shop; Sheffield Div., Giddings & Lewis Inc.; Superior Electric Co.

Conventional Machine Tools

Carborundum Co.; Carr Lane Mfg. Co.; Cincinnati Milacron Inc.; Clausing Industries Inc.; Dayton Progress Corp.; DoAll Co.; Dorian Tool International; Everett Industries Inc.; GE Superabrasives; Kelmar Associates; Moore Special Tool Co.; National Broach & Machine Co.; Society of Manufacturing Engineers; South Bend Lathe Inc.; Sunnen Products Co.

Cutting Tools/Holders

Carborundum Co.; Cincinnati Milacron Inc.; Cleveland Twist Drill Co.; DoAll Co.; Dorian International; General Abrasive Co.; GE Superabrasives; Greenfield Industries Inc.; Hertel Carbide Ltd.; Kelmar Associates; Kennametal Inc.; KPT Kaiser Precision Tooling Inc.; Niagara Cutter Inc.; Thriller Inc.; Union Butterfield Corp.

Dies, Jigs, Fixtures

Arntech Publishers; Carr Lane Mfg. Co.; Dayton Progress Corp.; National Tooling & Machining Association; Society of Manufacturing Engineers.

Hand Tools/Operations

Brown & Sharpe Mfg. Co.; Cleveland Twist Drill Co.; Dayton Progress Corp.; Hewlett Packard Co.; Kelmar Associates; L.S. Starrett Co.; Northwestern Tools Inc., Union Butterfield Corp.

Hydraulics/Pneumatics

General Motors Corp.; Kelmar Associates; Mobil Oil Corp.; Sun Oil Co.; Vickers Inc.

Manufacturing Processes

Association for Manufacturing Technology; Bridgeport Machines Inc.; Carboloy Inc.; Cincinnati Milacron Inc.; Dayton Progress Corp.; Deere & Company Mfg.; DoAll Co.; Giddings & Lewis Inc.; Ingersoll Milling Machine Co.; Makino Inc.; KPT Kaiser Precision Tooling Inc.; Society of Manufacturing Engineers.

Measuring & Inspection/Quality Control

Brown & Sharpe Mfg. Co.; DoAll Co.; Engineering Drawing & Design—Glencoe/McGraw-Hill; Fadal Engineering Div., Giddings & Lewis Inc.; Juran Institute; Kelmar Associates; LaserMike Div., Techmet Co.; Powertrain Div., General Motors Corp.; Praxair Inc.; Quality Assurance—Glencoe/McGraw-Hill; Sheffield Div., Giddings & Lewis Inc.; L.S. Starrett Co.; Superior Electric Co.; Taft-Peirce Mfg. Co.

Metallurgy, Heat Treating, Testing

American Iron & Steel Institute; Carboloy Inc.; GE Superabrasives; Kelmar Associates; Society of Manufacturing Engineers; U.S. Steel Corp.; Wilson Instrument Div., American Chain & Cable Co.

Metalcutting Fluids/Coolants

Cincinnati Milacron Inc.; DoAll Co.; Exair Corp.; Master Chemical Corp.; Modern Machine Shop.

Nontraditional Machining Processes

Cinncinati Milacron Inc.; Gerard Industries Pty. Ltd.; Giddings and Lewis Inc.; Kelmar Associates; Modern Machine Shop; QQC Inc., Div. Turchan Industries; Society of Manufacturing Engineers; 3-Dimensional Services; 3D Systems; Vaga Industries.

Plastics

DuPont Co.; Eastman Chemical Products Inc.; Firestone Plastics Co.; Hartig Plastic Machinery; HPM Div., Koehring Co.; Metco Inc.; Michigan Panelyte Molded Plastics Inc.; Owens-Corning Fiberglass Co.; Polytech Ltd.; Quantum Chemical Co.; Rohm & Haas Co.; Society of Manufacturing Engineers; The Society of the Plastics Industries Inc.; U.S. Industrial Chemicals Inc.

Lasers/Robotics

ABB Robotics Inc.; Association for Manufacturing Technology; Bausch & Lomb; Cincinnati Milacron Inc.; Giddings & Lewis Inc.; Kelmar Associates; LaserMike Div., Techmet Co.; Manufacturing Engineering Magazine; Modern Machine Shop; Robomatic International Inc.

Welding Processes

American Welding Society; Emco Maier Corp.; Hobart Brothers Co.; Lincoln Electric Co.

About the Authors

Arthur R. Gill served an apprenticeship as a tool and die maker. After 10 years of working at the trade, he entered the Ontario Community College system as a professor and coordinator of precision metal trades and apprenticeship training. During his 28 years at Niagara College in St. Catharines, he has been a member of the Ontario Precision Metal Trades college curriculum committee for apprenticeship training and head of Apprenticeship for Ontario. Art Gill has coauthored the textbook *CNC Technology and Programming* with Steve Krar. In 1991 he was invited by the People's Republic of China to assist in developing a Precision Machining and Computer Numerical Control (CNC) training facility at Yueyang University in Hunan Province.

Steve F. Krar spent 15 years in the machine tool trade and later graduated from the Ontario College of Education, University of Toronto, with a Type A Specialist's Certificate in Technical Education. After 20 years of teaching, he devoted full time to researching and coauthoring over 50 books on machine tools and manufacturing technology. The text *Technology of Machine Tools*, now in its fifth edition, is recognized as one of the leading texts in the world on the subject; it has been translated into four languages. During his years of research, he has studied under Dr. W. Edwards Deming, and has been associated with GE Superabrasives, and countless other leading machine tool manufacturers. He was invited twice to China to teach and share his knowledge about modern machining and manufacturing technology. Steve Krar, the former Associate Director of the GE Superabrasives Partnership for Manufacturing Productivity, is a life member of the Society of Manufacturing Engineers.

George C. Ku is a professor of Technology and Vocational–Technical Education at Central Connecticut State University, New Britain, Connecticut. He has taught material and manufacturing technology courses at CCSU for the last 26 years. George Ku holds B.S. and M.S. degrees in Industrial Technology from Southern Illinois University and an Ed.D. in Industrial and Technical Education from Utah State University. Prior to his current assignment, he taught industrial education subjects at LaSalle High School, South Bend, Indiana, and Logan High School, Logan, Utah. He also worked as a mechanic, machinist, and welder in modern industry. Dr. Ku has been invited to China and Taiwan to teach and share his knowledge about modern manufacturing technology and technical education programs in the United States. He has published a number of articles in professional journals and his publications range from machine tool operations to international programs.

Ken (Kun) Li holds B.S. and M.S. degrees in Manufacturing Engineering from Hunan University, China, and M.S. degrees in Industrial Engineering and Engineering Science from Louisiana State University. He has coauthored over 10 peer-reviewed technical papers on ceramic grinding, CNC/FMS, and cellular manufacturing for international journals and conferences. Ken Li is a member of the Industrial Diamond Association's Partnership for Manufacturing Productivity and has considerable expertise in the use of superabrasive grinding and machining tools. He is a member of the Society of Manufacturing Engineers, a Certified Manufacturing Technologist, and a Registered Engineer–Intern. Mr. Li is currently employed as a manufacturing engineer in aerospace industry.

J. William Oswald served an apprenticeship in machine shop; after 16 years in the trade he attended the Ontario College at the University of Toronto. After graduation, he received a Type A Specialist's Certificate in Machine Shop Practice and taught machine shop work for 25 years. During this time, he attended many upgrading courses in the operation of the latest machine shop and testing equipment offered by leading machine tool manufacturers. For several years, Mr. Oswald served on the teacher-training staffs at the University of Toronto and the University of Western Ontario. He has also served on many technical educational committee and organizations. Bill Oswald is a coauthor of the following publications: *Technology of Machine Tools, Machine Shop Training, Machine Tool Operations, Machine Shop Operations*, and of the following transparency kits: *Machine Tools, Measurement and Layout, Threads and Testing Equipment*, and *Cutting Tools.*

Paul J. Wanner has an 18-year history in the machining industry. During his 8-year apprenticeship in the job shop environment, he began working in various machine shops to broaden his skills while continuing his education through night school in injection mold design and mechanical engineering. He has worked in the primary metals industry, small parts manufacturing, and aerospace industries. Paul Wanner is a faculty member of Clackamas Community College/Advanced Technology Center in Wilsonville, Oregon. His areas of expertise are print reading, dimensional inspection. GDT, CAD/CAM, CNC, and CMM. He also teaches statistical process control, metrology, and calibration. In 1997 Paul advised a team of students for the National Skills Championships sponsored by Vocational and Industrial Clubs of America (VICA), and they won first place in the automated manufacturing competition. Paul Wanner is certified as a Mechanical Inspector and Quality Technician through the American Society for Quality (ASQ). He is also a Certified Manufacturing Technologist through the Society of Manufacturing Engineers (SME). Paul is a senior member of SME and received the Educator of the Year award in 1997 from his local chapter.

Dr. Steve C. Yuen is Professor of Technology Education at The University of Southern Mississippi. He holds a Ph.D. in Vocational–Industrial Education from Pennsylvania State University. He was a Fulbright Scholar to Taiwan in 1992-1993. Steven Yuen has published extensively in international and national journals, written grant proposals, conducted seminars and workshops, and presented numerous papers at national and international conferences. He has taught over 30 different graduate and undergraduate courses on a wide variety of technically advanced subjects. He has developed several software packages, numerous Web pages, and other innovative Web-based resources. Dr. Yuen serves on the Editorial Board of the *International Journal of Vocational Education and Training* and is a member of the Editorial Advisory Board of Occupational Education Forum. Steve Yuen is an active member of many honorary societies and professional organizations, including the American Technical Education Association, the Association for Educational Communications and Technology, the Fulbright Association, the International Society for Technology in Education, and the Society of Manufacturing Engineers.

SECTION 1

Computers and the Internet

Dr. Steve C. Yuen
Professor of Technology Education
University of Southern Mississippi
Hattiesburg, Mississippi

The use of computers, to control machine tools and manufacturing processes, has progressed from a humble beginning in the late 1950s and early 1960s to assume a major role by the late 1980s and early 1990s. It is the key to producing quality products and keeping U.S. manufacturing processes competitive with industries throughout the world. The computer today plays an essential role in everyday life, and will ensure that the present standard of living is maintained.

Computer Terms

absolute address Refers to a fixed location in the computer's memory.

□ In a spreadsheet program, an absolute address is a cell address that refers to a fixed location that does not change when the formula is copied or moved. An example is A3.

accelerated video adapter A type of video card that uses graphics coprocessor chips to perform video functions at very high speed.

access To get or retrieve data from a computer system.

access mechanism A device that moves the read/write heads of a storage disk over a particular track.

access time The average time, in milliseconds, required for a computer to locate data on the storage medium and transfer it into main memory.

accumulator A special-purpose storage location or register in the central processing unit (CPU) where a computer stores the results of arithmetic or logical operations.

acoustic coupler A type of modem equipped with cups to fit over a standard telephone handset that sends and receives its signals directly through the mouthpiece and earpiece of the phone.

Ada A high-level programming language developed by the U.S. Department of Defense in the late 1970s.

□ Based on Pascal, Ada is a sophisticated structured language that supports concurrent processing.

add-in card A circuit board or card that is inserted in the motherboard to enable the computer system to work with a hardware component. *See also* EXPANSION BOARD.

address Refers to a unique location in computer memory.

algorithm A prescribed set of well-defined rules for the solution of a problem in a finite number of steps.

alpha version The first version of testing of a new software product, usually carried out by the manufacturer's own staff.

alphanumeric Computer data consisting of letters, numbers, or other special characters such as punctuation marks and symbols. *(See figure.)*

alphanumeric. *(Courtesy Deckel-Maho Inc.)*

ALU *See* ARITHMETIC/LOGIC UNIT.

American National Standards Institute (ANSI) ANSI code is an 8-bit code used to represent letters, numerals, and special characters. It is a popular alternative to ASCII.

American Standard Code for Information Interchange (ASCII) A code representing characters in numeric form.

□ The ASCII code for the letter *a* is 01000001.

□ The ASCII file can be displayed on any computer.

analog computer A computer that processes constantly varying data rather than discrete values. For contrast, *see* DIGITAL COMPUTER.

analog device A machine that operates on continuously varying data.

analog transmission A form of data transmission over communication channels in a continuous waveform.

ANSI *See* AMERICAN NATIONAL STANDARDS INSTITUTE.

antivirus software A program used to detect and remove computer viruses.

APL *See* A PROGRAMMING LANGUAGE.

Apple Computer A computer company founded by Steve Jobs and Steve Wozniak at Dartmouth in 1976. Apple is best known for developing the Apple II and Macintosh computers.

AppleTalk A network protocol developed by Apple Computer for communication between Apple Computer products and other computers.

application programmer A person responsible for writing and maintaining computer application programs. The job duties involve testing, debugging, documenting, and implementing programs.

application software Software designed to accomplish a specific task on a computer system, such as word processing, database, spreadsheet, graphics, computer-aided design, or communication software.

A Programming Language (APL) A high-level programming language designed to handle complex operations on arrays that requires a specially designed terminal with Greek letters and some other special symbols.

archive To store infrequently used files on floppy disks or tape.

argument A value provided to a function.

□ For example, in Beginner's All-Purpose Instruction Code (BASIC) the string Ready is an argument to the PRINT statement: PRINT Ready.

arithmetic/logic unit (ALU) The part of the CPU in which arithmetic and logical operations are performed.

artificial intelligence (AI) A branch of computer science that deals with the use of computers to simulate human thinking. *(See figure.)*

ASCII *See* AMERICAN STANDARD CODE FOR INFORMATION INTERCHANGE.

assembler A program that translates assembly language into machine language.

assembly language A low-level programming language that uses convenient abbreviations called *mnemonics* rather than the binary commands of machine language.

Association for Computing Machinery (ACM) The oldest and largest educational and scientific society in the computing industry. Intended to advance the science of information processing, promote the exchange of ideas, and develop and maintain the integrity and competence of individuals in the computing field.

Asynchronous Transfer Mode (ATM) A form of high-speed digitized data transmission based on fixed-length cells that can carry voice, video, and data at 100 Mbits/s.

Asynchronous Transmission A transmission protocol that uses start and stop bits to mark the beginning and ending of each individual character (byte).

Atanasoff-Berry Computer (ABC) The first electronic digital computer developed by John Vincent Atanasoff and Clifford Berry at Iowa State University in 1939.

ATM *See* ASYNCHRONOUS TRANSFER MODE.

attribute A characteristic field that provides essential information about an entity for a database file.

authoring language A high-level language that allows a user to create computer-based, multimedia programs without having much knowledge of a computer language.

autoexec.bat A Microsoft Disk Operating System (MS-DOS) startup file that contains commands to be executed whenever the computer boots up.

auxiliary storage Also called *external storage* or *secondary storage.* A storage device that supplements the main memory storage of a computer.

□ Hard disks and tape drives are examples of auxiliary storage devices.

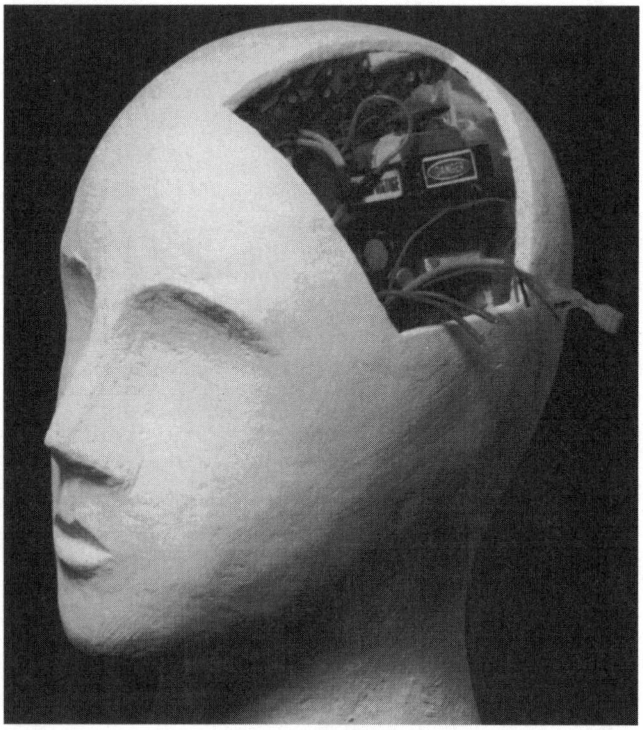

artificial intelligence.
(Courtesy Society of Manufacturing Engineers.)

□ Auxiliary storage is unlimited and nonvolatile.

backup A duplicate copy of a file kept on another type of storage medium, such as floppy disks or tapes, in case the original is damaged or destroyed.

backward compatibility The ability to handle hardware and software designed for earlier models or versions of computers and operating systems.

bad sector A defective area on a floppy disk or hard disk that the operating system marks as unusable.

band printer A line printer that uses a rotating band containing characters rotating horizontally past the print position, where they are pressed by hammers against a carbon ribbon to produce a line of print on paper.

bar code A pattern of wide and narrow bars imprinted on a label that can be read with a scanning device.

□ The most common bar code is the Universal Product Code used in labeling retail products. *(See figure.)*

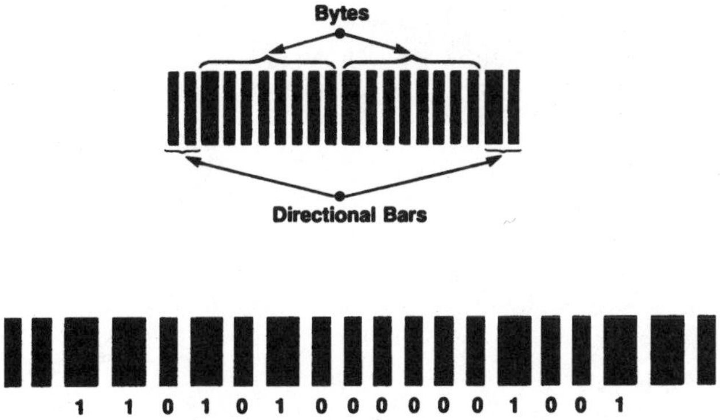

bar code. *(Courtesy Hewlett-Packard Co.)*

□ A device used to read a bar code by means of reflected light is called a *bar-code reader.*

BASIC Beginner's All-Purpose Symbolic Instruction Code. A high-level programming language developed by John Kemeny and Thomas Kurtz in the 1960s.

□ Considered to be the easiest computer language to learn; it is used in different versions by almost all microcomputers.

basic input/output system (BIOS) A part of the operating system that controls input and output.

□ On IBM PC–compatible machines, BIOS is a program stored on a ROM chip inside the computer and is loaded before the operating system. BIOS controls how the CPU interacts with input/output devices.

batch file A file whose name ends with .BAT, and which contains a list of DOS commands to be executed in sequence.

batch processing Computer programs or data are fed into the computer and executed in batches, with little or no interaction between the computer and the user.

baud A measure of modem speed. One baud can be regarded as one bit per second.

benchmark testing A test to measure the performance characteristics of a computer or a piece of software.

Bernoulli box A kind of removable media drive that holds high-capacity disk cartridges.

beta version The prefinal version of a software product, which is released to a large group of users for testing.

bidirectional Refers to a type of parallel port that can both send and receive data.

binary A numbering system represented in base 2, consisting of the digits 0 and 1. *(See figure.)*

ARABIC	BINARY	POWERS OF 2
0	0	
1	2	2^0
2	10	2^1
3	11	
4	100	2^2
5	101	
6	110	
7	111	
8	1000	2^3
9	1001	
10	1010	
11	1011	
12	1100	
13	1101	
14	1110	
15	1111	
16	10000	2^4
17	10001	
18	10010	
19	10011	
20	10100	
21	10101	
22	10110	
23	10111	
24	11000	
25	11001	
26	11010	
27	11011	
28	11100	
29	11101	
30	11110	
31	11111	
32	100000	2^5
64	1000000	2^6
128	10000000	2^7

binary. *(Courtesy Modern Machine Shop.)*

□ A binary digit is represented either by the digit 1 when electric current is transmitted or 0 when no electricity may be transmitted.

□ Computers store data internally as binary numbers.

binary-coded decimal (BCD) A system for representing decimal numbers and characters in binary.

BINHEX *Bin*ary *hex*adecimal A method for converting nontext (non-ASCII) files into ASCII.

BIOS *See* BASIC INPUT/OUTPUT SYSTEM.

bit Acronym for *bi*nary digi*t*; the smallest unit of digital information consisting of the number 0 or 1.

bitmapped The representation of a character or graphic as a series of dots or pixels (picture elements).

block A group of data records stored on magnetic tape or punch tape.

□ Each block is treated as a unit and is separated from other blocks by an end-of-block (EB) character.

BNC connector Standard connector used to connect coaxial cable to the T-shaped connector on a network interface card.

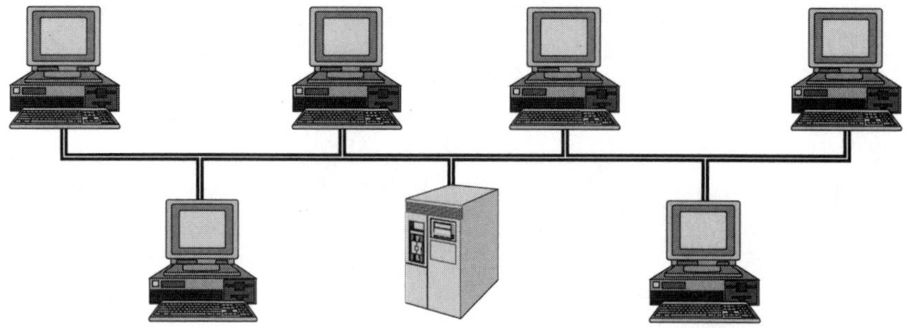
bus network. *(Courtesy Greco Systems.)*

board Also known as *printed-circuit board.* Many microcomputers contain expansion slots where you can add additional boards or cards to enhance the capability of the machine. *See also* EXPANSION BOARD.

boldface A darker version of the regular type formed from heavier strokes.

boolean An expression that evaluates to the logical value of true or false.

boot Short for *bootstrap,* which starts up the computer and loads the operating system of a computer into main memory.

□ Cold start or cold boot is the operation of booting a computer that has been completely shut down.

□ Warm start or warm boot is the operation of restarting an already running computer by pressing the Reset button or simultaneously pressing Ctrl, Alt, and Del keys.

bootable disk Also called *boot disk.* A floppy disk that contains the files needed to load the operating system into memory to start up a computer.

boot record A hidden file that contains information about the version of the operating system and the physical characteristics of the disk.

branch A statement used to alter its direction and branch to a new section of the program.

□ For example, a GOTO statement in a BASIC program is a branch statement.

bubble memory A memory medium where data is represented by magnetized spots or bubbles resting on a thin film of garnet wafer.

□ Bubble memories are more expensive than disk and are used in very lightweight, notebook computers.

buffer A temporary memory that is capable of storing incoming data for later transmission.

□ Often used in networking to compensate for differences in processing speed between network devices.

□ Also found on printers to allow the printer to receive information faster than it prints it.

bug An error in software that causes the software to malfunction.

bus An electronic pathway that transmits information in the form of electrical signals between CPU components.

□ The two most common types of bus are the data bus and the address bus.

bus network A network configuration in which all computers and peripheral devices are connected to a common cable called a *bus.* All communications travel the length of the bus cable. *(See figure.)*

button bar A panel containing buttons in a graphical user interface (GUI) software application.

byte The amount of memory space needed to store one character of information, which is normally 8 binary digits or bits.

C A powerful programming language developed at Bell Laboratories in the 1970s, based on two earlier languages: B and BCPL.

□ C is popular because it is highly efficient and portable and is implemented on various computer systems.

cache A place in the computer for storing data not yet ready to be processed.

□ A disk cache is a reserved section of random-access memory (RAM) that stores copies of frequently used disk sectors in RAM so that they can be read without accessing the disk.

□ A memory cache is a high-speed buffer between the CPU and memory.

CAD *See* COMPUTER-AIDED DESIGN.

CAD/CAM *See* COMPUTER-AIDED DESIGN/COMPUTER-AIDED MANUFACTURING.

CAE *See* COMPUTER-AIDED ENGINEERING.

CAI *See* COMPUTER-ASSISTED INSTRUCTION.

CAM Computer-aided manufacturing.

camera-ready copy High-resolution output that is ready to be photographed and offset-printed without further modification.

canned program A prewritten program available for purchase.

CAPP *See* COMPUTER-AIDED PROCESSING PLANNING.

card reader A machine that reads punched cards into a computer.

cascade A way of organizing information on a computer screen so that each open window is arranged in an overlapping pattern to the right and below the previous window.

CASE *See* COMPUTER-AIDED SOFTWARE ENGINEERING.

cathode-ray tube (CRT) Commonly known as a *monitor* or *video display unit.* Similar to a TV set, the

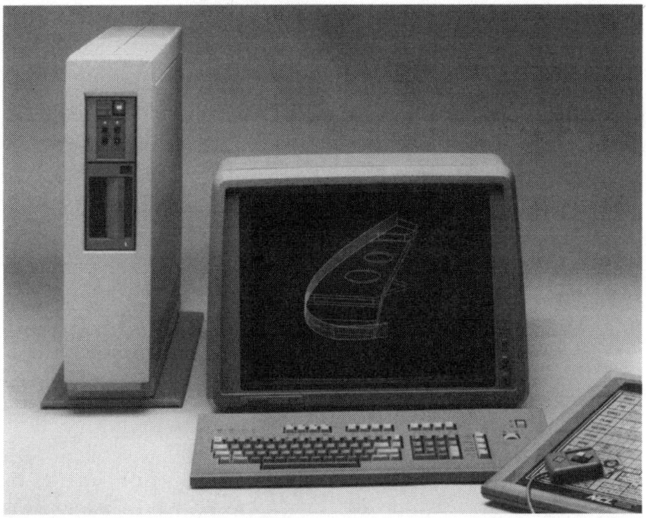

cathode-ray tube.
(Courtesy Numerical Control Computer Sciences.)

CRT is an output device containing an electron gun to emit a beam of electrons, illuminating phosphors on the screen to display images. *(See figure.)*

CD-ROM *See* COMPACT DISK–READ-ONLY MEMORY.

CD-I *See* COMPACT DISK–INTERACTIVE.

cell The intersection of a row and column in a spreadsheet in which data, labels, and formulas are entered.

cell address The location of a cell in a spreadsheet; for example, cell C7 is found at the intersection of column C and row 7.

central processing unit (CPU) The brain of the computer; composed of the arithmetic/logic unit (ALU) and the control unit.

- □ The CPU controls the operation of the computer.
- □ CPUs are often designated by a number, such as Intel series 8088, 8086, 80286, 80386, and 80486 for IBM-compatible computers and Motorola 68000, 68020, 68030, 68040, 68060, 601, 603, and 604 for Apple Macintosh computers.

Centronics Interface A standard format for parallel data transmission to and from microcomputer equipment, especially printers for IBM-compatible computers.

CGA *See* COLOR GRAPHICS ADAPTER.

chain printer A line printer in which the character set is mounted on a continuous metal chain, horizontally rotating past all print positions. A hammer presses an inked ribbon, forming the correct character on paper.

channel A communication path. The term *channel* is also used to describe the specific path between large computers and attached peripherals.

character An instance of a numeral, letter, or other linguistic, mathematical, or logical symbol. Characters are represented by bytes in computer storage.

character printer A form of impact printer that prints one character at a time.

character recognition A form of software that scans handwritten or printed characters and accepts them as input.

characters per second (CPS) A unit for measuring printer speed. The number of characters printed in one second. (The abbreviation *cps* is used also for cycles per second.)

checksum A technique often used by antivirus software in which a value is calculated by combining all the bytes in a file each time a program is run and comparing the result to the original sum of the bytes.

chip A very small wafer of silicon that contains hundreds of thousands of transistors and circuits to make up the entire central processing unit of a computer. *(See figure.)*

computer chip.

CIM *See* COMPUTER-INTEGRATED MANUFACTURING.

CISC *See* COMPLEX INSTRUCTION SET COMPUTER.

click To point the cursor at an on-screen object, and then press and release the mouse button. *See also* DOUBLE-CLICK.

client A node or software program that requests services from a server.

client/server network A type of network in which a central computer (server) provides file, printer, and communication support to other computers on the network.

clip art Collections of artwork created by professional artists that may be free or purchased and incorporated into your own documents.

clipboard An electronic holding place for the most recent cut or copy made from a document. Whatever is on the clipboard can be pasted into the current document.

clock speed The cycles per second that a microprocessor processes, measured in megahertz (MHz).

- □ A computer with a faster clock speed can perform more operations per second.

clone A computer that is an exact imitation of another and performs the same functions.

cluster A group of disk sectors; it is the smallest storage unit the computer can access.

CMOS *See* COMPLEMENTARY METAL OXIDE SEMICONDUCTOR.

CNC *See* COMPUTER NUMERICAL CONTROL.

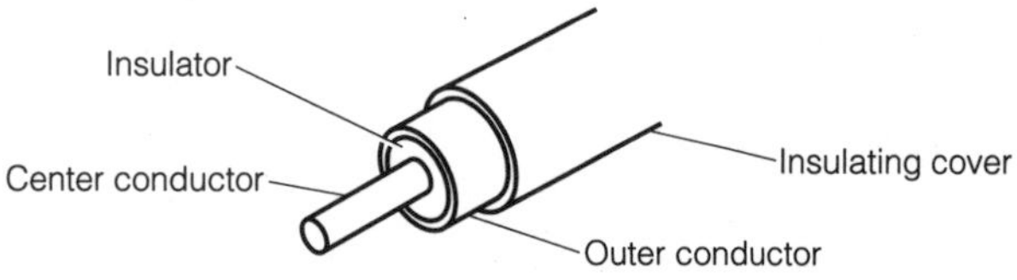

coaxial cable. *(Courtesy Kelmar Associates.)*

coaxial cable A communication cable consisting of a copper wire conductor surrounded by insulation and conductive shield. *(See figure.)*

COBOL Common Business-Oriented Language. A high-level programming language that is commonly used in business applications.

code The list of instructions comprising the computer program.

coding The process of writing an algorithm or other problem-solving procedures in a programming language.

cold site An empty alternate space for a user to install a computer system in case the main installation is inoperative due to a disaster.

cold start *See* BOOT.

Color Graphics Adapter (CGA) The original IBM standard for color adapters that display graphics in a matrix of 320 pixels horizontally and 200 pixels vertically (320×200 resolution).

command An instruction that tells the computer what to do.

COMMAND.COM An MS-DOS operating system file responsible for interpreting commands entered at the command prompt and for locating and loading an application or utility.

command-line interface A type of user interface where the user types commands in order to interact with the operating system and specify that it perform a certain task.

common carrier A company that provides communication services for the public.

communication The process of transmitting information from one computer to another, usually over telephone lines or direct-cable connections.

communication channel The communication link used to carry data between devices.

communication control software Software that monitors and controls the transmission of data among many computer systems.

communication networks Electronic links consisting of hardware, software, and communication lines that provide data transmission between one or more computer systems.

communications protocol The rules for sending and receiving transmissions in a computer network to ensure the orderly and accurate transmission and reception of data.

communications software Application software that enables your computer to communicate with other computers.

compact disk–interactive (CD-I) A combination of CD-ROM and computer technology designed for interactive presentation of multimedia information.

compact disk–read-only memory (CD-ROM) A computer storage medium that uses the same technology as musical compact disks to store data on a disk as a series of pits that are read using a laser beam.

□ A single CD-ROM can store up to 680 Mbytes of text, pictures, sound, and video.

compatibility The ability to use peripheral devices or software produced by one manufacturer on a computer produced by another manufacturer.

□ *Hardware* compatibility refers to the ability of peripheral devices to work together.

□ *Software* compatibility refers to the ability to run the same software on a variety of computers.

compiler A language translator that converts a program written in a high-level language such as FORTRAN or COBOL to assembly language.

complementary metal oxide semiconductor (CMOS) A special kind of memory chip that stores vital information about a computer system configuration even when the computer power is turned off.

Complex Instruction Set Computer (CISC) A CPU that processes longer and larger instructions such as Intel's 486, Motorola's 6040, and their predecessors.

com port A communication, or serial, port in a computer system.

composite video A single video signal encoding color (red, green, and blue) data.

compression The process of shrinking files to free up more disk space.

computer An electronic device that manipulates and processes data according to specified instructions or commands. *(See figure.)*

computer. *(Courtesy Hewlett-Packard Co.)*

□ Basic computers consist of a central processing unit (CPU), input/output devices, and memory.

computer-aided design (CAD) The application of computers to assist or enhance the design process.

computer-aided design/computer-aided manufacturing (CAD/CAM) The application of computers to improve productivity in design and manufacturing.

computer-aided engineering (CAE) The application of computers, special software, and various peripherals to accomplish such engineering tasks as analysis and modeling.

computer-aided manufacturing (CAM) In general, the use of computers to assist in any or all phases of manufacturing.

computer-aided processing planning (CAPP) The application of computers to develop detailed plans for manufacturing a part or assembly.

computer-aided software engineering (CASE) A software package that helps programmers write programs.

computer architecture The design, construction, or structure of a computer system.

computer-assisted instruction (CAI) The use of computers to assist in instructional activities such as drill and practice, tutorial, demonstration, simulation, problem-solving, and instructional games.

computer axial tomography (CAT) A scanning technique used to produce a cross-sectional x-ray of the human body.

computer crime An illegal act involving the use of a computer.

computer ethics The moral conduct and behavior in computer use.

computer-integrated manufacturing (CIM) The total integration of such individual concepts as CAD, CNC, robotics, and materials handling into one large automated system.

computer literacy Knowing how to use a computer for one's own applications and recognizing other potential uses for the computer.

computer network A collection of computers and peripheral devices that use communication channels to share hardware, software, and data.

computer numerical control (CNC) Using a computer and programmed instructions to control manufacturing machines and systems.

computer operator Individual responsible for running and monitoring the computer equipment.

computer phobia Fear of computers and their effects on society.

computer security The protection of hardware, software, and peripherals against the hazards to which computer systems are exposed and to control access to information.

computer virus A program designed to infect and disrupt a computer system by attaching itself to other software and spreading from one computer to another via networks or exchange disks.

concentrator A device that provides a central connecting point for the cables on a network.

concurrent user license A software license that allows a certain number of copies of a software program to be used at the same time.

CONFIG.SYS A special, MS-DOS startup file loaded into memory during the boot process that configures your computer system by loading device drivers and specifying the number of open files.

control key A special function key found on most computer keyboards that allows the user to perform specialized operations.

controller An expansion board that connects hard drives and floppy-disk drives to a computer system.

Control Program for Microcomputers (CP/M) An operating system developed by Digital Research for microcomputers.

control unit The section of the CPU that directs and coordinates the sequence of computer operations.

conventional memory The first 640K of memory.

coprocessor Also called *math coprocessor*; works in conjunction with the main microprocessor to increase the speed of mathematical calculations and processing of graphics.

copy-protected Refers to software that includes safeguards to prevent unauthorized duplication.

corrupted file A program or file whose contents have been altered as the result of a hardware, software, or power failure.

courseware Educational software designed for in-school use.

CP/M *See* CONTROL PROGRAM FOR MICROCOMPUTERS.

cracker A person who gains unauthorized access to other computer systems in order to perpetrate illegal activity in them. *See also* HACKER.

crash A system shutdown caused by hardware or software malfunction.

crash conversion A method of system implementation in which the old system is abandoned and the new one implemented at once.

cursor The prompting symbol usually displayed as a blinking marker that shows you where the character you type will appear.

cursor-control keys Keys that allow the user to move the cursor around the screen with the keyboard. These keys include the arrow keys, tab key, home key, or keys incorporated in the numeric keypad.

cylinders A conceptual slice of a disk made up of a vertical stack of tracks on a hard disk.

daisy wheel printer An impact printer that prints letter-quality characters with a removable print wheel. *(See figure.)*

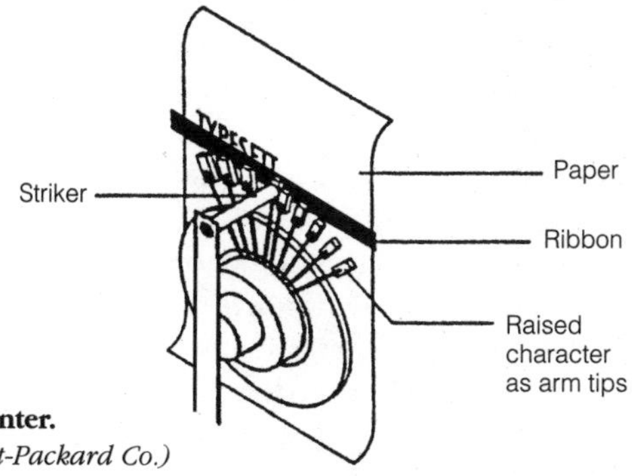

daisy wheel printer.
(Courtesy Hewlett-Packard Co.)

data Basic element of information that can be processed by a computer.

database A collection of related information that can be manipulated by a computer program.

database management software A productivity software that manages the collection, storage, access, and organization of data.

database management system (DBMS) A computer program that allows a user to enter, update, sort, query, and retrieve information in a computer database.

data bus The electronic circuitry used to send and receive data from other hardware components.

data communication The electronic transmission of data from one computer to another.

data compression The use of logical and mathematical methods to reduce the amount of storage space that software or data occupies.

data encryption The process of translating data into secret code as a security measure.

data flow diagram An illustration showing how data moves through the system.

data integrity The accuracy of the data in a database management system.

data packet A transmission packet that consists of a start bit, the character to be transmitted, a parity bit, and a stop bit.

data processing Also known as *electronic data processing* (EDP). Manipulation of large amounts of data by a computer.

Data Processing Management Association (DPMA) An international organization for computer professionals.

data security The protection of data from accidental or malicious destruction, disclosure, or modification.

data structure A system that defines the location of and the relationship between the data elements.

debugging The process of finding and correcting errors in a program.

decimal number A number expressed in ordinary base-10 notation.

decision support system (DSS) A computer information system designed to help managers make unstructured and semistructured decisions.

DECnet A proprietary network protocol developed by Digital Equipment Corporation.

decoding The reversal of a coding process in which data is converted by reversing some previous coding.

decryption The process of decoding encrypted data and restoring it to its original form.

default The factory settings, unless different instructions are given by the user.

default drive The drive on which external files are assumed to reside, and the drive to which files will be saved, unless otherwise specified.

defragmentation utility A program that rearranges files on a disk so that all sectors of each file are stored in consecutive sectors and empty spaces are removed between files.

delete A DOS, word-processing, or spreadsheet command to delete files, text and graphics, or cells, respectively.

demodulation A process of converting an analog signal to a digital signal.

desktop The background screen of a graphical operating environment, which serves as a graphics-based work area.

desktop model A computer chassis that is designed to lie flat on a desktop.

desktop organizers Pop-up utilities on a microcomputer such as phone dialer, calculator, card files, and notepads.

desktop publishing (DTP) A computer application that provides an electronic pasteboard for layout of text, graphics, and illustrations to produce professional-quality publications. *(See figure.)*

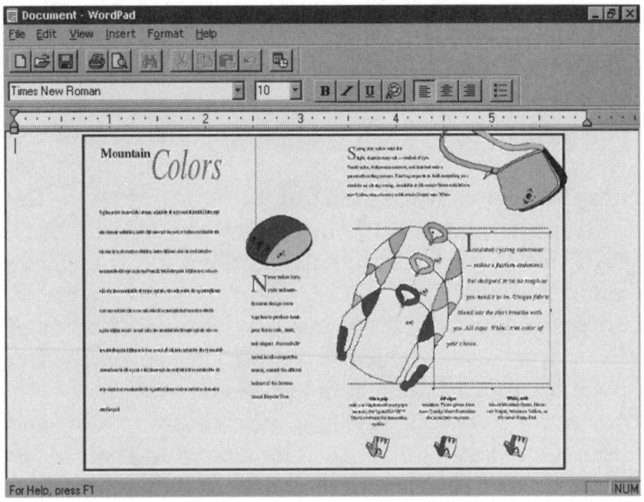

desktop publishing. *(Courtesy Corel Corp.)*

device A component within a computer system, such as mouse, keyboard, disk drive, monitor, or printer.

device address A unique number assigned to each peripheral device in a larger computer system.

device driver A program that extends the operating system in order to support a specific device.

dialog box A small window that appears on the screen to prompt the user to select options, or display useful information, warnings, and error messages.

digital Data stored in discrete units such as 0s and 1s.

digital audio Sound encoded as digital information rather than the analog signal used in conventional sound recordings.

digital computer A type of computer that represents information in discrete form. All modern general-purpose computers are digital computers.

digital device A device that works with discrete numbers such as 0 and 1.

digital image Any image that has been digitized on a digitizer.

digital signal processor (DSP) A special kind of processor that converts analog signals such as audio and video to digital signals.

digital transmission The transmission of data as distinct **ON** and **OFF** pulses.

digital video The representation and storage of video information by numbers or digits.

digital video–interactive (DVI) Optical storage medium that delivers full-motion, full-screen video, three-dimensional graphics, and high-quality audio capabilities.

digitize To convert analog input, such as photographs or sound, into digital values.

digitizer An input device (e.g., a scanner) that converts visual data into digital information.

direct-access storage device (DASD) Secondary storage device that allows data to be stored and accessed. Also known as *magnetic disks.*

directed memory access (DMA) An expansion board or other device that accesses system memory directly, without involving the CPU.

directory A listing of files on a disk, usually containing the filename, the file extension, the date and time the file was created, and the file size for every file in the directory.

disaster recovery plan A plan of action designed to return a computer system to full working order in case a disaster destroys the system.

disk, diskette An interchangeable data storage medium either $3\,^1/_2$ or $5\,^1/_4$ in in size.

disk access time The average amount of time required for a disk drive to move into position to read from or write to the disk, measured in milliseconds.

disk address The method used to identify a data record on a disk; consists of the disk surface number, track number, and record number.

disk cache *See* CACHE.

disk crash Contact between the read/write head and the hard-disk surface that results in loss of data.

disk drive A peripheral device that stores data onto or retrieves data from storage media, such as a hard disk or diskette.

diskless workstation A networked computer that has its own processing components, keyboard, and monitor, but lacks any storage device of its own.

disk map A visual representation of the organization of clusters on a floppy disk or hard disk.

Disk Operating System (DOS) A set of programs that allow interaction between the user, the computer, and peripheral devices.

disk optimization Reorganization of a disk so that the operating system can efficiently and quickly locate files and directories.

disk pack A stack of magnetic disks that are generally used for supercomputer, mainframe, and minicomputer systems.

disk server A high-capacity hard disk in the file server that stores programs and data for other computers on the network.

display terminal Also known as *display screen.* A peripheral that allows for the visual output of information for the computer.

distributed data processing (DDP) A system in which various users possess and control their own computer hardware and software, instead of using a central computer.

DMA *See* DIRECTED MEMORY ACCESS.

document A file or structural unit of text that can be stored, retrieved, and exchanged among systems and users as a separate unit.

documentation The materials that explain the proper use and possible applications of a given piece of software or hardware. Typically a printed manual.

DOS *See* DISK OPERATING SYSTEM.

dot matrix printer A type of impact printer that prints by using a matrix of dots to produce each character. *(See figure.)*

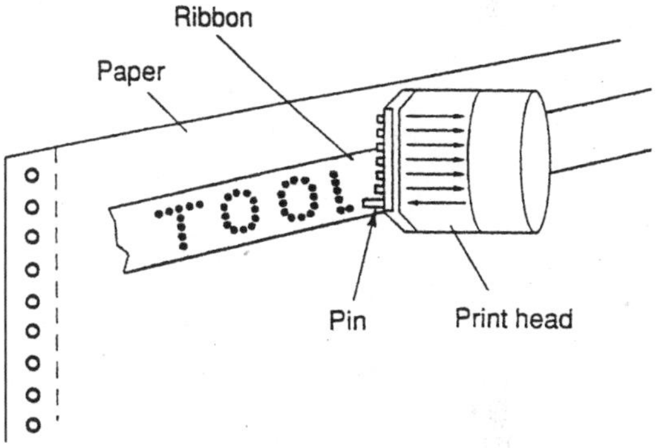

dot matrix printer. (*Courtesy Hewlett-Packard Co.*)

□ The quality of print depends on the number of pins (9 to 24) used.

□ This type of printer is generally speedy and inexpensive.

dots per inch A gauge of visual clarity, both on printouts and on the screen, measured by the number of dots per linear inch.

dot pitch The distance between the three colored dots that make up a single pixel on a color monitor.

□ Smaller dot pitches yield sharper images and higher resolution.

double-click A mouse technique of pressing the mouse button twice very rapidly.

double density A $3\,^1/_2$-in floppy disk with a storage capacity of 720K, or a $5\,^1/_4$-in floppy disk with a storage capacity of 360K.

double-side disk A disk that stores data on both of its sides.

download Transmitting a file or program from a host computer to your own computer.

downtime Time during which the computer is not available or is not working.

drag A mouse technique of holding down the mouse button on an object while moving the pointer on the screen.

drag and drop A mouse technique that allows you to move a selected object quickly by dragging it from one spot and dropping it in another.

DRAM *See* DYNAMIC RANDOM-ACCESS MEMORY.

drum plotter A plotter consisting of a rotatable drum and a plotting head that can only move parallel to the drum's axis of rotation.

dumb terminal Display terminal without an internal CPU that requires host computers for operation.

dump To transfer data from one place to another without regard for its significance.

duplexing The method that permits two computers to transmit data to each other simultaneously.

DVI *See* DIGITAL VIDEO–INTERACTIVE.

dynamic link A cross-reference between data files that allows a single data item to be used in several software applications.

dynamic random-access memory (DRAM) A type of RAM memory chip that requires a refresh signal to be sent to it periodically.

- □ The most common type of memory chip used for a personal computer's main memory.

EBCDIC *See* EXTENDED BINARY CODED DECIMAL INTERCHANGE CODE.

Echoplex A communication protocol in which information is echoed on a terminal screen on return of the appropriate signal from the other end of the line indicating that the information was received correctly.

EGA *See* ENHANCED GRAPHICS ADAPTER.

EIA *See* ELECTRONIC INDUSTRIES ASSOCIATION.

EISA *See* EXTENDED INDUSTRY STANDARD ARCHITECTURE.

Electronic Industries Association (EIA) An organization that specifies electrical transmission standards.

Electronic Numerical Integrator and Calculator (ENIAC) The first general-purpose electronic digital computer developed by Presper Eckert and John Mauchly in 1946 at the University of Pennsylvania. Used for calculating ballistics tables for the U.S. Army.

electrostatic printer A nonimpact printer that prints by using an electric charge to deposit toner on paper.

embedding The process of copying an object from a document in a software program to another document created by another software program.

Encapsulated PostScript (EPS) A high-quality, PostScript-based file format used in many drawing and illustration programs.

encryption A technique used to code or scramble the data so that it cannot be used until it is changed back to its original form.

end user The computer user, as opposed to the person involved in developing and marketing the computer.

Enhanced Graphics Adapter (EGA) An IBM graphics standard that produces 16 different colors on a screen with 640×350 resolution.

ENIAC *See* ELECTRONIC NUMERICAL INTEGRATOR AND CALCULATOR.

Enter key Also known as the *Return key.* Pressing the Enter key instructs the computer to execute the last command entered.

enterprise network A network connecting most major points in a company.

EPS *See* ENCAPSULATED POSTSCRIPT.

erasable optical disk An optical disk storage device that allows the storage and alteration of data.

erasable programmable read-only memory (EPROM) ROM chips that are programmed by the user and may be erased by exposure to ultraviolet light and reprogrammed again.

ergonomics Also called *human factors engineering.* The science of designing equipment, furniture, and working environments to suit human needs.

error message A message displayed or printed to notify the user of an error or problem in program execution.

even parity *See* PARITY BITS.

executable file A file that contains the instructions that tell a computer how to perform a specific task.

execution The actual processing of a program.

execution cycle The process of interpreting and then executing an instruction during the machine cycle.

expanded memory The paged memory that enables the program to recognize memory over 640K RAM under DOS.

expansion board A printed-circuit board that provides a hardware interface between a peripheral and the motherboard.

expansion slot A slot in the motherboard of a computer in which an expansion board is inserted.

expert system A form of artificial intelligence that is capable of solving complex problems at the competency level of a human expert. *(See figure on next page.)*

Extended Binary Coded Decimal Interchange Code (EBCDIC) An 8-bit code for character representation used by IBM mainframes.

Extended Industry Standard Architecture (EISA) A 32-bit bus used for microcomputers.

extended memory The memory above 1 Mbyte.

external command A DOS command that is not memory-resident. Examples are FORMAT, FIND, and CHKDSK.

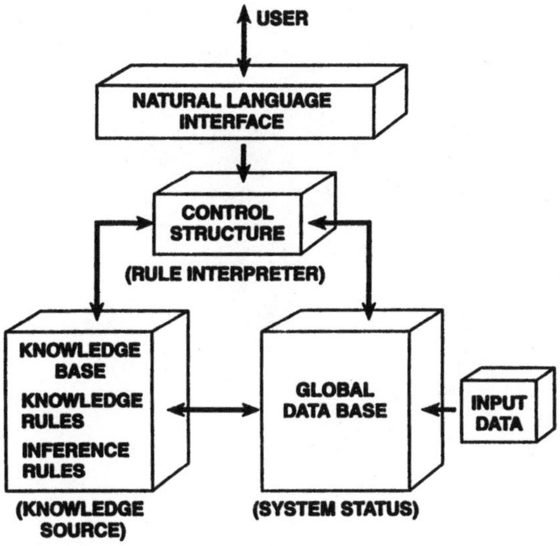

expert system. *(Courtesy Society of Manufacturing Engineers.)*

external storage Also known as *secondary storage.* Permanent storage of data on magnetic or optical media. *See also* AUXILIARY STORAGE.

facsimile machine (fax) A communication device that transmits copies of paper documents over telephone lines.

FAT *See* FILE ALLOCATION TABLE.

fax/modem A computer peripheral device that allows the user to send and receive faxes over the telephone system.

fiber-optic cable A high-bandwidth communication cable made of tiny threads of insulated glass. *(See figure.)*

□ Compared with other transmission media, fiber-optic cable is more expensive, but is not susceptible to electromagnetic interference, and is capable of higher transmission speed.

field A group of characters that identify data such as a name. Also the location in a record or database in which this group of characters is entered.

field name The label linked to a field in a database.

file A collection of information stored under one name on a storage device.

File Allocation Table (FAT) A table maintained by DOS which stores information on the size and physical location of files on the disk.

file compression A technique that reduces the size of a file by encoding it in a more compact format using a program such as PKZIP or Stuffit.

file exchange The transfer and conversion of files from one computer system to another.

file extension The addition of one to three characters, separated from the filename by a period, to further describe the file contents.

file manager A software application package designed to organize and manage files and directories.

filename The name assigned to a file stored on disk.

file server A computer dedicated to the local area network to ensure other computers in a network have access to data storage and retrieval.

file specification The complete description of a file, consisting of the drive name, path, and filename.

file transfer A network application allowing files to be moved from one network to another.

filter A process or device that screens incoming information for certain characteristics, allowing only a subset of that information to pass through.

Finder A part of the Apple Macintosh operating system that manages files.

firmware Software stored permanently in read-only memory (ROM).

first-generation computers Computers built in the early 1950s, using vacuum tubes.

fixed disk *See* HARD DISK.

flash memory A type of memory chip for nonvolatile storage that can be electrically erased in the circuit and reprogrammed.

flatbed plotter A plotter with a flat display surface fully accessible by the plotter head.

flat-panel display A compact, lightweight, energy-efficient monitor that does not use a picture tube; commonly used in portable and notebook computers.

flexible disk *See* FLOPPY DISK.

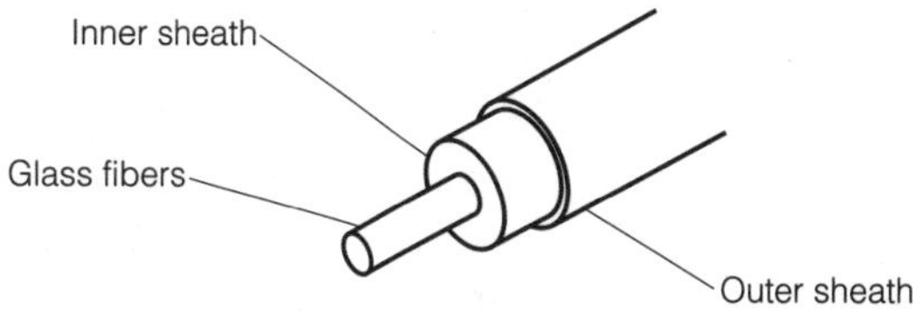

fiber-optic cable. *(Courtesy Kelmar Associates.)*

flexible manufacturing system (FMS) A group of processors or workstations connected by an automated materials-handling system and operated as an integrated system under computer control. *(See figure on the next page.)*

floppy disk Removable disk made of Mylar and encased in plastic.

□ Commonly used on microcomputers.

□ Disk diameters $3^1/2$ or $5^1/4$ in.

flowchart A graphical representation of the activities and logic flow in a program that facilitates program design or documentation.

flow control A method for ensuring that a transmitting entity does not overwhelm a receiving entity.

Folder A subdirectory on the Apple Macintosh computer.

font A collection of characters with a specific size and style (e.g., 12-point Helvetica bold).

font cartridge Cartridges inserted into non-PostScript laser printers, consisting of printer-resolution bitmaps of type characters.

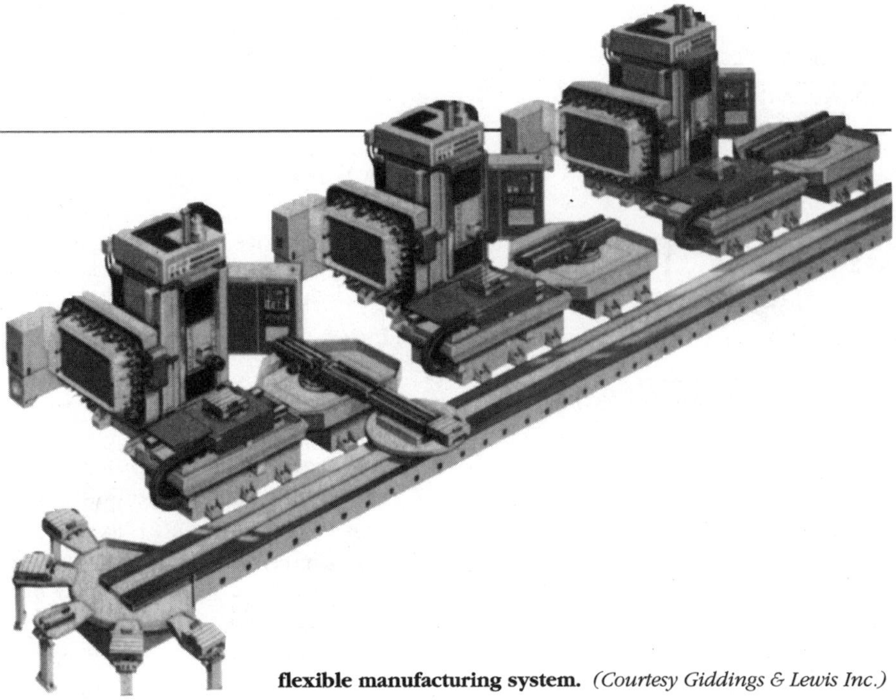

flexible manufacturing system. *(Courtesy Giddings & Lewis Inc.)*

Font Manager A utility program that controls screen fonts and printed fonts in a computer.

footer A repeating block of text at the bottom of each page that may contain a chapter heading or page number.

footprint The space a computer takes up on your desk.

format The process of preparing a disk for use by mapping into tracks and sectors.

formula A mathematical equation entered in a spreadsheet cell that tells the computer how to perform a calculation.

FORTH A high-level programming language, developed by Charles Moore in the 1970s, that is the standard language used in astronomical observatories around the world. It allows direct control over hardware devices.

FORTRAN (FORmula TRANslator) A high-level programming language, developed by John Backus at IBM in 1957; used primarily for mathematical, scientific, and engineering applications.

fourth-generation computer A computer that is built around integrated circuits with large-scale integration. It offers significant price and performance improvements over earlier computers.

fragmented file A file stored in noncontiguous, or nonadjacent, sectors on a disk.

Frame Relay A protocol used across the interface between the hosts and the network equipment.

freeware Software that can legally be copied and given free of charge to other users, usually available through a bulletin-board system (BBS) or online service.

front-end processor A device that provides network interface capabilities for a networked device.

full-duplex A communication channel that allows a device to send and receive data simultaneously and be used for high-speed computer-to-computer data transmission.

- Contrasts with half-duplex and simplex which permit alternate or one-way-at-a-time communications.

full-motion video TV-quality, or better moving video images displayed in full screen on the computer monitor.

function keys Programmable keys labeled F1 through F12, usually lining the top of the keyboard. Their function depends on the software being used.

functions Specialized formulas in a software application that allow the user to perform various operations.

garbage in, garbage out (GIGO) Serves as a reminder that the meaningfulness of computer output is only as valuable or accurate as the input.

gas plasma display A flat, thin screen filled with an inert gas that glows when exposed to an electric current.

- Used for the screens of some laptop computers.
- Easier to read than liquid crystal display (LCD) screens, but more expensive and consume more power.

gate An electronic switch used for mathematical or logic operation with several entrances but only one exit.

general-purpose computer A computer designed to perform a wide variety of tasks.

general-purpose language A programming language that is sufficiently flexible for use in a variety of applications.

gigabyte (Gbyte) 1024 megabytes (Mbytes), or approximately 1 billion bytes.

global setting A default setting that affects every subsequent entry in a document.

global variable A variable that can be recognized anywhere in a program.

grammar checker A program that checks a text file for errors in punctuation, format, grammar, spelling, usage, and style and offers suggestions for correcting the errors.

graphical user interface (GUI) An interface that provides a more visual method for interacting with the operating system through the use of icons, windows, menus, and dialog boxes.

graphics adapter Also known as a *graphics card* or *video card*. A video display card that controls the display of text and graphics on the monitor.

graphics coprocessor chip A microchip used on accelerated video adapters that is designed to perform video functions at very high speed.

graphics software Application software that allows the user to create illustrations, diagrams, graphs, charts, three-dimensional images, and animation.

gray scale Refers to a monochrome monitor that can create different shades of one color.

gray-scale graphic A bitmap graphic created by using shades of gray to produce images.

grid A feature in various draw, paint, and desktop publishing programs that guides the positioning of text and graphics. *(See figure.)*

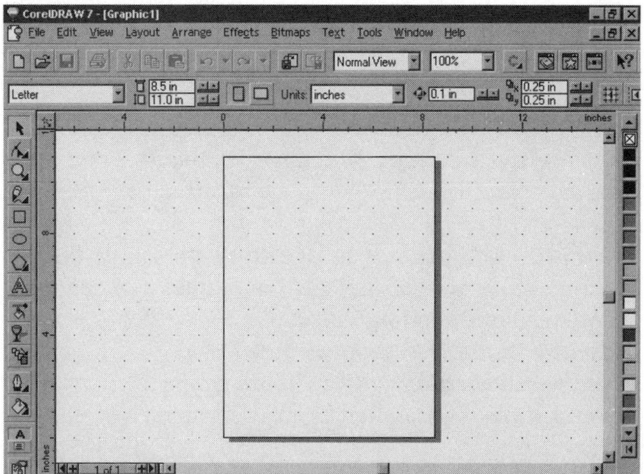

grid. *(Courtesy Corel Corp.)*

GW-BASIC An interpreter for the same dialect of BASIC as is implemented in ROM on the IBM or compatible computers.

hacker A person who gains unauthorized entry into a computer system, either for malicious reasons or just to prove that it can be done.

half-duplex A form of communication in which data travels in only one direction at a time. Contrasts with *full-duplex*.

handheld scanner A small, portable input device that converts drawings or text into digital form.

handshake A sequence of messages exchanged between two computers to ensure transmission synchronization.

hard card A type of hard disk designed in a card form that can be plugged into a slot inside the computer.

hard copy A paper copy of the computer's output. Contrasts with soft copy, which is displayed on a monitor.

hard disk Also known as *fixed disk*. A storage medium using rigid aluminum platters coated with iron oxide and capable of storing large amounts of data.

hardware The physical components of the computer, such as the disk drives, monitor, mouse, keyboard, CD-ROM drive, printer, modem, and scanner.

hardware interrupt A signal transmitted from a hardware device to the microprocessor when the device is ready to send or accept data.

hashing A mathematical operation that uses a formula to create a unique identifying number for the key field.

Hayes-compatible Refers to a modem that uses the standard dialing commands created by Hayes Microcomputer Products, Inc.

head crash *See* DISK CRASH.

header A repeating block of text at the top of each page in a document. Might contain title, author, issue date, and page number.

help features Files, built into the software package, that assist the user with a program task.

Hercules Graphics Card A card that provides high-resolution monochrome text and graphics at 720×350 pixels for IBM-compatible computers.

Hewlett-Packard Graphics Language A vector-based printer language used primarily by Hewlett-Packard plotters.

hexadecimal (HEX) A numbering system based on 16 rather than 10; consisting of the digits 0 to 9 and the letters *A* to *F*.

□ Used to display binary data because the hexadecimal form is more compact. For example, the number 2ECF in hexadecimal means 11,983.

hierarchical database A database structure in which the data elements are arranged to one another as "parents" and "children," following a family tree structure.

hierarchical network A network configuration in which workstations interact with the file server but not with each other. *(See figure on the next page.)*

hierarchical structure Also known as *tree structure*; the data structure in which one primary element may have numerous secondary elements linked to it at lower levels.

high-memory area (HMA) The first 64K of extended memory from 1024 to 1088K.

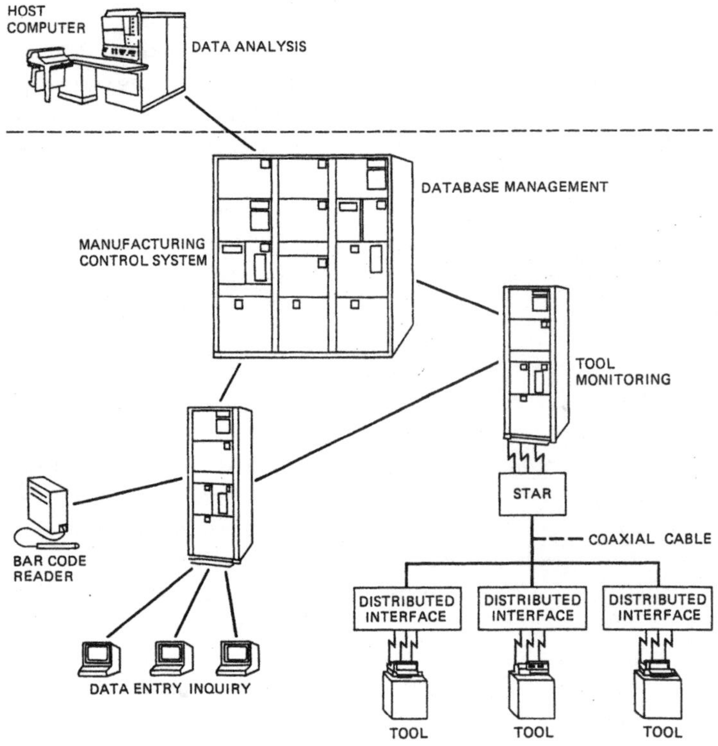

hierarchical network. *(Courtesy International Business Machines.)*

HIMEM.SYS An MS-DOS device driver that manages extended memory, the high-memory area, and the upper-memory area.

high-speed buffer *See* CACHE.

high-level languages Computer programming languages such as BASIC, Pascal, Logo, COBOL, and FORTRAN that allow the user to write programs in English-like commands without having to understand the inner workings of the computer.

high-resolution monitor A monitor with a high number of tiny dots (pixels) that resolve together to form a very sharp, clear image.

host computer A central computer that is in charge of operations of a group of computers linked in a network.

hot site An alternate oomputer facility for computer operations in case the primary site is inoperable because of a disaster.

hybrid computer A computer that combines features of both digital and analog computers and is used in a small number of engineering and scientific applications.

hybrid network An internetwork made up of more than one type of network technology.

HyperCard A Macintosh authoring tool, developed by Bill Atkinson in 1984, that implements many features of hypertext and hypermedia and allows users to create their own customized programs.

hypermedia An extension of hypertext to include graphics, sound, animation, and motion video in a manner accessible in multiple ways.

IBM Corporation International Business Machines Corporation, the largest manufacturer of mainframe computers and one of the major personal computer producers.

icon A small on-screen picture used by a graphical user interface to represent a task or command such as opening, saving, or printing a file.

impact printer An output printing device that physically contacts the paper in order to print output.

incremental backup A backup procedure that only copies the files that have changed since the last backup.

indexing The process by which records in a file are arranged in a logical sequence, allowing for both sequential and random retrieval.

Industry Standard Architecture (ISA) A 16-bit bus used for Intel-based personal computers.

information Data that has been organized and processed into a more useful form for the decision maker.

information processing The process of transforming raw data into useful information with the use of a computer.

□ The four basic functions of information processing are input, processing, output, and storage.

information-processing system A system consisting of hardware, software, people, data, and procedures that processes data to provide information.

information service An online company that allows the subscriber access to information on a large range of subjects.

initialize To prepare a blank disk for use on your computer. *See also* FORMAT.

inkjet printer A nonimpact printer that creates hard-copy images in dot matrix form by shooting a stream of tiny droplets of electrostatically charged ink at the page. *(See figure.)*

inkjet printer. *(Courtesy Hewlett-Packard Co.)*

input Data gathered and entered into the computer for processing.

input device A device used to enter information into a computer. The keyboard and the mouse are the most commonly used input devices.

input/output (I/O) Usually refers to one of the slots in a microcomputer to which peripheral devices may be connected.

insert mode The mode in a word processor in which characters that you type on the keyboard are inserted into whatever text already exists in the document.

insertion point A blinking vertical bar that marks the position where text will be inserted when you type on the keyboard. *See also* CURSOR.

instruction A command that tells the computer to perform a specific operation.

instruction execution cycle The time in which the CPU retrieves and executes an instruction.

instruction register A special holding area of the CPU that contains instructions that are to be executed by the ALU.

instruction set The list of logical and arithmetic procedures that a processor can perform, such as addition, subtraction, and comparison.

integrated circuit (IC) An electronic circuit consisting of many miniature transistors and other circuit elements on a single silicon chip.

integrated software Software package that integrates several computer applications, such as a word processor, spreadsheet, database, graphics, and communications, into one comprehensive program.

- Allows for easy exchanges of information from one application to another.

intelligent terminal Also known as "smart" terminal. A terminal with a built-in microprocessor that can perform some processing on its own. *See also* DUMB TERMINAL.

interactive Two-way communication between the user and the computer program.

interactive processing A computer process involving interaction with a user during processing. Such interaction may be via a keyboard, mouse, or other interactive input device.

interactive programming language A computer language that allows user input while a program is executing. Examples of interactive programming languages are BASIC and APL.

interblock gap (IBG) A blank space that separates records on a magnetic tape.

interface A device that allows two or more devices to pass information.

interface card A circuit board used to connect a computer to peripheral devices.

internal command A DOS command that is memory-resident. Examples are DIR, COPY, and ERASE. By the way of contrast, *see* EXTERNAL COMMAND.

internal font A font built into the printer.

internal memory The random-access memory (RAM) or read-only memory (ROM).

internal storage The computer built-in memory. By the way of contrast, *see* AUXILIARY STORAGE.

International Organization for Standardization (ISO) An international organization that is responsible for a wide range of standards, including those relating to computers and networking.

Internetwork Packet Exchange (IPX) A network protocol developed by Novel for use in NetWare networks.

interpreter A language processor that translates a high-level programming language into machine language one line at a time.

interrupt An instruction that tells a microprocessor to temporarily suspend normal processing operations.

IPX *See* INTERNETWORK PACKET EXCHANGE.

ISA *See* INDUSTRY STANDARD ARCHITECTURE.

job An individual program run by a computer system.

job control language (JCL) The command language used in batch jobs to tell the computer what to do.

job control program A part of the operating system that allows users to line up several jobs to enable the computer to immediately begin executing the next job as soon as the current job is completed.

joystick An input device that uses a movable lever to control a cursor or object on a computer screen; typically used in computer games.

justification The alignment of lines of text relative to the margins in word-processing documents.

- There are four forms of justification: left, center, right, and full.

K *See* KILOBYTE.

Kermit A popular file-transfer and terminal-emulation program, developed at Columbia University in 1981, for IBM-compatible computers.

kerning The process of adjusting the space between characters.

key A field by which a data file is sorted or searched.

keyboard Typewriterlike keys on a computer that consist of alphanumeric and command keys. *(See figure.)*

keyboard.
(Courtesy Numerical Control Computer Services.)

key field A field that is used as the criterion for identification of a record.

keypunch A keyboard device that punches holes in 80- or 90-column cards to represent data.

key-to-tape equipment A device used to record data directly onto magnetic tapes.

kilobyte (K) Storage capacity of 1024 bytes in memory or on disk that is often rounded off to 1000 bytes.

kiosk An online service's information booth.

labels Headings that describe the data in a specific row or column in spreadsheets.

landscape A horizontal page orientation that is wider than it is tall.

language translator A language processor that translates high-level languages or assembly language into machine language.

laptop computer A small, lightweight, portable lap-size microcomputer; often used interchangeably with *notebook computer*.

large-scale integration (LSI) Refers to integrated circuits that contain more than 100 logic gates. *See also* INTEGRATED CIRCUIT.

laser disk Also called *videodisk*. An optical disk storage device that uses lasers to store and read data.

laser printer A type of nonimpact printer that uses a laser beam to form images on a photostatic drum to produce a printed copy.

LCD *See* LIQUID CRYSTAL DISPLAY.

leading The distance from the baseline of one line of text to the next as measured in points.

letter-quality printer A printer that produces print equal in quality to that of a typewriter.

library A collection of files, computer programs, or subroutines.

light pen A light-sensitive, pen-shaped input device that is used to write on or select menus from a computer screen.

line printer An impact printer that prints an entire line at a time.

link To insert the location of an object file in another file. The object can be updated to reflect the changes in the linked object's source file.

liquid crystal display (LCD) A type of flat-screen display commonly found on portable computers.

liquid crystal display panel An LCD device that projects the contents of a computer screen via an overhead projector.

LISP A high-level programming (*lis*t-*p*rocessing) language, primarily used in artificial intelligence research, that processes data in the form of lists.

load The process of getting the program into the computer's memory.

local area network (LAN) A group of computers that are located within the same building or floor of the building and connected to each other through cables, allowing them to share files and peripheral devices. *(See figure on next page.)*

LocalTalk Apple Macintosh proprietary network protocol.

local system Peripheral devices connected directly to the CPU.

logic error A programming mistake that causes the wrong processing to take place despite correct syntax.

logic data structure A structure that defines how data items are related or organized so as to have meaning.

logic gate A device that accepts binary digits as inputs and produces an output bit according to a specified rule.

logical field A type of database field that indicates a logical condition, true or false.

Logo A high-level, graphics-oriented, interactive programming language, developed by Seymour Papert of MIT, designed to help young children develop problem-solving skills.

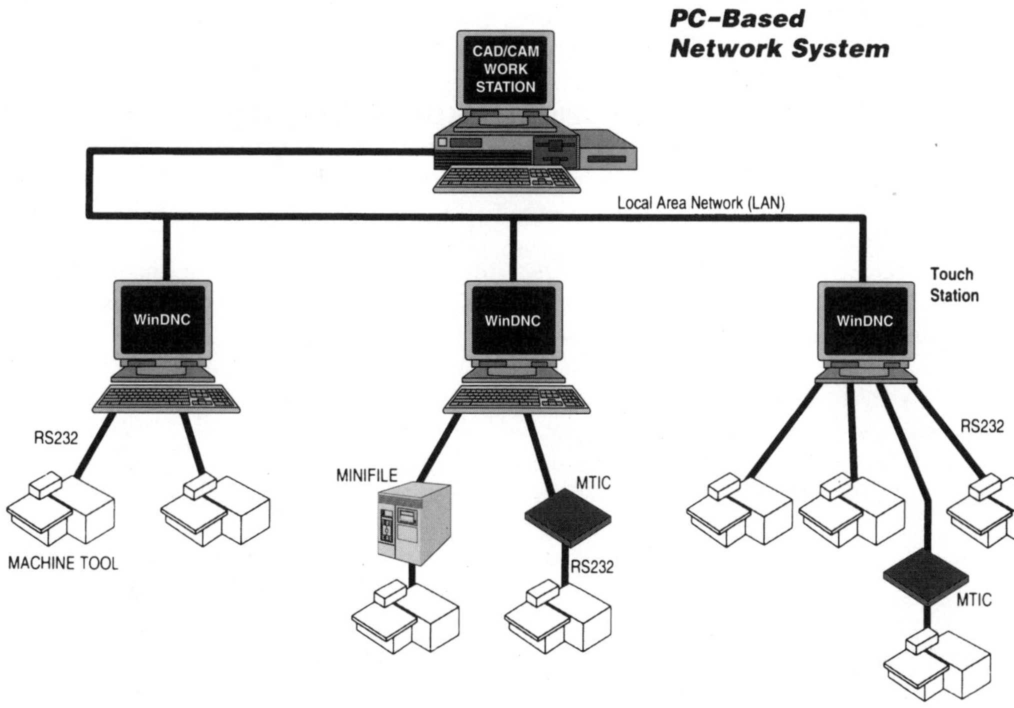

local area network. *(Courtesy Greco Systems.)*

logon The process of establishing a connection from your terminal to a multiuser computer and identifying yourself as an authorized user.

loop A logic structure that allows a set of instructions to be executed repeatedly until a given condition is true or false.

lost cluster A group of disk sectors that are not marked as free but are not allocated to a file.

Lotus 1-2-3 A popular spreadsheet program developed by the Lotus Development Corporation.

low-level language A programming language such as machine language and assembly language that can be executed with no translation.

machine cycle The process of executing a single instruction in the CPU.

machine-independent program The ability of a program to be run on many different types of computers.

machine language A low-level language containing instructions in a binary code that a computer can execute directly.

Macintosh Apple Computer's most successful line of computers, released in 1984, was the first personal computer to use a graphical user interface with a mouse.

macro Refers to the use of a simple command to execute a sequence of complex commands that helps the user customize an applications program. *(See figure.)*

magnetic disk Also known as *direct-access storage devices*. A storage medium consisting of a metal platter coated with a magnetized substance on which data is stored in the form of magnetized spots.

magnetic-ink character recognition (MICR) A process that involves reading characters imprinted with magnetic ink; commonly used in bank check routing numbers for rapid processing.

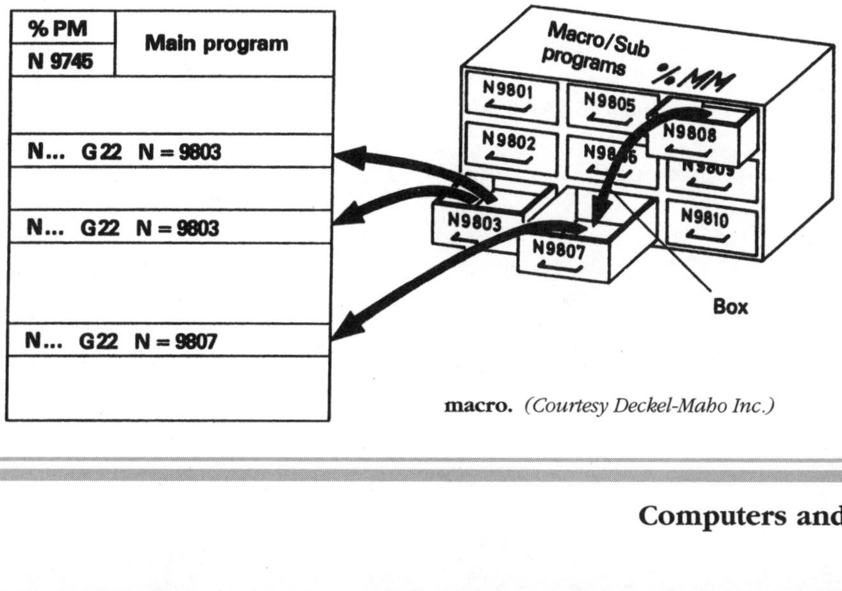

macro. *(Courtesy Deckel-Maho Inc.)*

magnetic tape A sequential storage medium consisting of a thin plastic or Mylar tape that is coated with magnetic material used to store information.

mail merge A word-processing feature used to combine a form letter with a secondary data file to produce personalized form letters.

main memory Also called *system memory.* Primary storage location that holds data and instructions for current use by the CPU.

mainframe A large, general-purpose computer system that stores large amounts of data and accommodates many users performing a variety of computer tasks. Generally used by business, institutions, and government.

management information system (MIS) A computerized information system designed to provide instant data to management for effective and efficient business operation.

Mark I Also known as *Automatic Sequence Controlled Calculator.* An electromechanical computer developed at Harvard University in 1944 by Howard Aiken.

mass storage system Special-purpose secondary storage devices capable of storing enormous quantities of data.

math coprocessor *See* COPROCESSOR.

maximize To fill the entire screen with one window.

mean time between failures (MTBF) A specification of the reliability of computer components, measured in hours.

megabits per second (Mbits/s) A million bits per second. Unit used to measure the data transmission speed on a network.

megabyte (Mbyte) Storage capacity of 1,048,576 (2 to the 20th power) bytes in memory or on disk. It is often rounded off to 1 million bytes.

megahertz (MHz) A unit of measurement for clock speed representing 1 million cycles per second.

memory Usually refers to the computer system's capacity for storing information in terms of RAM, but it also sometimes refers to the ROM and disk storage capacity.

memory chip An integrated circuit that contains memory.

memory manager A program that manages the use of a specific region of memory.

memory-resident program Also called a *terminate-and-stay-resident* (TSR) *program.* A program that remains in memory after being run and can be called up later.

menu A list of command choices or options presented by a program.

menu-driven interface A type of user interface in which the user can control the software by selecting options from several choices that are presented on screen.

metropolitan area network (MAN) A network that spans a metropolitan area.

□ A MAN spans a larger geographic area than a local area network (LAN) but a smaller geographic area than a wide area network (WAN).

Microcom Networking Protocol (MNP) A method for detecting and correcting errors in data transmission.

microcomputer Also called a *personal computer* (PC). A small computer dedicated to the single user for home or business use.

microprocessor A single integrated circuit containing the circuitry for both the arithmetic/logic unit and the control unit.

microsecond One-millionth of a second.

Microsoft One of the leading software companies. Founded by Bill Gates in 1975, the developer of DOS, Windows, and several popular application packages, including Microsoft Works, Word, Excel, and PowerPoint.

Microsoft Disk Operating System (MS-DOS) An operating system used by IBM and compatible computers. *See also* DISK OPERATING SYSTEM.

microwave system A data communications system that uses high-frequency electromagnetic waves to transmit data through the atmosphere.

MIDI *See* MUSICAL INSTRUMENT DIGITAL INTERFACE.

million instructions per second (MIPS) A unit used to measure the processing speed of a processor.

millisecond One-thousandth of a second.

minicomputer A class of general-purpose computers that are smaller and less expensive than mainframes but perform similar tasks. They are larger than a personal computer with corresponding power and memory capacity.

minimize To shrink a window so that it collapses to an icon.

MIPS *See* MILLION INSTRUCTIONS PER SECOND.

mnemonics Computer instructions written in a form that is easy for the programmer to remember but which must later be converted into machine language.

MNP *See* MICROCOM NETWORKING PROTOCOL.

modem *Mo*dulator-*dem*odulator. A communication device that enables computers to transmit information over telephone lines. *(See figure.)*

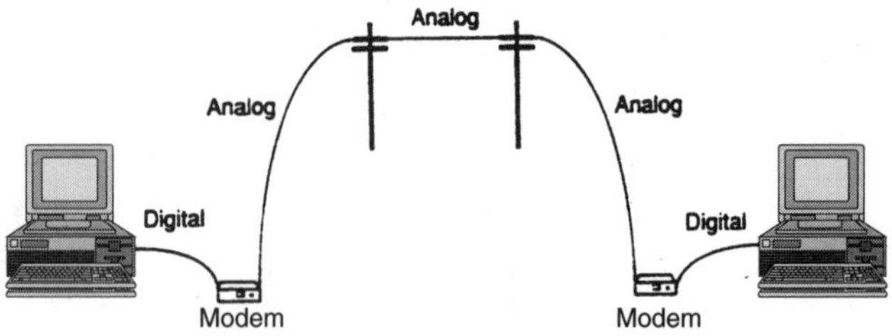

modem. *(Courtesy Hewlett-Packard Co.)*

□ Modems convert the digital signals used by computers into analog signals needed by voice telephone systems.

□ Modems are designed in either internally (plugged into an expansion slot inside a computer) or externally (attached to the outside of a computer by a cable).

monitor The display screen of a computer. *See also* CATHODE-RAY TUBE.

monochrome monitor A monitor which displays only one color in addition to black. Common monochrome combinations are green and black, black and white, and amber and black.

motherboard Also called the *system board.* The main circuit board in a microcomputer on which the CPU is mounted.

mouse A handheld pointing device that is rolled around on the desktop to aid interaction with the computer.

MPC Standard Multimedia PC Marketing Council Standard. The minimum hardware requirements needed to successfully run certain multimedia applications.

multimedia The integration of various media, such as text, graphics, sound, animation, photo images, and motion video, into a single computer application.

multiplexer A device that increases the efficiency of a network by permitting multiple low-speed devices to transmit data over the same communication channel.

multiprocessing The use of more than one CPU in a single computer system.

multiprogramming Concurrent execution of more than one program at a time on a single computer through timesharing.

multiscan Usually refers to a monitor that can handle more than one video mode.

multitasking A feature of an operating system that allows a user to run two or more applications at the same time.

multiuser system A computer system that accommodates many users at the same time.

music synthesizer A device that converts electronic signals into sounds, producing a high quality of music sound.

Musical Instrumental Digital Interface (MIDI) A protocol for communicating musical information between musical instruments, synthesizer, and computers.

nanosecond (ns) One-billionth of a second, often used to describe the speed ratings of memory chips.

narrowband channel The slowest catagory of communication channel that transmits data only at a rate between 45 and 90 bits/s.

□ Telegraph communications is an example of a narrowband channel.

natural language A developmental programming language using ordinary human language such as English.

near-letter-quality (NLQ) Refers to a printer that produces output that resembles the print of an old cloth-ribbon typewriter.

NetWare A popular network system software developed by Novell.

network A group of computers connected together to share hardware resources, software, and files. *(See figure.) See also* COMPUTER NETWORK and LOCAL AREA NETWORK.

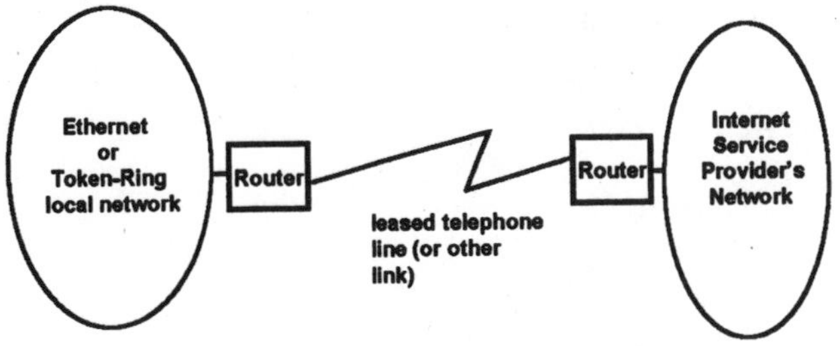

network. *(Courtesy McGraw-Hill Inc.)*

□ The two most common networking configurations are Ethernet and token ring.

network address A unique number assigned to each device on a network.

Network Communication Protocol The rules that specify the structure of packets transmitted over the network.

Network Control Program (NCP) A program that manages the flow of data of a communication network.

network interface card (NIC) An expansion card that plugs into the motherboard of the computer and is used to connect the computer to a network.

network server *See* FILE SERVER.

nibble or nybble Consists of 4 bits, or 1/2 byte.

node Any device on the network such as a computer, a terminal, a printer, or the file server.

noncontiguous sectors Nonadjacent sectors of file scattered across the surface of a disk.

nondestructive read The feature of computer memory that permits data to be read without erasing the information that is stored in that memory location. Contrasts with destructive read.

nonimpact printer A printer that forms images without striking the page. Thermal, inkjet, and laser-xerographic are examples of nonimpact printers.

nonvolatile memory Memory that retains data even after the power is shut off. ROM is an example of nonvolatile memory.

Norton Utilities A popular set of programs developed by Peter Norton that analyze storage allocation on disk, recover erased files, and correct other problems with disks.

notebook *See* LAPTOP COMPUTER.

NuBus The special kind of expansion slot commonly found in Macintosh computers.

null modem A small box or cable used to connect two computers together without a modem.

numerically controlled system A system in which actions are controlled by the direct insertion of

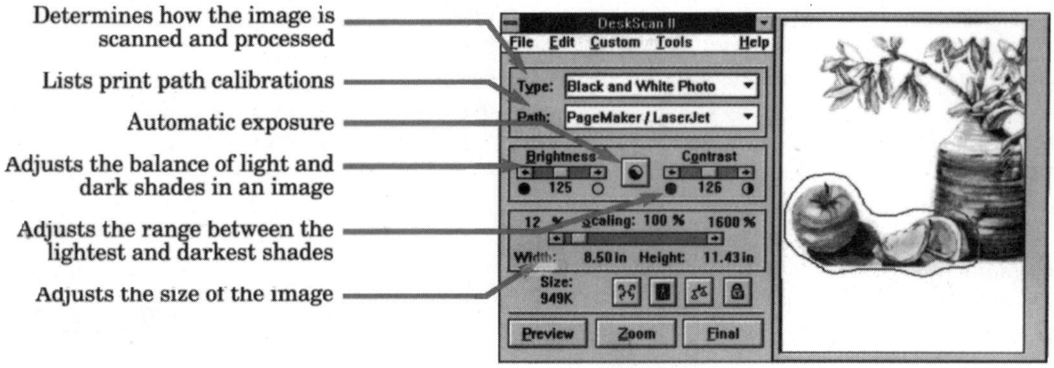

optical-character reader. *(Courtesy Hewlett-Packard Co.)*

programmed numerical data.

numeric keypad A separate set of keys, usually located at the right side of the keyboard, used for entering numbers quickly or for cursor movement.

object code The machine-language translation created by a compiler.

object linking and embedding (OLE) The process of connecting or inserting files created in one software package into a file created in another software package so that the OLE files can be updated and modified.

object program A program that has been translated into machine language and is ready to be executed.

octal number system A base-8 number system, consisting of the digits 0 through 7.

odd parity *See* PARITY.

online Referring to equipment or devices in a data-processing system that are under the direct and immediate control of the central processing unit. Contrasts with offline.

online help On-screen explanations displayed by a program to provide assistance for the user.

online service provider A company that provides dial-up access to information stored on its computer system for a service fee.

OP code Also called *operation code.* The part of a machine or assembly-language instruction that indicates what operation is to be performed.

open architecture A design that can be easily modified and expanded through the use of plug-in components. Such a computer system is seldom restricted to one manufacturer's components.

operating system (OS) A collection of programs that manage the computer and its peripherals, allowing the user to run programs and to control the movement of data to and from the memory and peripheral devices.

□ The operating system acts as an interface with application software.

□ Popular operating systems are Windows 3.1, Windows 98, Macintosh System 7, OS/2, and Unix.

optical-mark reader (OMR) A scanning device that reads optical-mark data and converts it into binary code for computer use.

□ Typically used in schools for gathering survey data or test answers using marks made on paper.

optical character A character printed in the OCRA font that is easily identified by optical scanners or the human eye.

□ Optical characters are used in postal and billing systems.

optical-character reader (OCR) Also called *optical scanner.* A device that reads numbers, letters, and other characters and then converts the optical images to digital form. *(See figure.)*

optical disk A high-density storage device that uses laser technology to store data on a thin coating of metal on the surface of a disk. CD-ROMs and videodisks are examples of optical disks.

OS/2 A multitasking operating system created by IBM and Microsoft that offers a graphical user interface and retains the ability to run DOS and Windows programs.

output Data that is produced and transmitted by a computer after it has been processed. The data can be transmitted to a device such as a printer, disk drive, or modem.

output device A hardware component used to display, print, or store information that results from processing data.

paging Also known as *virtual memory.* A virtual storage technique for managing the limited amount of main memory and a secondary storage by swapping program segments and data into high-speed internal storage.

page description language A special-purpose computer language used in conjunction with laser printers for page composition.

page layout software Software that supports online page layout of text and graphics into finished pages. *See also* DESKTOP PUBLISHING.

pages per minute (PPM) A unit for measuring printer speed.

parallel port A device interface that allows simultaneous transmission of 8 bits of data at a time.

parallel processing A type of processing in which computations are carried out at the same time on multiple processors.

parallel transmission The simultaneous transmission of 8 bits down eight separate data lines in a cable.

□ Sending data from a CPU to a parallel printer is a form of parallel transmission.

parity A transmission characteristic that is used to ensure error-free transmission.

□ Parity is designated as odd or even.

parity bit An extra bit added to a data block by the transmitting computer that enables computers to check the validity of data transmission.

partition To divide a hard disk into one or more logical drives.

Pascal A high-level structured programming language named after the French mathematician Blaise Pascal.

□ Designed to encourage programmers to write modular and well-structured programs.

□ Commonly used in a wide variety of applications such as business, science, and education.

password Secret word that is required to log onto a computer system, thus preventing unauthorized entry into the system.

path Describes the route to locate a document on disk.

□ For example, \word\letter\memo.doc means the file memo.doc is located in the subdirectory Letter inside the directory Word.

PC-compatible A computer based on the architecture of the IBM personal computers; hardware is interchangeable with the IBM PC standard.

PCMICA slot A special type of expansion slot developed by the Personal Computer Memory Card International Association for notebook computers that allows you to plug in a credit card–size expansion card.

peripheral device Any device that is connected to and controlled by the computer. Examples of peripheral devices are monitor, keyboard, mouse, printer, modem, and hard disk.

personal computer (PC) A computer designed to be used by only one person, either at home or in a business setting. *(See figure.)* The term often refers to IBM computers and compatibles.

286—a PC built around the 80286, a 16-bit processor from Intel.

386—a PC built around the 80386, a 32-bit processor from Intel.

486—a PC built around the 80486, a 32-bit processor from Intel.

Pentium—a PC built around the Pentium processor, a 32-bit processor from Intel.

pica A unit of measurement equal to $^1/_6$ in.

picoseconds One-trillionth of a second.

pixel The smallest unit on a display screen that can be illuminated.

□ The more pixels on a screen, the greater its resolution.

PL/1 A high-level programming language (Programming Language 1) developed by IBM in the early 1960s to combine the best features of COBOL and FORTRAN.

plotter An output device that uses pens to create a hard copy of high-resolution graphical images.

plug and play (PNP) A set of specifications for designing hardware so that the device is automatically configured when the computer boots.

point The basic measurement of type. A point is equal to $^1/_{72}$ in, and there are 12 points to a pica.

pointer The arrow-shaped cursor on the screen that is controlled by a mouse or other input device.

polling An access method in which the computer checks each device periodically to see if it has a message to send.

port A connection between the CPU and another peripheral device.

portable computer A computer that is small enough and light enough to be carried.

POST *See* POWER-ON SELF TEST.

PostScript A page description language developed by Adobe Systems and used by many laser printers.

power-on self-test (POST) A part of the boot process that tests the system component and diagnoses problems in the computer.

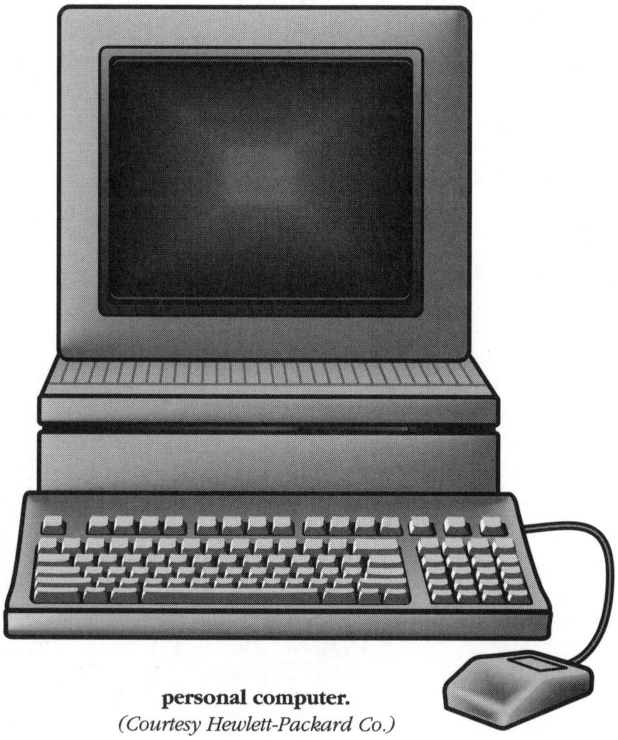

personal computer.
(Courtesy Hewlett-Packard Co.)

preemptive multitasking A multitasking strategy in which the operating system assigns priorities to the tasks being executed by several programs, so that the most important tasks are executed first.

presentation graphics A computer-based presentation that uses high-quality computer-generated graphics.

primary storage Also known as *internal storage* or *main memory*. The section of the CPU that holds instructions, data, and intermediate results during processing.

printer An output device that produces a hard copy. Examples are the line printer, daisy wheel printer, dot matrix printer, inkjet printer, thermal printer, and laser printer.

printer driver A type of software that translates files into a form that the printer understands.

print spooler Hardware and/or software that stores computer output in memory so that the computer can finish creating the output without waiting for the printer to print it.

printer server A networked computer system that manages the flow of traffic on shared printing devices.

PRN The device name for the first printer port in an IBM-compatible computer.

procedural language A programming language that requires a set of rules and procedures for problem solving.

processing The stage in which data is transformed into meaningful information.

processor *See* CENTRAL PROCESSING UNIT and MICROPROCESSOR.

productivity software Application software packages that can increase the productivity of the user.

☐ Types of productivity software include word processing, spreadsheet, and database management.

program A detailed set of instructions that tells a computer how to perform a task. Used interchangeably with software.

program file Also called an *executable file*. A file that contains the instructions that tell a computer how to perform a specific task.

programmable read-only memory (PROM) A type of ROM chip that can be programmed by the manufacturer or by the user.

☐ PROM can be programmed once, but not reprogrammed.

☐ EPROM can be erased and reprogrammed by a special process. *See* ERASABLE PROGRAMMABLE READ-ONLY MEMORY.

programmer The individual who writes programs or software.

programming The process of creating programs or software.

programming language The sets of characters and syntax used to create computer programs. Examples of programming languages include BASIC, FORTRAN, Pascal, and COBOL.

prompt A prompt is a symbol that appears on a computer screen to signal you that the computer is awaiting your input.

☐ For example, under IBM PC DOS, the prompt A> signals that drive A is the default drive and that the operating system is waiting for a response.

protocols A set of rules that governs the transmission of data between two devices.

pseudocode A useful programming aid consisting of a set of sentences or phrases that help the programmer write the code without worrying about proper syntax.

public-domain software Software that is available for use without copyright restrictions, and may be freely copied and distributed.

pull-down menu A menu that appears when a choice is made from the menu bar at the top of the screen. Often found in GUI operating systems and environments.

punched card A paper card on which data is represented by holes punched according to a particular pattern that can be read by a computer. *(See figure.) This method of representing data is no longer in wide use.*

☐ Standard punched cards are $7\frac{3}{8}$ in long and $3\frac{1}{4}$ in high. Each card contains 12 rows and 80 columns.

punch queue A special holding area with a list of files waiting to be printed.

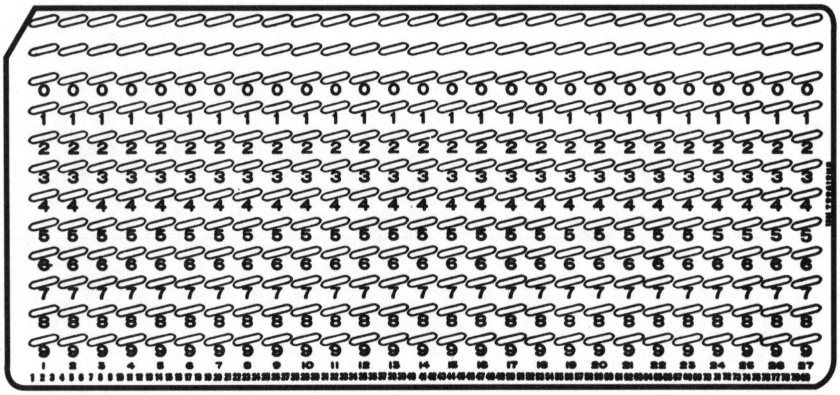

punched card. *(Courtesy Kelmar Associates.)*

query A request for data from a database that meets the criteria defined by the user.

query language A high-level language used to retrieve specific information from a database.

queue An ordered list of elements waiting to be processed.

QuickBASIC A high-performance BASIC compiler developed by Microsoft Corporation. In general, it accepts programs written in BASICA or GWBASIC.

QuickTime A video format, developed by Apple Computer, used to encode, compress, store, and play back video segments on computers.

random access The ability of a disk-based storage device to access data in any order, rather than sequentially.

random-access memory (RAM) Volatile memory that temporarily stores data and programs to which the CPU needs immediate access. Information in RAM is lost when the power is turned off.

RAM *See* RANDOM-ACCESS MEMORY.

RAM address A unique identifier assigned to each location within memory.

RAM disk A portion of random-access memory that temporarily functions as a diskette or hard disk to facilitate very rapid input and output operations.

RAM-resident program *See* MEMORY-RESIDENT PROGRAM.

read-only file A file that you can read from, but not write to.

read-only memory (ROM) Nonvolatile memory containing permanent data and instructions that cannot be changed by user programs.

read/write head The part of a disk drive that reads data or records data from a magnetic surface.

real-time processing Also known as *online transaction processing*. Refers to a computer system that processes input as soon as it is received.

record A collection of related fields, which relate to a single unit, such as a phone number.

□ A record in a database is equivalent to a card in a card file.

□ A collection of records is called a *file.*

Reduced Instruction Set Computing (RISC) A class of microprocessor that executes smaller, simpler instructions at a faster processing speed

register A special holding area of the CPU that stores temporary results or intermediate calculations and is used repeatedly by the CPU.

relational database A database structure in which some data items in one type of record refer to records of a different type.

relative cell address A spreadsheet cell address that indicates the position of a cell relative to another cell. An example is the address C5. Contrasts with absolute address.

remote job entry (RJE) An application that is batch-oriented, as opposed to interactive.

□ In RJE environments, "jobs" are submitted from a remote terminal to the central computer, and output is received later.

removable storage A storage medium, such as a diskette or hard disk, that can be inserted or removed from a storage device.

Report Program Generator (RPG) A problem-oriented high-level language developed by IBM in the 1960s to prepare business reports.

resistor An electronic component that resists the flow of electric current.

resolution A measure of the sharpness of the images that can be produced on a monitor or a printed page, expressed as dots per inch.

RF modulator Radio-frequency modulator. A device that converts video signals generated by the computer to signals that can be displayed on a TV.

RGB monitor A computer monitor that displays eight or more colors using a combination of three primary colors (red, green, and blue).

ring network A network configuration in which each computer is linked to two others to form a ringlike or circle pattern. *(See figure.)*

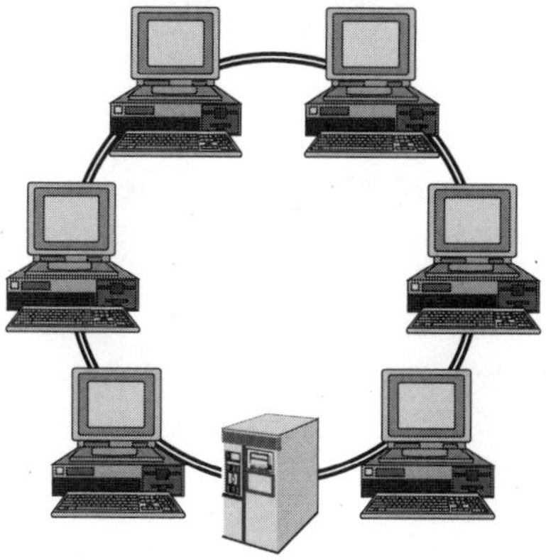

ring network. *(Courtesy Kelmar Associates.)*

RISC *See* REDUCED INSTRUCTION SET COMPUTING.

ROM *See* READ-ONLY MEMORY.

root directory The topmost or main directory on a disk.

RS-232C Interface Electronic Industries Association (EIA) Recommended Standard 232C. A standard for serial transmission devices. *See also* SERIAL PORT.

run-time error An error that occurs when a program is being executed.

sampling The process of turning sound into a digital audio file.

sampling rate The number of times per sound that the sound board measures the amplitude of incoming sound. Typical sampling rates range from 8 to 48 kHz.

sans serif A typeface that lacks serifs, or small crosslines at the end of the main letter strokes. Arial and Helvetica are examples of sans serif typefaces.

save To store a program on a storage device.

scanner An input device that converts text, photographs, or drawings on paper documents into digital form that the computer can display, print, and store. *(See figure.)*

□ The scanned images can be saved, edited, or manipulated on the computer.

screen saver A utility program that either blanks a screen or displays geometric patterns or moving pictures when no input has been received for a certain period of time. As soon as input is received, the screen redisplays whatever was on it before the screen saver was activated.

scrolling A process by which you adjust the screen view with the scroll bars or scroll boxes.

SCSI *See* SMALL COMPUTER SYSTEM INTERFACE.

search and replace A word-processing command that finds a specific character string in a document, and substitutes another character string chosen by the user.

secondary storage *See* AUXILIARY STORAGE and EXTERNAL STORAGE.

sector A portion of a track on a floppy or hard disk that stores 512 bytes of data.

seek time The average time it takes in milliseconds for the read/write head to position over the desired track of the disk.

semiconductor A material that is neither a good conductor nor a good insulator.

sequential access The ability to retrieve data only in the order in which it was stored. A tape storage device is an example of a sequential access device.

serial port Also known as *COM port.* A device interface capable of transmitting data one bit in each direction at a time.

serial transmission Transmission that proceeds one bit at a time.

serif typeface A typeface that has small crosslines at the end of the main letter strokes. Times Roman is an example of a serif typeface.

server A machine on a network that provides a particular service to other computers connected to the network.

shareware Software marketed under a "try before you buy" policy. If you like it, you should send in a registration fee to the author.

SIMM *See* SINGLE IN-LINE MEMORY MODULE.

simplex transmission Data transmission in only one direction, as found in commercial radio or television.

single in-line memory module (SIMM) A circuit board that holds RAM chips.

□ SIMMs are available in two forms: 30-pin (also known as 9-bit) and 72-pin (also known as 36-bit).

site license A software license that allows a software product to be used on a specified number of computers at a location.

scanner. *(Courtesy Hewlett-Packard Co.)*

small computer system interface (SCSI) A standard bus for connecting devices such as disk drives and scanners to computers.

□ A standard interface for Macintosh computers.

□ Provides a connection for up to seven peripheral devices.

smart terminal *See* INTELLIGENT TERMINAL.

soft copy A temporary output displayed on a computer screen. Contrasts with hard copy.

software A computer program (or programs) enabling a computer to perform a specific task.

software piracy Unauthorized duplication or use of commercial software without permission.

Software Publishers Association (SPA) An organization of the PC software industry that acts as a watchdog on software piracy.

SPA *See* SOFTWARE PUBLISHERS ASSOCIATION.

sorting A method of ordering the records in a database.

speech recognition The ability of an input device to translate human speech and use it as input media.

spelling checker A program or word-processing feature that compares the words in a document with an online dictionary to determine if the words are spelled correctly.

spreadsheet A program that provides worksheets with rows and columns for calculating and preparing charts and reports.

star network A network configuration in which each computer is connected directly to a central device called a *hub.*

startup disk A disk that contains operating system files and auxiliary files and is used for booting a computer.

stop bit Used in asynchronous protocol to indicate the end of a data block.

storage device A hardware component that stores and retrieves data from a storage medium.

storage media Materials on which data is stored. Floppy disks, hard disks, and optical disks are examples of storage media. *(See figure on next page.)*

storage media. *(Courtesy Hewlett-Packard Co.)*

structured programming A programming approach that produces structured programs by using a modular design that makes programs more comprehensive with less programming errors.

Structured Query Language (SQL) A standard language for database queries, supported by all client/server and most database programs.

stylus A penlike input device that is often used with a drawing tablet.

subdirectory A directory that is subordinate to another directory.

subroutine A subpart of a program, given a particular name, that can be called by another part of a program.

supercomputer The largest, fastest, most expensive type of computer capable of executing 100 million instructions per second; used to solve large, complex mathematical problems.

super VGA (SVGA) A video display adapter for the PC capable of displaying graphics from 640×480 pixels at 16 colors, to 1280×1024 pixels or higher at 256 or more colors.

surge suppressor Also known as *surge protector.* An electronic device that stops power surges and sudden voltage spikes from entering the computer and damaging sensitive electronic components.

SVGA *See* SUPER VGA.

swapping The process of exchanging segments of program and data from secondary storage into RAM for virtual memory.

synchronous transmission Transmission between two computers at predefined time intervals.

syntax The rules that specify how language symbols can be put together to form meaningful statements.

syntax error An error in a program code that violates the rules of the programming language.

system programmer An individual who is responsible for creating and maintaining system software.

system software A collection of programs that manage the basic operations of a computer. *See also* OPERATING SYSTEM.

tape drive A device used for making backup copies of a hard disk on magnetic tape.

telecommunications The transmission of data over communication media such as telephone lines or wireless channels.

telecommuter An employee who works at home through a link between a home computer and the company's central computer.

teleconferencing A method of holding "electronic" meetings from distant locations through the use of a computer network.

terminal An input/output device consisting of a keyboard and monitor linked to a multiuser system such as a mainframe.

terminal emulation The use of a personal computer and communications software to simulate a computer terminal.

terminate and stay resident (TSR) A program that remains in memory and is active in the background while you use other programs.

text editor A program for creating and editing simple text, or ASCII files.

timesharing The concurrent use of one computer by several users. In general, timesharing is the connection of multiple users at different terminals to a shared computer.

toggle key A key that switches back and forth between two modes of operation. Examples are Num Lock and Caps Lock, which affect the use of the numeric keypad and capitalization of alphabet keys.

token ring network A network standard developed by IBM that uses a token (a special message) to regulate the flow of data.

topology The physical arrangement and relationship of interconnected nodes and lines in a network. Common network layouts are bus, ring, and star. *See also* BUS NETWORK, RING NETWORK, and STAR NETWORK.

touch screen An input/output device designed for use with a menu-driven user interface. Users make choices by touching different parts of the screen.

track A concentric ring in which data is stored on a disk. *See also* SECTOR.

trackball A pointing device used by rolling the ball to position the pointer on the screen.

□ Unlike a mouse, a trackball does not move on the desk and therefore requires less space.

Trojan Horse A program that pretends to be a genuine shareware product but is actually designed to do mischief such as erasing the disks on a specified date.

typeface A complete set of printed characters that are created with a single style.

type family A complete set of characters in a particular style, such as Times Roman or Helvetica.

type size The height of a typeface measured in points.

type style An attribute of a typeface such as normal, bold, and italics.

turnkey system A complete computer system of hardware, software, and service sold by one vendor that is ready to perform a particular task with no further preparation.

twisted-pair cable A relatively low-speed transmission medium consisting of two insulated copper wires twisted together. *(See figure.)*

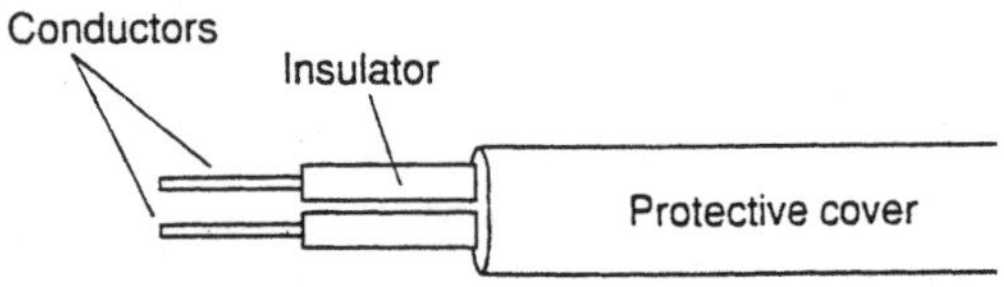

twisted-pair cable. *(Courtesy Kelmar Associates.)*

uninterruptible power supply (UPS) A device that contains a rechargeable battery to keep the computer up and running in the event of a power failure.

UNIX A multiuser operating system developed at Bell Laboratories in 1969. UNIX is the most popular operating system used on workstations.

upgrade Buying a newer version or more powerful hardware or software to increase the performance or capacity of a computer.

upload To send a file to another computer via a modem or network.

upper-memory area (UMA) Memory between 640 kbytes and 1 Mbyte, reserved for drivers that control peripheral devices.

UPS *See* UNINTERRUPTIBLE POWER SUPPLY.

upward compatibility A situation in which software works not only on the computer for which it was designed but also on newer models.

user-friendly Refers to hardware or software that is relatively easy for people to use.

user interface The hardware and software that provide a means by which humans and computers communicate.

utility A program that helps you perform maintenance tasks on the computer, such as backing up a hard disk, searching for files, checking for virus, or tuning up the system.

vaccine A computer program that offers protection from viruses by checking the integrity of the operating system and files.

vector graphics Graphics based on lines, curves, and shapes described by mathematical formulas.

very large-scale integration (VLSI) Refers to integrated circuits containing 10,000 or more logic gates.
□ VLSI is replacing large-scale integration (LSI) because it is smaller, faster, and less costly. *See also* LARGE-SCALE INTEGRATION.

VESA *See* VIDEO ELECTRONICS STANDARDS ASSOCIATION.

VGA *See* VIDEO GRAPHICS ARRAY.

video capture board An expansion board that is used to convert analog video from a VCR or camcorder into digital video and store it on disk.

videoconferencing A telecommunications application that uses a two-way, full-motion video and a two-way audio system for the purpose of conducting meetings between two distant locations.

Video Electronics Standards Association (VESA) An organization of graphics hardware manufacturers who devised and published the popular *VESA Local Bus* specification.

Video Graphics Array (VGA) The most widely accepted IBM graphics standard announced in 1987, with resolution of 640×480 pixels at 16 colors.

virtual reality Computer systems that simulate reality using computer-controlled three-dimensional video imaging.

virtual storage Also known as *virtual memory*. A technique of extending the size of a computer's memory by using a disk file to simulate additional memory space.

virus *See* COMPUTER VIRUS.

VISICALC The first spreadsheet program developed for the Apple II in the late 1970s.

VL Bus (VESA Local Bus) A popular expansion bus specification designed by VESA to provide fast graphics performance for 486 PCs.

voice-grade channel A wide-bandwidth communication channel that can transmit data at rates varying between 300 and 9600 bits/s.
□ Telephone lines are a form of voice-grade channel.

voice recognition *See* SPEECH RECOGNITION.

volatile memory Memory that loses its contents when the power is off. Volatile memory is sometimes called *read-and-write* memory.

wait state A brief delay in the CPU's operation as it waits for a slower component such as memory or the expansion bus to finish its operations.

warm boot *See* BOOT.

wide area network (WAN) A network spanning a wide geographic area.

wildcard A DOS symbol (? or *) that substitutes for all or part of a filename in a file specification.

Winchester disk *See* HARD DISK.

window A defined rectangular area on the screen for a special purpose or application.
□ Windows can be opened, closed, sized, and overlapped.

Windows A graphic user interface designed by Microsoft for IBM-compatible computers.

WIN.INI A windows configuration file that contains information on settings not directly related to hardware.

word processing The use of a computer and appropriate software to write, edit, format, store, and print documents.

word size The total number of bits that a register can store at one time.

word wrap A word-processing feature that automatically places a word on the next line as soon as the current line becomes full.

workstation A powerful microcomputer typically used for scientific research and engineering design. Many workstations are designed around Reduced Instruction Set Computing (RISC) architectures. *(See figure.)*

workstation. *(Courtesy Hewlett-Packard Co.)*

□ *Workstation* also refers to a personal computer connected to a network.

WORM *See* WRITE ONCE, READ MANY.

write once, read many (WORM) Refers to a type of optical disk that can be written to only once. Once written, data cannot be erased or changed.

write-protected Refers to a disk that can be read from but not written to.

write-protect notch The rectangular opening in the jacket of a floppy disk that allows the read/write head to access the disk.

WYSIWYG (what you see is what you get) Refers to on-screen displays that are close or identical to the same displays sent to the printer.

XModem A protocol for transmitting files from one computer to another and detecting transmission errors if they occur.

X-ON/X-OFF A communication protocol that uses ASCII characters to control the flow of data to ensure that the receiving computer and sending computer stay in sync.

YModem A communication protocol based on XModem, but faster. YModem is capable of transmitting more than one file at a time.

zip file A file compressed with the PKZIP utility; usually has the zip file extension.

ZIP socket Zero-insertion-force socket. A socket for CPUs that makes it easier to replace a chip, or upgrade the motherboard.

ZModem A fast and widely supported communication protocol that offers high-speed transmission and batch transfer.

INTERNET INTRODUCTION

The Internet is a network made up of tens of thousands of networks throughout the world that allow computer users to exchange and share information. It is virtually a storehouse of information contained in the most famous libraries, research centers, universities, government agencies, businesses, and industries throughout the world. This *global network* or *information highway* allows a person to exchange electronic mail (e-mail) with another person anywhere in the world instantaneously, or in a few seconds or minutes.

Internet Terms

10 base-T A cabling option for Ethernet LAN that uses unshielded twisted-pair cable to connect a computer to a network.

56K line A digital phone-line connection capable of carrying 56,000 bits/s.

acceptable-use policy (AUP) A policy statement from a network or organization that defines the acceptable uses of the network for local use and accessing the Internet.

ACK An abbreviation for acknowledgment. A response sent by a receiver that indicates information has been received.

Acrobat Software developed by Adobe Corporation that allows users to translate PostScript files into a Portable Document Format (PDF) and to view them on common computer platforms.

address A unique string of text that identifies an individual on the Internet or the location of a Web page on the World Wide Web, for example, e-mail: steve.yuen@usm.edu

Address Resolution Protocol (ARP) An Internet protocol used to convert an Internet address into the corresponding physical address.

administrative domain (AD) A collection of hosts and routers and the interconnecting networks, managed by a single administrative authority.

advanced digital network (ADN) Often refers to a 56-kbit/s leased line.

AFAIK Common abbreviation used in newsgroup postings and e-mail, meaning "as far as I know."

alias A user-created alternative name for a command or string. Often used for long e-mail addresses.

America Online (AOL) The largest commercial online service in the United States that has access to the Internet.

Anonymous FTP A scheme that allows users to log onto a remote computer without the need for the assigned user IDs and transfer files by logging on anonymously. *See also* FILE TRANSFER PROTOCOL.

AOL *See* AMERICA ONLINE.

API *See* APPLICATION PROGRAMMING INTERFACE.

Application Programming Interface (API) A specification of a function that defines how a service is invoked through a software package.

Archie Short for *archive*. A network system for finding files stored on anonymous FTP sites. *(See figure.)*

backbone A high-speed line and a primary conduit for traffic that link major computer centers together. Subnetworks attach to the backbone.

bandwidth The difference, in hertz (Hz), between the highest and lowest frequencies of a transmission channel.

□ In data-communication systems, high bandwidth allows fast transmission or transmission of many signals at one time.

BBS *See* BULLETIN BOARD SYSTEM.

Berkeley Internet Name Domain (BIND) Implementation of a DNS name server developed and distributed by the University of California at Berkeley.

BITNET ("because it's time" network) A low-cost, low-speed academic and research network consisting primarily of IBM mainframe computers running the VMS operating system.

□ BITNET separates from the Internet, but e-mail is freely exchanged between the Internet and BITNET.

bits per second (bits/s) A unit for measuring the speed of data transmission.

bookmark A tool for collecting favorite home pages and addresses in World Wide Web (WWW) browsers and in Gopher services for future references.

bounce The return of an e-mail because of an error in its delivery.

bridge A device that connects and passes packets between two networks. *See also* ROUTER.

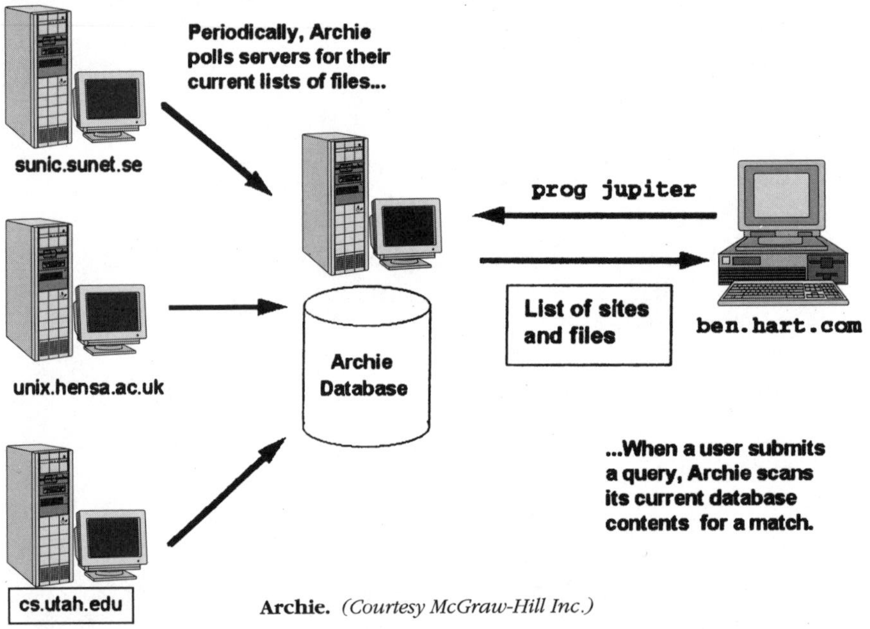

Archie. *(Courtesy McGraw-Hill Inc.)*

ARPANet Advanced Research Projects Administration Network. The precursor to the Internet. Established by the U.S. Department of Defense in 1969 as an experiment in wide area networking that would withstand nuclear attack.

authentication The verification of the identity of a person or process.

broadband channel A communication channel that can transmit data at high speed.

□ Coaxial cables, microwaves, satellites, and fiber optics are forms of broadband channels.

broadcast A packet sent to all network destinations. *See also* MULTICAST.

browser A client application software that allows

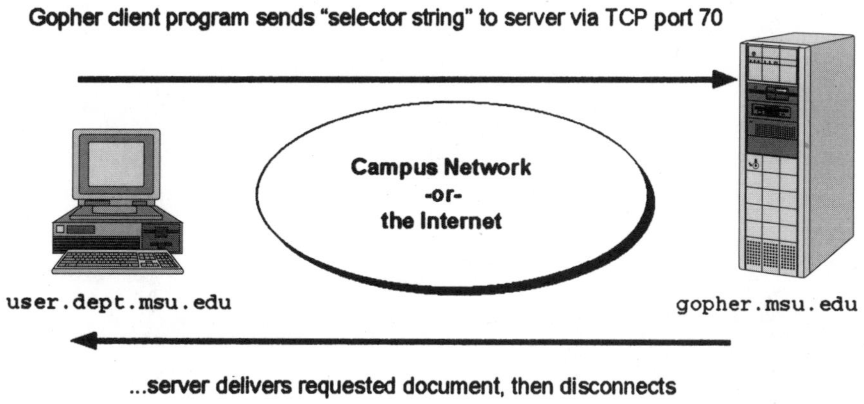

Campus Wide Information System. *(Courtesy McGraw-Hill Inc.)*

users to navigate and view WWW documents. Internet Explorer, Mosaic, and Netscape are examples of browsers.

BTA A common abbreviation in e-mail and newsgroup postings, meaning "but then again."

BTW A common abbreviation in e-mail and newsgroup postings, meaning "by the way."

bulletin board system (BBS) A computer system that provides an electronic forum, message center, archives of files, and any other services to its members.

Consultative Committee for International Telegraphy and Telephony (CCITT) An international organization responsible for the development of many communications standards such as X.25 and X.400.

Campus Wide Information System (CWIS) An information system for making information available (e.g., course schedules, current events, online library catalog, and job openings) to a campus or organization. *(See figure.)*

CCITT *See* CONSULTATIVE COMMITTEE FOR INTERNATIONAL TELEGRAPHY AND TELEPHONY.

CGI-BIN The most common name of a directory on a Web server in which CGI programs are stored.

chat The electronic version of Citizens Band (CB) radio that provides person-to-person, real-time conferencing.

Clearinghouse for Networked Information Discovery and Retrieval (CNDIR) An organization that promotes and supports the use of networked information discovery and retrieval software applications such as WAIS, Gopher, and WWW.

client A software program used to contact and obtain data from a server software program at a remote computer.

COM Refers to a commercial domain.

commercial information service A for-profit computer network such as CompuServe, Prodigy, or America Online that provides access to a wide range of financial, informational, and recreational services for a monthly or per-minute charge.

CompuServe A commercial online communications network that offers network services primarily through a telephone system.

connect time The amount of time your computer is connected to a communications network.

cookie A mechanism by which server side operations (such as CGI scripts) can store and retrieve information on the client side of the connection.

CU-SeeMe An Internet-based, real-time, multiparty, videoconferencing program developed by Cornell University that runs on Macintosh- and Windows-based systems.

- ☐ Provides a one-to-one conference or—by use of a reflector—a one-to-many, a several-to-several, or a several-to-many conference depending on user needs and hardware capabilities.

cyberspace A term originated by William Gibson in his novel *Neuromancer* to describe the world of computers and the society that gathers around them.

datagram Also known as *IP datagram*. A packet of data transmitted over the Internet.

- ☐ A message sent across the Internet is transmitted in one or more IP datagrams.
- ☐ Each datagram consists of the IP addresses of the sending and receiving computers and the data being sent.

dial-up access To connect to an Internet service provider through a phone line and a modem. *(See figure on next page.)*

digest A collection of messages from a Listserv or mailing list sent at regular intervals.

discussion list A form of group discussion and information sharing carried on electronic mail.

DNS *See* DOMAIN NAME SYSTEM.

domain name The unique name that identifies an Internet site.

- ☐ Usually has two or more parts separated by dots.
- ☐ A suffix that indicates the type of organization or country of origin. Examples of suffixes: com (commercial), edu (education), gov (U.S. government),

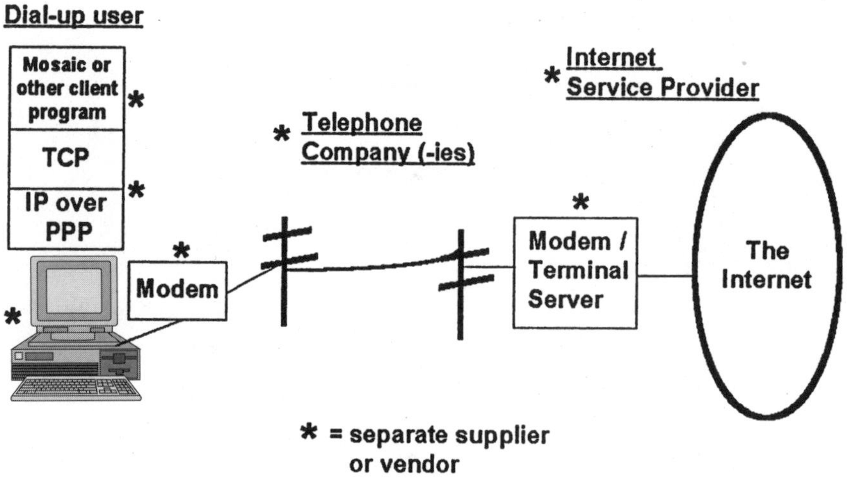

dial-up access. *(Courtesy McGraw-Hill Inc.)*

org (organization), mil (U.S. military), au (Australia), ca (Canada), and hk (Hong Kong).

- ☐ Examples of domain names: www.microsoft.com, evergreen.cc.usm.edu, and www.pbs.org.

domain name system (DNS) A system used to translate the name of a site (such as www.netdoor.com) into its numeric Internet address (e.g., 204.255.16.8) and vice versa.

- ☐ DNS allows users to use the Internet without remembering long lists of numbers.

dot address Also known as *dotted decimal notation*. Refers to the common notation for Internet addresses in the form of <n.n.n.n.>.

EBONE An acronym for European backBONE. The wide area network facilities that interconnect many European countries to one another and to the Internet.

electronic mail (e-mail) A system that allows users to send and receive messages electronically via a communications network; e-mail is one of the most popular uses of the Internet.

e-mail *See* ELECTRONIC MAIL.

e-mail address An electronic mailbox address assigned to a user.

- ☐ A user must have the e-mail address of the recipient in order to send an e-mail.
- ☐ An e-mail address is comprised of UserID@node. An example is jdoe@ocean.usm.edu, where jdoe is the user name, ocean is the machine name, usm is the regional node, and edu is a large domain.

Ethernet A local area network specification developed by Xerox Corporation that transmits data at 10 Mbits/s over coaxial cable.

Eudora A popular e-mail program for Windows and Macintosh computers.

FAQ *See* FREQUENTLY ASKED QUESTION.

FDDI *See* FIBER DISTRIBUTED DATA INTERFACE.

Fetch A popular Macintosh FTP program developed by Dartmouth University that allows users to transfer files on the Internet.

Fiber Distributed Data Interface An ANSI standard for transmitting data on optical fiber cables at a rate of 100 Mbits/s.

- ☐ FDDI is 10 times faster than Ethernet and about twice as fast as T3.

File Transfer Protocol (FTP) A program (and protocol) for transferring files between two Internet sites. *(See figure.)*

- ☐ Anonymous FTP allows anyone to log on to a system and retrieve files by logging in anonymously.
- ☐ Other FTP systems require the user to enter a password.

Finger An Internet software tool that displays information about a particular user, or all users, logged on the local system or a remote system.

firewall A combination of hardware or software that separates a LAN into two or more parts to protect against unauthorized access.

flame An emotional, obnoxious, or inflammatory mes-

File Transfer Protocol. *(Courtesy McGraw-Hill Inc.)*

sage delivered by e-mail or public posting on a newsgroup or electronic bulletin board.

□ Also applied to messages that contain strong criticism or disagreement with a previous message or article.

freenet A community-oriented network that allows users to access the Internet free of charge or for a small membership fee.

frequently asked question (FAQ) A document, often associated with mailing lists and newsgroups, that provides answers to commonly asked questions on a specific topic, technology, or Internet service.

□ FAQs are usually the first places users should look to find answers to questions or to get information on a specific topic.

FTP *See* FILE TRANSFER PROTOCOL.

FYA A common abbreviation of "for your amusement" used in e-mail and newsgroup postings.

FYI An abbreviation often used in e-mail and news, meaning "for your information."

gateway A hardware or software setup that transfers information between different types of networks.

□ A gateway may connect two different networks, like DECnet and the Internet, or it may allow two incompatible applications to communicate over the same network.

GIF *See* GRAPHICS INTERCHANGE FORMAT.

Gopher A hierarchical menu system that allows users to access files, documents, and other Internet services on the Internet. Developed at the University of Minnesota, Gopher is characterized by relative simplicity of design and ease of implementation.

□ Gopher is a client/server program that requires the user to have a Gopher client program.

Gopher server A computer that allows Internet users to access and use Gopher.

Gopherspace The network of Gopher servers available on the Internet.

Graphics Interchange Format (GIF) The most common graphic file format developed by CompuServe that is used to store and transfer graphic images on the Internet.

hacker Originally referred to a person who spent many hours to refine a computer program or system; now refers to someone trying to break into computer systems illegally.

header The portion of a packet, preceding the actual data, containing source and destination addresses and error-checking fields.

hit A count that is recorded anytime someone connects to a remote computer, site, or home page.

hog The passage of a packet through a router.

home page A page or screen of a site accessible through a World Wide Web browser.

host Any computer on the Internet that allows users to communicate with other computers.

□ Technically, any computer connected via Internet Protocol is considered a host.

hotlist A list of favorite or important links to sites, resources, and services compiled during WWW sessions for future use. *See also* BOOKMARK.

hostname The name given to a computer. *See also* DOMAIN NAME.

HTML *See* HYPERTEXT MARKUP LANGUAGE.

HTTP *See* HYPERTEXT TRANSPORT PROTOCOL.

hub Center of a star topology network that houses the network software and direct communications within the network. It can also act as the gateway to another network.

hyperlink Using HTML tags to reference a Web location.

hypertext A system for supporting embedded links within documents. Clicking on a hypertext word on a Web page may take you to another location, page, document, or links to sounds, graphics, or movies.

□ Hypertext links are not limited to the local computer network and can take you to information located on another computer on the Internet.

HyperText Markup Language (HTML) A formatting language consisting of a series of tags that are read by the World Wide Web browser to determine text styles, page layout, and links to other documents and files.

HyperText Transport Protocol (HTTP) The communication protocol (rules or procedures) established for the World Wide Web that tells computers how to handle and send hypertext documents and data from one computer to another. HTTP is the most important protocol used to access a World Wide Web document.

HYTelnet A hypertext database of the Internet that is accessible by Telnet. It presents a hypertext interface to an organized list of Telnet sites that are arranged in categories by the type of service, such as a library catalog, bulletin board, electronic book, and network information.

image map A special type of graphics file that allows Web users to click on an area of an image that jumps to other documents, images, graphics, or Web pages on the Internet.

image tag An HTML tag used to insert a graphical image in a World Wide Web page.

IMHO A common abbreviation in e-mail and newsgroup postings meaning "in my humble opinion."

inline image A graphics file in a Web page that is aligned with text using HTML tabs.

Integrated Service Digital Network (ISDN) A set of high-speed transmission standards designed to permit telephone networks to carry data, voice, and other source material.

□ A standard 128-kbit/s ISDN line is about 5 times faster than a 28.8-baud modem and one-twelfth the speed of a 1.55-Mbit/s T1 line.

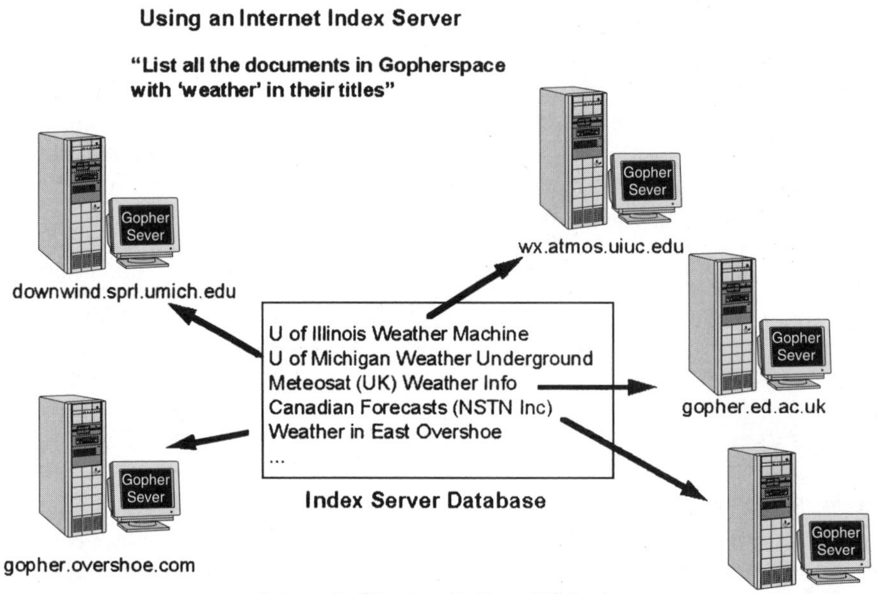

Internet. *(Courtesy McGraw-Hill Inc.)*

interest group Group discussion and sharing of information carried on by electronic mail.

Internet The worldwide network of networks, initially linking academic, research, and government facilities but now available to anyone through online services and network gateways. Also called *information highway,* the *Net, cyberspace,* or *data superhighway.* *(See figure.)*

Internet Architecture Board (IAB) The technical body that oversees the development of the Internet standards and protocols.

Internet Engineering Task Force (IETF) A part of IAB that is responsible for designing and testing new protocols to be used on the Internet.

Internet Multicasting Service An Internet service that sends audio information across the Internet like a radio station.

Internet Network Information Center (InterNIC) An organization funded by the National Science Foundation (NSF) to provide information about the Internet, registration, and database services and other documentation.

Internet Protocol (IP) A specification for the format of packets that computers use to route information from one computer to another on the Internet. IP makes sure that the packets arrive at the correct destination. *See also* TRANSMISSION CONTROL PROTOCOL and TCP/IP.

Internet Protocol: The Next Generation (IPng) Refers to the successor of IP being planned by the Internet Engineering Task Force.

Internet Relay Chat (IRC) An Internet tool that enables users to chat in real time using a keyboard on the Internet.

□ IRC is similar to CB radio channels, where many users can converse on one channel.

□ Any user can create a channel, and anything that a user types in a given channel is seen by all other users in the channel.

Internet service provider (ISP) A company or organization that provides access to the Internet for individuals or businesses.

Internet Society (ISOC) A nonprofit, professional organization established to facilitate and support a worldwide information network.

Internet telephone Also called *cyberphone.* An Internet application that allows real-time voice communication over the Internet.

InterNIC *See* INTERNET NETWORK INFORMATION CENTER.

intranet A private network inside a company or organization that uses the same kinds of software for internal use as found on the public Internet.

IP address Also called *IP number, dotted quad,* and *Internet address.* A unique number assigned to each computer on the Internet.

□ IP addresses consist of four numerals, each in the range of 2 through 255 separated by periods, for example, 135.9.5.16.

IPng *See* INTERNET PROTOCOL: THE NEXT GENERATION.

IRC *See* INTERNET RELAY CHAT.

ISDN *See* INTEGRATED SERVICE DIGITAL NETWORK.

Java A simple, object-oriented, platform-independent, multithreaded, dynamic, general-purpose programming language developed by Sun Microsystems.

□ Java is best for creating Applets and applications for the Internet, intranets, and any other complex, distributed network.

□ Using small Java programs called *Applets,* Web pages can include functions such as animations, interactive graphics, calculations, and other fancy tricks.

Joint Photographics Expert Group (JPEG) A standard used for the compression of color still photographs.

JPEG *See* JOINT PHOTOGRAPHICS EXPERT GROUP.

Jughead An Internet search tool named as a pun on the Archie comic strip.

leased line A permanently connected private phone line between two locations.

- ☐ Typically used to connect a LAN to an Internet service provider.

Listproc Software used to administer and manage a discussion list, interest group, or mailing list.

Listserv Software used to manage subscriptions to a mailing list. *(See figure.)*

- ☐ Listservs originated on BITNET, but they are now common on the Internet.

group, but all postings are sent directly to the subscribers by e-mail.

mail reflector A special mail address that allows automatic forwarding of e-mail to a set of other addresses. Often used to implement a mail discussion group.

Mailserve Software used to administer and manage a discussion list, interest group, or mailing list.

Majordomo Software used to manage and administer a mailing list.

map A search tool that displays locations of Web sites geographically. Examples of maps are the Virtual Tourist and the Wanderer, a directory of service sites organized by geographic location.

MAVEN A software program developed by Charley Kline at the University of Illinois to allow for audioconferencing on the Internet.

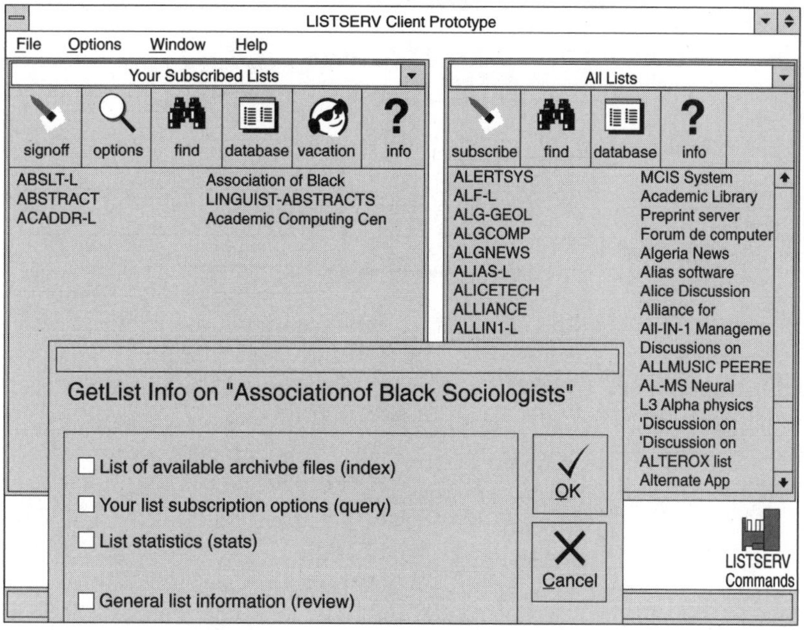

Listserv. *(Courtesy McGraw-Hill Inc.)*

login/logon The process of entering a user account ID and password to gain access to a host computer or Internet service provider.

logout/logoff The process of disconnecting your computer from the host computer.

LOL A common abbreviation in e-mail and newsgroup postings meaning "laughing out loud."

lurker A person who reads mailing lists, bulletin boards, online chats, or Usenet groups without contributing or posting a message.

Lynx A text-based World Wide Web browser that is usually available on UNIX-based hosts.

MacTCP An extension program that allows a Macintosh to connect to the Internet.

mailbox A storage area that holds incoming e-mail messages until a user reads the mail.

mail filter Software that screens incoming e-mail.

mailing list A discussion group similar to a news-

MBONE The multicast backbone of the Internet.

MIME *See* MULTIPURPOSE INTERNET MAIL EXTENSION.

Multipurpose Internet Mail Extension (MIME) An extension of Internet mail standards that provides the ability to transfer nontextual data, such as graphics, spreadsheets, formatted word-processor documents, and sound files.

mirror A service that replicates another service; for example, a site may mirror another site's Gopher or FTP offerings.

moderated A newsgroup or mailing list that has a person who screens all incoming messages before posting them to its readers.

MOO *See* MUD, OBJECT-ORIENTED.

Mosaic A multipurpose client program, developed by the National Center for Supercomputing Applications (NCSA), that provides a graphical user interface (GUI) to multiple Internet services, including Gopher,

WAIS, WWW, Usenet News, and FTP.

Motion Picture Experts Group (MPEG) A standard for the compression of color, full-motion video.

MUD The multiuser dungeon game programs.

MUD, object-oriented (MOO) One of several kinds of multiuser role-playing environments.

multicast The technique used to send a given packet of information to a selected computer.

multiuser simulated environment (MUSE) One kind of MUD, usually with little or no violence.

radio and atomic clocks located on the Internet.

newbie An individual who is new to the Internet and its use.

news article A message that appears on a Usenet newsgroup.

newsfeed An agreement by which one site sends copies of network news articles to another.

newsgroup A bulletin board system on the Internet that is dedicated to a particular topic of interest. *(See figure.)*

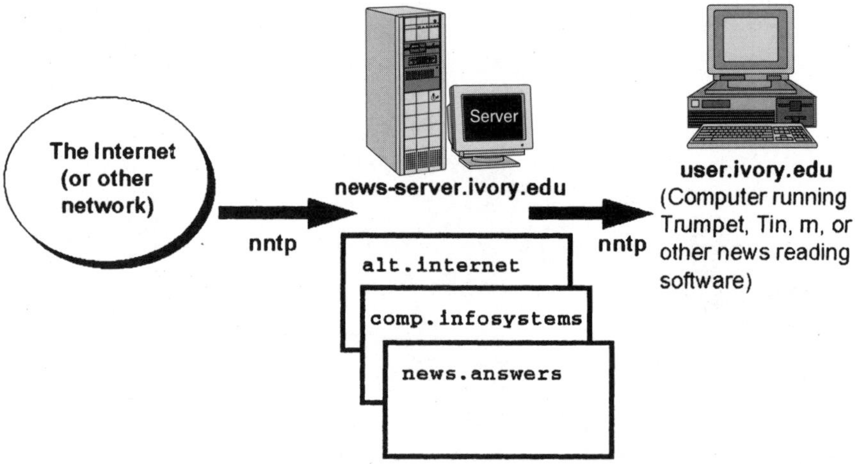

newsgroup.
(Courtesy McGraw-Hill Inc.)

name server A server provided on the Internet that resolves network names into IP addresses.

National Science Foundation Network (NSFNET) A very high-speed backbone network controlled by the National Science Foundation (NSF) that carries much of the noncommercial Internet traffic in the United States.

navigating Moving from one electronic site to another on the Internet.

Net, the *See* INTERNET.

NetFind An Internet service that allows users to find e-mail addresses based on information, such as a person's name, location, or domain name.

Netiquette A collection of social and behavioral rules for the Internet.

Netizen A "citizen" of the Internet or someone who uses networked resources.

NetNews Also called *Usenet News*. The collection of articles available to users on the Usenet network.

Netscape Navigator The most popular graphical Web browser developed by Netscape Communications Corporation.

Network Information Center (NIC) Any organization that serves the Internet community by providing user assistance, documentation, training, and other services. InterNIC is an example of the NIC.

Network News Transmission Protocol (NNTP) A protocol for the distribution, inquiry, retrieval, and posting of Usenet news articles.

Network Time Protocol (NTP) A protocol that assures accurate local timekeeping with reference to

newsreader A client software that allows users to read, post, and reply to newsgroup articles. Free Agent (for Windows), NewsWatcher (for Macintosh), and Tin (for Unix) are examples of newsreader.

NIC *See* NETWORK INFORMATION CENTER.

NNTP *See* NETWORK NEWS TRANSMISSION PROTOCOL.

NSFNET *See* NATIONAL SCIENCE FOUNDATION NETWORK.

Online Computer Library Catalog (OCLC) A nonprofit organization that offers online computer-based services to libraries, institutions, and their users.

- OCLC connects more than 10,000 libraries worldwide.

packet The unit of data sent across the Internet.

- On the Internet, files are broken into packets and each packet travels independently. When the packets reach their destination, they are reassembled.
- A packet includes a header containing control information (addresses of the sender and the recipient) and user data.
- Packet sizes can vary from roughly 40 to 32,000 bytes, depending on network hardware and media, but packets are normally less than 1500 bytes long.
- An Internet packet is also called an *IP datagram*.

Packet Internet Groper (PING) A program used to test the reliability of a network device. PING sends the computer a packet and waits for a reply.

packet switching The method used to move data around on the Internet that allows the most efficient use of the network.

□ On the Internet, data from the sending machine is broken up into chunks; each chunk has the address of where it came from and where it is going. This method allows chunks of data from many different sources to commingle on the same lines and be sorted and directed to different routes by special machines along the way. This way many users can use the same lines at the same time.

pager A program that displays the contents of a file one screen at a time.
□ In UNIX, examples of pagers are "more" and "pg."

password A code that must be supplied at login.
□ Hosts use passwords to provide security from unauthorized access.
□ Good passwords contain letters and nonletters and are not easy to guess.

PEM *See* PRIVACY-ENHANCED MAIL.

PICO The full-screen text editor commonly used with the PINE e-mail program in UNIX.

PINE A program for Internet news and e-mail in UNIX.

PING *See* PACKET INTERNET GROPER.

Point-to-Point Protocol (PPP) A popular method that allows a computer to connect directly to the Internet using a standard phone line and a high-speed modem. Similar to Serial Line Internet Protocol.

POP *See* POST OFFICE PROTOCOL.

port number A number that identifies a particular Internet application.

post To send an article or message to a Usenet newsgroup.

postmaster An individual who manages e-mail software on a computer.

Post Office Protocol (POP) A protocol designed to allow single-user hosts to read mail from a mail server. *(See figure.)*

remote login The ability to log in to a remote computer and become a user. *See also* TELNET.

RFCs Official documents (Requests for Comments) that define both proposed and adopted Internet protocol standards.
□ RFCs published by the IETF are available on the Internet.

Rlogin A terminal emulation program, similar to Telnet, offered in Unix.

round-trip time (RTT) A measure of delay on a network at any one time.

route A path that network traffic takes from its source to its destination.

router A device that connects different types of network.
□ Routers have the capability to forward packets from one network to another, based on the network protocol.

RTFM A common abbreviation in e-mail and newsgroup postings meaning "read the fine manual."
□ Often used to "flame" someone who has asked a question that could be answered easily by reading the available documentation.

search engine A program on the World Wide Web that allows users to search the Web for information, people, news, documents, software, etc. Alta Vista, Excite, Infoseek, Lycos, and WebCrawler are examples of search engines.

Serial Line Internet Protocol (SLIP) A standard for using a regular line and a high-speed modem to connect a computer as a real Internet site.
□ Similar to PPP, but SLIP is gradually being replaced by PPP.

service provider See INTERNET SERVICE PROVIDER.

SGML *See* STANDARD GENERALIZED MARKUP LANGUAGE.

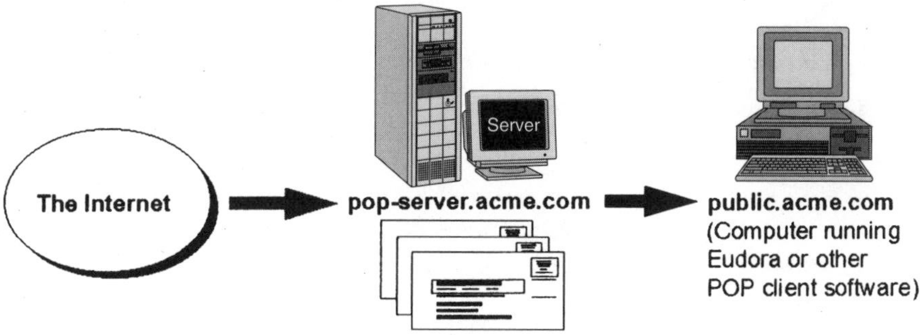

Post Office Protocol.
(Courtesy McGraw-Hill Inc.)

PPP *See* POINT-TO-POINT PROTOCOL.

Privacy-Enhanced Mail (PEM) A proposed standard for encrypting e-mail on the Internet.

protocol A set of rules governing the transmission of data between two devices.

RealAudio A commercial client and server product developed by Progressive Networks that uses proprietary data-compression schemes to move real-time audio across the Internet.

Shockwave A delivery platform developed by Macromedia Inc. for viewing interactive multimedia and high-impact graphics on the Web.

signoff A term used in a command to leave, quit, or unsubscribe from a discussion list, interest group, Listserv list, or mailing list.

signature A small text file often inserted at the end of e-mail messages or Usenet news articles that contains a name, mailing address, phone number, quote, etc.

Simple Mail Transfer Protocol (SMTP) An Internet protocol for sending e-mail messages between computers on the Internet.

Simple Network Management Protocol (SNMP) The Internet standard protocol developed to manage nodes on an IP network.

SLIP *See* SERIAL LINE INTERNET PROTOCOL.

SMDS *See* SWITCHED MULTIMEGABIT DATA SERVICE.

smilie Characters used in e-mail and news articles to indicate humor and irony.

□ For examples, :-) means a smiling face, ;-) means winky smilie, and :-(means a frowning smilie.

SMTP *See* SIMPLE MAIL TRANSFER PROTOCOL.

snail mail Refers to the mail sent by the U.S. Postal Service.

spam To send unwanted or commercial messages to many lists, Usenet newsgroups, or online addresses at the same time. Considered rude.

spider A program that visits Web pages to gather, compile, and index information.

Standard Generalized Markup Language (SGML) A language used to describe structural information embedded within a document.

□ SGML is a popular language in desktop publishing and is the predecessor of HTML used by the World Wide Web.

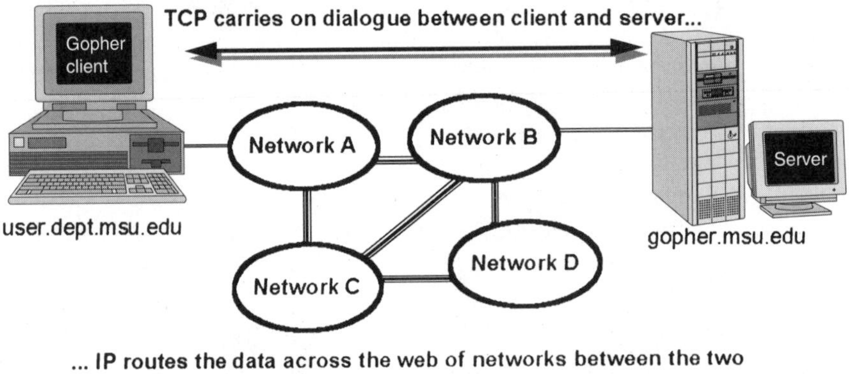

Transmission Control Protocol. *(Courtesy McGraw-Hill Inc.)*

strand A series of linked Web pages.

subscribe The process of joining a discussion list, interest group, Listserv list, or mailing list.

surfing Browsing or exploring the Internet more or less at random, hopping from document to document, host to host, etc.

Switched Multimegabit Data Service (SMDS) A standard for high-speed, packet-switched data transfer offered by the telephone companies.

SYSOP An individual (system operator) responsible for the physical operation of a computer system or network resource.

T1 A leased-line connection capable of carrying data at 1.544 Mbits/s.

□ T1 provides fast speed connection commonly used to connect networks to the Internet.

T3 A leased-line connection capable of carrying data at 45 Mbits/s.

□ T3 provides a very fast speed connection that is more than enough to do full-screen, full-motion video.

tags HTML codes.

□ Tags can be written in upper- or lowercase and are enclosed in brackets, < >.

TCP/IP An acronym for Transmission Control Protocol/Internet Protocol. A set of protocols developed by the U.S. Department of Defense in the 1970s to transfer packets of information around the Net.

□ All computers that use the Internet must have TCP/IP software.

Telnet An Internet tool that allows a user to log in to a remote server or a host computer and use its computing resources.

thread A series of linked messages on the same theme or topic in newsgroup postings.

time out Refers to a situation in which two computers are communicating and one computer fails to respond for whatever reason.

TIN A UNIX-based newsreader.

TN3270 A special version of the Telnet program that allows TCP/IP connections to IBM mainframes.

tour guide Also called *guide* or *travel guide.* A search tool used to find new, unusual, and outstanding Web pages. Spry City and Global Network Navigator (GNN) are examples of tour guides.

Transmission Control Protocol (TCP) One of the two major parts of TCP/IP protocols. TCP provides reliable transmission of data. *(See figure.)*

TTFN A common abbreviation of "ta ta for now" used in e-mail and newsgroup postings.

unicast A technique for sending a packet via the Internet from a single source to a single destination.

Uniform Resource Locator (URL) The standard used for specifying the server and path information for documents on the World Wide Web. A URL consists of one of the following:

http://www.abc.com/index.html
telnet://psupen.psu.edu

ftp://sumex-aim.standford.edu/info-mac

unsubscribed The process of removing a discussion list, interest group, Listserv list, or mailing list.

Usenet A collection of newsgroups or bulletin boards where Internet users can read and exchange news items electronically on a wide variety of topics. *(See figure.)*

□ Numerous Usenet newsgroups are on the Internet for a variety of topics.

UNIX-to-UNIX Copy Program (UUCP) A protocol used for communication between UNIX systems.

uudecode A program used to re-create binary files from the ASCII or text form to which they are converted by uuencode.

uuencode A program used to convert a binary file into ASCII form so that it can be sent by e-mail to another Internet user.

uunet A nonprofit organization that provides low-cost access to e-mail, netnews, source archives, public-domain software, and standard information.

VERONICA An acronym for Very East Rodent-Oriented Netwide Index to Computerized Archives. A tool that searches for keywords in gopher menus throughout the Internet.

vi A UNIX-based, full-screen, text editor.

videoconferencing A system that allows users to exchange video information over the Internet.

virtual library An online collection of electronic books and journals on the Internet.

Virtual Memory System (VMS) A Digital Equipment Corporation operating system.

Virtual Reality Modeling Language (VRML) A computer language that allows an individual to program in a virtual reality environment where the screen gives the illusion of being in a three-dimensional world.

VMS *See* VIRTUAL MEMORY SYSTEM.

VRML *See* VIRTUAL REALITY MODELING LANGUAGE.

VT 100 The dominant communications protocol for full-screen terminal sessions; commonly used with Internet services and programs.

WAIS *See* WIDE AREA INFORMATION SERVERS.

Web browser *See* BROWSER.

Web page A document on the World Wide Web that displays text, images, movies, and animation.

White Pages A directory listing of Internet users that are accessible through the Internet. It is set up similar to telephone white pages that are indexed to name of person, place, and resource.

Whois An Internet tool that allows users to query a database of people and other Internet entities, such as domains, networks, and hosts, kept at the NIC.

Wide Area Information Servers (WAIS) An Internet search service that allows users to locate documents that contain key words or phrases on the Internet.

Winsock A program that conforms to a set of standards called the *Windows Socket API*.

□ The Winsock program controls the link between Windows software and a TCP/IP program.

World Wide Web (WWW or W3) Also called the *Web*. A hypertext-based, distributed information system, developed at the Corporation for Research and Education Networking (CERN) in Switzerland, that allows users to access information organized by hypertext-linked pages.

X.25 An ISO communication protocol widely used in many wide area networks.

X.400 An ISO standard protocol for sending electronic messages from one network to another. It is widely used in Canada and Europe.

Yellow Pages (YP) A directory service often used by UNIX administrators to manage databases distributed across a network.

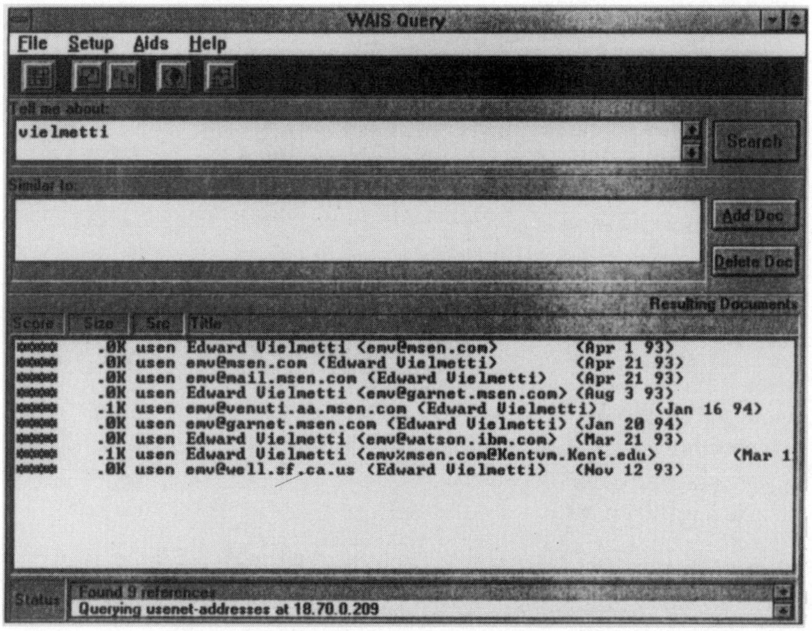

Usenet.
(Courtesy McGraw-Hill Inc.)

SECTION 2

CNC Machines and Programming

Ken "Kun" Li
Manufacturing Engineer
GEC Precision Corporation
Wellington, Kansas

The first numerical control (NC) machines developed in the 1950s were basic machine tools equipped with a laboratory-constructed control unit. Modern computer numerical control (CNC) technology with contouring capabilities and automatic tool changers has been applied to all types of machine tools, resulting in an increase in productivity and a reduction in manufacturing costs. The most common metalworking CNC machines are machining centers, chucking and turning centers, electrodischarge machines, coordinate measuring machines, and computer-aided design/manufacturing (CAD/CAM) design systems.

***A* axis** The axis of rotary movement of a CNC machine tool member or slide about the *X* axis. *See also* COORDINATE SYSTEM. It is used to
- □ Specify angles about the *X* axis.
- □ Distinguish positive from negative angular motions, allowing use of the "right-hand rule."

abort The stopping of a CNC part program or computer program caused by hardware, software, or the operator.

absolute accuracy Accuracy as measured from a specified reference.

absolute angle An angle measured from the polar reference line, which is positive when counterclockwise or negative when clockwise.

absolute coordinate/dimension Measurement made from an origin or fixed reference point in the coordinate system, rather than from another point or location. *(See figure.)*

absolute positioning Locating the tool in relation to a fixed datum (reference) point.

absolute programming A mode of CNC programming in which all axis movements are made in relation to a fixed datum point.
- □ The preparatory function G90 must be included in the CNC part program for absolute programming.
- □ Compare with *incremental programming.*

absolute system A CNC system in which all positional dimensions, both input and feedback, are taken with respect to the same common datum point (a fixed point of origin).
- □ The alternative is the *incremental system.*

ACCANDEC *Acc*eleration *and dec*eleration in feedrate.
- □ It provides smooth starts and stops when operating under CNC, and also when changing from one feedrate value to another.

access time The time interval between the instant at which information is
1. Called for from storage and the instant at which delivery is completed: the *read* time.
2. Ready for storage and the instant at which storage is completed: the *write* time.

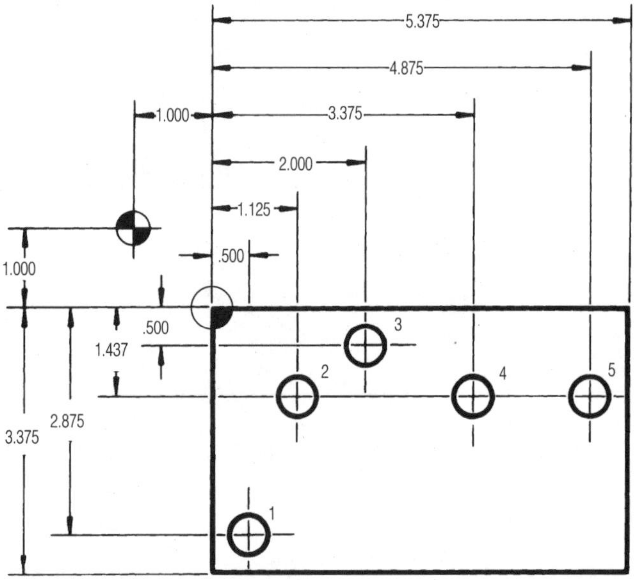

absolute dimensioning. *(Courtesy Kelmar Associates.)*

accumulative error An error that can build up in relative coordinate dimensioning if the user is not careful.

ACCUPIN A General Electric trademark for its linear feedback transducers.

accuracy The degree of closeness to correct value.

□ On CNC machine tools, it is a measure of the control system's ability to position the machine table at a desired or commanded location, which is defined by a set of axis coordinate values: accuracy=(control resolution)/2 + 3 (standard deviation of mechanical error).

□ Do not confuse *accuracy* with *precision,* which refers to the degree of preciseness of a measurement.

ADaptation of APT (AD-APT) An Air Force adaptation of the APT programming language with limited vocabulary but many features of APT.

□ It can be used on some small-sized computers for both positioning and contouring jobs.

adaptive control A means of automatically adjusting feeds and/or speeds of a cutting tool from sensor feedback (continuous monitoring) to maintain the best cutting conditions. *(See figure.)*

adaptive control. *(Courtesy Cincinnati Milacron, Inc.)*

□ Sensors may measure cutting forces, torques, cutting temperature, vibration amplitude, horsepower or spindle deflection, work material hardness, and width or depth of cut.

□ Adaptive control has the capability to respond to and adjust for these variations during machining.

□ It also provides long tool life and/or lower machining cost.

□ Current adaptive control machining systems are generally systems where certain limits are set on each process variable.

address (word address) **1.** A format of CNC part programming in which each word is preceded by a letter to identify its function. Several examples are *N* for block sequences, *G* for preparatory functions, and *M* for miscellaneous functions. **2.** A computer system location which can be referred to in a computer program. It can be a main memory location, a terminal, a peripheral device, a cursor location, or any other physical item in a computer system.

administrative data Information used to manage the manufacturing operation as a business on a daily basis, such as materials and labor costs.

AGV *See* AUTOMATICALLY GUIDED VEHICLE.

algorithm An ordered, logical sequence of finite, precise steps that describe the solution to a given problem.

□ CNC and computer programs are developed by this method.

alphanumeric coding A coding system in which the character set contains letters of the alphabet, numerals 0 to 9, punctuation marks, and special signs.

□ CNC code systems use alphanumeric characters.

analog (data/quantity/signal) Continuous variable physical quantities, such as the amplitude, phase, or frequency of a voltage; the amplitude or duration of a pulse; the angular position of a shaft; or the pressure of a fluid. *(See figure.)*

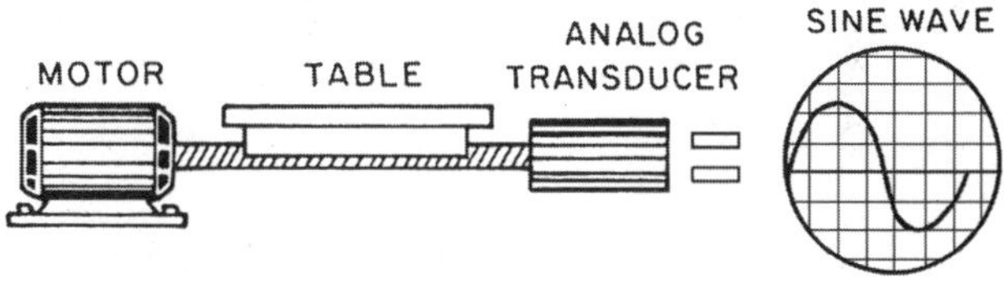

analog signal. *(Courtesy Kelmar Associates.)*

□ In CNC, analog information such as electrical voltage magnitudes or ratios is used to represent physical axis positions.

□ Analog data is contrasted with digital data represented in a discrete (discontinuous) form.

AND A logical operator that has the property such that the function X and Y is True only when both logic variables *X* and *Y* are true. It is usually represented by a centered dot "•," or by an asterisk "*" within a boolean expression.

AND gate A logic circuit which is designed to produce a resultant TRUE signal only when all the inputs are true.

application software Computer programs that are developed to perform specific problem-solving tasks such as inventory control, tool monitoring, or management reports.

APT *See* AUTOMATICALLY PROGRAMMED TOOLS.

arc, clockwise An arc produced by the coordinated motion of two axes, in which the tool movement is clockwise to the work movement when viewing the plane of motion from the positive direction of the perpendicular axis. *(See figure on next page.)*

arc, counterclockwise An arc produced by the coordinated motion of two axes, in which the tool movement is counterclockwise to the work movement when viewing the plane of motion from the positive direction of the perpendicular axis.

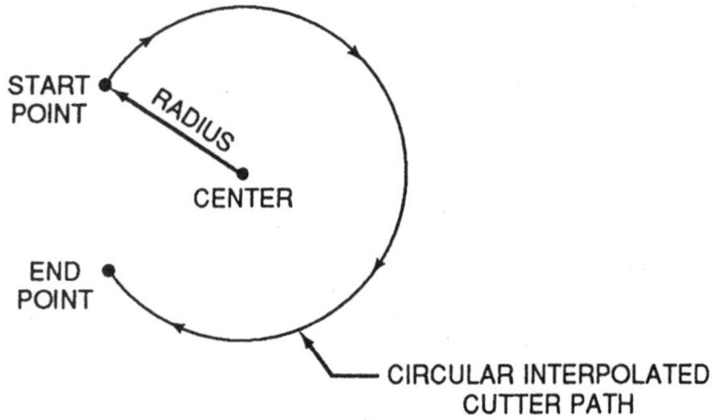

clockwise arc. *(Courtesy Kelmar Associates.)*

arithmetic/logical unit (ALU) A part of the CPU of a computer that performs the arithmetic and logical operations.

AUTO-MAP *see* AUTOMATIC MACHINING PROGRAMMING

AUTOmatic MAchining Programming (AUTO-MAP) A computer-aided part programming language system used for simple contouring and straight-line programming. It is a subset of APT.

automatic pallet changer (APC) An automatic system of changing workpieces on CNC machining centers, coordinate measuring machines, and other systems. Workpieces are loaded onto interchangeable subtables called *pallets*.

automatic storage and retrieval system/warehouse (ASRS/W) A computer-controlled, automated materials-handling system/warehouse in which parts, materials, or products are automatically stored and retrieved from storage locations using equipment such as stacker-cranes and narrow-aisle and miniload equipment.

automatic system for positioning tools (AUTOSPOT) A computer-assisted CNC part programming system for positioning and straight-cut operations.

automatic tool changer (ATC) A mechanical armlike device on the CNC machine tool for automatically changing cutting tools under the control of a part program. *(See figure.)*

automatically guided vehicle (AGV) A driverless vehicle, usually battery powered, that is used to load, unload, and transport various cargoes within an overall materials-handling system.

□ It has automatic guidance equipment and follows a prescribed guide path around the manufacturing floor. The route can be modified by remote or on-board computers.

automatically programmed tools (APT) A system for computer-assisted part programming with the objective of providing a means for the part programmer to communicate the machining instructions to the machine tool in simple English-like statements. Modern versions of APT can be used for both positioning and continuous-path programming in up to five axes. There are four types of statements in the APT language:

1. *Geometry statements* (*definition statements*)—define the geometric elements that constitute the workpart
2. *Motion statements*—describe the path taken by the cutting tool
3. *Postprocessor statements*—specify feeds and speeds and actuate other features of the specific machine tool and control system
4. *Auxiliary statements*—identify the part, tool, tolerances, and so on

automation Automatically controlled operation of an apparatus, process, or system by mechanical or electronic devices that take the place of human observation, effort, and decision.

auxiliary axis Any machine axis other than the main *X, Y,* and *Z* axes. It may be either linear or rotary.

auxiliary function A programmable function, usually ON/OFF-type operation, of a CNC machine tool other than the control of machine movement.

auxiliary storage Secondary data storage other than main memory storage, on high-memory-capacity magnetic tapes, disks, or CD-ROMs. *See also* SECONDARY STORAGE.

axis A principal direction of relative motion between cutting tool and workpiece, which may be either linear or rotary, and is identified by a letter of the alphabet. *(See figure on next page.)*

□ There are usually three linear axes at right angles to each other (*X, Y,* and *Z*).

axis inhibit A function of CNC machine tools that prevents movement of the selected slides with the power on.

axis interchange A capability of CNC machine tools for inputting the information concerning one axis into the storage of another axis.

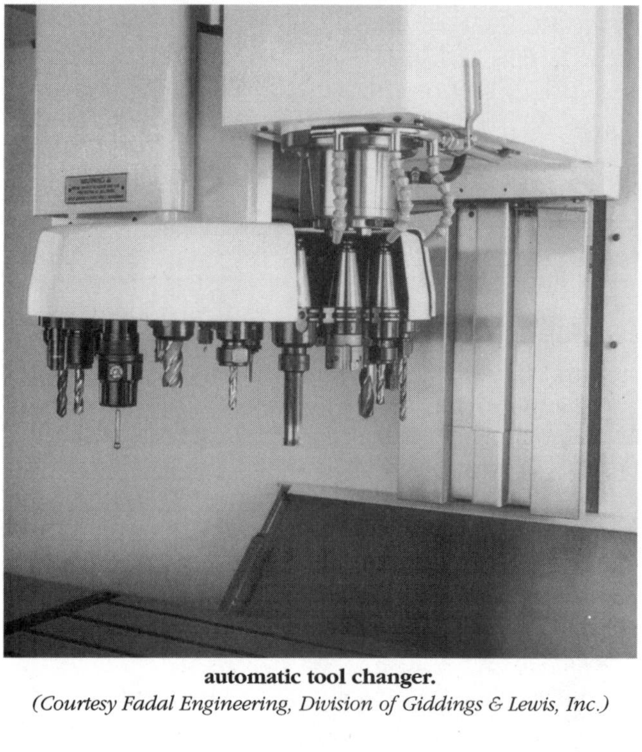

automatic tool changer.
(Courtesy Fadal Engineering, Division of Giddings & Lewis, Inc.)

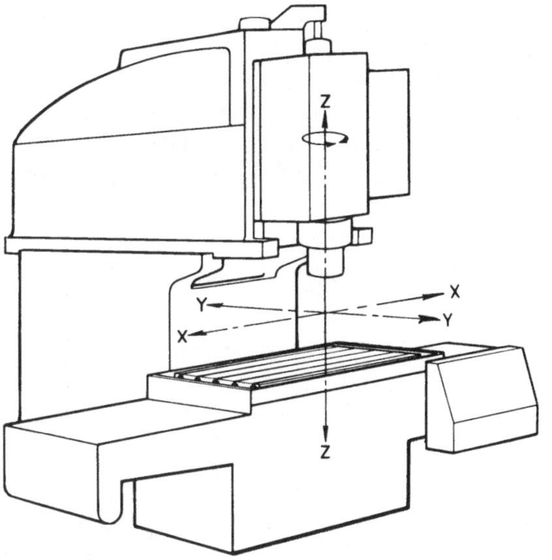

axis (machine). *(Courtesy Modern Machine Shop.)*

axis inversion A capability of a CNC system that allows the reversal of normal plus and minus values along an axis. It makes possible the machining of a left-handed part from right-handed programming or vice versa. Same as mirror image.

***B* axis** The axis of rotary motion of a machine tool member or slide about the *Y* axis.

backing storage Mass storage medium for the permanent storage of programs and data.

□ For CNC applications, punched tape, magnetic tape, or magnetic disk is generally used.

backlash The amount of mechanical play (relative movement) between interacting mechanical parts as a result of looseness.

bar code A standard identification system in which an array of rectangular, high-contrast bars and spaces of relative widths are printed on labels to identify parts or products *(see figure)*. These codes can be read automatically by a bar-code reader or scanner for data collection.

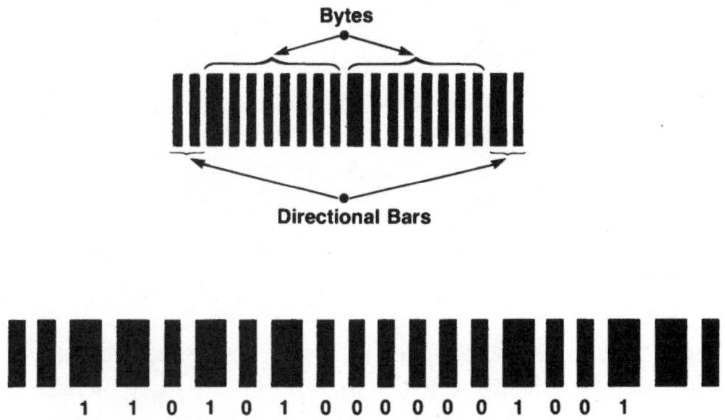

bar code. *(Courtesy Hewlett-Packard Co.)*

BCD *See* BINARY-CODED DECIMAL.

behind-the-tape-reader (BTR) system A DNC arrangement in which the computer is linked directly to the regular CNC controller unit, replacing the tape reader by the telecommunication lines.

binary-coded decimal (BCD) A binary representation system for decimal numbers in which each decimal digit 0 to 9 is represented by four binary digits 0000 to 1001.

□ Most CNC machine tools use binary-coded decimals in a form compatible with the ON/OFF switching signals of a digital computer.

black box An approach to the visualization of the operation of a system by considering only the input to and the output from that system. Since it is not necessary to know how the system works (only how it affects the input to produce the output), the system is known as a *black box.*

block A collection of words as a complete CNC instruction which consists of all the information required to carry out one machine operation. It is usually one line of a part program.

□ Each block is identified by a unique block or sequence number and ends with an "end of block" (EOB) character.

block delete A capability of CNC systems that allows selected blocks of a part program to be ignored by the control system, at the discretion of the operator and with permission of the programmer.

block diagram A simplified schematic drawing in which a system or computer program is represented by annotated boxes for the various elements to show the function and by interconnecting lines to show the relationship between the elements.

***b*oundary *rep*resentation (B-REP)** A type of solid modeling technique that defines surfaces around solid objects.

bulk memory A high-capacity auxiliary data storage device such as a disk, drum, tape, or CD-ROM.

bus A path or channel for transmitting electrical signals or data between computers and peripheral devices.

***C* axis** The axis of rotary motion of a CNC machine tool member, or slide, about the *Z* axis. *(See figure.)*

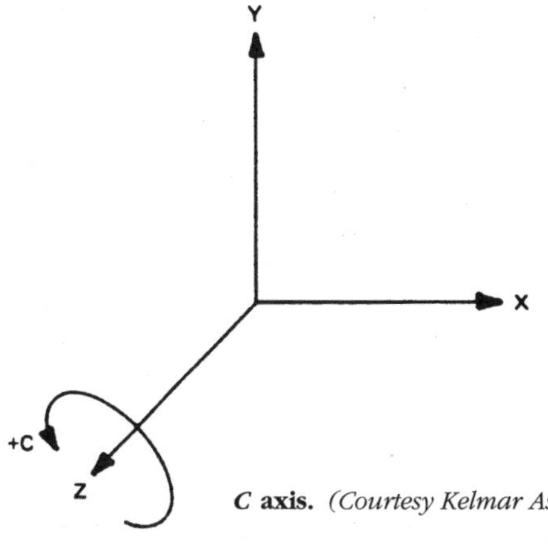

***C* axis.** *(Courtesy Kelmar Associates.)*

canned/fixed cycle A fixed sequence of machine operations initiated by a single preparatory G function.

For example, code G84 will perform tap cycle by CNC. Each fixed-cycle command must be accompanied by additional information.

□ Canned cycles decrease programming time and simplify the part program.

card punch A device for punching holes in cards in accordance with a standard code scheme.

card-to-tape converter A device that converts information directly from punched cards to punched or magnetic tapes.

cartesian coordinates A set of three numbers that define the location of a point within a rectilinear coordinate system which consists of three axes (*X, Y,* and *Z*) perpendicular to each other *(see figure).* The numbers represent distances from the origin, the intersection of the three axes.

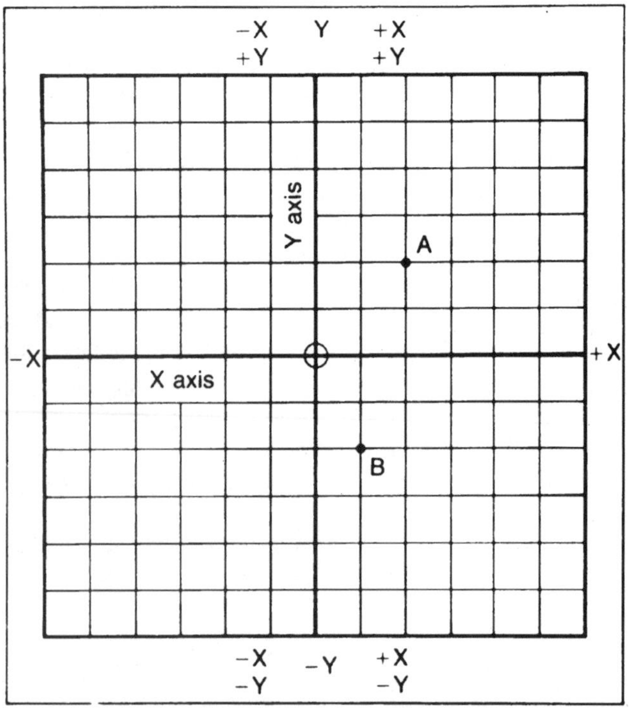

cartesian coordinates. *(Courtesy Allen Bradley.)*

cathode-ray tube (CRT) A special form of electronic vacuum tube in which a controlled, focused beam of electrons strikes a luminescent screen to display graphical images.

chad Tiny pieces of material removed in a card- or tape-punching operation.

check digit/checksum A digit attached to a numeric or coded data item for error checking in the recording of the coded data. Its value is based mathematically on the characters that make up the data item.

□ The system reading the data uses the same formula to compute the check digit.

□ Any difference between the two values indicates an error in transmission.

check surface The surface that stops (checks) the forward movement of the tool in its current direction in APT contouring motions.

circular interpolation An interpolation scheme for programming a circular arc by specifying the coordinates of its endpoints, the coordinates of its center, and its radius, as well as the direction of the cutter along the arc. *(See figure.)*

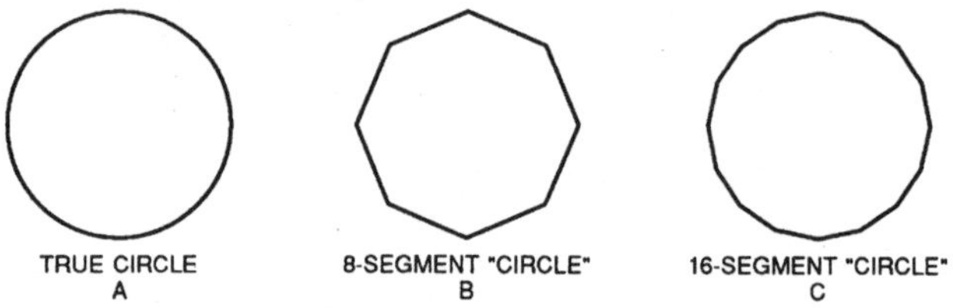

circular interpolation. *(Courtesy Allen Bradley.)*

□ A tool path consists of a series of straight-line segments, with the segments being calculated by the interpolation module rather than the programmer.

□ The cutter is directed to move along each line segment one by one in order to generate the smooth circular path.

CL data/file *See* CUTTER LOCATION DATA/FILE.

closed-loop control/feedback control system A machine control system in which outputs (e.g., slide position and velocity) are continuously monitored and fed back for comparison with the input commands (desired values).

□ Any difference between the input value and the measured value is used to regulate the system behavior toward a zero difference. The monitoring and comparing operations are collectively termed *feedback.*

□ The function of the feedback loop in a CNC system is to assure that the table and workpiece have been properly located with respect to the cutting tool.

□ Closed-loop CNC systems generally use DC servomotors or hydraulic actuators (less commonly used).

closed-loop MRP An enhancement to materials-requirements planning (MRP) which adds capacity-planning functions to adjust schedules or manufacturing resources in response to over- or under-scheduling conditions.

CMM *See* COORDINATE MEASURING MACHINE.

CNC *See* COMPUTER NUMERICAL CONTROL.

component/part family A collection of components or parts which possess similar design or manufacturing characteristics.

computer-assisted (-aided) part programming A part programming system in which the part programmer prepares the set of processing instructions in a high-level computer language (English-like statements) and the computer interprets that high-level language and commands and performs much of the tedious computational work and data processing required in manual programming.

□ When using one of the CNC programming languages, the part program usually consists of defining the geometry of the workpart and specifying the tool path and/or operation sequence.

□ The computer's job in computer-assisted part programming consists of the following: input translation, arithmetic calculations, cutter offset computation, and postprocessor.

computer-automated part programming The automatic generation of part programs using software capable of making logical, and even quasi-intelligent, decisions about how the part should be machined from the geometric model of a part that has been defined during product design.

computer numerical control (CNC) A CNC system that uses dedicated, stored program computers to carry out some or all of the basic numerical control functions.

□ CNC control features include storage of part programs, use of diskettes, program editing at the machine tool site, fixed cycles and programming subroutines, complex interpolation, positioning features for setup, cutter length compensation, online diagnostics, and standard communications interface.

computer programming language A defined set of representative characters or symbols, combined with specific rules and conventions necessary for their interpretation, which is used to communicate a set of instructions or program to the logic circuits of a computer.

constant surface speed (CSS) A CNC code (G96) used on lathes and turning centers that automatically adjusts the speed of the machine spindle as the workpiece diameter changes in order to keep a constant cutting speed.

constructive solid geometry (CSG) A type of solid modeling technique which uses a building-block approach to create a three-dimensional model out of simple shapes or primitives such as blocks or spheres.

contouring/continuous-path CNC A CNC system that can simultaneously control more than one axis movement, and continuously control the path of a machine tool to produce the desired geometry of a workpiece. *(See figure.)*

□ Straight or plane surfaces, circular paths, conical shapes, or almost any other forms are possible under contouring control.

□ This is the most complex, the most flexible, and the most expensive type of machine tool control.

control system A system of hardware and software that controls the operation of a CNC machine tool. For motion control, it may use either nonservo techniques that control endpoints only, or a servo control of the path and speed.

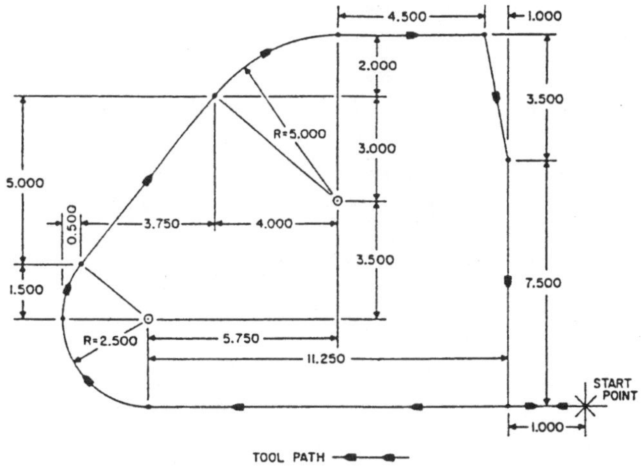

contouring. *(Courtesy Allen Bradley.)*

conversational programming A method of CNC part programming that allows a machine tool to be programmed by the operator who enters information in response to questions provided by the CNC system.

coordinate dimensioning word (*X*, *Y*, and *Z* words) A word in a block of a CNC part program that provides the coordinate positions of the tool.

coordinate measuring machine (CMM) A numerically controlled machine for measuring shapes and dimensions of solid objects, which consists of a contact probe and a means of positioning the probe in three-dimensional space relative to the surfaces and features of a workpiece to be measured. *(See figure.)*

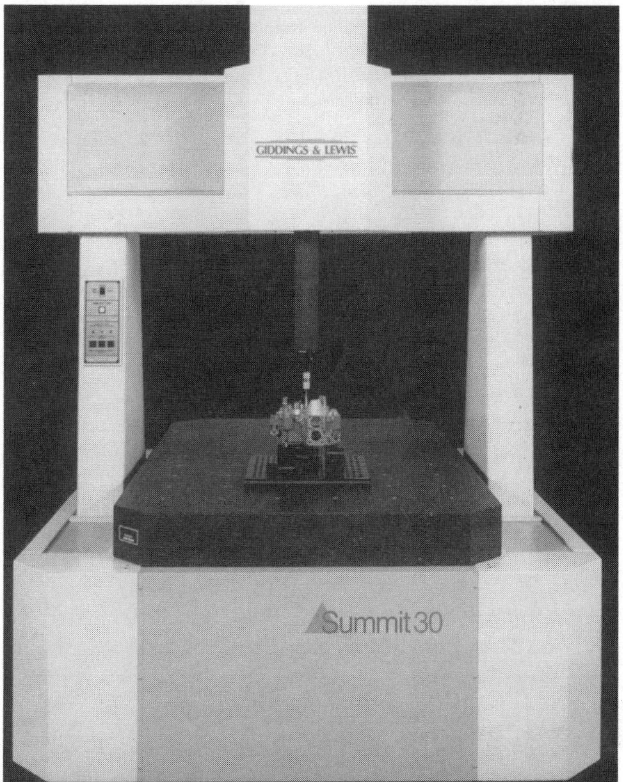

coordinate measuring machine. *(Courtesy Giddings & Lewis, Inc.)*

□ The locations of the tip of the probe can be accurately and precisely recorded to obtain dimensional data concerning the workpiece geometry.

coordinate system in CNC A standard axis system which describes the positions of the cutting tool with respect to the workpiece *(see figure)*. There are three *linear* axes—*X, Y,* and *Z*—and three *rotational* axes—*A, B,* and *C.*

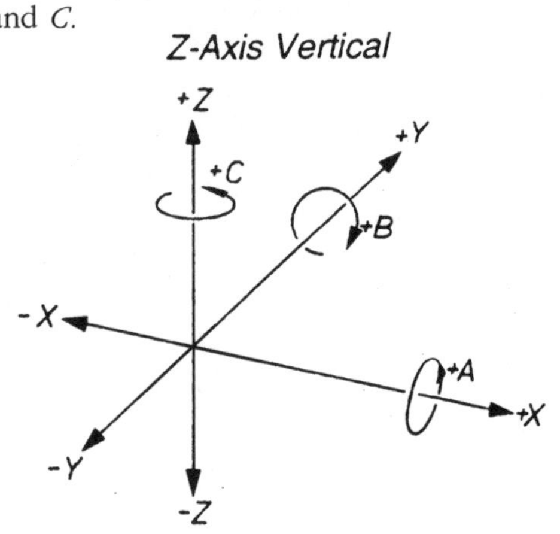

coordinate system. *(Courtesy Allen Bradley.)*

□ The *X* and *Y* axes are defined in the plane of the table; the *Z* axis is perpendicular to this plane, and movement in the *Z* direction is controlled by the motion of the spindle.

□ The *A, B,* and *C* axes are used to specify angles about the *X, Y,* and *Z* axes.

□ To identify positive from negative angular motions, the right-hand rule can be used.

crash zone An operator-defined boundary to restrict slide or table movement, so that the control system stops the machine movement and sounds an alarm signal to alert the operator if a programmed tool path attempts to enter this area.

CRT *See* CATHODE-RAY TUBE.

CSG *See* CONSTRUCTIVE SOLID GEOMETRY.

cubic interpolation An interpolation system that provides approximations of free-form curves using third-degree curves. It is often used in the automotive and aerospace industries to generate complex profiles with a small number of data inputs.

cutter diameter/radius compensation A capability of the CNC system that allows an automatic adjustment to the cutter centerline path, to account for the difference between actual and programmed cutter diameters. The net effect is to move the cutter center away from, or closer to, the edge of the workpiece.

cutter location data/file (CL data/file) Any *X, Y,* and *Z* coordinates and other CNC information describing a cutting tool centerline at the tip of the tool which is independent of any particular machine tool.

□ This is usually generated by a computer-assisted part programming system as a result of processing a geometric shape definition.

□ It is converted into part program codes for a specific machine by a postprocessor or link.

cutter offset A capability of the CNC system to offset the tool path (the axial center of a cutter) from the desired workpiece surface outline by the radius of the cutter.

cutter path/tool path The path taken by the center of a rotating cutter or tool or the tip of a lathe tool. *(See figure.)*

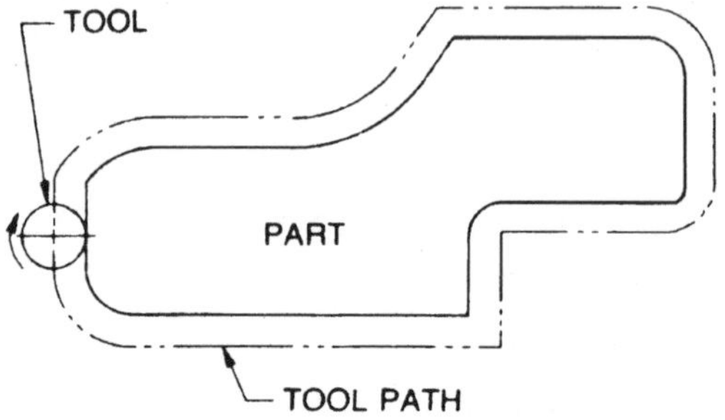

cutter/tool path. *(Courtesy Allen Bradley.)*

cycle time The period of time from starting one machine operation to starting another operation in a pattern of continuous repetition.

damping A characteristic built into electrical circuits and mechanical systems to provide resistance to motion introduced, reduce undesirable oscillations, and counter the effects of excessive overshoot or undershoot.

data Any facts, letters, numbers, or symbols that refer to or describe an object, concepts or instructions, and so on.

database A collection of related information that has a specific, predetermined structure and organization and is suitable for communication, interpretation, and processing by human or automatic means.

datum A reference point from which movement or measurements are made. *(See figure.)*

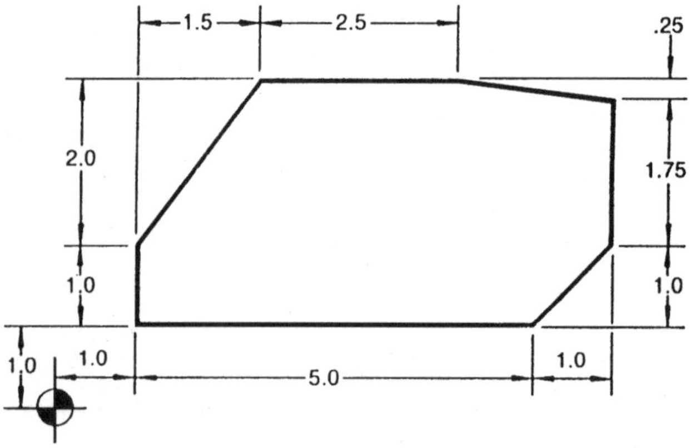

datum. *(Courtesy Kelmar Associates.)*

debugging The process of detecting, locating, and removing mistakes from computer software or hardware.

dedicated computer A computer system that is dedicated to performing a fixed, clearly defined task.

delete character A character which is used primarily to remove any unwanted characters on punched tapes.

diagnostic program/routine A special test program or routine that is used to locate and identify errors and malfunctions in hardware or software on a computer or machine control unit.

digital Representation of data in discrete numbers.

digital quantity A quantity that can be represented by one of the two mutually exclusive states, 1 and 0.

digital-to-analog converter (DAC) An electronic device that changes digital data into analog data.

digitize To obtain the digital representation of a measured quantity or analog signal.

digitizer A graphics tablet.

dimensional tolerance The amount of deviation (above and/or below) allowed from a stated dimension.

direct numerical control (DNC) The direct control of a number of separate CNC machine tools by a large central host computer. The part programs are downloaded from a host computer directly into the memory of a CNC machine tool as required. *(See figure.)*

- □ The system consists of four components: central computer, bulk memory which stores the CNC part programs, telecommunication lines, and machine tools.
- □ The advantages of DNC are timesharing, greater computational capability for functions such as circular interpolation, remote computer location, elimination of tapes and tape reader for improved reliability, and elimination of hard-wired controller unit on some systems.

display An input/output device that is able to display alphanumeric characters or graphics. It is typically a videomonitor, much like a television.

distributed numerical control A hierarchical control in which the machine tool control units are connected to a central plant computer and the controllers are themselves CNC units.

downtime The period of time during which the operation of production equipment, a computer, or a communications network is stopped because of mechanical or electric failure or lack of materials.

drift An undesired change in output over a period of time.

drive surface The surface that guides the side of the cutter in APT contouring motions.

drive unit A DC servomotor, stepping motor, or hydraulic actuator that is connected to the table by a leadscrew. Rotation of the motor causes the leadscrew to turn, producing linear movement of the table.

dry run Running a CNC part program in automatic mode with no component installed on the machine tool, to verify the programmed path of the tool under continuous operation.

dwell A CNC preparatory function (G04) that allows a short pause in a machining operation.

dynamic tool display A CAD/CAM device for graphically displaying a cutting tool and the tool path on a CNC machine tool, in order to verify and simulate the cutting process. *(See figure on next page.)*

dynamic visual feedback A computer graphics technique that allows the user to see the effects of actions to create and modify images such as rubberbanding or dragging.

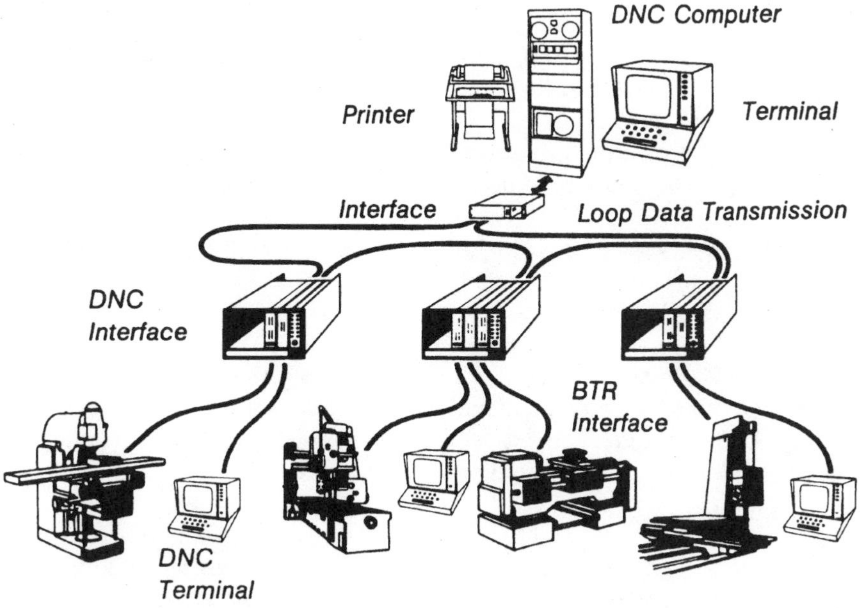

direct numerical control. *(Courtesy Modern Machine Shop.)*

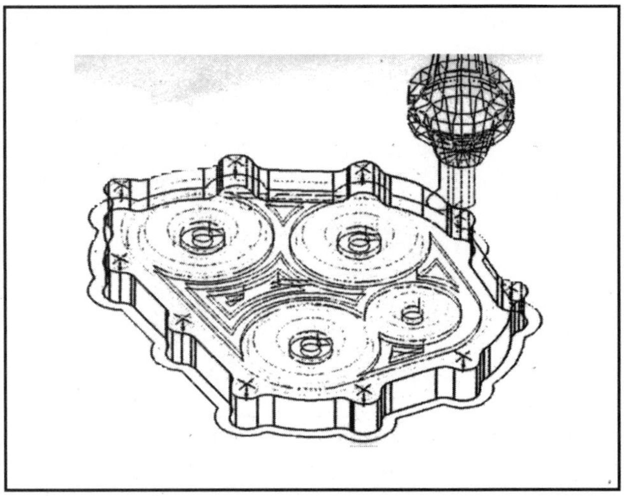

dynamic tool display. *(Courtesy Modern Machine Shop.)*

editing The rearrangement, addition, and deletion of data.

EIA coding system The U.S. Electronic Industries Association standard coding system for representation of characters on eight-track punched tapes, with track 5 used for parity checking.

electromagnetic interference (EMI) Unwanted electrical energy or noise that is produced in the circuits of a device by other electromagnetic fields.

encoder An electromechanical transducer that is used to change angular or linear position, or velocity into electrical signals.

- ☐ Encoders are commonly used on CNC machines because of their digital quality, which can be easily compared with the pulses produced by the CNC system.

end-of-block character A character that indicates the end of a block in CNC part programs.

end of program A CNC miscellaneous function (M30) indicating the end of a part program or the completion of a workpiece. It stops spindle, coolant, and feed after completion of all commands in the block.

error The difference between the measured, observed quantity and the desired, theoretically correct value.

EXAPT *See* EXTENDED SUBSET OF APT.

exclusive OR A logical operator between two logical variables which yields a result of 1 if one, and only one, of the variables has the value 1 and yields a result of 0 otherwise.

EXtended subset of APT (EXAPT) An APT-based system developed in Germany to compute optimum feeds and speeds automatically. There are three versions:

- ☐ *EXAPT I*—for positioning (drilling, and also straight-cut milling)
- ☐ *EXAPT II*—for turning
- ☐ *EXAPT III*—for limited contouring operations

F-word A CNC multiple-character code, containing the letter *F* followed by digits, that specifies the feedrate in a machining operation.

fault tolerance The ability of a computer system or program to continue to operate properly when a fault occurs. This can be accomplished by the hardware or by error-recovery software.

feed The amount of movement of the cutting tool into the workpiece per revolution of the workpiece in turning, per revolution of the drill in drilling, per tooth of the milling cutter in milling.

feedback A feature of closed-loop CNC systems in which a signal of the machine axis position or velocity is fed back to the controller and compared with the input command signal which specifies the demanded position. *(See figure.)*

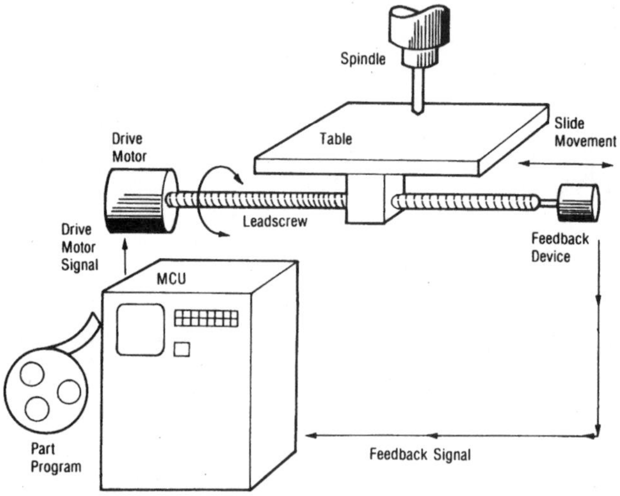

feedback system. *(Courtesy Modern Machine Shop.)*

- ☐ The result of the comparison of the error between the desired response and the actual response is used by the controller to adjust for differences.

feed engage point A point where the axis motion from rapid traverse to a programmed feed occurs.

feedrate bypass A function of CNC systems that directs the control system to ignore programmed feedrate and substitute a selected operational rate.

feedrate override A manual control function, usually a rotary dial, that can direct the CNC system to reduce or increase the programmed feedrate from a few percent to over 100 percent of the programmed feedrate.

first-off component The first machined component subject to dimensional, geometric, and physical inspection for the final proving of a CNC part program.

fixed automation The mechanization and control of a fixed or repetitive sequence of simple processing or assembly operations in order to achieve high production rates.

- ☐ Examples of fixed automation include mechanized assembly lines and machining transfer lines.

fixed-block format A format of a CNC part program where the words in each block must be in identical sequence and the characters within each word must be of the same length and in the same format.

□ The least flexible, and probably the least desirable, format of a CNC part program.

fixed cycle *See* CANNED/FIXED CYCLE.

fixed-sequence format A format of a CNC part program, in which words are written in a set order (fixed sequence) and no address letter is required.

fixed zero The origin of a CNC coordinate system that is always located at the same position on the machine tool from which all machine movements are referenced. *(See figure.)*

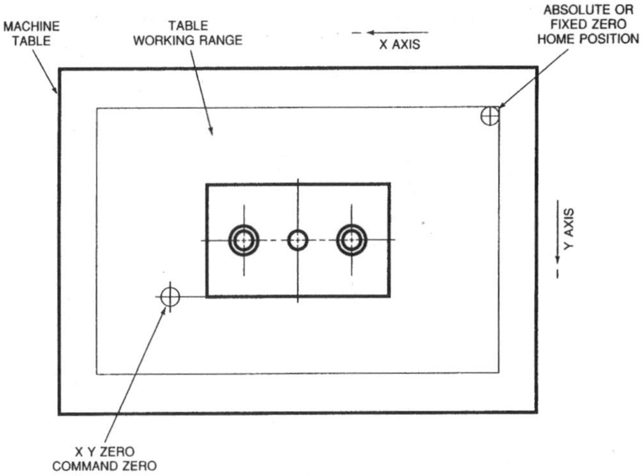

fixed zero. *(Courtesy Kelmar Associates.)*

□ Usually, it is the lower left-hand corner of the machine table, and all locations must be defined by positive *X* and *Y* coordinates from the fixed origin.

□ With fixed-zero systems, the part programmer and machine operator must reference the job to the machine's permanent zero point.

floating zero A common feature on modern CNC machines that allows the machine operator to set the zero datum at any position within the programmable area on the machine table.

□ The decision is based on part programming convenience; for example, the workpiece may be symmetrical, and the zero point should be established at the center of symmetry.

FORTRAN *For*mula *tran*slation. A high-level, algebraic-procedure-oriented computer programming language that uses common mathematical notation for programming engineering and scientific problems.

□ FORTRAN is used in handling the APT system.

G-word (code) A CNC word addressed by the letter *G* and followed by a numerical code, which defines preparatory functions to prepare the controller for instructions that are to follow.

gage height (plane) A predetermined height or plane above the workpiece surface along the *Z* axis to which the cutter retreats from time to time to allow safe *X-Y* table travel, or the rapid travel of the spindle changes to feed. *(See figure.)*

□ Also called the *reference plane* or *rapid level.*

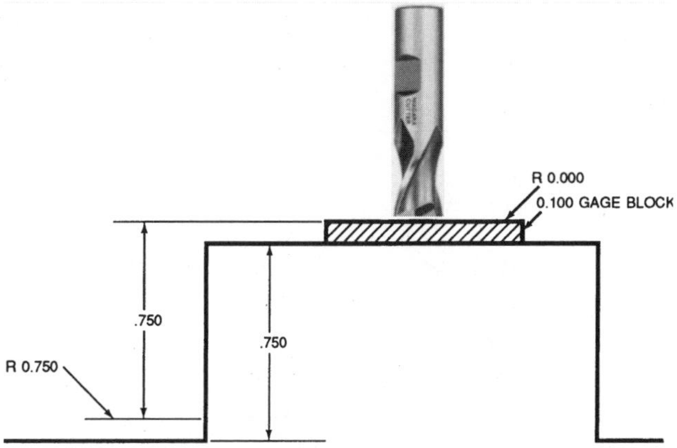

gage height. *(Courtesy Kelmar Associates.)*

geometric model A mathematical representation of the geometric form or properties of a part or product that is used by a computer to display and manipulate its images.

geometric tolerance The amount of allowable error, in shape or form, from true geometry.

grid plate A baseplate with accurately positioned dowel holes or machined slots, and tapped fixing holes in a grid or matrix formation at regular intervals. It is used for the quick and accurate location of components to be machined.

hard copy A visually readable form of data output on paper.

hardware All physical (mechanical, magnetic, electrical, and electronic) components of a computer system.

□ Contrast with *software.*

hard-wired A control system using physically wired circuitry to implement system function.

□ It is less flexible than software-programmed systems.

helical interpolation A form of interpolation that combines the circular interpolation for two axes with linear movement of a third axis; this permits the definition of a helical path in three-dimensional space. *(See figure.)*

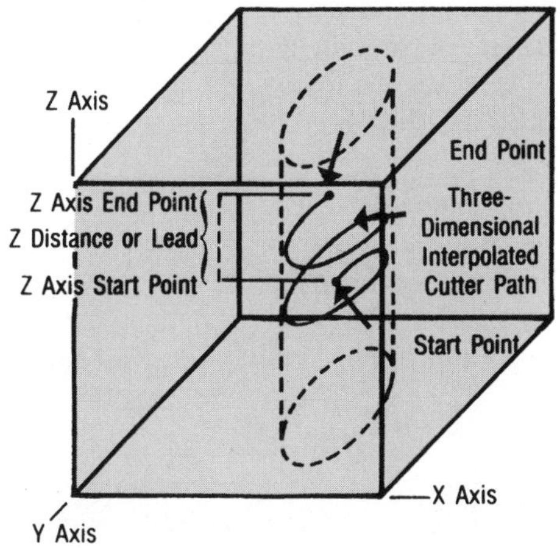

helical interpolation. *(Courtesy Modern Machine Shop.)*

host computer The primary or controlling computer in a multiple-computer system, usually equipped with a large memory and a variety of peripheral devices. It provides high-level services such as computation, database access, and special programs or programming languages for other computers in the network.

IGES *See* INITIAL GRAPHICS EXCHANGE SPECIFICATION.

incremental system A CNC system in which each coordinate location is taken from the previous position rather than from a fixed zero point *(see figure)*. It is initiated by a G91 command in the part program.

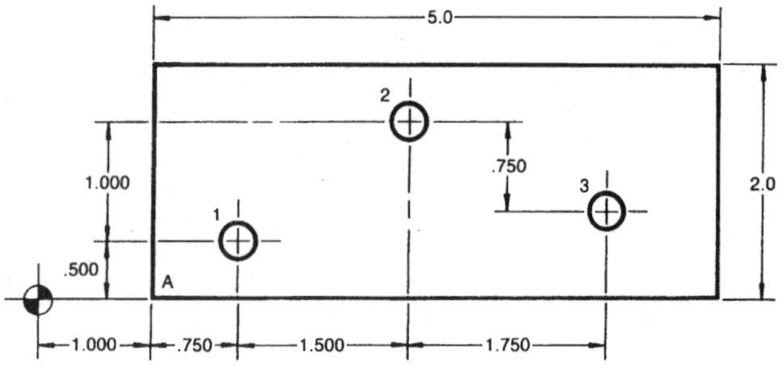

incremental dimensioning. *(Courtesy Kelmar Associates.)*

Initial Graphics Exchange Specification (IGES) A computer graphics standard that specifies a file structure and language format for communicating geometric data.

initialization The act of establishing fixed values in certain areas of memory, or setting counters, switches, or addresses to starting values at prescribed points in the execution of a program.

input Transfer of external information (data to be processed) by the keyboard or external storage device into a computer or machine control unit.

input/output (I/O) devices Computer equipment used to provide data to the computer (input devices such as a keyboard) or to people or other machines (output devices such as a display or printer).

input/output (I/O) routines Programs that interpret data transfer instructions, such as READ and WRITE, and control the movement of data between the I/O devices and computer memory.

instruction A statement to a computer or machine control unit that specifies an operation to be performed by the system and the values or locations of its operands.

interchange station The position in which the tool of an automatic tool-changing CNC machine awaits automatic transfer to either the spindle or the appropriate tool magazine slot.

interferometer An instrument that uses light interference phenomena to determine wavelength, spectral fine structure, indices of refraction, and very small linear displacements.

interlock A safety arrangement to control machines or devices to render their operation interdependent in order to ensure their proper coordination, or to ensure that a piece of apparatus will not operate until certain precautions have been taken.

intermediate transfer arm The mechanical device in automatic tool changing of CNC machines that grips and removes a programmed tool from the tool magazine slot and places it into the interchange station, where it awaits transfer to the machine spindle.

interpolation CNC routines produced by the interpolation module in the MCU to calculate the intermediate points the cutter must follow in order to generate a particular mathematically defined or approximated path. *(See figure.)*

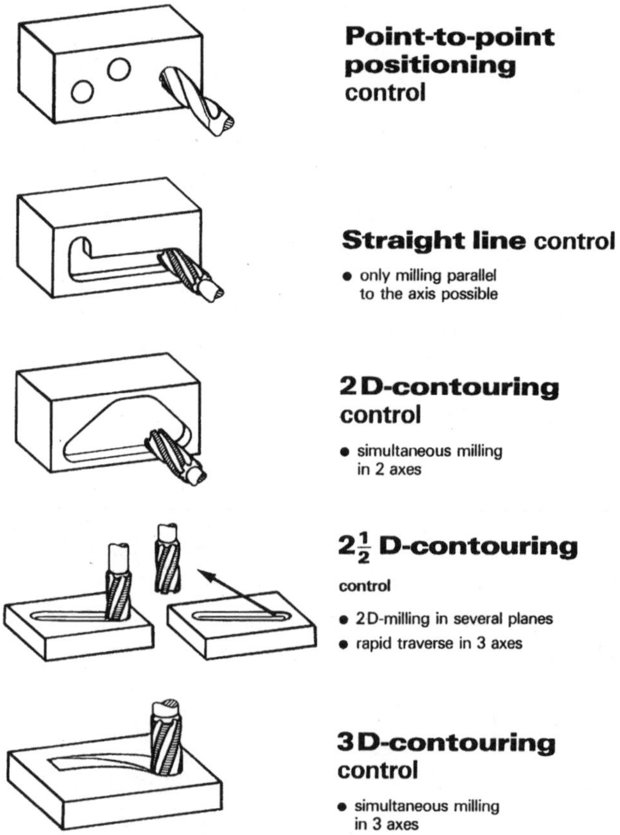

interpolation. *(Courtesy Deckel-Maho, Inc.)*

- ☐ There are interpolation systems for linear interpolation, circular interpolation, helical interpolation, parabolic interpolation, and cubic interpolation.
- ☐ Each of these systems allows the programmer or operator to generate machine instructions for linear or curvilinear paths, using relatively few input parameters.

interrupt A signal from a peripheral device or a request from a program to temporarily suspend the currently executing program and perform a specific service.

iteration A set of repetitive calculations in which the output of each step is the input to the next step with gradually narrowing results until an exact solution is found.

jogging A CNC function of manually controlling machine axis movement by depressing a jog button to move the selected axis by a fixed amount of movement.

ladder (logic) diagram A graphical method of showing the logic and, to some extent, the timing and sequencing of the control system. The various logic elements and other components are displayed along horizontal lines or rungs connected on either end to two vertical rails, resembling a ladder.

LAN *See* LOCAL AREA NETWORK.

large-scale integration (LSI) A classification for a scale of complexity of an integrated electron circuit chip. More than 100 equivalent gates manufactured simultaneously on a single slice of semiconductor material, forming the basis of microprocessor and minicomputer logic systems.

lead time The total time required between the receipt of an order and the delivery of that order to manufacture a product.

leading zeros All zeros to the left of the first nonzero digit of a number.

leading-zero suppression A feature of CNC systems in which the zeros before the first nonzero digit of a number do not have to be entered.

linear interpolation A basic CNC function requiring the part programmer to specify the beginning point and endpoint of the straight line, as well as the feedrate that is to be followed along the straight-line path. *(See figure.)*

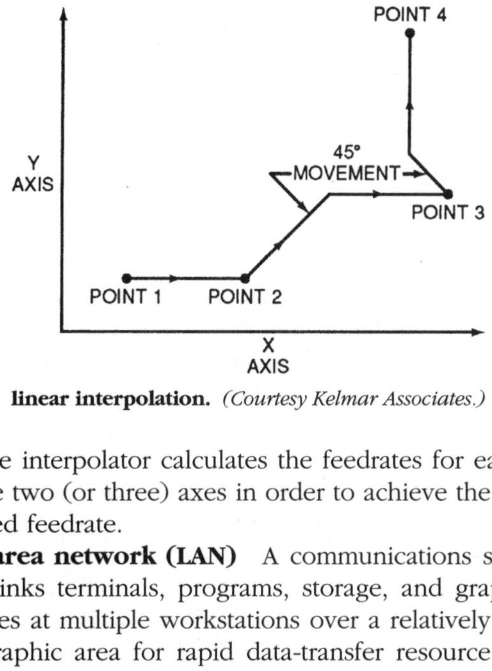

linear interpolation. *(Courtesy Kelmar Associates.)*

- □ The interpolator calculates the feedrates for each of the two (or three) axes in order to achieve the specified feedrate.

local area network (LAN) A communications system that links terminals, programs, storage, and graphical devices at multiple workstations over a relatively small geographic area for rapid data-transfer resource sharing.

logistical data Information controlling the flow of materials and products through the manufacturing process, such as quantities, schedules, and routings.

loop A structure of computer or part program language in which a sequence of instructions is executed repeatedly until a terminating condition is satisfied.

LSI *See* LARGE-SCALE INTEGRATION.

M-word A CNC word with a two-digit code preceded by the letter *M*. It specifies certain miscellaneous or auxiliary functions such as turning on and off coolant or operating power clamps, and is generally the last word in the block.

machine control unit (MCU) The electronics and control hardware of a CNC system that read and interpret the program of instructions and convert it into mechanical actions of the machine tool or other processing equipment.

machining cell An organization of a small number of machine tools that is designed to manufacture a defined set of components.

machining center A CNC machine tool, resembling a milling machine, that is capable of a variety of machining operations such as drilling, milling, boring, reaming, and tapping on a workpiece in the *X, Y,* and *Z* axes under program control.

- □ Some machining centers may have automatic tool changing, automatic workpart positioning, and a pallet shuttle.
- □ A vertical machining center has its spindle on a vertical (*Z*) axis in relation to the work table, and is generally used for flat work that requires tool access from the top.
- □ A horizontal machining center has its spindle (*Z* axis) on a horizontal plane and is used for cube-shaped parts where tool access can be best achieved on the sides of the cube.

macro (command/subprogram) The ability of a CNC programming language that allows a sequence of commands, in the form of a subprogram, to produce a series of tool paths within the part program.

manual data input (MDI) The manual programming of part programs to the memory of a CNC control unit by using the console keyboard at the site of the machine tool. Communication between the operator-programmer and the system is through the CRT display monitor and the keyboard.

manual override A feature on a CNC machine tool that allows the operator to use a dial to reduce or increase programmed feed and speed values without altering programmed values held in memory.

Manufacturing Automation Protocol (MAP) A set of communications protocol standards designed for use in a multivendor factory automation system. It uses a bus network configuration, broadband transmission, a token-passing access scheme, and a data-transmission rate of 10 Mbits/s.

manuscript A form in manual programming used by a part programmer for listing detailed machining instructions that can be directly fed to the machine control unit (MCU).

mirror imaging A CNC feature that reverses the sign of programmed dimensions in one or two axes by means of a switch, allowing opposite-hand geometries to be produced by a single programmed command *(see figure).* It can also be used to repeat geometric features programmed in a single quadrant in other quadrants.

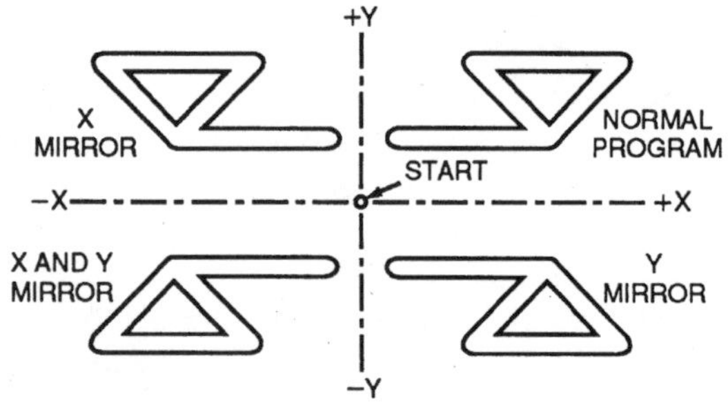

mirror imaging. *(Courtesy Superior Electric Co.)*

modal function A CNC programmed function that remains in effect until canceled or superseded by a function of the same type.

moiré fringe A repeating interference pattern of alternating light and dark bands that is produced as a result of two optical gratings moving relative to each other. The mathematical relationship between the number of lines on the gratings and the movement of the fringes for a given movement of the gratings enables accurate position measurement.

N-word Sequence number to identify the block number.

NOT gate A logic element with a single input which has an output of a 1 if the input is 1 0 and an output of a 0 if the input is a 1.

number base Radix, the fundamental number in a number system.

□ 10 is the base in the decimal system; 8, in the octal system; and 2, in the binary system.

numerical control (NC) A form of programmable automation in which the machine tool is controlled by means of a predetermined program of coded instructions and commands which consist of numbers, letters, and other symbols.

□ NC is suitable for low- and medium-volume production.

□ *See also* COMPUTER·NUMERICAL CONTROL.

offline Pertaining to operating software or hardware not under the direct control of a central processor, or operations performed while a computer is not monitoring or controlling processes or equipment.

offline programming The development of CNC part programs away from the machine console to be transferred to the MCU at a later time.

offset *See* TOOL OFFSET.

open-loop control A control system with no feedback to verify the axis position of a machine tool slide. Position and velocity are determined by built-in features of the driving mechanism.

operating program The master control program in ROM that directs the action of a CNC control system. It is provided by the control system designers and cannot be changed by the machine operator.

optical encoder A feedback sensor device used in CNC that measures linear or rotary motion by detecting the movement of markings past a fixed beam of light.

optional stop A Miscellaneous Function command (M01) similar to Programmed Stop, except that the control ignores the command unless the operator has previously set the validate switch.

OR A logic element which outputs a value of 1 if either of the inputs has a value of 1, and 0 otherwise.

overshoot The amount of overtravel beyond the commanded position.

parabola A plane curve generated by a point moving so that its distance from a fixed second point is equal to its distance from a fixed line.

parabolic interpolation A high order of interpolation which produces contoured shapes by the simultaneous and coordinated control of two axes of motion such that the cutting tool travels through parabolas or portions of parabolas.

□ Its application is concentrated in the automobile industry for fabricating dies for car body panels styled with free-form designs.

parametric programming A system of CNC part programming in which dimensional values are replaced by letters or symbols, with actual dimensions supplied when the program is executed. This enables the same program to be used to produce components of the same shape but of different dimensions.

part program A specific and complete set of CNC-coded instructions and data for the complete machining of a component on a CNC machine. *(See figure on next page.)*

part program format The order and form in which CNC part programs must be written to be accepted by a particular control system. The three formats are word address, tab sequential, and fixed-block.

part programmer A person who prepares CNC part programs.

part programming The planning and documentation of the sequence of steps to be performed on a CNC machine into a complete process sequence in the form of a CNC part program.

part programming using CAD/CAM An advanced form of computer-assisted part programming that uses an interactive graphics system equipped with CNC programming software.

□ The actions indicated by the commands are displayed on the graphics monitor, providing visual feedback to the programmer.

□ Certain portions of the programming cycle are automated by the CNC programming software to reduce the total programming time required.

Part No. CNB-140-1 | Rev. P | Note | Date 10-20-97 | Page 115

Program Number O(;) 1160 | (Note) End of block code (;) is CR in EIA or LF in ISO | Programmer M.C.

/	N	G	X	Y	Z	A/B/C	R/I	J	K	B	P	Q	L	H/D	F/S	T	M	;
	%																	
	Ō11	60	(CNB-	140-1)														
	N10		G0 G17	G20 G40	G49	G64	G80	G90	G54									
		G43	X.407	Y-7.094	Z5.									H1	S1000		M3	
	G81	G98			Z-.1		R.3								F3.		M8	
	N20															T2	M6	
		G43	X.407	Y-7.094	Z5.									H2	S1000		M3	
	G81	G98			Z-.1		R.3								F3.		M8	
	N30															T3	M6	
		G43	X1.5	Y-6.	Z5.									H3	S450			
	G82	G98			Z3.26		R3.9				P0700				F2.			
	G0	G80		Y-5.														
					Z3.25													
		G1		Y-7.														
	G0				Z5.													
			X6.141	Y-5.2														
					Z2.072													
	G1			Y-2.8														
		G0			Z5.													
				Y-5.2														
					Z2.062													
		G1		Y-2.8														
		G0	X7.16	Y-6.														
					Z.473													

part programming. *(Courtesy Kelmar Associates.)*

□ Advanced CAD/CAM systems have the capability to automate portions of tasks of both geometry definition and tool path specification.

part surface The surface on which the bottom of the cutter rides in APT contouring motions and may or may not be an actual surface of the workpiece.

□ The part programmer must define this plus the drive surface for the purpose of maintaining continuous path control of the tool.

PAU *See* POSITION ANALOG UNIT.

peck drilling A canned cycle for deep-hole drilling where the direction of the *Z*-axis feed is reversed at regular intervals for chip removal.

photocell/photodetector An electronic sensor in an optical encoder that emits an electrical signal when a flash of light falls on it.

planetary roller leadscrew A leadscrew system where threaded rollers revolve around a threaded leadscrew in a manner similar to the way planets revolve around the sun, whereby sliding friction is replaced by rolling friction.

PLC *See* PROGRAMMABLE LOGIC CONTROLLER.

point-to-point (PTP) control/positioning system A CNC system that controls motion of the cutter only to reach a predetermined location, but exercises no control over the tool's speed or path. Once the tool reaches the desired location, the machining operation is performed at that position.

□ Positioning systems are the simplest machine tool systems; NC drill presses are a good example of PTP systems.

polar coordinates A system of coordinates that locates a point in a plane with respect to its distance from a fixed point (origin or pole), and the angle this line makes starting from the polar reference line. *(See figure.)*

position analog unit (PAU) The unit which feeds back analog information about the position of machine slide to be compared with input positional information.

positional measuring system A measuring system (sensors or transducers, normally attached to the leadscrew of the machine) designed to monitor the position or movement of a controlled axis to produce

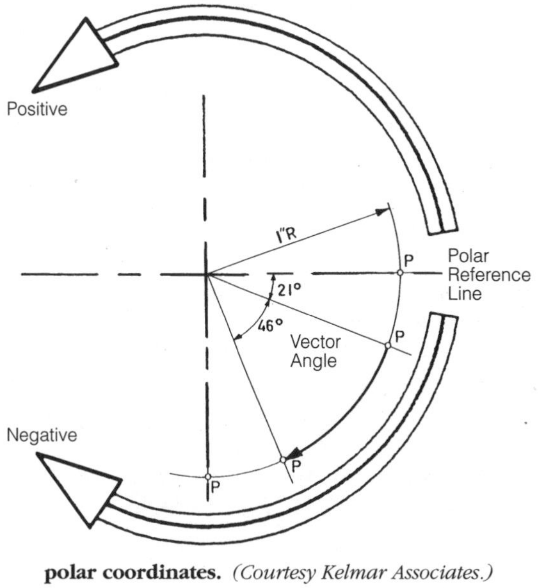

polar coordinates. *(Courtesy Kelmar Associates.)*

feedback in a closed-loop control system.

postprocessor A separate computer program that puts the cutter path coordinate data (output by a computer-aided part programming system) into a form suitable for the control system of a particular CNC machine tool that will machine the part.

precision Degree of preciseness of measurements.

preparatory function (G-word) A CNC command with a two-digit number preceded by the letter *G* that prepares the control system for a particular mode of operation. The preparatory word is needed to enable the controller to correctly interpret the data which follows it in the block.

□ Examples include absolute/incremental coordinates, inch/metric dimensions, rapid/feed axis movement, and linear/circular interpolation.

preset fixture A system of eliminating machine downtime by setting cutting tools away from the machine tool, on special equipment that duplicates the machine datum conditions.

printed-circuit board A self-contained board or card with electronic circuits. It may allow the simple replacement of damaged or faulty circuitry and is a simple means of updating or enhancing control system facilities.

program A sequence of coded instructions to be executed by a control or a computer to perform a given function.

program proving Verifying the safe and correct operation of part programs running on CNC machine tools.

programmable automation Using computer control to reprogram or change the sequence or control of a mechanized operation.

□ One of the most important examples of programmable automation is computer numerical control.

programmable logic controller (PLC) A digital electronic stored program device used in sequencing, timing, counting, and arithmetic to control—through digital or analog input/output modules—various types of machines or processes.

□ It uses a programmable memory for the internal storage of instructions, tests the state of input lines, and sets output lines in accordance with input state.

programmed acceleration A controlled increase in velocity to the programmed feedrate of a CNC machine.

programmed dwell A programmable length of time that the tool is stationary.

programmed stop A Miscellaneous Function (M00) command that stops the spindle, coolant, and feed after completion of the dimensional move commanded in the block. To continue with the remainder of the program, the operator must initiate a restart.

punched tape A storage medium made of paper, plastic, and polyester laminates, that is used for the permanent storage and loading of CNC part programs, on which characters are represented by combinations of holes.

□ Punched tape for CNC applications is 25 mm (1 in) wide, consisting of eight parallel tracks of holes along its length and holds 10 characters per 25-mm length.

□ The presence or absence of a hole in a certain position represents bit information, and the entire collection of holes constitutes the CNC program.

quadrant Any of the four parts into which a plane is divided by rectangular coordinate axes in that plane.

qualified tooling Cutting tools on which the position of the cutting edge is guaranteed to close limits of accuracy within ±.0003 in. or ±0.008 mm relative to the specified datum on the toolholder. *(See figure.)*

qualified tooling.
(Courtesy KPT Kaiser Precision Tooling, Inc.)

□ This makes possible the precise replacement of cutting tools in the machine tool.

read-only memory (ROM) A memory in which the data is stored permanently and can only be read out. ROMs are used primarily as internal memories to store small control programs that can be accessed frequently at high speeds.

ball screw. *(Courtesy Cincinnati Milacron, Inc.)*

real-time control A type of control system in which the calculation or control functions necessary for operation are performed at the time control is occurring, instead of preprocessing or predetermining the control responses.

recirculating ball screw A leadscrew system in which sliding friction is replaced by rolling friction *(see figure).* Both the leadscrew and the nut have a precision ground form in which ball bearings are allowed to run.

reference block A block within a CNC program which is identified by an *O* or *H* in place of the word address *N* and contains sufficient data to enable resumption of the program following an interruption.

- □ This block should be placed at a convenient point in the program to enable the operator to reset and resume operation.

register Specialized, temporary memory locations in the CPU for the arithmetic and logical operations.

repeatability The ability of the control system to return to a given location which was previously programmed into the controller. *Repeatability* refers to the capabilities of the CNC machine tool to produce parts that do not vary in dimensions from one part to the next.

reset To return a register or storage location to zero, or to a specified initial condition.

resolution The minimum positioning motion that can be specified by a machine control unit.

right-hand rule Using the right hand with the thumb pointing in the positive linear axis direction (*X, Y,* or *Z*); the fingers of the hand are curled to point in the positive rotational direction. *(See figure.)*

ring network A computer network in which each computer has a neighboring computer on either side, forming a continuous ring.

ROM *See* READ-ONLY MEMORY.

route sheet The document used to specify the process sequence. It lists the production operations and associated machine tools for each component (and subassembly) of the product.

S-word The specification of the cutting speed or the rate at which the spindle rotates in a CNC program block. Units are revolutions per minute (r/min).

secondary storage A memory outside the CPU used to store data that is waiting to be processed or to record data for future use. *See also* AUXILIARY STORAGE.

sequence number (N-word) Positive integer number preceded by the letter *N* that identifies the separate blocks, and the relative positions of those blocks within a CNC part program.

serial interface An interface that transmits data bit by bit.

servo amplifier The part of the servo system that amplifies the error signal and provides the power to drive the machine slides or the servo valve controlling a hydraulic drive.

servo control A negative-feedback (error-actuated) closed-loop control system that is applied to machine tool axis control. It is capable of controlling the velocity acceleration and path of motion as well as the endpoints.

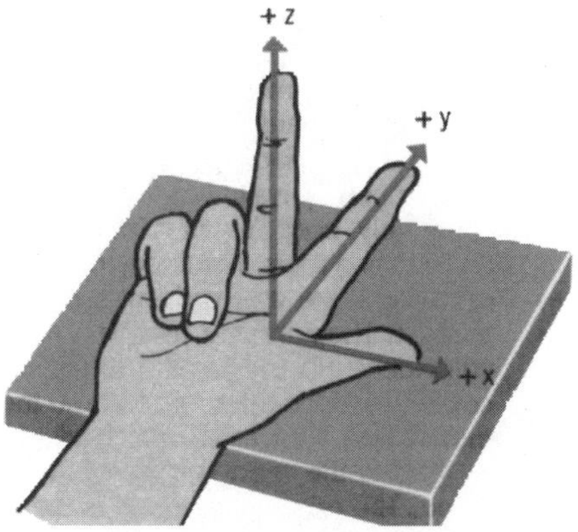

right-hand rule.
(Courtesy Modern Machine Shop.)

servomechanism An automatic control system used in CNC machine tools for axis positioning. It incorporates feedback and power amplification to enable an output quantity to follow an input (command) signal without error.

shaft encoder A transducer that converts the angular position of a rotating shaft into digital coded form using a digitally coded disk. *See also* OPTICAL ENCODER. *(See figure.)*

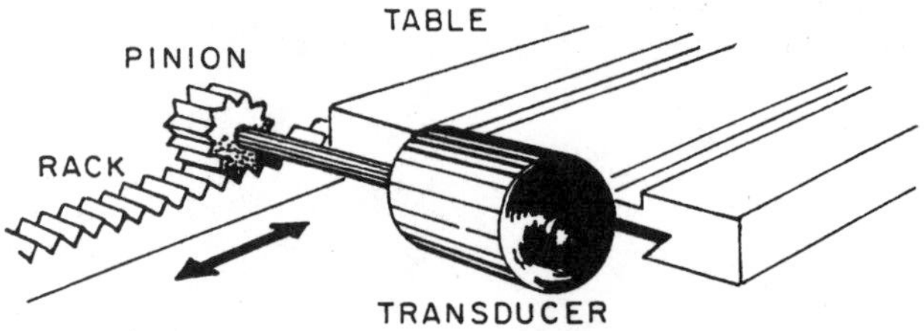

shaft encoder. *(Courtesy Kelmar Associates.)*

significant digit A digit that contributes to the precision of a number.

□ The most significant digit is the one contributing the most value; the least significant digit is the one contributing the least value.

single step A mode of operation of a CNC machine tool running a part program block by block. After a block has been executed, the machine stops and waits for further commands.

slowdown span A span of information that has the necessary length to allow the machine to decelerate from the initial feedrate to the maximum allowable cornering feedrate which maintains the specified tolerance.

software The collection of programs, procedures, routines, rules, and other documentation that directs the operation of a computer.

spindle speed (S-word) A multiple-character code containing the letter *S* followed by digits, which determines the number of revolutions per minute of the cutting spindle of the machine.

SPLIT *See* SUNDSTRAND PROCESSING LANGUAGE INTERNALLY TRANSLATED.

stepping/stepper motor A bidirectional, permanent magnet motor that is driven and controlled by an electrical pulse train generated by the MCU (or other digital device). Each pulse drives the stepping motor by a fraction of one revolution called the *step angle.*

straight-cut CNC system A CNC system that is capable of moving the cutting tool parallel to one of the major axes at a controlled rate suitable for machining.

subroutine/subprogram A separately defined part of a computer or CNC part program that can be called to execute from various points in the main program. *(See figure.)*

□ It is used to simplify and shorten programming by defining commonly used sequences once only, and calling them into the program as required.

Sundstrand Processing Language Internally Translated (SPLIT) A proprietary system intended for Sundstrand's machine tools that can handle up to five-axis positioning and also possess contouring capabilities.

□ The postprocessor is built into the program, and each machine tool uses its own SPLIT package, so there is no need for a special postprocessor.

synchroresolver An electromagnetic position transducer (shaft encoder) whose output voltage depends on the angular position of its rotor.

system software The programs that direct the internal operations of the computer while it is executing instructions from the user. They include functions such as translating languages, managing computer resources, and developing programs.

T-word/tool function A program code that specifies which tool is to be used in the operation for CNC machines with a tool turret or an automatic tool changer.

tab A nonprinting spacing action on tape preparation equipment which sets typewritten information on a manuscript into tabular form.

□ A tab code is used to separate words or groups of characters in the tab sequential format.

tab sequential format A format of CNC part pro-

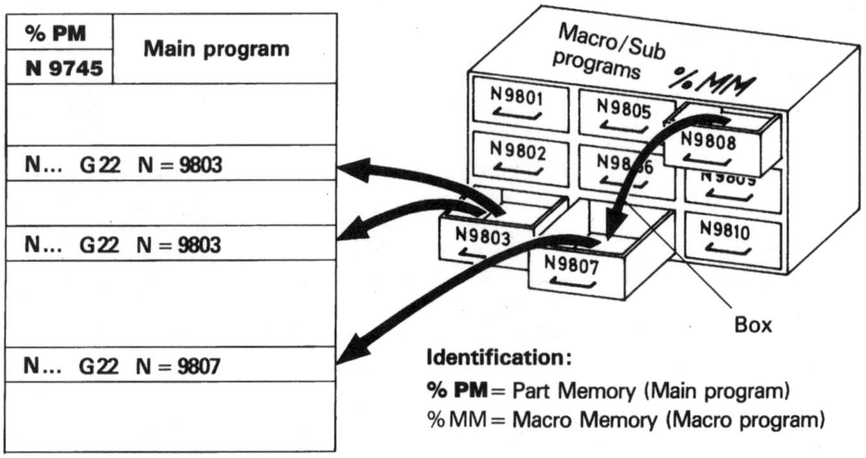

subroutine. *(Courtesy Deckel-Maho, Inc.)*

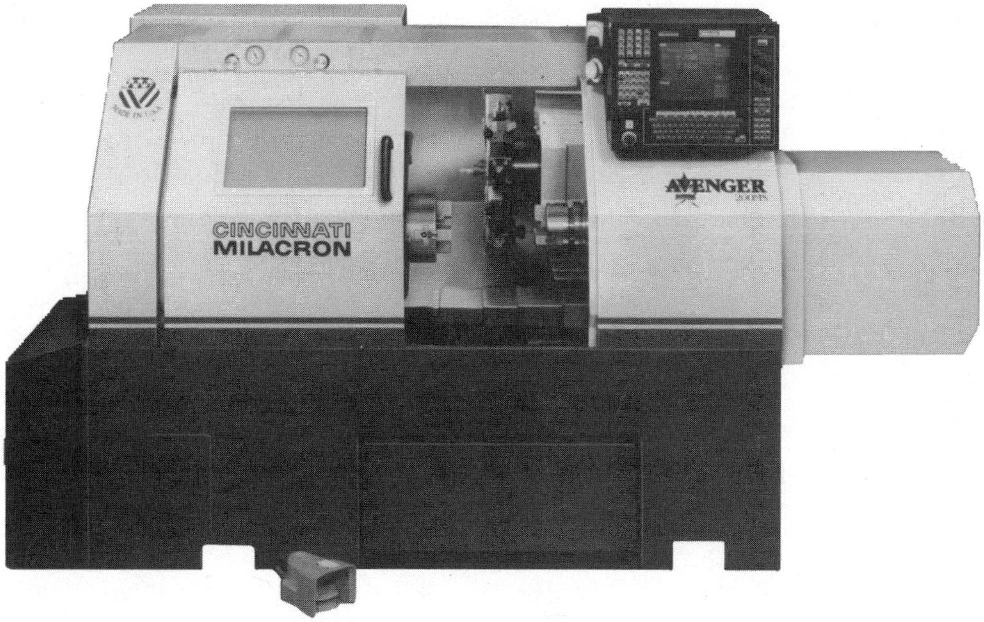

turning center.
(Courtesy Cincinnati Milacron, Inc.)

grams in which words are listed in a fixed sequence and separated by depressing the TAB key. Since the words are written in a set order, no address letter is required.

tape format The organization of words within blocks. There are three commonly used formats: word address, tab sequential, and fixed-block.

tape reader An electrical-mechanical device for winding and reading the punched tape containing the program of instructions.
- □ CNC tape readers operate by using photoelectric cells, electrical contact fingers, or the vacuum method; the photoelectric cells method is more common.

tool changer Mechanisms that automatically change cutting tools on a CNC machine under program control.
- □ It may take the form of a carousel with a variety of cutting tools for machining centers, or a special device at the end of a robot arm that provides for quick changes of the end-effector or tool.

tool length compensation/offset (TLO) A manual input to adjust for differences between the programmed length of a cutter and its actual length in order to avoid over- or undercut.

tool nose radius compensation (TNRC) Tool offset values that allow for small variations in tool point locations on CNC lathes. Although the tool point exists as a programmable point, it may not exist physically.

tool offset A correction for tool position parallel to a controlled axis, due to tool change or wear. Offset values are manually entered into the CNC control unit before starting to machine.

trailing-zero suppression *See* ZERO SUPPRESSION.

transducer A device that converts energy from one form to another such as temperature, pressure, and weight into electrical signals. Transducers are the basic mechanism used to provide sensing functions for machine tools.

truth table A tabulation that describes a logic function by listing all possible combinations of input values and the corresponding logical output values for each combination.

turning center A CNC machine tool, resembling a lathe, that is capable of automatically boring, drilling, turning outer and inner diameters, threading, and facing multiple diameters and faces of a part, at a single setup and in a number of axes simultaneously. *(See figure.)*
- □ Most turning centers are equipped with a system for automatically changing or indexing cutting tools, and a chip-removal system.

turnkey system A CAD/CAM, CNC, or computer system for which the supplier or vendor assumes total responsibility for building, installing, and testing; also the training of the user personnel.

unmanned machining/untended machining A mode of operation that allows machining to be performed under computer control with no operators present.

variable-block format A CNC part program format consisting of a combination of the word address and tab sequential formats to provide greater compatibility in programming.
- □ Words are interchangeable within the block, and the length of the block varies.

word address format A CNC part program format in which each word is preceded by a letter to identify its

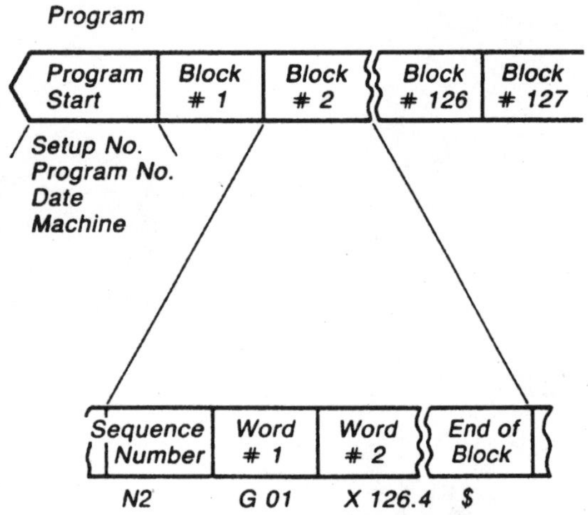

word address format.
(Courtesy Allen-Bradley Co.)

function and to address the data to a particular location in the controller unit. *(See figure.)*

□ The *X* prefix identifies an *X*-coordinate word; an *S* prefix identifies spindle speed, and so on.

□ The standard sequence of words for a two-axes CNC system is N-word, G-word, X-word, Y-word, F-word, S-word, T-word, M-word, and EOB.

word length The number of bits or characters in a word.

work in process (WIP) The amount of product that is currently in the factory either being processed or between processing stations. High WIP means that there is a large amount of working capital tied up in partly finished product.

***X* axis** Axis of motion that is always horizontal and parallel to the workholding surface in relation to the machine tool table.

***Y* axis** Axis of motion in the plane of a CNC machine tool table that is perpendicular to both the *X* and *Z* axes.

***Z* axis** Axis of motion that is always parallel to the main spindle of the machine tool.

zero offset A CNC feature that allows the zero point on an axis to be moved anywhere within the programmable range of a CNC machine tool. The control always keeps a record of the location of the permanent zero.

zero shift A CNC function that allows the machine zero to be shifted to any point within the programmable area of the machine. The control does not keep a record of the location of the "permanent" zero.

zero suppression **1.** *Leading-zero suppression* is the elimination of insignificant leading zeros to the left of significant digits usually before printing. **2.** *Trailing-zero suppression* is the elimination of insignificant trailing zeros to the right of significant digits usually before printing.

SECTION 3

Conventional Machine Tools and Accessories

J. W. Oswald
Consultant
Kelmar Associates
Niagara Falls, Ontario

There is a wide variety of conventional machine tools for performing many types of machining operations on different types and sizes of workpiece materials. To operate efficiently and achieve their full potential, these machine tools must be equipped with the proper tools, attachments, and accessories. This can give one machine the versatility to perform operations that would normally require another machine if these tools and accessories were not available. For example, a milling machine that is normally used for producing flat surfaces can, with the proper accessories and attachments, produce gears, sprockets, and helical forms.

abrasive cutoff machine, saw A machine that uses thin resin- and rubber-bonded cutoff wheels, impregnated with abrasive particles, to cut or part stock. *(See figure.)*
□ Used for cutting most metals, glass, and ceramics.
□ *See also* SAW, SAWING MACHINE.

abrasive-wire bandsawing A bandsawing operation that uses a small-diameter wire with diamond, cubic boron nitride, or aluminum oxide abrasives bonded to the surface as the cutting blade.
□ Abrasive-wire bandsawing may be used in place of electrical-discharge machining for producing dies, stripper plates, electrodes, and cams from difficult-to-machine materials.
□ *See also* BANDSAWING.

angle plate An L-shaped tool, usually made of cast iron, that has two faces at right angles to each other. It is used in the layout, machining, and inspection of finished workpieces.

arbor A spindle-mounted shaft used to support cutting tools for various machining operations.
□ In *grinding* it is the spindle for mounting the wheel; in *drill press and lathe work,* it may be used for supporting flycutters for boring operations; in *milling and other cutting operations,* it is the shaft for mounting the cutter.

automatic bar machine A production machine, similar to an automatic chucking machine, used for turning bar stock.
□ The stock size used is limited to through-the-spindle capacity and work is held by push, draw, or stationary collets rather than by chucks.
□ *See also* AUTOMATIC CHUCKING MACHINE *and* TURNING MACHINE.

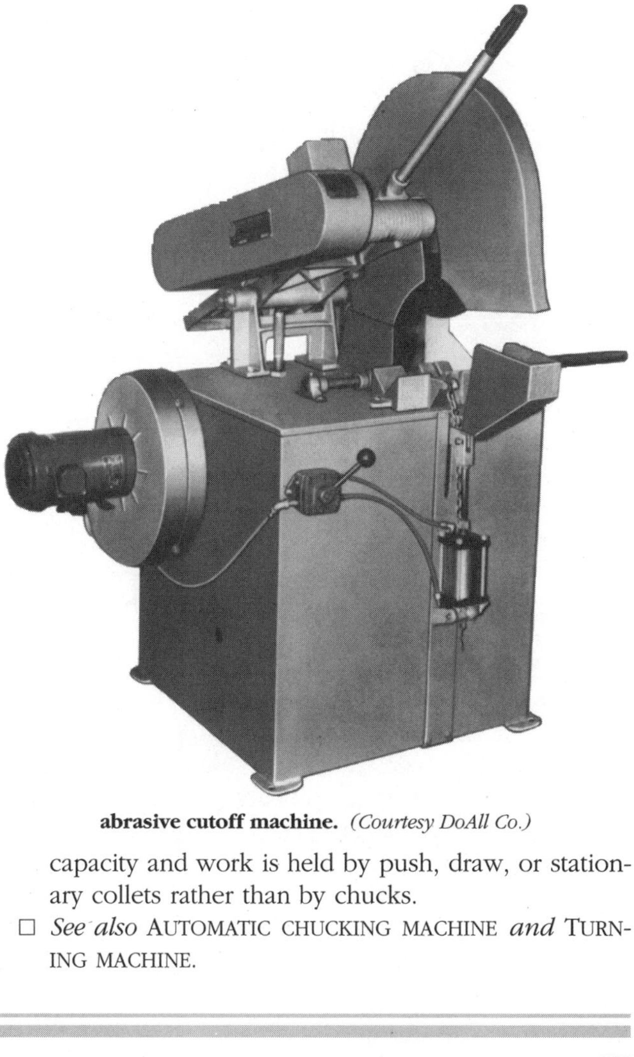

abrasive cutoff machine. *(Courtesy DoAll Co.)*

automatic chucking machine A machine with multiple chucks and toolholding spindles that permits processing several parts either simultaneously, or by multiple machining steps in one pass or one cycle of the machine.

□ *See also* AUTOMATIC BAR MACHINE.

automatic screw machine A precision turning machine, usually with a single spindle but sometimes multiple spindles, designed to produce parts automatically from coil or bar stock.

□ The two basic types are cam (mechanical) and programmable (computer-controlled).

□ *See also* LATHE *and* TURRET LATHE.

back rest A support mounted on a cylindrical grinder to prevent deflection and springing when long, small-diameter stock is being ground.

band polishing An operation similar to bandsawing that uses an abrasive band to smooth or polish parts previously sawed or filed.

□ *See also* BANDSAWING.

band welder A welder, usually mounted on the column of a contour band machine, that welds the bandsaw blade to allow internal sections to be cut out of a workpiece. *(See figure.)*

□ The saw band is cut to a specified length, and one end is fed through a hole drilled in part of the piece to be removed.

□ The two saw ends are welded together, and the saw blade is mounted over the pulleys of the machine.

bandsaw A sawing machine that uses an endless metal band with serrated teeth, for cutoff or contour sawing operations.

□ *See also* SAW, SAWING MACHINE.

bandsawing Power bandsawing, or band machining, uses an endless band with many small teeth traveling over two or more pulleys or wheels to cut material to length or size.

□ The lower pulley is the driven wheel, and the upper one is adjustable to provide for the tension and tracking of the saw band.

□ The saw band produces a continuous and uniform cutting action with evenly distributed, low individual tooth loads.

barrel finishing A production finishing process that produces a low-pressure abrasion action by tumbling workpieces in a hexagonal or octagonal barrel together with an abrasive slurry.

□ *See also* FINISHING.

bending machine A machine designed to bend a workpiece to a controlled, predetermined shape using various dies or forms and a bending arm.

blocks Workholding devices used on machine tools along with some type of strap clamps to fasten a workpiece to the machine table.

□ Common clamps include finger, gooseneck, step, telescoping, and quick-clamp.

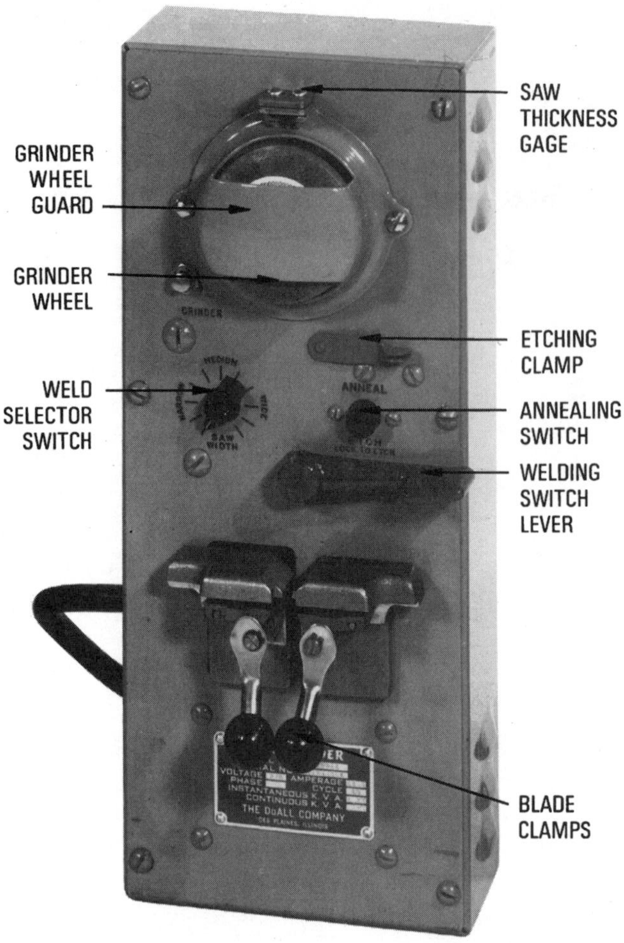

band welder. *(Courtesy DoAll Co.)*

boring The operation of enlarging a hole already drilled or cored, with a single-point lathe-type tool to true the hole, bring it to size, and improve the surface finish.

□ The boring operation is basically internal turning because a single-point cutting tool usually forms the internal diameter or shape.

boring machine A type of turning machine that is used to enlarge drilled or cored holes with a single- or multipoint cutting tool.

□ The three main types are the horizontal and vertical boring machines and the jig borer.

1. *Horizontal boring machine.* This machine is generally used for large parts and has a table that can be moved lengthwise and crosswise.

   The cutting tool is held in a horizontal revolving arbor that has an end support column.

2. *Vertical boring machine.* This machine has a horizontal circular work table that holds the work for facing, vertical turning, and boring operations.

   A swivel-type turret head can hold several tools for performing different machining operations.

   A slide arrangement allows the cutting tools to be moved horizontally and vertically.

3. *Jig borer.* This is a vertical boring machine gener-

jig borer. *(Courtesy Moore Special Tools.)*

ally found in toolrooms for precision locating, drilling, and boring of holes in jigs, dies, and fixtures. *(See figure.)*

The rectangular table can be moved accurately lengthwise and crosswise to within ± .0001 in. (0.0025 mm) by means of precision screws and measuring devices.

brake forming The use of a manual or power-operated brake to bend sheet metal to the required angle, form, or shape.

□ Heavy or tough material may require a progressive bending sequence in order to bring it to the required form or shape.

broaching An operation in which a cutter progressively enlarges a slot or hole, or shapes a workpiece exterior by using a multitoothed formed tool. *(See figure.)*

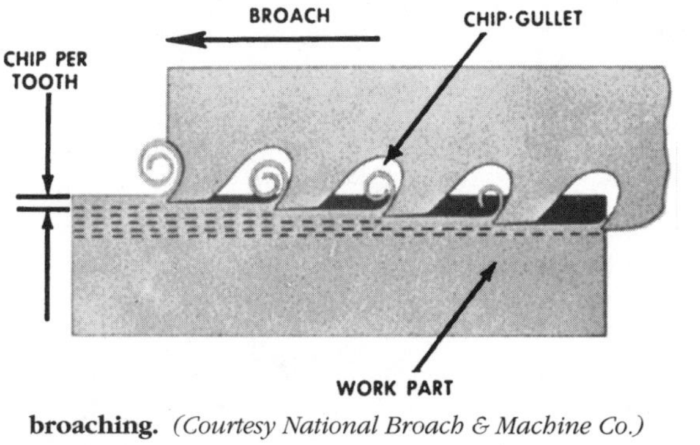

broaching. *(Courtesy National Broach & Machine Co.)*

□ The teeth of a broach are manufactured on a slight taper so that the end teeth start the cut, the intermediate teeth remove the majority of the material, and the final teeth finish the form.

□ Broaching is usually a one-step operation, but multiple passes may be used when removing much metal.

broaching machine A machine that uses a specially shaped broaching tool having a series of progressively larger teeth to cut internal or external shapes on a workpiece.

□ The broach is pulled or pushed through or over the part by mechanical or hydraulic means to reproduce its size and shape.

□ Broaching may also be performed on arbor presses (manual and powered).

brushing An operation that uses rapidly spinning wires or fibers to effectively and economically remove burrs, scratches, and similar mechanical imperfections from precision and highly stressed components.

□ Used extensively for the removal of sharp edges from gears and bearing races.

buffing Smoothing and polishing the workpiece surface by pressing it against an abrasive compound embedded in a soft wheel or belt.

cam-cutting attachment *See* LONG AND SHORT LEAD ATTACHMENT.

carriage stop A mechanical device fastened to the lathe head or ways to prevent overtravel of the carriage and cutting tool that might damage the machine or workpiece.

center drilling Drilling cone-shaped holes with a small straight hole at the bottom for mounting work between lathe centers.

□ Center holes are often used as "starter" holes when drilling larger holes in the same location.

□ *See also* DRILLING.

centering The process of locating the center of a workpiece to be mounted between lathe centers.

□ Also, the process of mounting and truing a workpiece in a chuck concentric to the machine spindle.

centerless grinding The workpiece, resting on a knife-edge support, rotates through contact with a regulating or feed wheel and is ground by a grinding wheel. *(See figure.)*

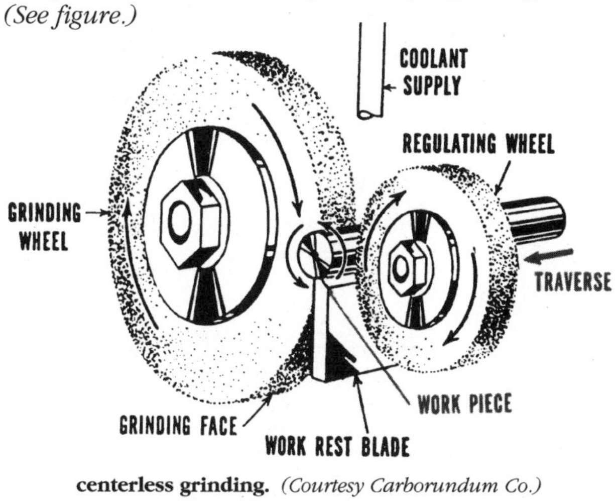

centerless grinding. *(Courtesy Carborundum Co.)*

☐ This method allows long, thin parts to be ground without the use of steadyrests and with less taper problems.

center rest A support provided at the center of a shaft during cylindrical grinding to stop the part from springing or deflection.

centers Tapered metal plugs that fit into the spindles of various machines such as lathes and cylindrical grinders to support the ends of a part during a machining operation.

☐ The lathe live center fits into the headstock spindle and revolves with the workpiece, while the dead center fits into the tailstock.

1. A solid tailstock dead center does not revolve and is rarely used today.
2. The revolving tailstock dead center, equipped with antifriction bearings, is most commonly used because it revolves with the work and can adjust to expansion of the part due to heat created during machining.

chamfering The operation of machining a bevel on a workpiece to reduce the 90° sharp edge that can easily become burred or damaged.

chuck A workholding device that is mounted on a mill, lathe, or drill-press spindle to hold and revolve a workpiece or cutting tool. *(See figure.)*

chuck. *(Courtesy Clausing Industrial, Inc.)*

☐ Two or more adjustable jaws may be actuated manually, pneumatically, hydraulically, or electrically, to hold the tool or part.

☐ *See also* COLLET *and* MAGNETIC CHUCK.

circular saw Any type of cutoff machine that uses a circular blade with serrated teeth to cut off unhardened metal.

☐ Large cutoff saws generally use blades having inserted teeth.

☐ *See also* SAW, SAWING MACHINE.

coining A closed-die squeezing operation in which all surfaces of the work are confined by the die and punch surfaces to produce a pattern or form onto the workpiece.

☐ Coining also refers to a press-brake bending operation in which the punch bottoms against the workpiece and the die.

☐ This process is limited to fairly soft metals and requires special high-pressure presses.

collet A spring- or flex-collet device that fits into the spindle of a machine tool to hold a tool or workpiece accurately when the draw bar is tightened.

☐ These collets, which have a small size range, are used to hold accurately sized workpieces securely.

☐ They are available to hold round, square, or hexagonal workpieces for light machining and finishing operations.

contouring attachment A workholding jaw, used on the contour bandsaw to hold and guide the workpiece into the saw blade when cutting contours. *(See figure on next page.)*

☐ The work is fed into the saw by a weight-type power feed that is attached to the jaw by a cable and pulley.

counterboring The process of enlarging one end of a previously drilled hole to a given depth to accommodate the head of a screw or nut.

☐ Counterboring is used to provide a flat seat for boltheads and nuts below the work surface.

countersinking Cutting a beveled edge at the entrance of a hole so that a screwhead can sit flush with the workpiece surface.

☐ *See also* COUNTERBORING *and* SPOTFACING.

creep-feed grinding A grinding operation in which the grinding wheel, set to the full depth required, is slowly fed into the work to produce a finished form in one pass of the wheel. *(See figure.)*

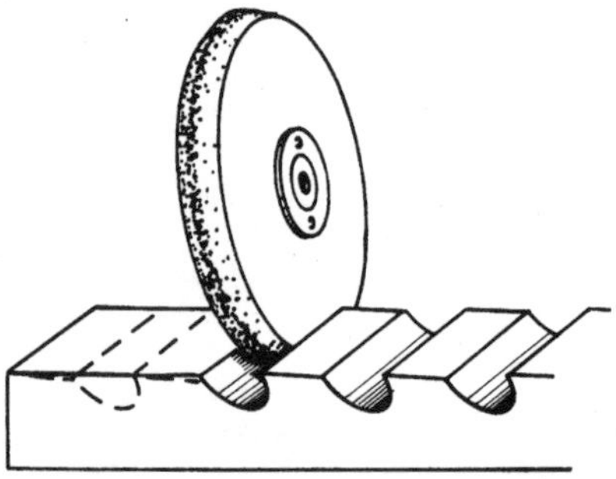

creep-feed grinding. *(Courtesy GE Superabrasives.)*

☐ *See also* GRINDING.

cutoff An operation performed on most types of turning machines that produces a slug, blank, or other workpiece for machining or other processing by separating it from the original stock.

☐ Also performed on milling machines with slitting

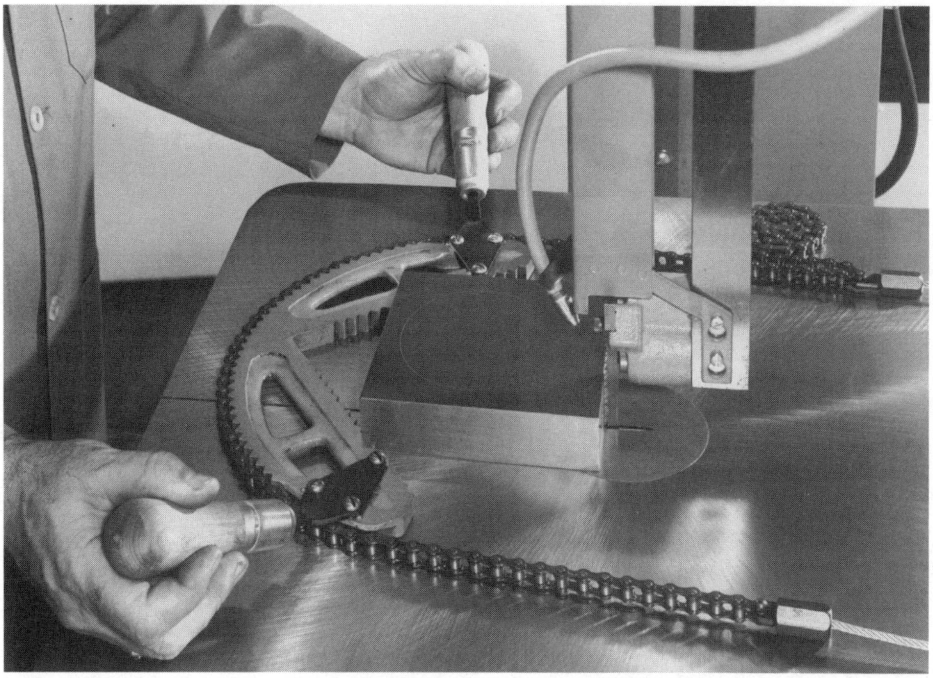

contouring attachment. *(Courtesy DoAll Co.)*

saws, circular sawing machines, hacksaws, bandsaws, or abrasive cutoff saws.

□ *See also* MICROSLICING, SAWING, *and* TURNING.

cylindrical grinding attachment A portable grinder, mounted in the toolpost of a lathe, for grinding round parallel, tapered, or angular surfaces.

diamond bandsawing An operation in which a saw band impregnated with diamond points is used to machine carbides, ceramics, and other extremely hard-to-cut materials.

disk-cutting attachment An adjustable device used on a contour bandsaw to position stock when sawing arcs and circular shapes.

disk grinding An operation in which the workpiece is placed against the side of a wheel rather than the wheel's periphery, such as grinding a vertical surface with the side of a wheel in surface grinding.

□ *See also* GRINDING.

dividing head A milling-machine table accessory, also known as an *indexing head,* used to accurately divide the circumference of a part into many equal spaces for the machining of gear and sprocket teeth, spline keys, serrations, and other components or elements. *(See figure.)*

□ The most 'common dividing head has 40 teeth around its spindle that mesh with a single-threaded worm driven by the indexing crank.

□ This 40:1 ratio is used to calculate the number of crank turns required to equally space gear teeth or move the workpiece an exact number of degrees.

drawing The production of deep, cup-shaped forms in a variety of steel parts such as automobile body panels, fenders, doors, and many other similar parts.

□ The cups can be drawn with or without a flange and in almost any desired cross section.

□ Internal resistance flow of the metal toward the punch and die sets up compressive stresses which, combined with tensile stresses created by the motion of the punch, produce two-dimensional shear.

drill-grinding gage A tool used to check the length and angle of a drill's lips or cutting edges when grinding or resharpening.

drilling The operation of using a rotating tool to produce a round hole in a workpiece where none existed before.

□ Drilling is the operation generally performed before the machining operations of boring, reaming, tapping, counterboring, countersinking, and spotfacing.

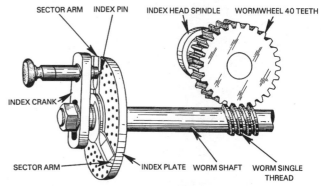

dividing head. *(Courtesy Kelmar Associates.)*

drilling machine, drill press A machine designed to rotate and feed end-cutting tools for drilling, reaming, tapping, countersinking, counterboring, spotfacing, and boring operations.

- □ Special drill presses are available in multiple drilling heads and multispindle models for production purposes.
- □ Radial drill presses, in which the arm carrying the drilling head can be swung around the column, are used for drilling-type operations on large workpieces.

drill jig A drill press accessory used to hold a workpiece securely while guiding a drill or other tool to an accurate and repeatable location. *(See figure.)*

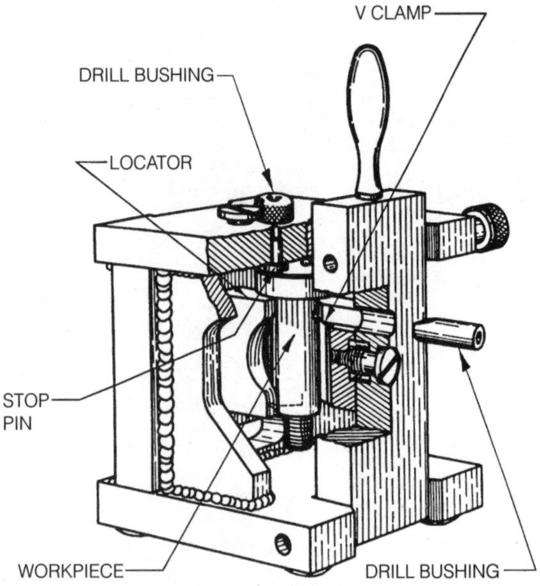

drill jig. *(Courtesy Kelmar Associates.)*

drive plate An accessory mounted on the lathe spindle that has a slot or slots that engage a driving dog to turn the workpiece mounted between lathe centers.
- □ *See also* CENTERS *and* DRIVING DOG.

driving dog A clamping device, equipped with a ring or clamp on one end, that slips over and is clamped to the workpiece to be turned.
- □ The opposite end (tail) of the dog fits into a drive plate slot and provides a drive for the workpiece.
- □ *See also* DRIVE PLATE.

edge finder A locating tool that is mounted in the spindle of a vertical mill or jig borer to align the part edge with the machine spindle center.

electroforming A process for making thin parts where an electrode deposits metal on a mandrel or mold.
- □ When the proper thickness is achieved, the mold is removed from the dielectric and the deposit is removed from the mold.
- □ This process is similar to electroplating, except that the deposit is usually thicker than an electroplated deposit.
- □ *See also* ELECTROPLATING.

electrohydraulic or electrospark forming A method of shaping hollow tubes or preforms by discharging electrical energy inside the workpiece, which is filled with water or another suitable medium that produces shock waves to deform the metal. *(See figure.)*

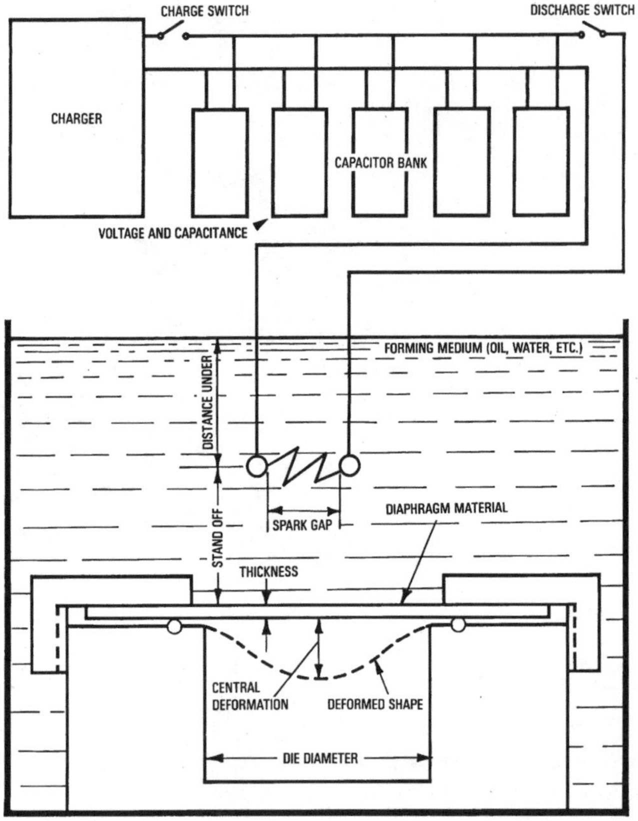

electrospark forming. *(Courtesy Kelmar Associates.)*

- □ Either the spark-discharge or exploding-bridge-wire method is used to generate the shock waves.
- □ Deformation is controlled by forming into external dies, regulating the energy released, or using shapers within the transfer medium.

electron-beam cutting (EBC) A process related to laser-beam cutting or welding, utilizing beam-spot intensities several times greater so that complete vaporization occurs along the beam's path of travel.

electron-beam machining (EBM) The use of electrical energy to produce thermal energy for removing material.
- □ A pulsating stream of high-speed electrons, produced by a generator, is focused by electrostatic and electromagnetic fields to concentrate energy on a very small area of work.
- □ High-power beams are used with electron velocities exceeding half the speed of light.
- □ As the electrons strike the work surface, their kinetic energy is transformed into thermal energy and melts or evaporates the material locally.

electroplating An electrical process that uses dc electric current to deposit a metal coating on the conductive surface of an object (cathode). *(See figure on next page.)*
- □ A plate of metal being deposited serves as the anode.

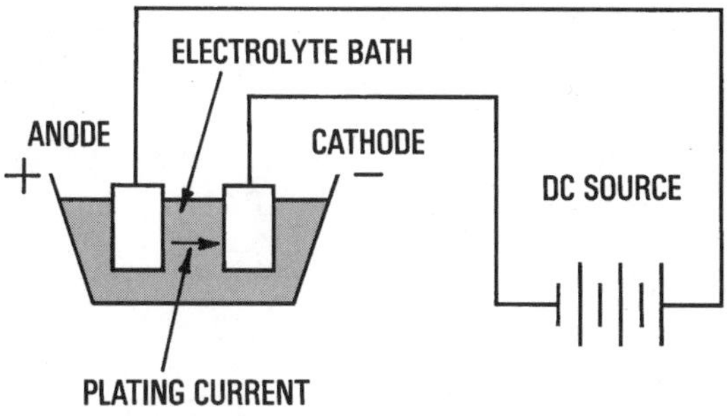

electroplating. *(Courtesy Kelmar Associates.)*

□ Both are immersed in a suitable electrolyte solution containing a salt of the material being deposited.

end milling A milling operation where an end mill is held in the machine's spindle for machining flat and vertical surfaces, grooves, slots, keyways, dovetails, T-slots, and circular slots.

face milling A type of operation that uses a slab or shell milling cutter, having teeth or inserts on both its periphery and its end face, to produce a flat surface generally at right angles to the axis of the cutter. *(See figure.)*

face milling. *(Courtesy Cincinnati Milacron, Inc.)*

faceplate A flat, round plate having a series of slots and mounted on the lathe spindle; regular- or irregular-shaped parts are clamped to its surface for machining.

□ If a part is irregular in shape, counterbalances should be used to prevent vibration that can be caused by centrifugal force.

□ *See also* DRIVE PLATE.

facing A cleanup operation to produce a true, flat surface that acts as a reference point for other machining operations which may have to be performed.

filing An operation in which a hardened cutting tool with numerous small teeth is used manually to round off sharp corners and remove burrs and nicks.

□ Filing can also be performed on a power filer or contour band machine with a special filing attachment.

filing attachment An accessory mounted on a bench filer or contour bandsaw for power-filing operations.

finishing Any process that is used for surface, edge, and corner preparation, as well as conditioning, cleaning, and coating of a part. It is usually the final operation in the manufacture of a part or component.

fixture A holding and locating device that is made to suit a specific part for some type or types of machining operations. *(See figure.)*

fixture. *(Courtesy Cincinnati Milacron, Inc.)*

□ *See also* JIG.

flame cleaning A method of cleaning metal parts by playing an oxyacetylene flame over the surface to burn off oils, dry water residues, or heat scale rapidly to cause it to flake off.

flattening die A die used to flatten and close the seams in light-gage sheet-metal products.

□ It consists of a top and bottom die with a flat surface that can close one section (flange) to another (hem, seam).

fluting The operation of cutting or grinding straight or spiral grooves in drill, end mill, reamer, and tap blanks to improve cutting action and chip removal.

follower rest A device equipped with two or three soft-faced jaws and attached to the lathe carriage to support the workpiece immediately behind the cutting tool to prevent it from springing during machining.

□ *See also* BACK REST *and* STEADYREST.

forging The plastic deformation of metals, generally at elevated temperatures, into complicated shapes by compressive forces exerted through a die.

□ The simplest forging operation is upsetting or compressing metal between two flat, parallel platens.

forming A process in which thin sheet material is usually stamped, stretched, bent, or given a new shape without removing material.

form-rolling machine A cold-forming machine used to roll splines, gears, worms, and threads for production processing of previously machined parts.

friction sawing A fast method that uses carbon steel

bands to saw hard ferrous alloys up to 1 in thick at saw-band speeds of up to 15,000 surface feet per minute (sf/min). *(See figure.)*

friction sawing. *(Courtesy DoAll Co.)*

- □ The high band speed creates frictional heat that softens the metal; then the teeth scoop out the molten material.
- □ This process cannot be used for cutting aluminum alloys, brass, or other materials that may load the saw teeth.
- □ *See also* SAWING.

gang cutting, milling, and slitting An operation used in production where machining occurs with several cutters mounted on a single arbor to produce the desired work shape.

gear shaper A machine that reciprocates the form-shaped tool to cut the gear as compared to mills and hobbing machines, which use rotating cutters.

grinding An operation in which material is removed from the workpiece by a powered abrasive wheel, stone, belt, paste, sheet, compound, slurry, or other agent to improve the surface finish and bring the work to size. *See also* GRINDING MACHINE.

- □ *Surface grinding*—used to produce flat and squared surfaces
- □ *Cylindrical grinding*—used to produce external and internal cylindrical parts
- □ *Centerless grinding*—used for production grinding of cylindrical, tapered, and multidiameter parts
- □ *Tool and cutter grinding*—used for sharpening cutting tools and milling cutters

grinding machine A machine that removes metal using a rotating abrasive wheel for the rough and finish grinding of workpieces.

- □ Rough grinding (abrasive machining) is generally done on large, heavy grinders, some capable of removing $^1/_2$-in depth of cut in one pass.
- □ Finish grinding is usually performed on lighter machines that can produce fine finishes and close tolerances.
- □ Following are some of the most common grinding machines:
  1. *Cutter and tool grinders*—used mainly for sharpening milling cutters and cutting tools. *(See figure.)*

cutter and tool grinding.
(Courtesy Cincinnati Milacron, Inc.)

  2. *Cylindrical grinders*—used to produce cylindrical work on long lengths of bar stock either between centers or by centerless process. *(See figure.)*

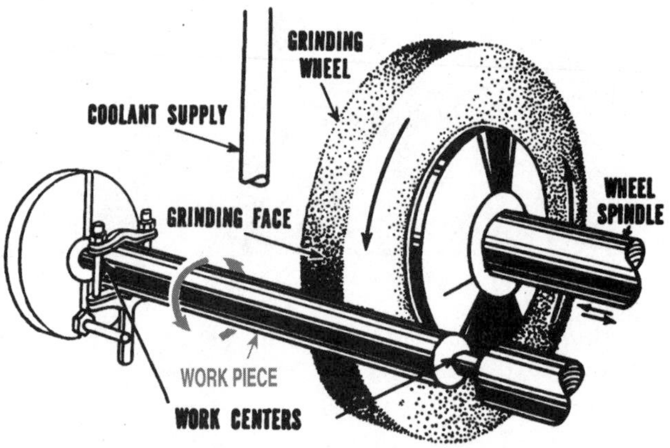

cylindrical grinding. *(Courtesy Carborundum Co.)*

  3. *Internal grinders*—used for finish grinding of holes and internal forms. *(See figure.)*

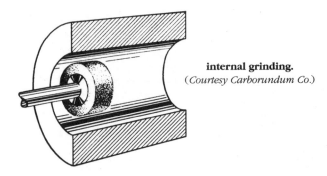

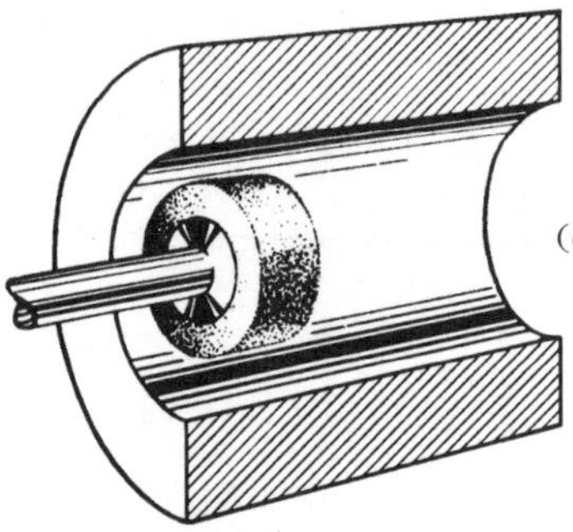

internal grinding.
(Courtesy Carborundum Co.)

4. *Surface grinders*—generally used to produce fine finishes and close tolerances on flat surfaces. *(See figure.)*

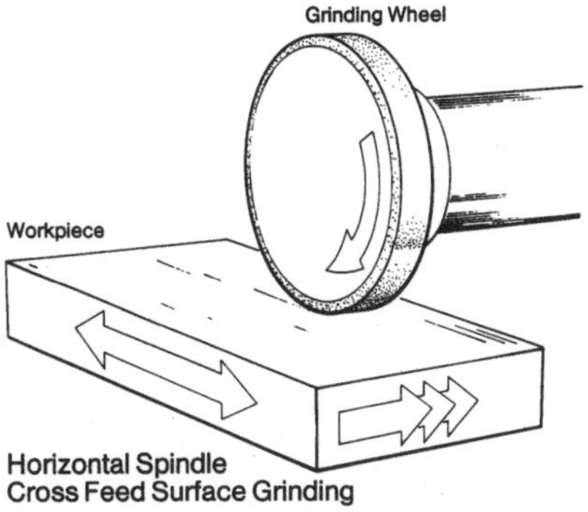

surface grinding. *(Courtesy GE Superabrasives.).*

5. *Pedestal and bench grinders*—used for off-hand grinding of workpieces and the sharpening of cutting tools such as lathe toolbits and drills.

□ *See also* CENTERLESS GRINDING, GRINDING, HONING, OFFHAND GRINDING, *and* SURFACE GRINDERS.

grooving The operation of producing square, round, or V-shaped forms at the end of a thread or shoulder.

□ Used in thread cutting, it provides a runout for the cutting tool or die.

□ At the shoulder, it is used to break a sharp edge or provide strength where the diameter meets a shoulder.

gun drilling A process using a single-lip, self-guiding drill to produce deep, precise holes.

□ High-pressure coolant is fed to the cutting area, usually through the gun drill's shank, to cool the drill point and flush out the chips.

hacksaw, power A machine fitted with a serrated blade held tightly in a reciprocating frame that cuts in one direction, on either the forward or return stroke.

□ *See also* SAW, SAWING MACHINE.

helical milling A milling operation in which the workpiece is rotated and fed lengthwise into the cutter at precalculated rates to create a spiral form. *(See figure.)*

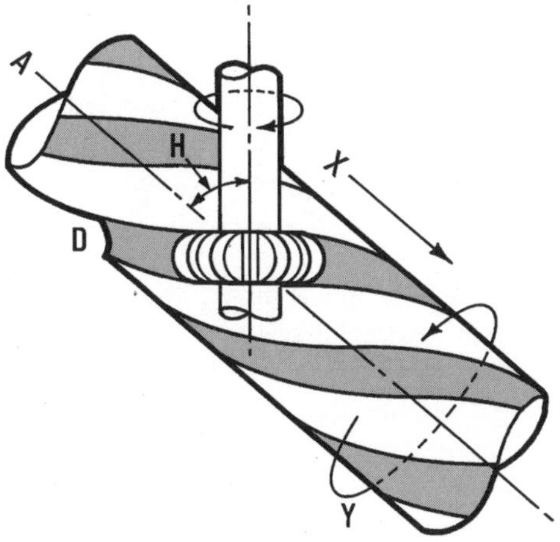

helical milling. *(Courtesy Kelmar Associates.)*

□ The table of the milling machine must be swung to the helix angle of spiral form, and the dividing head must be geared to the leadscrew of the machine.

□ Often used to mill flutes on end mill, milling cutter, and twist-drill blanks.

high-speed milling attachment An attachment generally used on universal milling machines that has a higher speed range than the machine spindle for operating small-diameter end mills at high speeds.

□ *See also* UNIVERSAL MILLING ATTACHMENT.

hobbing **1.** A gear-tooth-generating process that involves rotating and advancing a fluted-steel worm cutter (hob) past a revolving blank. *(See figure.)*

hobbing. *(Courtesy Gleason Works.)*

□ The speed ratio of the hob and blank depends on the number of teeth to be generated on the gear, and whether the hob is single- or multithreaded.

□ The hob cutting speed is controlled by change gears that vary the speed of the hobbing machine's main drive shaft.

2. A method of producing mold cavities in soft steel by pressing a hardened steel form or hob into the soft steel using a hydraulic press.

□ This process can produce multiple exact cavities with highly polished surfaces.

hobbing machine A machine in which a hob and a blank rotate in precise relation to each other to create special forms such as worms, spur and helical gears, and splines.

□ *See also* GEAR SHAPER.

hold-down This usually consists of a T-slot bolt, strap clamp, or other device for clamping parts to a machine tool table.

holemaking Using a cutting tool such as a drill, reamer, punch, liquid medium, or electrode to produce holes in the workpiece. Usually the first step before a machining and finishing operation.

honing A low-speed, internal finishing operation that uses aluminum oxide or silicon carbide abrasive sticks, mounted on a mandrel, to smooth and bring a hole to size. *(See figure.)*

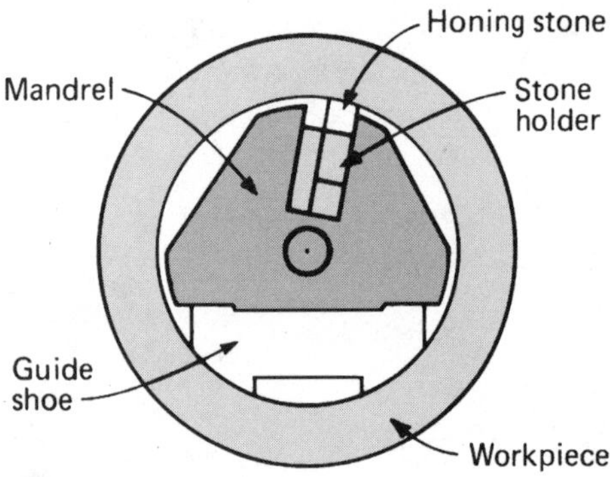

honing. *(Courtesy Sunnen Products Co.)*

- □ Because of the low speed, heat and pressure are minimized, resulting in excellent size and geometry control.

indexing head *See* DIVIDING HEAD.

jig A device used to hold the workpiece and locate and guide a cutting tool such as a drill, reamer, and tap for the manufacture of many similar parts.
- □ A hardened bushing, located at the exact spot required, aligns and guides the cutting tool.
- □ Jigs are also used in the subassembly or final assembly of parts to act as assembly aids.
- □ *See also* FIXTURE.

jig borer A precision boring machine, with precision table leadscrews and locational measuring devices, that was originally developed to accurately locate and bore holes in jigs, dies, fixtures, and gages.
- □ It is now used for production machining of precision parts, extremely accurate hole location, and similar tasks.
- □ Basic types include fixed-bridge, open-side or C-frame, and adjustable-rail.
- □ *See also* BORING MACHINE *and* MILLING MACHINE.

jig boring A high-precision machining operation used for locating and boring holes at precise locations for the manufacture of jigs and fixtures.
- □ Other jig boring operations include centering, drilling, reaming, through and step boring, counterboring, and contouring.
- □ Numerical control machining centers have greatly reduced the amount of work still performed on jig borers.

knockout A mechanism, sometimes called an *ejector* or *shedder,* used for releasing workpieces from a die.

knurling Generally a lathe process that consists of impressing diamond-shaped or straight-line patterns on the periphery of a workpiece to improve the appearance and provide a better grip.
- □ In automotive applications, it is used to enhance clearances and help pistons and valve guides retain oil.

lancing An operation performed on a punch press where a hole is partially punched through a piece of sheet metal and the cut side is bent down to form a louvred opening. *(See figure.)*

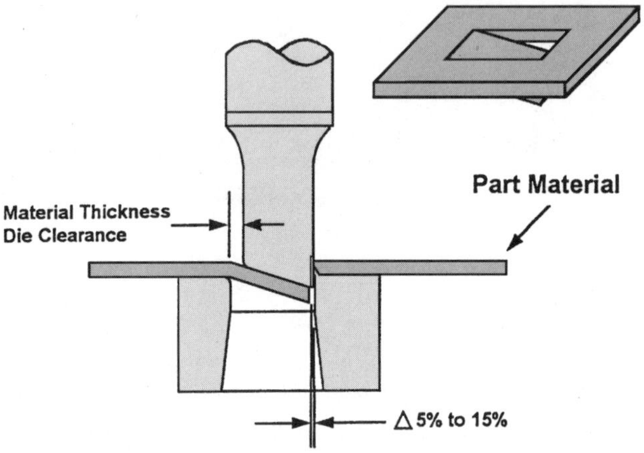

lancing. *(Courtesy Dayton Progress Corp.)*

lapping A finishing operation where a loose, fine-grained abrasive flour in a liquid medium abrades and polishes the material.
- □ An accurate process that produces very fine surface finishes to allow a close fit between mating surfaces.

lathe A turning machine capable of producing round diameters by rotating a workpiece against a stationary single-point cutting tool.
- □ Common lathe operations include straight and form turning, facing, chamfering, grooving, knurling, taper cutting, drilling, reaming, boring, and threading.
- □ Common types of lathes are engine *(see figure on the next page)*, turning and contouring, turret, and computer numerical control (CNC).
- □ Toolroom and bench lathes are generally used for precision work and small parts.
- □ Special lathe types include through-the-spindle, camshaft and crankshaft, brake drum and rotor, spinning machines, and gun-barrel.
- □ Modern lathes are often equipped with digital readouts and CNC.

lathe turning A machining operation in which a workpiece is rotated in a lathe against a stationary cutting tool that can be moved lengthwise or crosswise to produce external or internal round surfaces.

layout The use of layout dye, scribers, prick punches, rules, combination sets, surface gages, and vernier height gages to create a part outline that machinists use to visually check the part shape during machining.

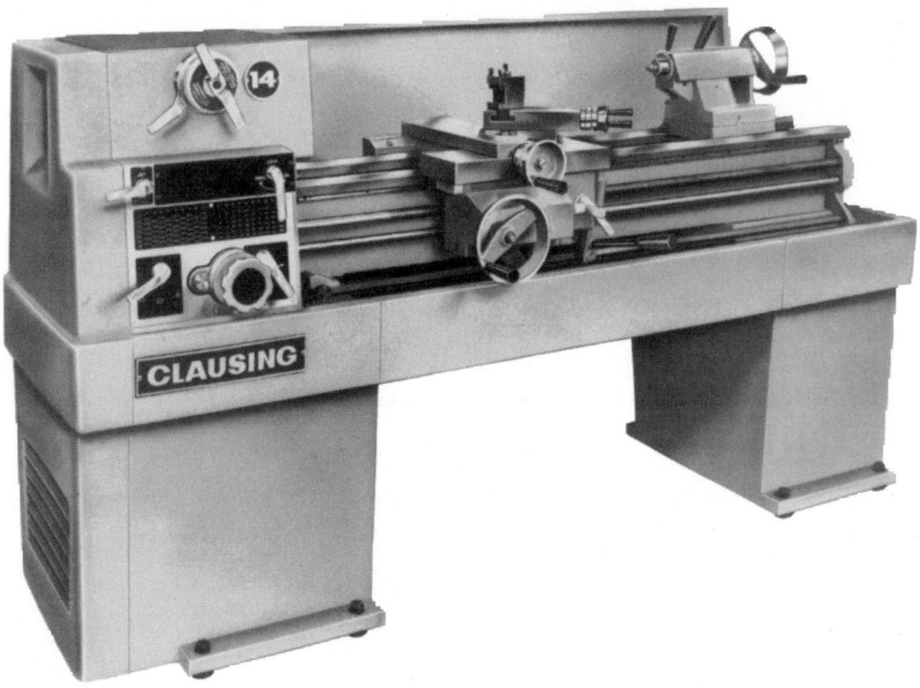

engine lathe. *(Courtesy Clausing Industrial, Inc.)*

long and short lead attachment A milling machine attachment that provides a calculated drive from the longitudinal leadscrew of the machine to a table-mounted dividing head for machining cam forms.

magnetic chuck A workholding device, commonly used on surface grinders, that contains permanent magnets or an electromagnetic system for holding ferrous parts with large, flat sides on the machine table. *(See figure.)*

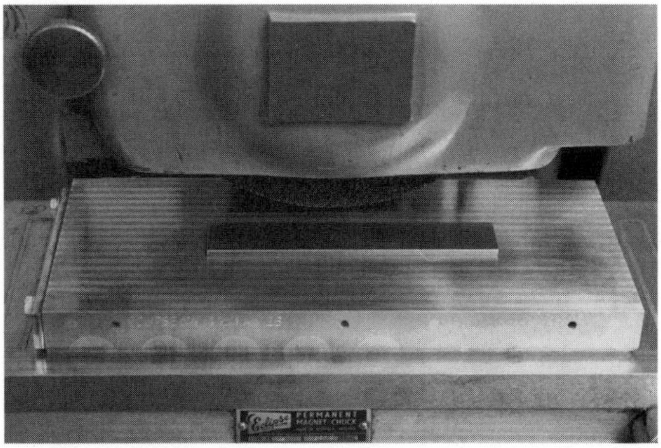

magnetic chuck. *(Courtesy Kelmar Associates.)*

□ Magnetic chucks may also be used on lathes or milling machines for light machining operations.

□ *See also* CHUCK.

mandrel A tool that fits workpieces having an internal bore to hold and provide a drive so that other machining operations are concentric with the bore.

□ Common types available are the standard, expanding, pin, and threaded.

microslicing Cutting very thin slices from a base material using a special machine with a thin, tensioned blade that makes a very fine cut.

□ It is used primarily to cut expensive materials such as silicon, germanium, and other computer-chip materials to keep the amount of material cut to chips to a minimum.

milling A machining operation in which metal or other material is removed by bringing the workpiece into contact with a horizontally or vertically mounted cutter.

□ In *horizontal milling,* the cutter is mounted horizontally on the spindle or arbor and machining can be done either by conventional milling (the cutter rotates opposite the direction of table feed) or by climb milling (the cutter rotates in the same direction as the feed).

□ Common milling operations include plane or surface, end, face, side, angle, form, profiling, straddle, gang, sawing, and slitting.

□ In *vertical milling,* the cutting tool or end mill is mounted into a vertical spindle.

milling arbor A shaft that is inserted in the milling machine spindle to hold and provide a drive for peripheral- or face-milling cutters.

□ The other end of the arbor is aligned and supported by the arbor support.

milling machine There are two common types of milling machines: horizontal and vertical. Horizontal mills are generally larger and more powerful, whereas the vertical mill is more versatile and easier to set up.

□ *Horizontal milling machines.* These machines are available as plain and universal and are used to produce flat and angular surfaces, grooves, contours, gears, racks, sprockets, and helical grooves.

1. The horizontal spindle, mounted in the column, provides a means of holding and driving a wide variety of cutting tools.
2. The universal type has a table that may be swiveled to machine helical surfaces or grooves.
3. Modern milling machines may be equipped with a digital readout measuring system and computer numerical control for greater accuracy.

□ *Vertical milling machines.* These machines are similar to horizontal mills except that the spindle is positioned vertically; they are used for basically the same operations as horizontal mills and are easier to set up.

1. The head of this machine can be swiveled to permit the machining of angular surfaces.
2. The common types of tools used are end mills, shell end mills, form cutters, drills, and boring tools for flat and angular machining, drilling, and boring operations.
3. With the use of a rotary table, it is possible to cut circular grooves.

modular fixturing A system in which workholding fixtures are built from standard, reusable components that can be put together or taken apart quickly. *(See figure.)*

□ Common modular fixturing components consist of tooling plates and blocks, mounting, locating, and clamping elements, and special-purpose accessories.

□ The part to be machined can be held in place by manual or power-operated clamping devices.

multifunction machines Machines and machining/turning centers capable of performing a variety of tasks, including milling, drilling, boring, turning, and cutoff, usually in just one setup.

nontraditional machining A variety of chemical, electrical, mechanical, and thermal processes used to machine workpieces.

□ The term *nontraditional machining* usually applies to new or nonconventional processes developed since 1955.

notching A stamping operation that uses a punch to cut V-shaped pieces from the edge of a sheet strip to gradually make a blank contour before it is detached by a cutoff or parting step. *(See figure.)*

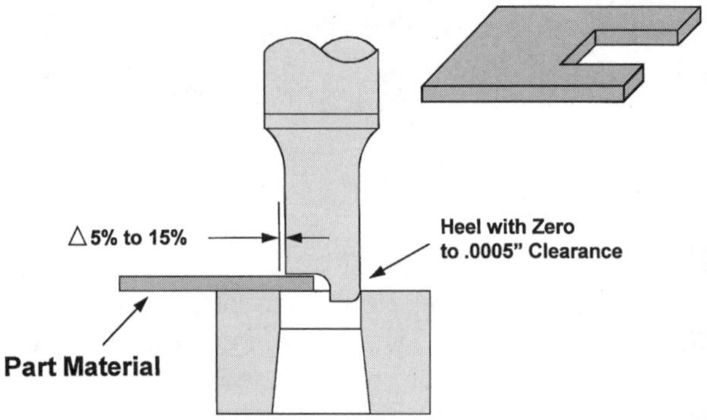

notching. *(Courtesy Dayton Progress Corp.)*

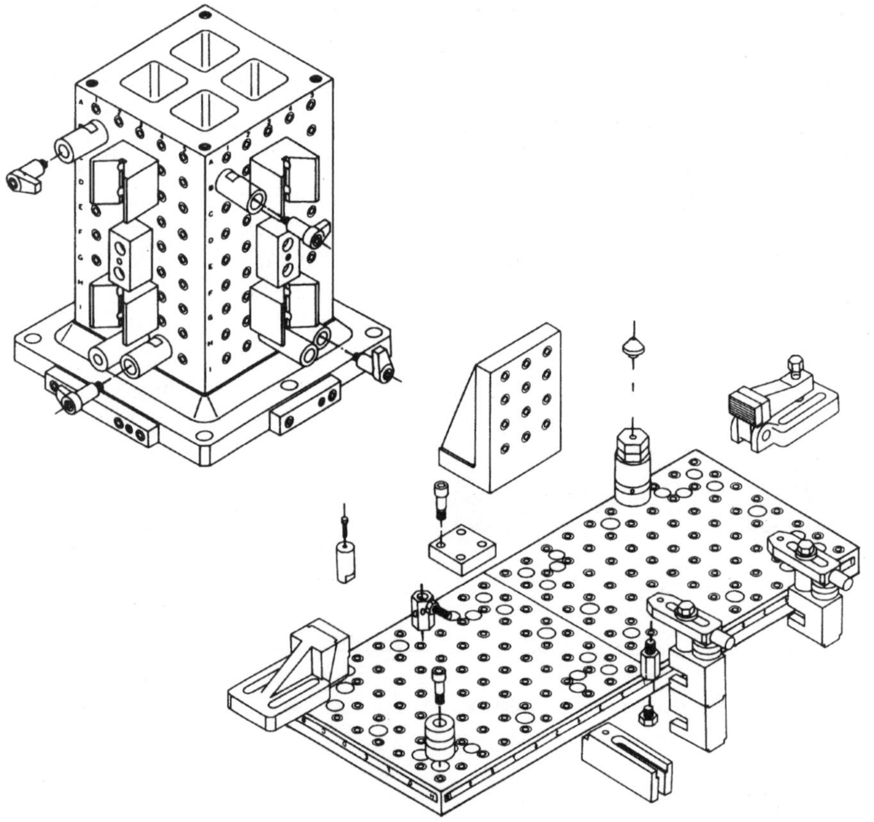

modular fixturing. *(Courtesy Carr Lane Mfg. Co.)*

offhand grinding Hand-feeding a workpiece into a bench or pedestal grinder to rough grind metal and resharpen tools.

parallel A strip or block of precision-ground stock, usually made in matched pairs, that is used to raise and provide a solid seat for a workpiece during layout and machining operations.

parting An operation used in lathe or screw machines to separate a completed part from chuck- or collet-held stock using a narrow, flat-end cutting tool (parting tool).

□ In sheet-metal forming, parting is a stamping operation that performs two cutoff operations in the same stroke of the press.

piercing The general term for shearing or punching openings such as holes and slots in sheet metal, plate, or parts.

□ This operation produces a slug, the shape of the punch, that is scrap.

plain milling A milling operation that produces a flat finished surface usually in a plane parallel to the axis of a cutter which has teeth or inserted tools on its periphery. *(See figure.)*

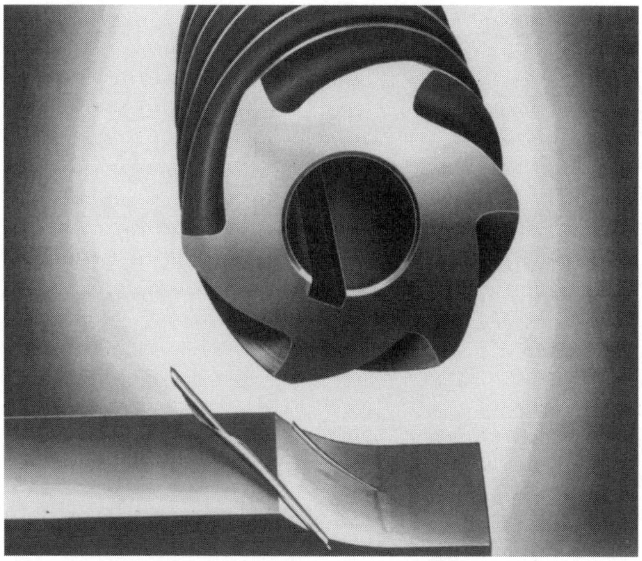

plain milling. *(Courtesy Cincinnati Milacron, Inc.)*

□ *See also* MILLING.

planer, planing machine Planer-type machines are used for machining large, flat surfaces that are too large for milling machines. The cutting tool is generally in a stationary position, and the work, mounted on the reciprocating machine table, is moved under the cutting tool.

□ The cutting tool may be of the single-point type or consist of a milling-type head equipped with a large-diameter milling cutter with inserted blades.

□ There are various types of planers, including double-housing, open-side, convertible and adjustable open-side, double-cut, and milling.

planing A planer operation that produces flat surfaces and contour shapes by reciprocating the workpiece, attached to the machine table, under one or more single-point cutting tools.

point-to-point system A CNC locating system used to position cutting tools rapidly for drilling, reaming, and other types of machining operations.

□ The cutting tool is usually positioned above the work surface and *is not* in contact with the workpiece during this locating operation.

polishing An abrasive process that improves surface finish on flat and contour surfaces by using some form of abrasive particles attached to a flexible backing to abrade the workpiece.

polishing band A continuous loop of usually 1-in-wide abrasive cloth that is mounted on a special polishing attachment of a contour bandsaw to finish work surfaces. *(See figure.)*

polishing band. *(Courtesy DoAll Co.)*

□ Coarse-grained abrasive belts are used for roughing; fine-grained belts are used for finishing.

power brushing Any process that uses a power-driven, rotary industrial brush to deburr, clean, or finish a metal part.

□ Brush fibers may be made of abrasive-filled plastics, metal wires, animal hairs, and other materials, and may be fiberglass-coated.

power hacksawing A machine-sawing process that uses a reciprocating motion of a short, straight-toothed blade to cut off the workpiece while cutting fluid may be applied to the saw blade.

press A machine with a stationary bed and a slide (ram) that has a controlled reciprocating motion toward and away from the bed surface.

□ Presses are usually used with punches and dies for forming, punching, and shearing operations to produce a wide variety of thin metal parts.

□ *See also* PRESS BRAKE.

press brake A type of gap-type press that was developed for making long bends in sheet metal.

□ Complex shapes can be bent from long stock with reasonably simple dies.

□ This type of press is ideal for the fabrication of long structural parts that must be made from sheet metal, such as aircraft and other transportation parts.

press forming Any forming operation performed on a mechanical or hydraulic press using a die and related tooling.

profiling The operation of machining the vertical edges of a workpiece having irregular contours with an end mill on a vertical milling machine or with a profiler that is following a master form or pattern.

punching A punch- or stamping-press operation in which holes are cut in ferrous, nonferrous, and nonmetallic materials with the proper-shaped punch that produces scrap slugs.

□ Other operations similar to punching include marking, nibbling, notching, perforating, piercing, pointing, and slotting.

quick-change toolholder An indexable-lathe toolholder, holding four or more different cutting tools, that permits rapid and accurate tool changes for various machining operations. *(See figure.)*

□ *See also* TOOLHOLDER.

rack-indexing attachment A milling attachment used to index and provide accurate spacing for cutting gear teeth on a flat bar or rack.

□ This attachment, when used along with a universal dividing head, can also be used to machine worms.

radial drill A large drilling machine with an arm that can be rotated around the column to provide positioning flexibility, greater reach, and stability.

□ This type of machine is used for large workpieces that would be difficult or awkward to position for hole locations.

□ *See also* DRILLING MACHINE, DRILL PRESS.

reaming A machining process using a multiedge, fluted cutting tool to smooth and accurately size a previously cored, drilled, or bored hole.

□ Reaming is commonly performed on drill presses, lathes, and milling machines.

□ *See also* DRILLING.

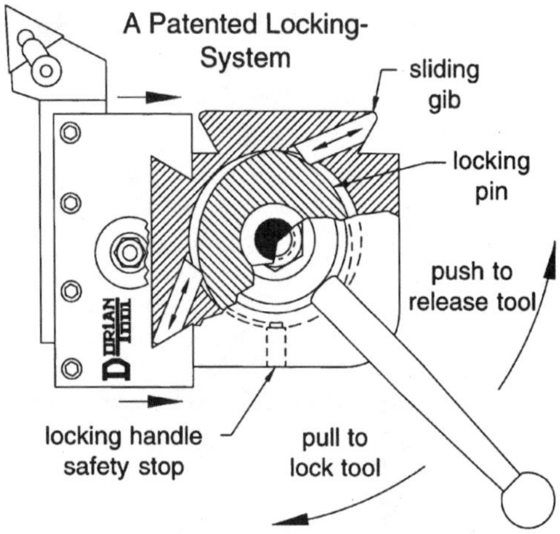

quick-change toolholder.
(Courtesy Dorian Tool International.)

roll bending A process of forming cylindrical or semicylindrical shapes from sheet metal or plate stock. *(See figure.)*

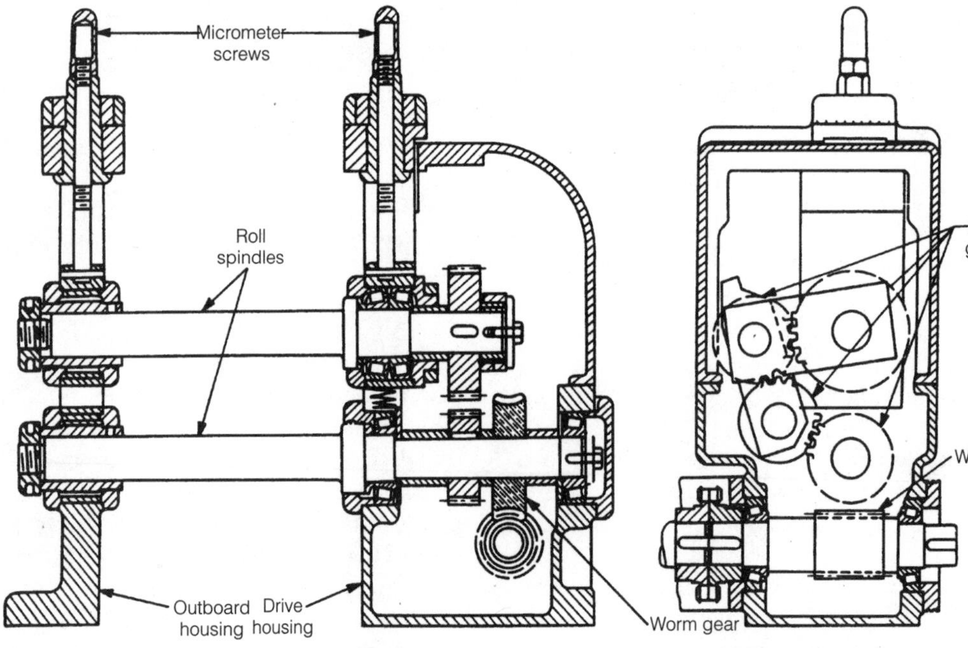

roll bending. *(Courtesy Society of Manufacturing Engineers.)*

□ The material is fed edgewise into the gap between a cluster of at least three relatively small-diameter straight rolls that progressively form the piece to the desired shape.
□ The size of the gap between rolls and its relation to the thickness of the sheet determine the bend radius of the finished cylinder.

roll forming A process used for forming metal parts from sheet, strip, or coiled stock into shapes of uniform cross section. The material is fed into a continuous set of rolls that progressively change its shape until the desired cross section is formed.
□ Forming usually starts at the center and gradually progresses to the outer edge with little or no change in the metal thickness.

rotary table A milling accessory that can be bolted to the machine table for machining circular slots, bolthole circles, cams, and other forms.

saw, sawing machine A type of machine that uses a hardened serrated-tooth blade to cut metal or other material to length or shape.
□ They are available in a wide variety of types and sizes, such as
1. The *hacksaw,* a simple, rugged machine that uses a reciprocating motion to cut metal or other materials.
2. The cold or circular saw (*see figure*) that uses a circular blade to cut structural materials.

cold or circular saw. *(Courtesy Everett Industries, Inc.)*

3. The two types of bandsaws that use an endless saw band:
   ▸ The *cutoff saw,* usually a horizontal machine, is used to cut long bars to the desired length.
   ▸ The *contour bandsaw,* usually a vertical machine, is used to cut intricate contours and shapes.
   ▸ The *abrasive cutoff saw* uses an abrasive disk rotating at high speeds instead of a blade with serrated teeth to cut off metal.

sawing A cutting off or sawing operation that uses a blade with hardened cutting teeth to cut material into lengths or shapes. Sawing may be divided into five categories:
1. *Abrasive bandsawing.* This is actually a grinding operation that uses a saw blade with abrasive particles, generally diamond, fused to the teeth edges to cut superhard materials.
2. *Abrasive cutoff sawing.* A thin abrasive wheel revolving at high speed is used to cut hardened or soft steel as well as glass and ceramics.
3. *Bandsawing.* A flexible, endless blade mounted on pulleys is fed into a workpiece by hand, or mechanical feed, to cut straight or contour forms. *(See figure.)*

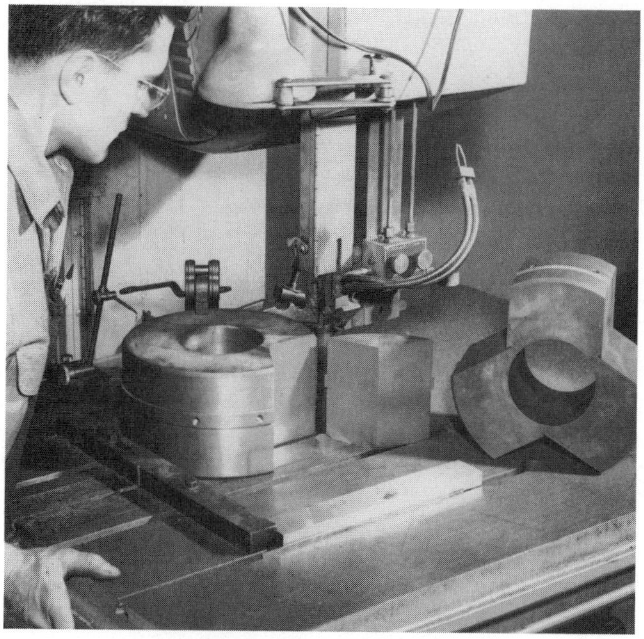

bandsawing. *(Courtesy DoAll Co.)*

4. *Circular sawing.* A power-driven, shaft-mounted disk, having teeth on its periphery, is rotated slowly and fed into the workpiece.
5. *Hacksawing.* A hand- or power-driven reciprocating blade with hardened teeth on one side is fed into the workpiece.

shaper, slotting machine A machine that uses a single-point, reciprocating cutting tool to machine flat surfaces and cut slots in a workpiece.
□ This machine is a rarely used machine because it cuts only on the forward stroke of the tool and it cannot compete with the metal-removal rates of other machines.
□ It is sometimes used to machine special, unusual, and intricate shapes on low-volume runs.
□ *See also* BROACHING MACHINE, MILLING, *and* MILLING MACHINE.

shaping An operation used to produce flat horizontal, vertical, or angular surfaces on a shaper.
□ Cutting takes place when a tool is reciprocated along the workpiece and metal is removed only on the forward stroke.
□ This process is seldom used since shapers cannot compete with the metal-removal rates of milling machines and machining centers.

shaving A secondary shearing operation using dies with very small clearances to remove very small amounts of metal from a sheet-metal part.

shear forming (flow turning) A power-spinning operation performed in a lathe where a metal plate is held firmly against a form mandrel and a power roller presses against the side of the revolving plate to form it to the mandrel shape. *(See figure.)*

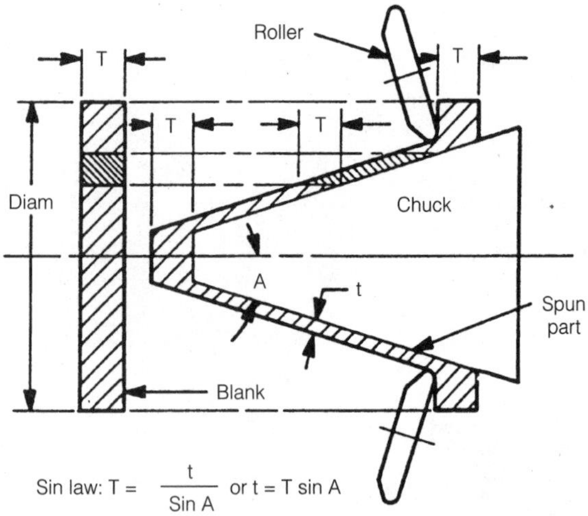

shear forming.
(Courtesy Society of Manufacturing Engineers.)

shearing The process of cutting large sheets of material into smaller pieces of specified length and width in a shearing press.

□ These pieces may be used in further operations such as blanking, forming, lancing, notching, punching, shaving, and trimming to produce the final product.

slitting The operation of making incomplete or single-line cuts in sheet-metal parts or sheets using a gang of circular blades.

□ *See also* MICROSLICING.

slotting A milling operation that produces grooves, slots, and forms such as T-slots and dovetails in workpieces.

□ This is also a punch-press operation in which rectangular and elongated slots are cut.

slotting attachment An attachment, mounted on the milling machine spindle, to convert the rotary motion of the spindle into a reciprocating motion for slotting tools to machine keyways and slots.

spade drilling The operation of using a spade drill consisting of a rigid shaft or holder equipped with a flat interchangeable end-cutting blade to produce large holes beyond the range of twist drills.

□ *See also* DRILLING *and* TREPANNING.

spindle adapters Any device that provides a means of fitting the shank of a cutting tool or arbor into the spindle of a machine.

spinning A chipless production method of forming axially symmetrical metal shapes similar to shear forming where a metal disk is plastically deformed by bringing it into contact with a rotating chuck (mandrel) by axial or axial-radial motions of a tool or rollers. *(See figure.)*

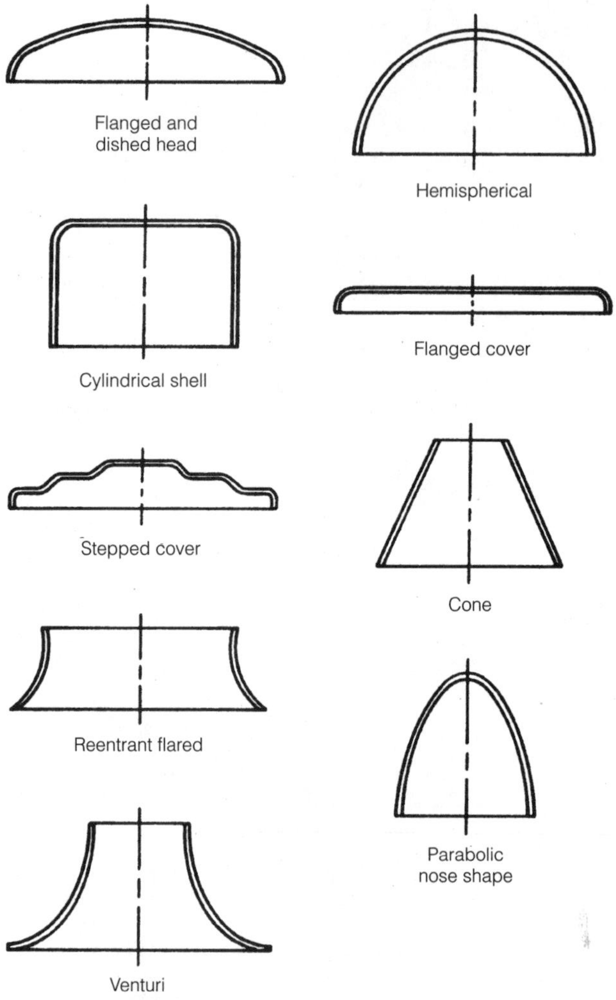

spinning. *(Courtesy Society of Manufacturing Engineers.)*

□ Common shapes produced include cones, cylinders, tubes, and other hollow symmetrical parts.

spotfacing The operation of smoothing and squaring the surface around a hole with a counterbore or boring tool to provide a flat spot for a cap screw or nut.

□ *See also* COUNTERBORING *and* COUNTERSINKING.

steadyrest A lathe accessory that can be fastened to the lathe ways to support and prevent long, thin, or flexible work from springing while being machined.

□ *See also* FOLLOWER REST.

stretch forming The process of forming large sheet-metal pieces of thin metal over a tool or form block by applying tensile forces in a metal stretch press.

□ The forming punch moves vertically as the jaws move laterally to stretch metal over the curve of the punch face and then the punch sides.

□ The metal is stretched to its elastic limit, causing a slight thinning of the metal with very little springback.

□ Used extensively in the aircraft industry to produce parts of large radii or curves.

surface grinders Grinding machines that produce fine surface finishes and cut flat-plane surfaces to accurate size using a grinding wheel as the cutting tool.

- □ The work is generally held on a magnetic chuck for grinding; however to accommodate a certain workpiece size and material, vises, angle plates, and fixtures are sometimes used.
- □ There are four types of surface grinders:

1. *Horizontal spindle, reciprocating-table surface grinder. (See figure.)*

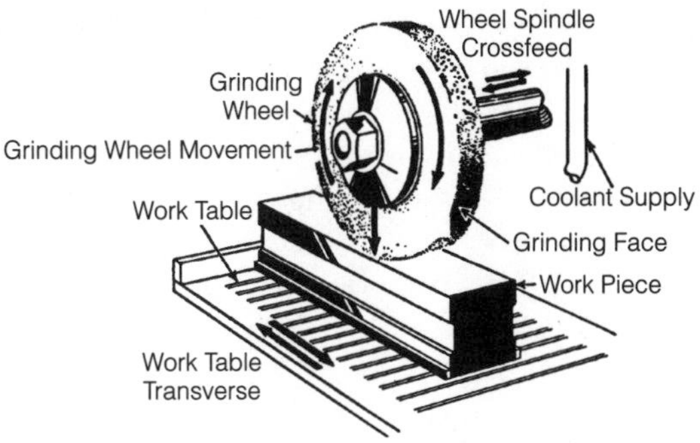

horizontal spindle reciprocating table.
(Courtesy Carborundum Co.)

- ▶ The workpiece is ground by the periphery of the grinding wheel while it is reciprocated under the revolving grinding wheel.
- ▶ The crossfeed movement of the table provides the feed, and the depth of cut is set by the downfeed wheel.
- ▶ This machine is used to grind flat surfaces with the periphery of the wheel or vertical surfaces with the side of the wheel.

2. *Horizontal spindle grinder with a rotary table. (See figure.)*

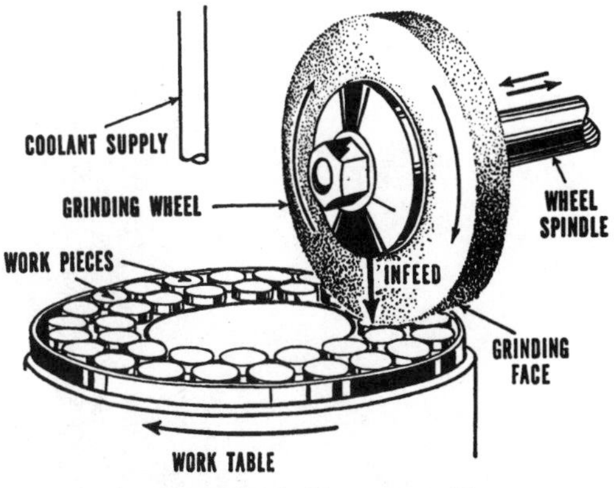

horizontal spindle rotary table.
(Courtesy Carborundum Co.)

- ▶ Work is mounted on a rotary magnetic table that revolves under the periphery of a grinding wheel.
- ▶ This machine is used for production grinding the faces of numerous small parts quickly and accurately.

3. *Vertical spindle with a rotary table. (See figure.)*

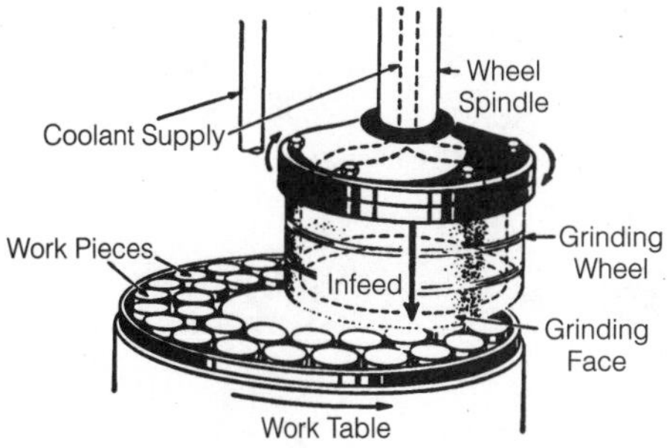

vertical spindle rotary table. *(Courtesy Carborundum Co.)*

- ▶ Work is ground by the face of a vertical-spindle grinding wheel while the work on the rotary table is being revolved.
- ▶ A very efficient and accurate grinder that produces a surface finish consisting of a series of intersecting arcs.

4. *Vertical spindle with reciprocating table. (See figure.)*

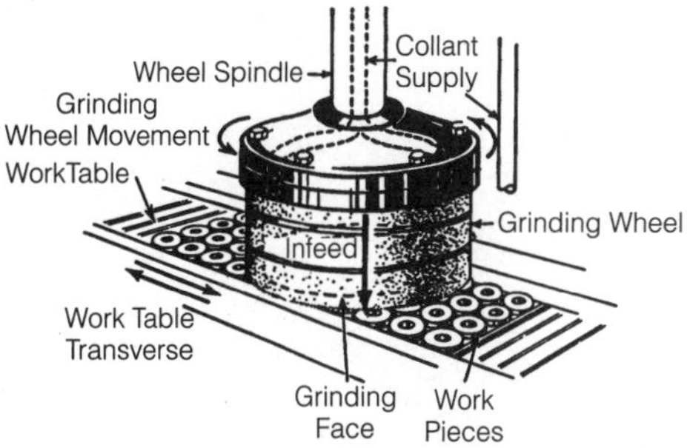

vertical spindle reciprocating table.
(Courtesy Carborundum Co.).

- ▶ The work reciprocates under the wheel face that may be tilted a few degrees to permit cuts as deep as .500 in.
- ▶ This grinder is capable of deep cutting and high material-removal-rate operation because of the greater area of contact between the wheel and work.
- ▶ The surface pattern produced is a series of intersecting arcs.

tailstock drill and tapholder An accessory mounted in a lathe tailstock that can be used for center drilling and tapping work held in a chuck.

- □ *See also* CHUCK.

taper-turning attachment An attachment, to which the lathe cross slide is connected, for machining internal or external tapers in inches or degrees without disturbing the alignment of the tailstock.

tapping An operation in which a cutting tool called a *tap,* having teeth on its periphery, cuts internal threads in a predrilled hole having a smaller diameter than the tap diameter.

- □ Threads are formed as the helical form of the rotating tap pulls it through the workpiece.
- □ Tapping can be performed manually or in a drill press with a power tapping attachment.

□ *See also* TAPPING MACHINE *and* THREADING.

tapping attachment An accessory that fits into a drill-press spindle to slowly advance a tap into a previously drilled hole to produce an internal thread under power. *(See figure.)*

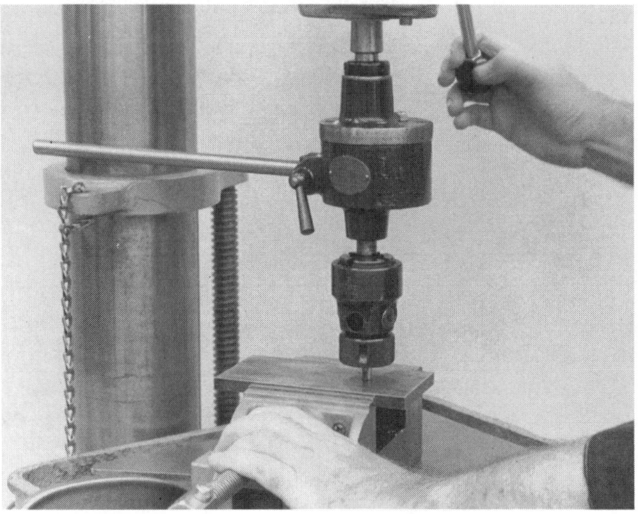

tapping attachment. *(Courtesy Kelmar Associates.)*

□ *Down pressure* applied to the downfeed handle of the machine causes the tapping attachment to turn clockwise and feed the tap into the hole.

□ *Up pressure* reverses the spindle direction to bring the tap out of the hole.

tapping machine A machine used for high-volume production and cost-effective tapping that features repeatability, high production rates, and reduced tap breakage.

□ Machine accessories can include indexing units with multiple tapping spindles, precise stroke-depth settings, and automatic features to increase the productivity of these machines.

thread grinder A form-grinding machine that has precision gears and leadscrew so that thread may be formed or finished by grinding.

□ A grinding wheel dressed to the exact form of the thread required is used; the wheel must be dressed occasionally during use to maintain the required thread form.

□ Most threads are formed in two or more passes, with the final pass removing only a few thousands of an inch to eliminate distortion.

threading The process of cutting, turning, or rolling various thread forms into material. *(See figure.)*

□ Standard specifications such as thread angle, depth, and width are available for each type of thread so that the thread cut will fit mating parts.

□ Threading is performed on lathes and specialized machines designed for threading.

threading machine A single- or multispindle universal threading machine uses dieheads and thread chasers to cut threads automatically or semiautomatically.

toolchanger An attachment fitted to a machining center that holds a variety of tools and inserts each one into the machine spindle when directed by the computer program. *(See figure on next page.)*

toolholder Any device used to hold a solid or indexable cutting tool firmly for machining operation.

toolroom lathe A high-precision lathe built to hold tighter tolerances than regular, general-purpose lathes.

□ *See also* LATHE *and* TURNING MACHINE.

tooth rest A specially ground piece of metal that contacts and supports the edge of a cutter's tooth while it is being resharpened on a tool and cutter grinder.

tracer attachment A device containing a stylus connected to a servo mechanism that follows the form of a template or sample workpiece and hydraulically guides the movements of the cutting tool to produce a complex workpiece.

trepanning The operation of drilling deep holes that are too large to be drilled by high-pressure coolant drills or gun drills.

□ The trepanning drill has one straight flute and no dead center.

□ The cutting chips form a solid core of metal between the flute and the hole wall that guides the drill.

□ *See also* BORING, DRILLING, *and* SPADE DRILLING.

turning The workpiece, which may be held in a chuck, mounted on a faceplate, or held between lathe centers, is rotated against a single-point cutting tool that may be fed along its periphery or across its end or face. *(See figure on next page.)*

□ Feeding along its periphery, called *straight turning,* produces a cylindrical shape; feeding across its end (*facing*) produces a true flat surface.

□ Taper turning, step turning, chamfering, facing, and thread cutting are other operations performed on a lathe.

threading. *(Courtesy Kelmar Associates.)*

toolchanger. *(Courtesy Cincinnati Milacron, Inc.)*

☐ Turning operations may also be performed on turning centers, chucking machines, automatic screw machines, and turret lathes.

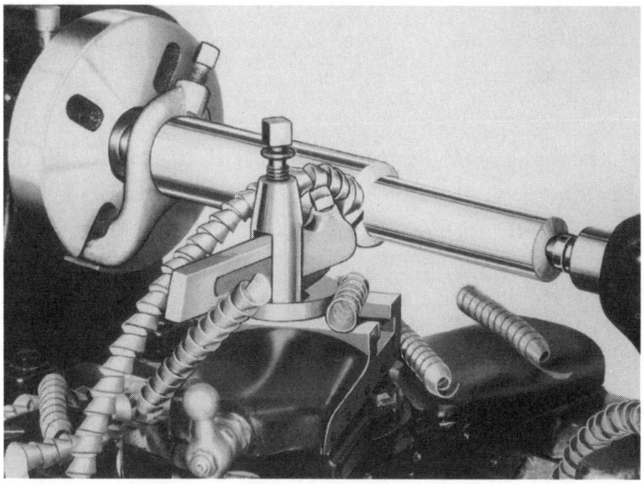

turning. *(Courtesy South Bend Lathe, Inc.)*

turning machine Any machine that rotates a workpiece while feeding a stationary cutting tool into it to produce round workpieces.
☐ *See also* LATHE.

turret lathe A production machine that differs from an engine lathe by the normal compound rest being replaced with a square, multitool indexable turret mounted on the cross slide and a hexagon turret mounted in the tailstock.
☐ Used for the production of many similar parts.

undercutting
☐ In *lathe work,* often called *necking* or *grooving,* the cutting of a groove next to a shoulder on a workpiece of work. This provides a runout for the thread-cutting tool at any point along a diameter, or a sharp corner when the diameter must be ground.
☐ In *welding,* a groove formed in the base metal next to the toe or root of a weld that is not filled by the weld metal. This is a washout usually caused by improper heat (voltage) and improper welding procedures.

universal helical-milling attachment An attachment used on a universal mill to cut helixes, screw threads, worms, and other helical shapes. *(See figure.)*
☐ *See also* RACK-INDEXING ATTACHMENT.

helical milling attachment.
(Courtesy Cincinnati Milacron, Inc.)

universal milling attachment A table-swivel housing, found on a universal milling machine, that allows the table to be swiveled to any angle for machining angular and helical forms.

V-block A workholding device having a V-shaped slot for holding round stock for layout, machining, or inspection operations.

vertical-milling attachment An attachment mounted on a horizontal mill so that vertical and angle milling operations can be performed.

vise A workholding device mounted on various machine tools to hold smaller workpieces for machining.

- □ Vises are available with fixed or quick-change jaws and can be manually or hydraulically operated.
- □ Common vise types are plain, angle, gang, swivel, quick-change jaw, self-centering, clusterlock, and indexable.

wheel-balancing stand An accessory used to test a grinding wheel to ensure that it is accurately balanced on its spindle adapter before it is mounted on a grinder.

- □ Proper balancing will ensure that the wheel does not chatter during grinding and will produce good surface finishes.

wheel dresser A single diamond or multidiamond tool used to true the face of a grinding wheel to make it concentric with its axis, and dress it to clean the wheel and make it cut better. *(See figure.)*

wheel dresser. *(Courtesy Kelmar Associates.)*

wheel flange The collars of a grinding-wheel adapter that form the side support for a grinding wheel when it is mounted on a machine spindle.

work-rest blade The part of a centerless grinder that guides, supports, and keeps the workpiece in contact with the regulating and grinding wheels. *(See figure.)*

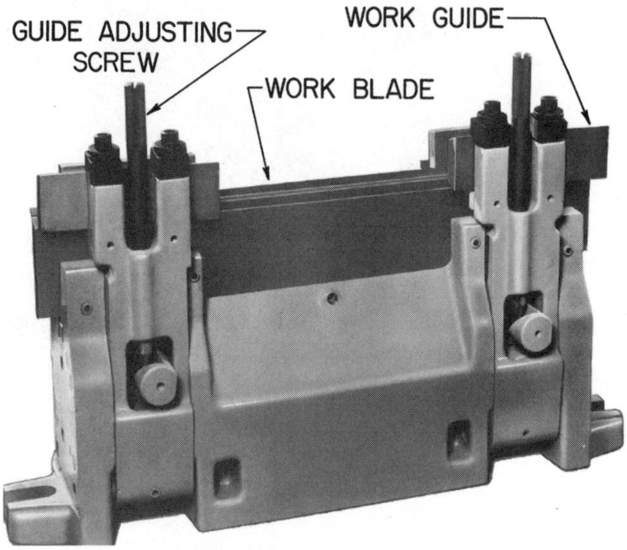

work-rest blade. *(Courtesy Cincinnati Milacron, Inc.)*

- □ Most work, up to 1 in in diameter, is ground with the top of the work-rest blade set above the wheel-center height, about half the diameter of the work.
- □ For larger-diameter parts, the blade should rarely be set more than $1/2$ in above the wheel-center height.

work-squaring bar An accessory that can be mounted to the contour bandsaw table to square the work to the blade.

SECTION 4

Cutting Tools and Holders

Arthur R. Gill
Coordinator, Precision Metal Trades
Niagara College—Glendale Campus
Niagara-on-the-Lake, Ontario

Steve F. Krar
Consultant, Kelman Associates
Welland, Ontario

An important component of the machining or grinding process is the type of cutting tool selected and how the tool is used. The new metal alloys developed to perform at high temperatures and high pressures are very difficult to machine and grind. Diamond and cubic boron nitride superabrasive grinding wheels and cutting tools are well suited to cutting these superalloys. Combination cutting tools–modular toolholders have helped increase the accuracy and productivity of many machining and grinding operations.

abrasive Natural or manufactured abrasive particles, each acting as a small single-point cutting tool, that produces a good finish on a metal surface. *(See figure.)*

abrasives. *(Courtesy General Abrasive Co.)*

abrasive band Continuous abrasive-coated band.

abrasive belt Continuous abrasive-coated belt used for deburring, polishing, and similar operations.

abrasive cutoff wheel A thin grinding wheel made with a rubber bond, sometimes reinforced, is used for cutting through a workpiece or cutting slots and grooves in a workpiece.

abrasive dressing stick An abrasive stick such as boron carbide, used to dress the face of a grinding wheel.

acme tap A cutting tool used to produce an internal screw thread that has an included angle of 29°.

adapters Machine tool accessories used for mounting straight or taper shank tools in a machine spindle.

adjustable reamer A cutting tool with inserted blades that can be adjusted approximately .015 in (0.04 mm) over or under the nominal reamer size to finish a hole to size and produce a good surface finish.

angle cutters A type of milling cutter with teeth that are neither parallel nor perpendicular to the cutting axis. They are used for milling angular surfaces such as grooves, serrations, and chamfers. *(See figure on next page.)*

arbors Milling accessories that are used to mount milling cutters. They are inserted and held in the spindle by a draw bolt or a quick-change adapter.

ball-nose end mill An end-milling cutter with a spherical radius on the end of the cutter; the cutter end is the same size as the cutter diameter.

angle cutter. *(Courtesy Union Butterfield Corp.)*

band file A steel band, used on a contour bandsaw, that has a number of short interlocking file segments riveted to it to form a continuous loop. File bands are available in a variety of cuts, widths, and cross-sectional file segments.

bandsaw blade Continuous bands of various widths with different sizes and styles of cutting teeth; also knife-edge blades used as the cutting tools for the bandsaw.

bonded abrasive Abrasive products that are held together by a suitable bonding material which holds the abrasive grains together to form grinding wheels, sharpening stones, and other abrasive products. *(See figure.)*

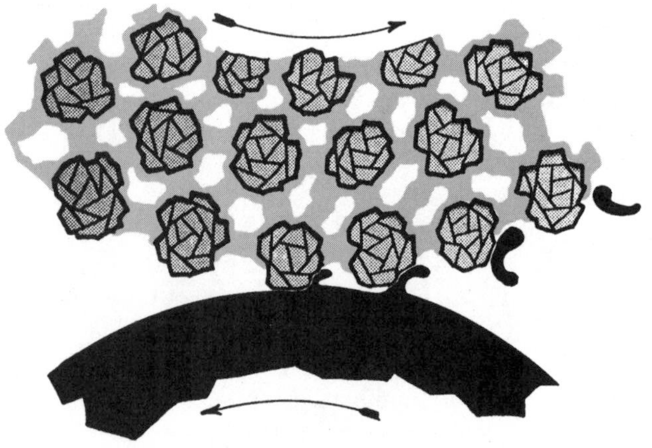

bonded abrasive. *(Courtesy Carborundum Co.)*

boring bar Usually a round bar, used mainly on lathes, mills, and boring and drilling machines, that holds one or more cutting tools during a boring operation. It can be held stationary and moved axially while the workpiece revolves around the cutting tool, or it can be revolved and moved axially while the workpiece is held stationary.

boring tool A cutting tool mounted in the boring bar that is used to true and enlarge a previously drilled or cored hole.

brazed tools Carbide inserts that are brazed onto steel shanks. They are available in a variety of sizes and shapes and may be ground to resharpen or change the shape of the tool to suit the machining operation.

broach A tapered tool having a series of teeth increasing in size and/or shape that is pushed or pulled into a workpiece. *(See figure.)*

broach. *(Courtesy Kelmar Associates.)*

☐ Each successive tooth removes a small amount of metal to enlarge a hole or slot to a size or shape.

built-up edge Small metal particles that form on the edge of a cutting tool as a result of high temperature, high pressure, and high frictional resistance during machining.

burr *See* ROTARY BURR.

carbide A wide variety of hard metals produced by powdered metallurgy technology. Cutting tools made of cemented carbide can increase cutting speeds to about 3 to 5 times faster than that used for high speed steel tools.

carbide-tipped tools Cutting tools with carbide tips brazed to their cutting edges. They are used when abrasion or high heat in machining are considerations. *See* BRAZED TOOLS.

center drill Combination drill and countersink used for predrilling operations and producing mounting holes for work to be turned between centers in a lathe. *(See figure on next page.)*

center drill. *(Courtesy Cleveland Twist Drill Co.)*

ceramics A class of hard, brittle, and high-melting nonmetallic cutting tool materials made from aluminum oxide, zirconium oxide, and beryllium oxide.

- □ Ceramics are generally used for machining hard ferrous materials and cast iron at cutting speeds higher than those possible with the use of high-speed steel tools.
- □ They lower manufacturing costs and increase productivity.

ceramic-tipped tools Cutting tools with ceramic inserts cemented to the tool shank with an epoxy glue.

cermets Inserts composed of solid titanium carbide (TiC) and titanium nitride (TiN) with a superalloy metal binder; used for high-velocity precision machining of cast irons and steels where high temperatures are encountered.

chamfering tool A cutting tool or abrasive wheel that will create a chamfer or beveled edge on a workpiece or tool.

chatter Surface imperfections on a workpiece that are usually caused by vibrations transferred from the cutting tool or grinding wheel to the workpiece.

chip A small piece of material removed from the workpiece by the cutting tool.

chipbreaker A groove or other tool feature that breaks chips into small pieces as they come off the workpiece. They are designed to prevent chips from becoming so long that they are difficult to control, catch in turning parts, or cause safety problems.

chip clearance A space provided in the body of the cutting tool to allow for the formation of chips.

chucking reamer A type of machine reamer that is generally held in some type of chuck for the reaming operation. The most common types are the *rose* and *fluted* reamers.

claw tooth A hook or claw-shaped tooth form that is commonly used on saw blades. It has a positive rake on the cutting face that provides a cutting rate.

clearance A space behind a tool's cutting edge that allows the tool to cut and prevents it from rubbing. Too much clearance will reduce the support for the cutting edge and result in rapid tool failure.

climb milling A form of milling, also called *down-cutting,* where the rotation of the cutter and the feed of the workpiece are going in the same direction. *(See figure.)*

- □ Climb milling should only be performed on a machine that has a backlash eliminator to prevent the work from being pulled in too quickly and cause cutter breakage.

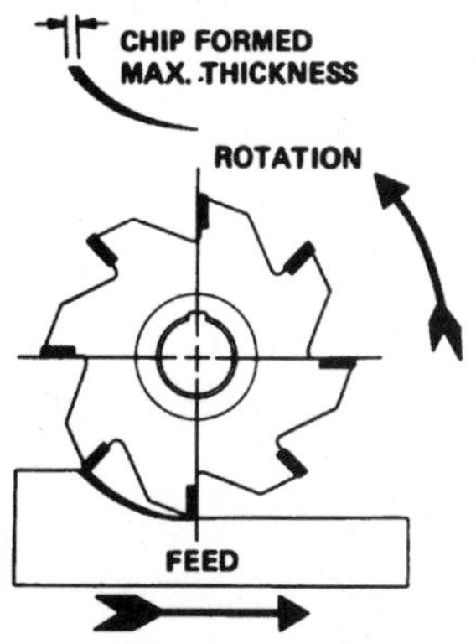

climb milling. *(Courtesy Hanita Cutting Tools.)*

coated abrasives A flexible backing of cloth or paper to which abrasive grains have been bonded.

coated carbide tools Cemented carbide tools are coated with a film of titanium carbide, titanium nitride, or aluminum oxide to increase tool life and manufacturing productivity while reducing machining costs. *(See figure.)*

coated carbide tool. *(Courtesy Kennametal Inc.)*

- □ Various coatings are applied singly, or in a combination of two or more materials, to give a cutting tool special properties. Coatings provide high wear and abrasion resistance at moderate speeds for rough and finishing operations.
  1. *Titanium oxide coating* has excellent lubricating properties and good crater resistance, and reduces edge buildup. It is used for heavy roughing cuts at high speed.
  2. *Aluminum oxide coating* provides chemical stability and maintains hardness at high temperatures. It is used for high speed roughing and finishing operations.
- □ *Triple-coated* tools use
  1. A strong wear-resistant titanium carbide coating as the innermost layer.
  2. A thick layer of aluminum oxide is then applied on top of the titanium carbide layer.
  3. A thin coating of titanium nitride (the third layer) which provides a lower coefficient of friction and reduces the chances of a built-up edge forming.

cobalt An element added to alloy steel to increase wear resistance and red hardness in cutting tools. It allows a tool to maintain a cutting edge at elevated temperatures.

combination tool Tools capable of performing more than one operation with the same tool, such as drilling, chamfering, and threading. These tools are generally used on machining centers capable of helical interpolation. *(See figure.)*

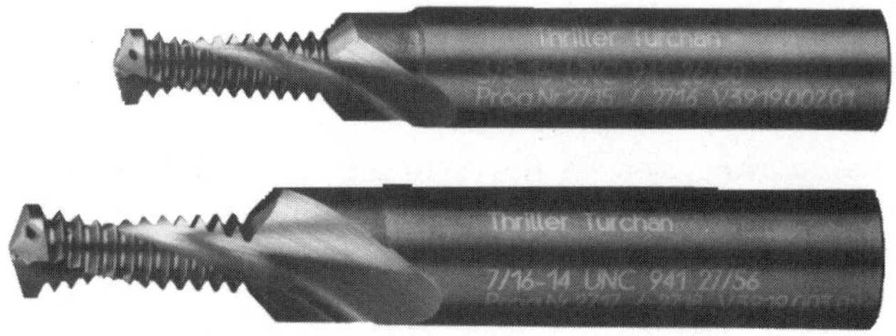

combination tool. *(Courtesy Thriller, Inc.)*

conventional milling A form of milling, also called *up-cutting,* in which the rotation of the cutter and the feed of the workpiece are going in an opposite direction. *(See figure.)*

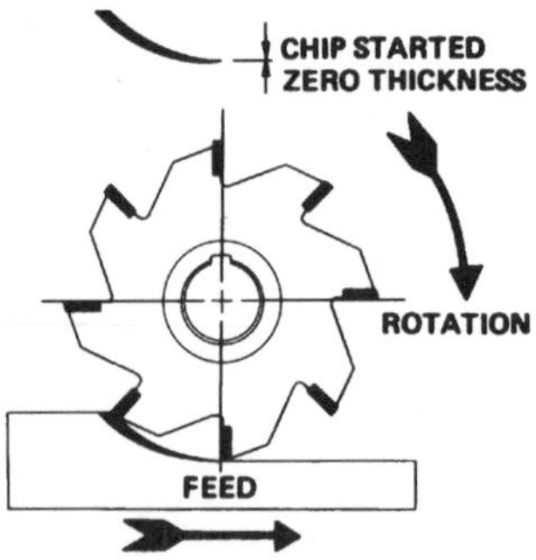

conventional milling. *(Courtesy Hanita Cutting Tools.)*

core drill A type of twist drill that has three or four flutes, generally used to enlarge cored, drilled, or punched holes. Core drills are available in sizes from $1/4$ to 3 in (6 to 76 mm) in diameter.

counterbore An end-cutting tool with a pilot on its end to keep the tool in line with the hole being counterbored. Some types of counterbore tools have interchangeable pilots to suit various-size holes.

countersink A tool with an angular cutting surface used to enlarge the start of a hole to allow a screwhead or other object to sit flush with the surface of the workpiece.

cratering The grooves or depressions created behind the cutting edge of a tool by chips sliding across the top of the tool or insert.

crush-form dressing A wheel-dressing operation in which hardened preformed rolls are forced into a slowly revolving grinding wheel to produce a desired shape or contour. This type of wheel dressing is used for complex profiles in production runs. *(See figure.)*

crush rolls Hardened tool steel or carbide rolls, used for crush-form dressing, that have the desired form or contour of the finished workpiece.

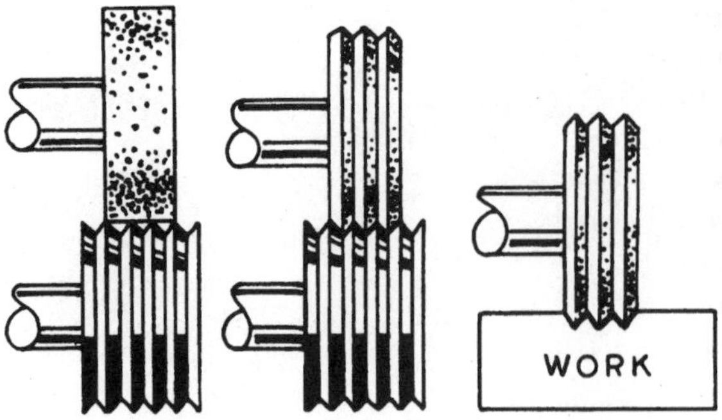

crush-form dressing. *(Courtesy DoAll Co.)*

cubic boron nitride (CBN) A manufactured superabrasive developed by the General Electric Company in 1969, with hardness properties just below that of diamond.

□ CBN cutting tools and grinding wheels are used to cut hard, abrasive ferrous materials.

cutting speed The speed, in feet or meters per minute, at which the cutting tool passes the work, or vice versa.

cutoff tool *See* PARTING TOOL.

cutoff wheel *See* ABRASIVE CUTOFF WHEEL.

diamond The hardest known substance, available in its natural form, is used primarily for gems; its manufactured form is generally used for industrial purposes. *(See figure.)* In 1957, the General Electric Company began the commercial production of manufactured diamond by subjecting a form of carbon, along with a catalyst, under high pressure and high temperature.

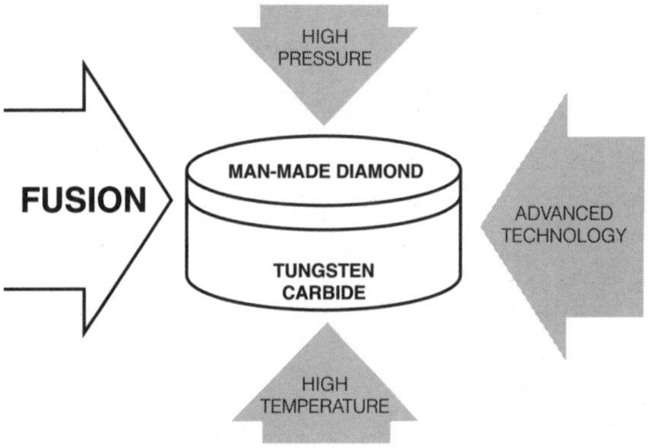

diamond tool manufacture. *(Courtesy GE Superabrasives.)*

□ Diamond is used to manufacture cutting tools, grinding wheels, and wheel-dressing tools.

diamond-tipped tools Cutting tools used to machine hard and abrasive nonferrous and nonmetallic materials that require high surface finish and extremely close tolerances.

diamond wheel Commonly used to grind cemented carbides or high vitreous materials such as glass and ceramics. Wheels $1/2$ in (13 mm) or less in diameter have diamond particles throughout (impregnated); larger metal-bonded wheels have a diamond surface bonded to the grinding face.

diamond wheel dresser This usually consists of a round steel pin with one or more diamonds mounted in the end. It is held in a holder or radius wheel dresser for dressing and truing the grinding wheel.

dies *See* THREADING DIES.

dovetail cutter A milling cutter that has a flat bottom and either a 45° or 60° included angle that is used to form the side of a dovetail slot.

dressing The operation of removing metal particles from a wheel and restoring sharp abrasive edges to the abrasive grain, also to create the desired form on a wheel. This operation improves the wheel's cutting action and produces a better surface finish on the workpiece.

drill, drill bit, drilling tool *See* TWIST DRILL.

electrobandsawing A low-voltage high-amperage current is fed into the saw blade, which is traveling at 6000 sf/min (1827 m/min). This sawing method is used to cut thin-wall tubing, stainless steel, and aluminum and titanium honeycomb sections. *(See figure.)*

electrobandsawing. *(Courtesy DoAll Co.)*

end mill A milling cutter with teeth on its end as well as on the periphery. End mills are manufactured in a variety of sizes and styles to suit a wide variety of milling operations.

expansion reamer A type of reamer that is similar to an adjustable reamer which is expanded by a threaded plug fitted in the end. The amount that it can be expanded is very limited; for example, a 1-in. (25.4-mm)-diameter reamer only expands up to .005 in (0.1 mm).

facemill A milling cutter usually over 6 in. (152.4 mm) in diameter that has a series of inserted teeth blades. It is used for producing large flat surfaces quickly and accurately.

feedrate The distance in inches or millimeters per minute that the workpiece moves into the cutter when milling; when drilling, the distance the drill advances into the work for each revolution; when machining in a lathe, the distance the cutting tool advances for each work revolution.

finishing tool The cutting tool used for the finish machining operation on a workpiece. It generally produces a smooth surface finish and brings the work to an accurate size.

flat, screw flat A flat surface machined or ground on a straight-shank cutting tool. A hardened screw is tightened on this flat to hold the tool in an adapter.

fluteless tap A forming tool used to produce internal threads in ductile materials such as copper, aluminum, brass, and leaded steel. *(See figure.)*

fluteless tap. *(Courtesy Greenfield Industries Inc.)*

flutes A groove cut into the body of a cutting tool that produces cutting edges, provides chip clearance, and allows cutting fluid to reach the cutting edges of the tool.

flycutter A single- or multiple-pointed cutting tool with the cutting end ground to the desired shape. It is held in an adapter and used for milling large flat surfaces.

form cutters Milling cutters that incorporate the exact shape of the form that is to be produced on a part. This permits the exact duplication of irregularly shaped forms on many similar parts.

form dressing The operation of producing a desired shape or form on a grinding wheel with a radius wheel dresser or a hardened form roll.

fraction drills A size designation for twist drills in sizes ranging from 1/64 to 4 in. Fractional drills usually vary in 1/64-in increments from one size to the next.

friction sawing A contour bandsaw operation used to cut hard or tough ferrous materials up to 1 in (25.4 mm) in thickness. The sawband generally has a velocity of up to 15,000 sf/min. (4572 m/min) and removes a thin kerf of metal through a combination of heat and friction.

gang milling Two or more milling cutters mounted on an arbor to produce a desired shape on the surface of a piece of metal. It is commonly used in production work where many parts must have the same form or steps. *(See figure on next page.)*

gear cutter Involute gear cutter.

gear hob A formed gear cutter used for continuous cutting of gear teeth on gear-generating machines.

gang milling. *(Courtesy Cincinnati Milacron, Inc.)*

helical cutter. *(Courtesy Cleveland Twist Drill Co.)*

grinding wheel A cutting tool composed of natural or manufactured abrasive grains that are held together in a desired form with a suitable bond material. Grinding wheels are manufactured in a variety of sizes and shapes to suit various grinding operations or different types of work material. The most commonly used grinding wheel is made of aluminum oxide grains that are held together with a vitrified bond.

gullet The curved area between two saw blade teeth that provides a space for chips to form.

gun drill A self-guided drill for producing holes as deep as 20 ft (6.09 m). It consists of a round tubular stem with a flat two-fluted drilling insert on the end. Cutting fluid is forced through the center of the tubular stem for flushing out the chips.

hacksaw blade A thin metal saw blade made of high-speed molybdenum or tungsten alloy steel that is hardened and tempered. Saw blades are manufactured in various pitches (number of teeth per inch) to suit the sawing operation and the type and size of material to be cut.

half-side milling cutter A cutter with teeth on only one side and around the periphery. The cutter is made with flat interlocking faces so that two cutters may be placed side by side for slot milling.

helical cutter A milling cutter having right- or left-handed spiral flutes cut at helix angles of 45° to 60°. This gives the teeth a shearing action, reduces chatter, and produces a good surface finish. These cutters are suited for milling wide and intermittent surfaces. *(See figure.)*

helix angle The angle that the tool's leading edge makes with the plane of its centerline.

hob *See* GEAR HOB.

hole saw A thin cylindrical-diameter cutter with a twist drill in its center that provides a guide for the cutting teeth of the hole cutter. They are used for drilling holes in thin material, pipe, and sheet metal. Hole saws are available in various diameters and saw teeth pitches to suit various types of materials.

honing tool A tool bit that has abrasive inserts incorporated in its body to produce a fine finish and accurate size on internal holes and surfaces.

indexable insert A replaceable geometric-shaped tool, made of carbide, ceramic, cermet, polycrystalline cubic boron nitride, or polycrystalline diamond that is mechanically held in special holders. Indexable inserts usually have multiple cutting edges that can be used and resharpened when all the edges are dull. *(See figure.)*

indexable insert toolholders Specially designed toolholders to suit the wide variety of indexable inserts manufactured, such as turning, boring, threading, and parting tools.

involute gear cutter A formed cutter having the exact shape of the space between the teeth of the gear to be cut. *(See figure on next page.)*

kerf The width of the slot left after a saw blade or tool passes through the workpiece.

knife sawband A thin metal band available with knife, wave, or scalloped edges, used to cut soft, fibrous materials such as cloth, cardboard, cork, and rubber.

knurling tool A lathe tool used for impressing a diamond or straight-line pattern into the workpiece to improve grip, increase the diameter, and improve its appearance.

lathe tool Usually a single-point cutting tool held in some type of toolholder. It is used for producing round forms using machining operations such as turning, facing, threading, and form turning.

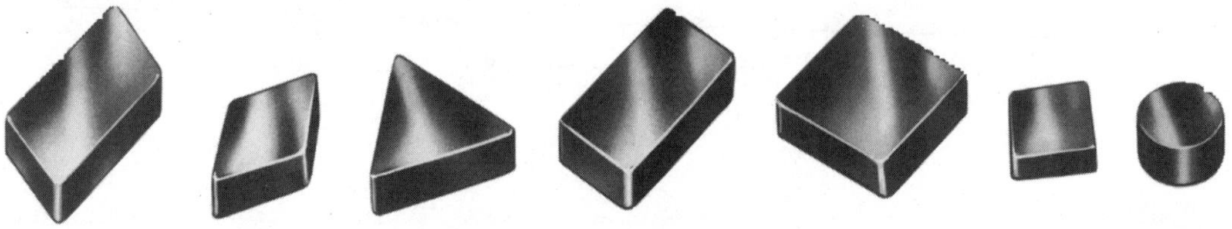

indexable inserts. *(Courtesy Hertel Carbide Ltd.)*

involute gear cutter. *(Courtesy Union Butterfield Corp.)*

letter drill An alphabetical size designation for twist drills ranging from letters *A* to *Z*. The A-drill is the smallest, at .234 in diameter; the Z-drill is the largest, at .413 in diameter.

line grinding bands These are flexible steel bands that have an abrasive bonded to their thin edge. They are run at high speeds 3000 to 5000 sf/min (914 to 1524 m/min) and are used to cut hardened steel alloys, brick, marble, and glass. *(See figure.)*

line grinding bands. *(Courtesy DoAll Co.)*

machine reamers Also called *chucking reamers*; used in lathes, drill presses, milling machines, and other equipment to bring a hole to an accurate size and produce a good surface finish. They are available in a variety of types and sizes.

millimeter (metric) drills A metric size designation for twist drills. Drill sizes range from miniature metric drills 0.04 to 0.09 mm in 0.01-mm increments and straight-shank standard drills ranging from 0.5 to 20 mm. Taper shank drills are available in sizes of 8 to 80 mm.

milling cutter Any type of rotary cutting tool with one or more teeth; used on a milling machine to remove material as the workpiece is moved past the rotating cutter.

milling cutter (horizontal) A cylinder made of hardened tool or high-speed steel with teeth cut on the periphery.

milling cutter (vertical) A rotary cutting tool with teeth on its end as well as on the periphery. These tools are generally held in a vertical spindle by a suitable adapter.

modular tooling A complete tooling system that provides the flexibility and versatility to build a series of tools from a stock of component parts. Modular tooling systems combine high accuracy and quick-change capabilities to increase productivity. *(See figure.)*

modular tooling.
(Courtesy KPT Kaiser Precision Tooling Inc.)

number drills A size designation for twist drills with drill sizes ranging from no. 1 measuring .228 in in diameter, to no. 97, measuring .0059 in in diameter.

oil-hole drills Twist drills with one or two oil holes running from the shank to the cutting points through which compressed air, oil, or cutting fluid can be forced during drilling. *(See figure.)*

oil-hole drill. *(Courtesy Cleveland Twist Drill Co.)*

- ☐ This type of drill is used for long hole drilling where it is difficult to get cutting fluid to the drill's cutting edges and to remove chips.

parting tool A thin cutting blade that is held in a cutting-off or parting toolholder. Both of its sides are tapered from top to bottom to prevent the tool from rubbing when work is being grooved or parted off.

pitch The number of threads (thread points) per inch on a screw, nut, or threaded object, or the number of teeth per inch on a saw blade.

plain milling cutter A hardened or ground cylinder made of high-speed steel with teeth cut on the periphery. It is probably the most common cutter used to produce a flat surface on a horizontal milling machine.

planing bit A single-point cutting tool similar in appearance to a turning tool or shaper toolbit but with a longer shank.

point angle The included angle at the point of a twist drill or similar tool. For general-purpose drilling, the included point angle is 118°.

- ☐ For drilling hard materials, the included point angle ranges from 135° to 150°, depending on the hardness of the material being drilled.
- ☐ For drilling soft materials, the included point angle ranges from 60° to 90°.

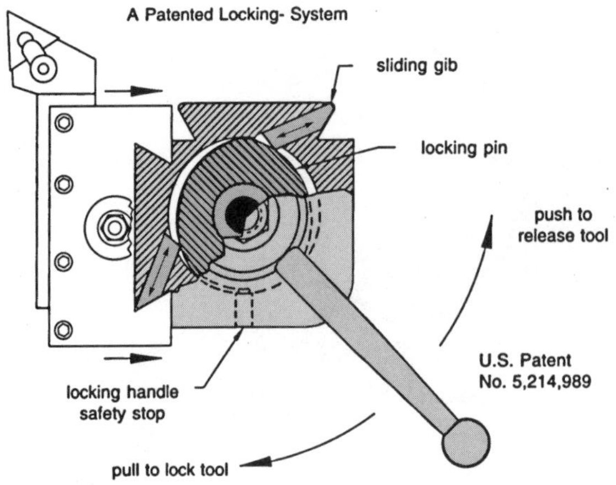

quick-change toolholder. *(Courtesy Dorian Tool International.)*

polycrystalline cutting tools *See* SUPERABRASIVE TOOLS.

quick-change toolholder A style of toolholder capable of quickly and accurately clamping holders with preset tools on a dovetailed tool post. It is used when multiple machining operations and quick-tool changes are required to increase productivity. *(See figure.)*

quick-change tooling *See* MODULAR TOOLING.

raker set A type of set pattern in which one tooth is offset to the right, one tooth is offset to the left, and the third tooth is straight. It is the most commonly used set pattern for most sawing applications.

reamer A rotary cutting tool with straight or helical cutting edges along its body. It is used to accurately size and smooth a hole that has been previously drilled or bored.

rose reamer An end-cutting tool where the cutting is done on the 45° chamfer at the end of the reamer. These reamers are usually .003 to .005 in (0.07 to 0.12 mm) under the nominal size. They are used for rapid metal removal rate and close sizing of holes. *(See figure.)*

rose reamer. *(Courtesy Cleveland Twist Drill Co.)*

rotary burr A shank-mounted high-speed steel or carbide cutter with uniformly ground teeth used where it is necessary to remove only a minimum amount of metal. They are available in a wide range of sizes and shapes.

rotary file A shank-mounted high-speed steel or carbide cutter whose teeth are cut and form broken lines to dissipate heat. They are available in a wide range of sizes and shapes and are used to remove a minimum amount of material.

rubber wheel *See* ABRASIVE CUTOFF WHEEL.

sawband A continuous band used on a contour or cutoff bandsaw for sawing metal. The most commonly used blades are made of carbon-alloy steel, high-speed steel, and tungsten-carbide-tipped.

set The amount of displacement that the teeth of a saw blade are offset on either side of center to produce clearance for the back of the band or blade. *(See figure.)*

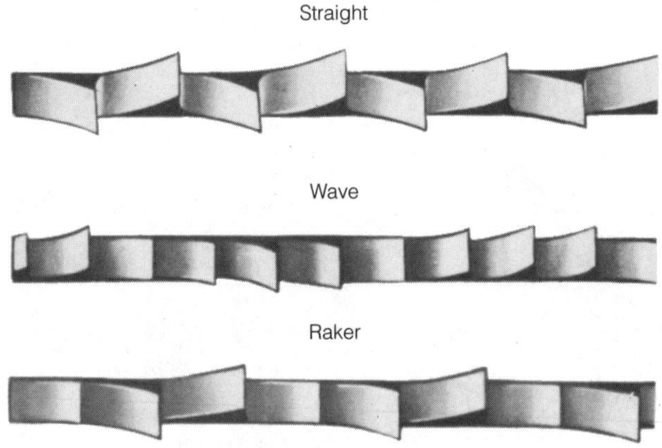

set. *(Courtesy DoAll Co.)*

sf/min *See* SURFACE FEET PER MINUTE.

shank The portion of a cutting tool that is held in a chuck, collet, or similar holding device.

shell end mill A multitooth cutter with teeth on the face and the periphery. Shell end mills are fitted on a stub arbor and can be changed quickly when the cutter becomes dull or must be replaced with a different size. *(See figure.)*

shell end mill. *(Courtesy Cleveland Twist Drill Co.)*

shell reamer A large reamer head that is mounted on a driving arbor which may be straight or tapered. It is economical to use for large holes, especially where a number of different large holes must be reamed, as the reamer head can be changed quickly from one size to the next.

side-milling cutter A narrow cylindrical milling cutter with teeth on each side as well as the periphery. It is used for cutting slots and for face and straddle milling operations.

slitting saw A thin plain-milling cutter with sides relieved or dished to prevent rubbing or binding during machining. They are available in widths ranging from $1/32$ to $3/16$ in. (0.8 to 5 mm). *(See figure.)*

slitting saw. *(Courtesy Niagara Cutter Inc.)*

slot drill This cutting tool is similar to a two-flute end mill except that one flute is larger than the other. This tool provides plunge milling capabilities because the larger flute overlaps the center of the cutter.

spade drill This is similar to a gun drill in that the cutting end is a flat blade with two cutting lips. They are available in a wide range of sizes, ranging from very small microdrills to 12 in (304.8 mm) in diameter.

spiral sawband A round sawband manufactured in diameters of .020, .040, .050, and .074 in (0.51, 1.02, 1.27, and 1.88 mm) that are used for making narrow cuts and sharp turns through various materials. *(See figure.)*

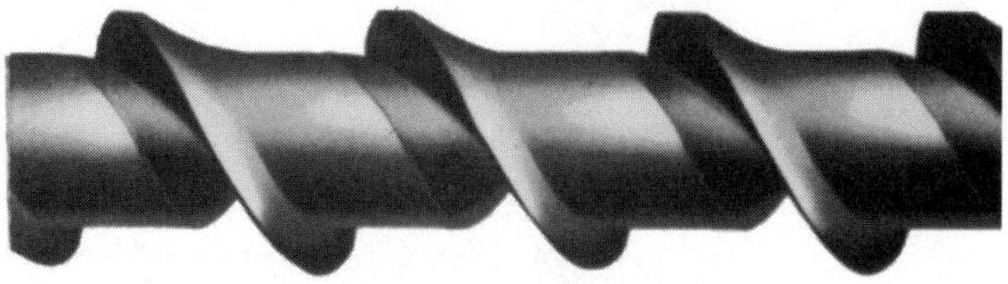

spiral sawband. *(Courtesy DoAll Co.)*

- ☐ Spring-tempered blades are used for cutting soft materials such as plastic and wood; all-hard blades are used for cutting light metals.

spotfacer A cutting tool equipped with a pilot to guide it in a previously drilled hole. It is used to smooth and square the surface around the top of a hole for the head of a cap screw or nut.

stellite A cast alloy material that contains 25 to 35 percent chromium, 4 to 25 percent tungsten, and 1 to 3 percent carbon; the remainder is cobalt. It is used to make toolbits because it has excellent red hardness qualities, good wear resistance, and hardness.

stellite-tipped tools Cutting tools in which the cutting edges are tipped with stellite inserts. They are capable of speeds and feeds that are 2 to $2\frac{1}{2}$ times higher than that of high-speed steel on uninterrupted cuts.

step drill A drill that has two or more diameters ground on it. It is used to drill and countersink, or drill and counterbore in one operation.

straddle milling A milling operation in which two side-milling cutters are used to machine opposite sides of a workpiece parallel in one operation. The cutters are separated on an arbor by spacers of the required spacing to suit the surfaces being cut. *(See figure.)*

straddle milling. *(Courtesy Cincinnati Milacron, Inc.)*

straight-flute drill A type of drill that has flutes running straight along the length of the body to prevent the drill from drawing itself into the work material. They are often recommended for the drilling of soft materials such as copper, brass, and bronze.

straight set A set pattern on a saw blade where one tooth is set to the right and the next to the left. It is used for cutting nonferrous materials, thin sheet metal, and tubing.

superabrasive tools Cutting tool inserts made from diamond or cubic boron nitride (CBN) that can be used for the high-speed cutting of hard, abrasive materials.

- ☐ *Diamond* insert tools are used for machining hard, abrasive nonferrous or nonmetallic materials.
- ☐ *Cubic boron nitride* (CBN) insert tools are used for machining hard, abrasive ferrous materials through a chemical vapor deposition (CVD) process.

surface feet per minute (sf/min) The cutting speed of a metal may be defined as the speed in surface feet or meters per minute at which the metal may be machined efficiently and most cost effectively.

T-slot cutter A milling cutter that consists of a small side-milling cutter with teeth on both sides and the periphery, mounted on an integral shank. *(See figure.)*

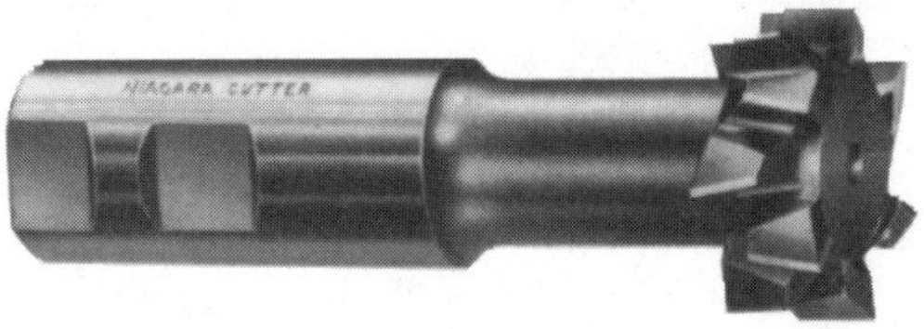

t-slot cutter. *(Courtesy Niagara Cutter Inc.)*

tang The flat portion on the end of a tapered drill shank that fits into a slot in a drill press spindle to prevent the drill from slipping. It is used primarily for milling T-slots for bolt heads and clamping devices.

□ Also the tapered end of a file that fits into a handle.

tap A cylindrical fluted, hardened cutting tool that is used to produce internal threads.

□ Taps are manufactured in many different types of thread forms and are available in a wide variety of sizes.

□ Hand taps are designed for hand tapping, and a standard set consists of a taper, plug, and bottoming tap.

□ Machine taps are designed for use on machine tools such as lathes, drill presses, and milling machines. The most common are the stub and spiral-flute taps.

tap extractor A tool having three or four fingers that fit into the flutes of a tap broken in a hole of the workpiece. The extractor is turned counterclockwise gently and carefully to remove a right-hand tap.

taper-pin reamer A reamer designed to ream tapered holes to suit 1/4-in taper per foot taper pins. They are available in sizes from 7/0 to 14, with straight, spiral, or helical flutes.

taper reamer A hardened tapered cutting tool available with straight or spiral flutes for all standard-size tapers. *(See figure.)*

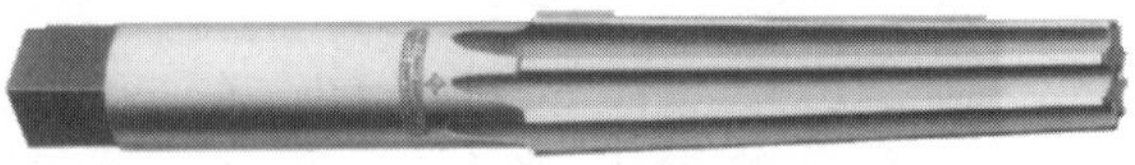

taper reamer. *(Courtesy Cleveland Twist Drill Co.)*

□ It is used to accurately size and finish a tapered hole.

threading dies Hardened cutting tools used to produce external threads on round work. The most common types are solid, adjustable, and adjustable screw-plate dies.

titanium-coated tools *See* COATED CARBIDE TOOLS.

toolholder A holding device used to hold a cutting tool during a machining operation. They are available in a variety of styles and shapes such as straight, left-, and right-hand lathe toolholders and parting, threading, and boring toolholders.

□ Modular and quick-change tooling systems are finding increasingly wider use because their flexibility and quick-change potential helps increase productivity.

truing The operation of using a dressing tool to make a grinding wheel round and concentric with its driving spindle. *(See figure.)* Automatic balancing devices are also used to true the grinding wheels on modern grinders.

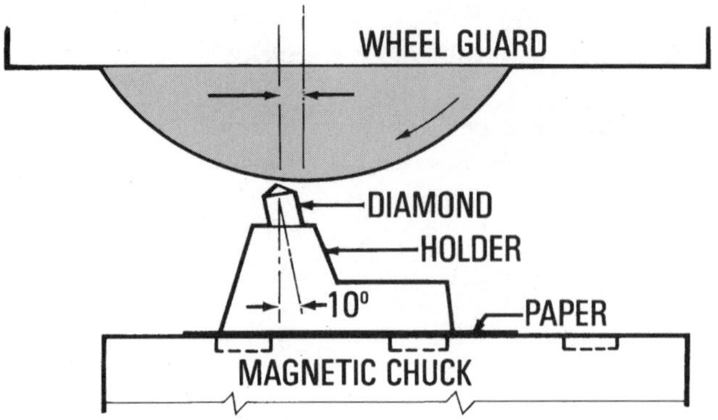

truing. *(Courtesy Kelmar Associates.)*

twist drill A rotary end-cutting tool used to produce holes in most types of materials. Two helical flutes are cut lengthwise around the body of a standard drill to provide cutting edges and space for the cuttings to escape while drilling.

wave set A set pattern on a saw blade in which a group of teeth are offset to the right and the next group to the left, producing a wavelike pattern. They are used to cut structural shapes or pipes where the cross section of the workpiece changes.

web The thin portion in the center of the drill that extends the full length of the flutes. This part forms the chisel edge at the cutting end of the drill. *(See figure.)*

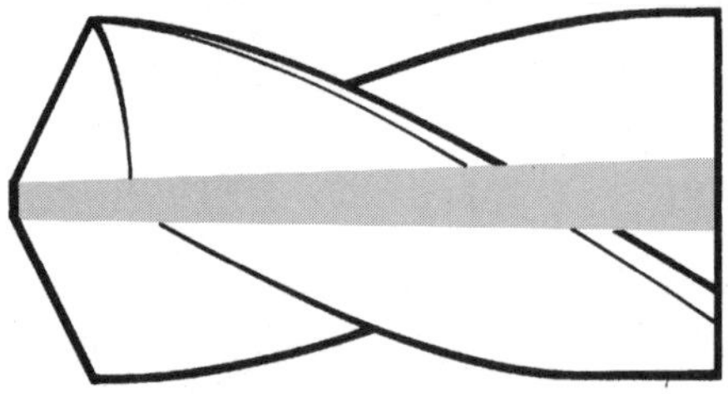

web (drill). *(Courtesy Cleveland Twist Drill Co.)*

□ The web gradually increases in thickness toward the shank to give the drill strength.

wheel dresser A device used to true and dress a grinding wheel manually or automatically. *See also* DIAMOND WHEEL DRESSER.

Woodruff keyseat cutter A shank-mounted milling cutter that is similar to a side-milling cutter with teeth cut only around the periphery of the cutter. *(See figure.)* It is available in a wide variety of sizes and is used for cutting semicircular keyseats.

woodruff keyseat cutter. *(Courtesy Niagara Cutter Inc.)*

Superabrasives

Superabrasives are superhard, super-wear-resistant tools that increase productivity, produce consistently high-quality products, and reduce machine downtime. Since their introduction in the mid-1950s and the early 1960s they have revolutionized the machining and grinding of hard, abrasive, and difficult-to-cut materials throughout the world.

AA *See* ARITHMETIC AVERAGE.

abrade The rubbing or wearing away caused by the friction or scraping between two moving surfaces.

abrasion The wearing away of the solid surface of a material by frictional action of hard particles or protuberances.

abrasion resistance The ability of a material to resist wearing away by friction, rubbing, or grinding.

abrasive A small, nonmetallic hard particle having sharp edges and an irregular shape that is used for grinding, honing, lapping, superfinishing, polishing, pressure blasting, or barrel finishing.

- □ Common abrasives are alumina, silicon carbide, boron carbide, diamond, cubic boron nitride, garnet, and quartz.
- □ They are capable of removing small amounts of material from a softer surface by a cutting process that produces tiny chips.

abrasive belt A coated abrasive product, in the form of a belt, used in production grinding and polishing.

abrasive machining The process in which chips are formed by the small cutting edges on abrasive particles. Grinding is a typical abrasive machining process.

- □ Abrasive machining processes have two unique characteristics:
  1. Each abrasive particle cutting edge is very small, and many of these edges can cut simultaneously;
  2. Because abrasive particles can be produced, hard materials such as hardened steel, glass, carbides, and ceramics can be machined very readily.

abrasive wear Unintended and unwanted removal of material when a hard and rough surface, or a surface containing hard, protruding particles, slides across a surface under pressure. This type of wear removes particles by forming microchips or slivers, thereby producing grooves or scratches on the softer surface.

abrasive wheel A grinding wheel composed of abrasive grits and a bonding agent to hold the abrasive together.

alumina Also referred to as *aluminum oxide* (Al_2O_3), alumina is the most widely used oxide ceramic. The natural crystalline mineral is called *corundum,* but the synthetic materials used for abrasive and ceramic materials are obtained by heating hydrates of alumina.

- □ Most synthetic aluminum oxide is obtained by fusing bauxite, iron filings, and coke. Alumina abrasives, in a form of grain crushed from fused alumina, are used to grind ferrous materials.
- □ The recently developed unfused alumina (also known as *ceramic aluminum oxides*) can be harder than fused alumina, the purest (free of flaws) form of which is seeded gel. The seeded gel abrasives can grind cool and resist loading and wear while cutting more aggressively.

amborite The DeBeers trade name for polycrystalline cubic nitride (PCBN) products that are used for machining hard, abrasive metals and superalloys.

amorphous carbon Carbon in which there is no uniform arrangement of carbon atoms as in the case of graphite or diamond.

angle-approach grinding A cylindrical grinding process in which the grinding wheel is plunged into the workpiece at an angle. The form dressed on the grinding wheel can grind a number of side faces and diameters in one plunge.

arc of contact The portion of wheel circumference that is in direct contact with the workpiece during grinding. *(See figure.)*

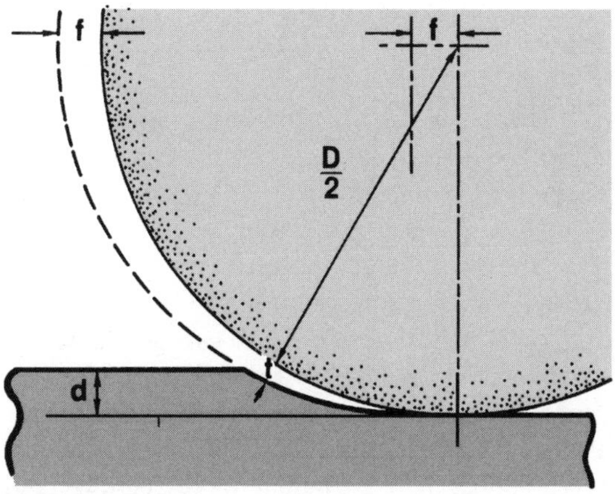

arc of contact. *(Courtesy GE Superabrasives.)*

area of contact This is often referred to as the *arc of contact* multiplied by the width of the grinding wheel. The true area of abrasive grain in contact with the workpiece is typically 0.5 to 5 percent of that area.

arithmetic average (AA) The mean between a number of values. It is used in the United States and corresponds to the British Centre Line Average (CLA), or the European arithmetic roughness average R_a value.

attritious wear The dulling of the cutting edges of a sharp abrasive grain, developing a wear flat.

balance (dynamic) The balancing of a grinding wheel in two planes usually to balance out both radial and axial deflections.

balance (static) The balancing of a grinding wheel in one plane only, usually to balance out radial deflections.

balancing Reducing or eliminating the unbalance of a grinding wheel; the unbalance results from an irregular mass distribution within the grinding wheel.

bauxite A naturally occurring, whitish to reddish mineral, rich in hydrous (water-containing) alumina. It is the most important ore (source) of aluminum, alumina abrasives, and alumina-based refractories.

belt apparatus The punch-and-die system designed to withstand the high pressure and high temperature required to manufacture superabrasives such as diamond and cubic boron nitride.

binder A material used as a cementing medium in the manufacture of abrasive products. It is added to promote adherence between particles of a solid, and between the particles and the surface. Binders are usually resins, ceramics, and low-melting metals.

black diamond A variety of crystalline carbon that is related to diamond but shows no crystal form.

- □ It is found only in Brazil and because of its hardness finds wide use in abrasives.

blotter A disk of compressible material, usually blotting-paper stock or preferably plastic washers, that must be placed between the sides of the grinding wheel and the flanges of the arbor (wheel flanges) to avoid concentrated stresses and prevent the wheel from cracking.

boart or bort Diamond grit obtained by crushing natural diamond crystals of a quality not suitable for gem use.

body *See* CORE.

bond The material that holds the abrasive grain in place in the form of a grinding wheel or other relative rigid abrasive product. Standard grinding wheels may use one of the following bonds: vitrified, resinoid, rubber, shellac, silicate, or metal.

borazon CBN Trademark of GE Superabrasives for its cubic boron nitride (PCBN) abrasive products.

boron carbide A black crystalline powder of high hardness, the composition of which is either B_6C or B_4C.

- □ It is produced by dehydrated boric acid mixed with high-quality coke.
- □ Applications include loose abrasives, cheap substitutes for diamond dust, and blast nozzles and other wear-resistant components.

brake-controlled truing device The most common device used for truing straight-faced resin- and metal-bond superabrasive grinding wheels.

- □ It is equipped with a 60-grit, L-hardness SiC wheel.
- □ It contains brake shoes that expand as the centrifugal force increases to cause a braking action that slows the truing wheel so that truing can occur.
- □ It allows effective removal of the bond matrix without creating damaging frictional heat or severe crushing of the superabrasive particles.

brittle fracture Fracture without first undergoing significant plastic deformation.

brittleness The property of a material that permits no permanent distortion before breaking. Cast iron is a brittle metal; it will break rather than bend under shock or impact. Brittleness is opposite of ductility.

built-up edge (BUE) Chip material adhering or welding to the tool cutting face next to the cutting edge during machining. Low-carbon machine steel and many high-carbon alloyed steels, when cut at a low cutting speed with a high-speed steel cutting tool and without the use of cutting fluids, generally produce a continuous-type chip with a built-up edge.

- □ A BUE changes the geometry of the cutting edge, produces poor surface finish, and shortens the cutting-tool life.
- □ Built-up edge formation can be eliminated or minimized by increasing the cutting speed, reducing the depth of cut, increasing the rake angle, applying an effective cutting fluid, or changing a cutting tool material.

burning of workpiece The blackened discoloration on ground steel surfaces caused by excessive temperature during grinding.

- □ The discoloration indicates surface oxidation due to extremely high, localized surface temperatures caused by poor grinding practice.
- □ The surface layers may undergo phase transformations, with martensite forming in higher-carbon steels from rapid cooling (metallurgical burn).
- □ This condition will influence the surface properties of ground parts, reducing surface ductility and toughness.

burnishing Finish sizing and smooth finishing of surfaces (previously machined or ground) by plastic displacement, rather than removal, of minute surface irregularities with smooth-point or line-contact fixed or rotating tools.

- □ The softer surface has been flattened, gives a shiny appearance, and is highly stressed in a compressive mode.

burr The ragged material on the edge of a workpiece created by a dull cutting tool. Burrs may interfere with the assembly of parts, reduce the fatigue life of components, and be a safety hazard to personnel.

bushing The insert sometimes found on the bore of a grinding wheel to make the arbor hole fit the required spindle size. The material may be lead, babbitt, wax, or plastic.

BZN tool blanks A trademark of GE Superabrasives for its polycrystalline cubic boron nitride (PCBN) tool blanks.

cam grinding The machining of the cam profile by rocking the camshaft back and forth with the grinding wheel in a stationary position. The latest CNC cam grinding machines use a different concept where it is the grinding wheel that is moved and the camshaft which is stationary.

carbon The element that provides the backbone for all organic polymers. Graphite is a crystalline form of carbon, while diamond is the densest crystalline form. Glassy carbon or carbon black is amorphous carbon.

carbon affinity The tendency of a substance to react with carbon.

□ Grinding steel (0.03 to 1.5 percent carbon) with a diamond wheel (100 percent carbon) will cause the carbon of the diamond to enter into the steel surface.

carbon solubility potential A chemical characteristic of ferrous metals that describes their potential for further reaction with carbon. *(See figure.)*

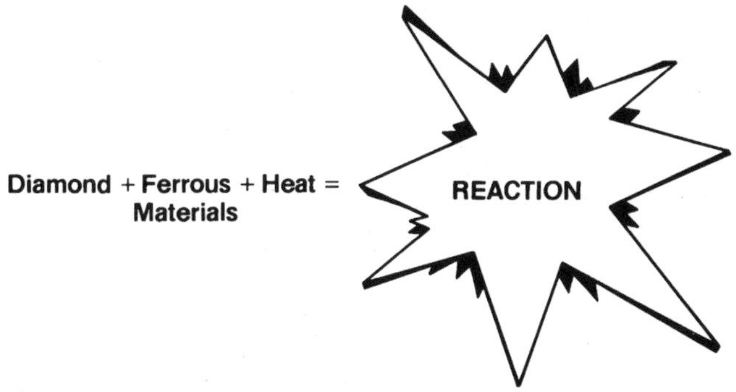

carbon solubility potential. *(Courtesy GE Superabrasives.)*

□ Most steels, especially low-carbon steels, are ready to react with any source of free carbon and absorb this carbon into their surface.

□ Such a reaction is easily triggered under temperature and pressure conditions that normally occur in the grinding and machining process.

catalyst A substance capable of changing the rate of a reaction without undergoing any net change itself. The use of a molten-metal solvent-catalyst in combination with a pure source of carbon such as graphite produced the first diamond ever manufactured.

CBN *See* CUBIC BORON NITRIDE.

CD *See* CONTINUOUS DRESSING.

CDA-M A DeBeers multigrain abrasive in which each grain is composed of many fine diamond particles bonded together in a metal-eutectic matrix.

□ Used in resin or vitrified wheels for high metal removal rates and cool, free-grinding performance.

cemented tungsten carbide A hard, wear-resistant, refractory material manufactured by sintering finely divided carburized tungsten metal powder in a cobalt metal binder. It is used in a wide range of applications, including metal cutting, mining, construction, rock drilling, metal forming, structural components, and wear parts.

center (dead and live) A 60° angle cone, either solid (*dead center*) or rotating on a bearing (*live center*), used to support a cylindrical workpiece.

centerless grinding External cylindrical grinding in which the workpiece is supported on a workrest blade between a grinding wheel and a regulating wheel that is mounted at an angle to the plane of the grinding wheel and revolves at a much lower speed. It is a high-production process for roller bearings, piston pins, engine valves, camshafts, and similar components.

CGS-II Trademark of GE Superabrasives for a diamond abrasive, consisting of nickel-coated, blocky crystals of medium toughness, designed for grinding composite cemented carbide and steel.

Centre Line Average (CLA) A measure of surface quality given in micrometers or microinches. It is defined as the average value of the departure of the profile from its centerline, whether above or below it, throughout a certain length.

chatter Self-excited vibration caused by the interaction of the chip-removal process and the structure of the machine tool, producing a wavy surface on the work. The tendency for chatter in grinding can be reduced by using soft-grade wheels, dressing the wheel frequently, changing dressing techniques, reducing the material-removal rate, and supporting the workpiece rigidly.

chatter marks Pattern of surface imperfections on the workpiece, usually caused by vibrations transferred from the wheel-work interface during grinding. It is particularly common in surface peripheral or external cylindrical grinding.

chip The piece of material removed from the workpiece by one cutting edge or abrasive grain.

chip breaker A piece of metal clamped to the rake face of the tool, or a notch or groove ground in the face of the cutting tool behind the cutting edge to bend the chip and break it.

chip curl In all cutting operations on metals and nonmetallic materials, chips develop a curvature as they leave the workpiece surface.

chip flow A metal-cutting process where a cutting tool contacts the workpiece to remove metal in the form of a chip.

□ The force of the cutting tool compresses the metal in ahead of the tool causing stress lines to be concentrated at the cutting-tool edge and radiating into the workpiece.

□ This concentration of stresses causes the chip to shear from the workpiece and flow along the chip-tool interface by either plastic flow or rupture.

chip-tool interface The area where the contact between the face of a cutting tool and the chip being removed from the workpiece occurs. *(See figure on next page.)*

CLA *See* CENTRE LINE AVERAGE.

cleavage The fracturing of a crystal by crack propagation across one or more of its principal crystallographic planes (cleavage planes).

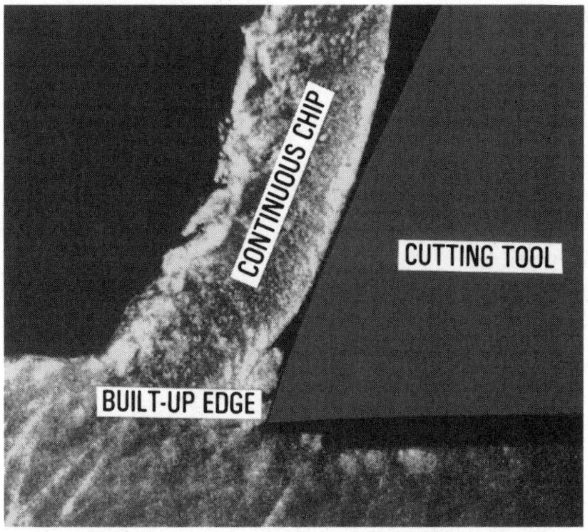

chip-tool interface. *(Courtesy Cincinnati Milacron Inc.)*

cleavage plane A characteristic crystallographic plane or set of planes in a crystal on which cleavage fracture occurs easily.

compact A generic name for sintered polycrystalline superabrasives.

compax blanks A trademark of GE Superabrasives for its polycrystalline diamond (PCD) tool blanks.

composites A mixture or mechanical combination of two or more physically and chemically distinct, mutually insoluble materials (reinforcing elements, fillers, and composite matrix binder) which are solid in the finished state. Composite materials provide properties and/or performance that are superior to those of the individual materials.

compressive strength The maximum stress in compression that a material will take before it ruptures or breaks.

□ Superabrasive tools have 2 to 10 times higher compressive strength than do conventional tools.

compressive stress A stress that causes an elastic body to deform (shorten) in the direction of the applied load.

concentration A designation for the amount of superabrasive contained per unit volume of grinding wheel rim. A wheel with 100 concentration contains 25 percent by volume of superabrasive.

concentricity The condition in which the grinding wheel periphery and its geometric center are on a common axis.

conductivity, electrical A measure of the ease with which a material conducts electric current.

conductivity, thermal The ease with which heat flows within and through the material. Metals generally have higher thermal conductivity, whereas ceramics and plastics have poor conductivity. For metals, thermal conductivity is directly proportional to electrical conductivity.

contact length The length of the portion of the rim in contact with the workpiece (depends on the type of grinding operation).

continuous dressing (CD) The process used with conventional wheels on high-performance machines where a diamond dressing tool is dressing the grinding wheel all the time the wheel is grinding. This ensures perfectly predictable wheel wear, perfect profile, and optimal freeness of cut as the grits do not have time to develop wear flats.

conventional abrasives A term referring to aluminum oxide or silicon carbide abrasives in contrast to the much harder superabrasive grits.

conventional wheels Wheels with conventional abrasives (aluminum oxide or silicon carbide) as removal agents.

coolant *See* CUTTING FLUID.

core The center or body of the grinding wheel to which the superabrasive is bonded. It is usually made of some form of metal.

corundum Natural pure alumina occurring as rhombohedral crystals and also in masses and variously colored grains. Corundum and its artificial counterparts are especially suited to the grinding and polishing of metals.

cost-effective A description of a grinding or machining operation where the increase in productivity and quality through the use of superabrasives more than justifies their cost.

cost per part The total cost of producing a part when all costs such as superabrasive tools, materials, labor, and overhead are included.

crater wear An undesirable form of cutting-tool wear that occurs just behind the cutting edge on the rake face of the tool due to friction of the chip sliding over the face of the tool. *(See figure.)*

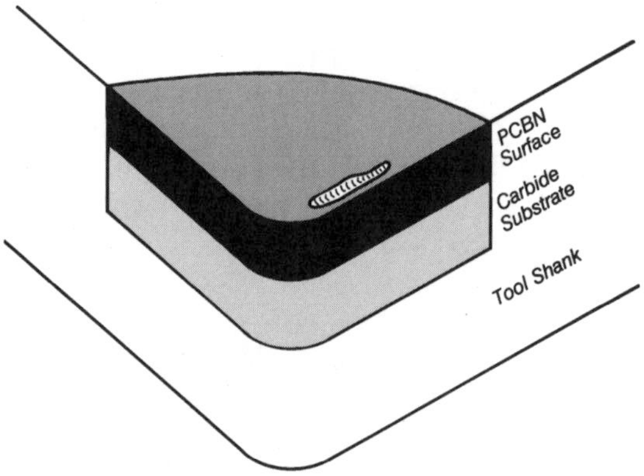

crater wear. *(Courtesy GE Superabrasives.)*

creep-feed grinding A high-stock-removal method of grinding where the table speeds are very low and the wheel is fed down to full depth of cut in one or two passes to produce an accurate form. To be contrasted with oscillating or reciprocating grinding, with shallow infeeds and a large number of passes.

crosshatch Pattern on the workpiece surface in face grinding or honing.

crush dressing/crush forming A technique using a hardened steel or carbide roll, having the same contour of the part to be ground, to produce special forms or profiles on the face of a grinding wheel which is usually vitrified bonded. *(See figure.)*

crush dressing. *(Courtesy GE Superabrasives.)*

- □ The form roll is forced against a slowly revolving grinding wheel to produce its reverse profile on the wheel.
- □ The crushing action fractures and dislodges some of the abrasive grains, exposing fresh sharp edges.
- □ This process is especially valuable when complex forms must be ground on a large number of parts.

crystal A solid composed of atoms, ions, or molecules, arranged in orderly configuration with geometrically arranged cleavage planes and internal faces that assume a group of patterns associated with the atomic structure.

cubic boron nitride (CBN) An extremely hard material synthesized by high-pressure, high-temperature sintering of hexagonal boron nitride. It is harder than any other material except diamond.

- □ CBN is used in the machining and grinding of ferrous materials such as tool steels, cast irons, hardfacing alloys, and surface-hardened steels.

cutting edge The leading edge of a cutting tool (such as a lathe tool, drill or milling cutter) where a line of contact is made with the work during machining.

cutting fluid A fluid applied over the cutting tool and work to cool, lubricate, prevent corrosion, and wash away chips and debris to improve finish, tool life, or dimensional accuracy.

- □ The most common cutting fluids are water, water solutions or emulsion of detergents and oils, mineral oils, chlorinated or sulfurized mineral oils, fatty oils, and mixtures of those oils. For water-based fluids, antiwear additives and rust inhibitors are essential.

cutting speed The linear or peripheral speed of relative motion between the tool and workpiece in the principal direction of cutting measured in feet or meters per minute.

cylindrical grinding A grinding process used to grind the outer cylindrical surface of a rotating part.

D *See* DIAMOND.

deburring The removal of burrs from the corners or edges of workpieces.

detritus The residue from the metal-removal process, typically the chips, commonly referred to as *swarf.*

diamond (D) A highly transparent mineral composed entirely of carbon (allotropic form of carbon) having a cubic structure. The hardest material known, it is used as a gemstone and as an abrasive in cutting and grinding applications.

- □ Natural diamonds are produced deep within the earth's crust at extremely high pressures and temperatures.
- □ Manufactured diamonds are synthesized by subjecting carbon, in the form of graphite, to high temperatures and pressures using large special-purpose presses.

diamond dresser (nib) A single-point, multipoint, or cluster-type diamond dresser, mounted in a suitable holder, used to dress precision grinding wheels.

diamond-impregnated nib A truing device, generally 3/8 in (9 mm) in diameter, with 100/120-mesh diamond evenly distributed throughout a metal-bond matrix on the end of the steel post. It should be used only to true resin- and vitreous-bond wheels up to 8 in (250 mm) in diameter and up to 1 in (25 mm) wide. *(See figure.)*

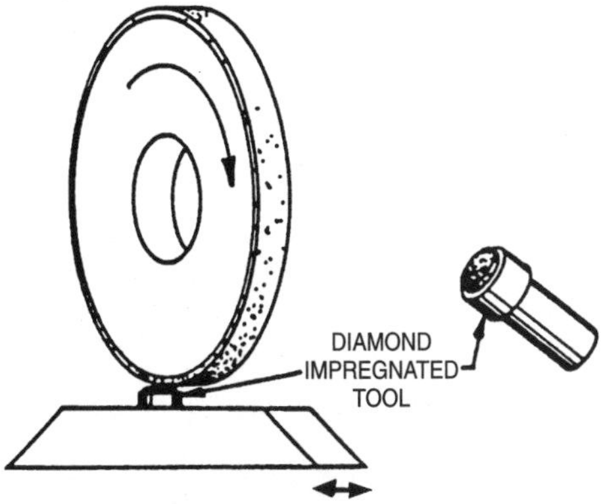

diamond (multipoint). *(Courtesy GE Superabrasives.)*

diamond roller dresser A wheel that has a peripheral surface impregnated with diamonds, by either plating or sintering. The surface of the roller may be formed into a very detailed and complex shape for profiling grinding wheels.

diamond tool **1.** A diamond, shaped or formed to the contour of a single-point cutting tool, for use in precision machining of nonferrous or nonmetallic materials. **2.** An insert made from polycrystalline diamond compacts.

diamond truing block A metal block that may be either diamond-impregnated or electroplated with diamond abrasive in the 100/180-mesh range. It is most commonly used for truing resin- and metal-bond straight superabrasive wheels.

dielectric fluid A fluid such as light lubricating oil, transformer oil, silicon-based oil, or kerosene that is used in EDM electrospark machining to control arc discharge in the erosion gap and wash away chips.

down-cutting A mode of grinding where the grinding wheel is rotating in a direction that, at the arc of contact, is the same as the motion of the workpiece.

downfeed *See* INFEED/DOWNFEED.

dressing A process of removing some bond material from the surface of a trued wheel to expose sharp abrasive grits to make the wheel grind efficiently. Dressing generates sharp, protruding grits and chip clearance space.

DTG Abbreviation for "difficult to grind."

ductility The ability of a material to deform plastically without fracturing.

□ Soft metals are generally ductile; hard metals are not.

Dutch finish Pattern on the surface of workpieces that have been ground in the face grinding mode with a cup wheel (related to the crosshatch pattern as found in honing).

elasticity The property or quality of a material where deformation caused by applied stress disappears on removal of the stress.

□ A perfectly elastic body completely recovers its original size, shape, and bulk after release of the compression, extension, or other distortions.

electroplated wheel A superabrasive wheel with a single layer of abrasive bonded to the machined surface of a metal wheel core in a nickel electroplating bath. The electroplated nickel matrix provides outstanding retention of the abrasive. *(See figure.)*

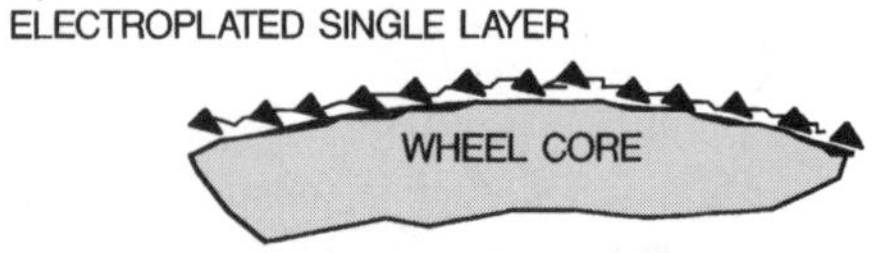

electroplated bond. *(Courtesy GE Superabrasives.)*

□ The wheel is generally used in form grinding, where an accurate form must be produced in a workpiece.

emery A naturally occurring, fine-grained, impure form of corundum used as an abrasive material. Its hardness varies according to the amount of iron oxide present.

ETG Abbreviation for "easy to grind."

external (cylindrical) grinding The grinding of the outside diameter of cylindrical or tapered components.

face grinding Grinding with the wheel's face where the workpiece and the wheel touch each other on a plane perpendicular to the wheel's axis. If the workpiece traverses the wheel radially, the surface of contact is minimal and is given simply by the product (rim width times workpiece height).

facing In machining, generating a surface on a rotating workpiece by the traverse of a tool perpendicular to the axis of rotation.

finish(ing) grinding The final grinding action on a workpiece to obtain final dimensions with acceptable tolerance and surface finish.

fixture Any device designed to hold each particular workpiece as firmly as possible during the grinding operation. The most common is the magnetic chuck used on surface grinders.

flank wear The wear occurring on the flank of the cutting tool due to rubbing contact between the workpiece and the tool and high temperature during cutting. *(See figure.)*

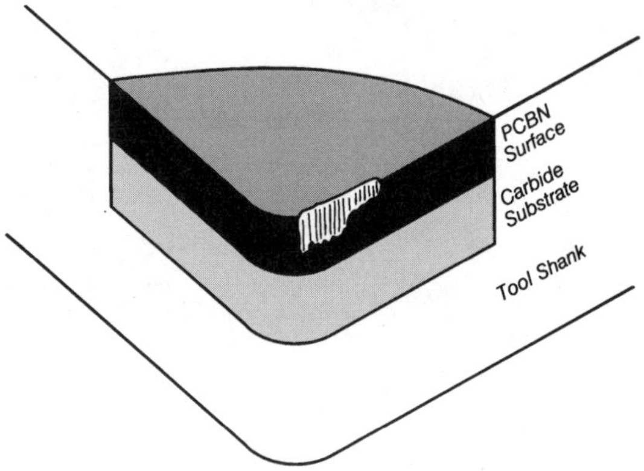

flank wear. *(Courtesy GE Superabrasives.)*

floor-to-floor time The time taken to perform an operation from the time the part is picked up from the floor to when it is completed and put down. This time includes the cutting time and the loading and unloading time.

flour wheel A wheel with extremely fine grains in the region of 4000 grain size.

form cutter/form tool Any cutter, profile-sharpened or cam-relieved, shaped to produce a specified form on the work.

form grinding/profile grinding Grinding with a wheel having a contour on its cutting face that is a mating fit to the desired form.

free-cutting wheel A softer-grade wheel that grinds with little effort, little noise, low spindle power, and low heat generation. There is no quantitative definition available, although spindle power under load is often a usable approximation.

free machining Pertaining to the machining characteristics of an alloy to which one or more ingredients such as sulfur or lead to steel, lead to brass, lead and bismuth to aluminum, and sulfur or selenium to stainless steel, have been added to produce small broken chips, lower power consumption, better surface finish, and longer cutting-tool life.

friability The ability of abrasive grains to fracture (break down) into small pieces under stress. This property gives abrasives self-sharpening characteristics; as the cutting edges become dull, part of the grain breaks off and presents new cutting edges.

frictional heat The heat created when a wheel or cutting tool rubs the workpiece surface.

functional costing A new costing technique that is used to justify large expenditures on new tools and major manufacturing equipment.

glazing A term describing the shiny appearance of the grinding wheel surface caused by the excessive wear or roundness of abrasive grits.

grade A measure of the strength with which the bond holds the abrasive particles in the wheel. This is also referred to as the *hardness* of the wheel.

graded abrasive An abrasive powder in which the sizes of the individual particles are confined to certain specified limits.

grain size The number of openings per linear inch in the screen through which the abrasive particles just pass. Grain sizes ranging from 4 to 240 grit are the most common sizes used in the manufacture of grinding wheels. *(See figure.)*

grain size. *(Courtesy Carborundum Co.)*

graphite A natural form of crystalline carbon that generally occurs as a black, soft mass.

□ It is a chemically inert variety of carbon with a metallic luster and a slippery texture.

green state A grinding wheel is said to be in a green state after it is formed and before it is fired.

grindability The relative ease by which a material can be ground. In a practical sense, the term "relative ease" refers to the amount of tool wear and the amount of power required to grind or machine a material.

grinding One of the most commonly used machining operations that involves cutting by abrasive wheels and other shapes, or by abrasive belts.

□ The grinding operation generates close-tolerance dimensions or produces smooth finishes.

□ Abrasive materials include aluminum oxide, silicon carbide, diamond, and cubic boron nitride; the one selected depends on the material to be ground.

□ Grinding operations include surface grinding, creep-feed grinding, cylindrical grinding, internal grinding, centerless grinding, gear grinding, and thread grinding.

grinding cracks Shallow cracks formed in the surfaces of relatively hard materials because of excessive grinding heat or the high sensitivity of the material.

grinding fluid *See* CUTTING FLUID.

grinding interface The area in which the surface of the wheel rim, workpiece, and the wheel penetration depth are removing the material.

grinding ratio (g ratio) The ratio of the volume of work material removed to the volume of wheel worn during grinding under a specific set of conditions.

grinding stress Residual stress, generated by grinding, in the surface layer of work. It may be tensile or compressive, or both.

grinding wheel A cutting tool of circular shape made of abrasive grains bonded together.

grit size Nominal size of abrasive particles in a grinding wheel, corresponding to the number of openings per linear inch in a screen through which the particles can pass.

hardness Resistance of a material to forceable penetration or plastic deformation. It can be measured by Brinell, Rockwell, Vickers, Knoop, or Scleroscope hardness tests.

HBN *See* HEXAGONAL BORON NITRIDE.

heat flux The amount of heat generated over the area of contact between the grinding wheel and the workpiece.

heat sink A material that absorbs or transfers heat away from a critical element or part.

hexagonal boron nitride (HBN) A white, powdery substance, sometimes known as *white graphite,* used in the manufacture of cubic boron nitride.

high-speed grinding Any type of grinding mode characterized by a very high stock removal rate, high workpiece speed (in the meters-per-minute range), and high wheel speed (in the 100-m/s range). It requires machines built with high spindle power and efficient coolant and lubricant installations.

honing An abrading operation which aims at obtaining both internal and external cylindrical high-quality surface finish with high accuracy.

□ Its characteristic feature is the honing tool, similar to a grinding wheel that reciprocates entirely through the length of the bore and rotates slowly at the same time.

hub *See* CORE.

imbalance Lack of balance caused by an irregular mass distribution within a grinding wheel.

impact strength A measure of the resiliency or toughness of a solid. The maximum force or energy of a blow (given by a fixed procedure) that can be endured without fracture, as opposed to fracture strength under a steady applied force.

impregnated bond A method of attachment of superabrasive grains to the rim or face of a grinding wheel by uniformly distributing the grains through the thickness of the bond material (resin, metal, or vitrified). *(See figure on next page.)*

indexable insert Individual cutting tools with a number of cutting edges that are usually clamped on the tool shank with various locking mechanisms.

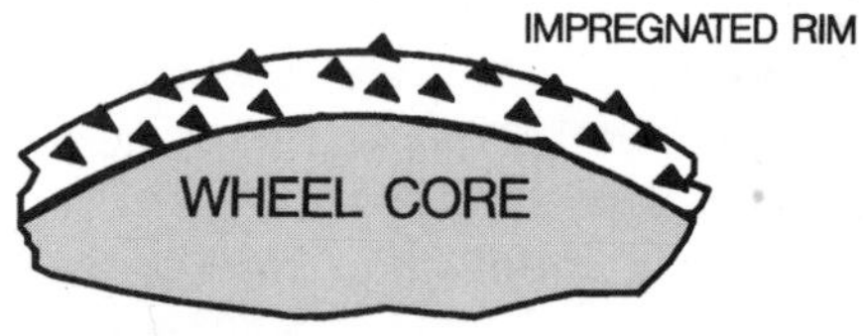

impregnated bond. *(Courtesy GE Superabrasives.)*

☐ After one edge is worn, it is indexed (rotated in its holder) to present another cutting edge. A square insert, for example, has eight cutting points, and a triangular insert has six.

infeed/downfeed The feed motion of the grinding wheel into the workpiece.

in-process gage A mechanism that measures the size of a part during the machining process to achieve very accurate sizing.

interface The boundary or surface between two different, physically distinguishable media.

☐ The chip-tool interface is the surface of the cutting tool that a chip slides over as it separates from the workpiece.

internal-diameter (ID) grinding. Grinding of the inside diameter of holes or profiles.

lattice **1.** A *space lattice* is a set of equal and adjoining parallelopipeds formed by dividing space by three sets of parallel planes; the planes in any one set are equally spaced.

2. A *point lattice* is a set of points in space located such that each point has identical surroundings.

loading Occurs when the porosities on the grinding surfaces of the wheel become filled or clogged with chips.

☐ Loading can occur in grinding soft materials or by improper selection of grinding wheels, such as with low porosity and improper process parameters.

☐ A loaded wheel does not cut very well, generates much frictional heat, and causes surface damage and loss of dimensional accuracy.

machinability The relative ease with which a metal can be machined with a cutting or abrasive tool.

macrofracture A type of wear of an abrasive grain where cleavage takes place along a principal plane. *(See figure.)*

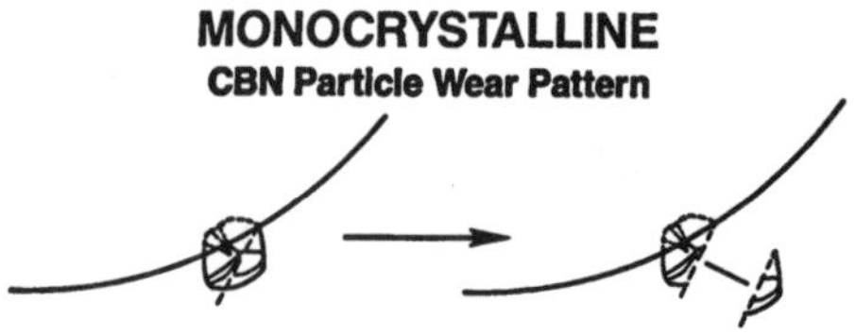

macrofracture. *(Courtesy GE Superabrasives.)*

macroscopic Anything that is visible at magnifications at or below 25×.

macrostructure The structure of metals as revealed by macroscopic examination of the etched surface of a polished specimen by eye or at low magnification.

MAN-MADE Diamond A trademark of GE Superabrasives for its manufactured diamond abrasive products.

MDA 100 A DeBeers diamond abrasive used in metal bond grinding wheels that has high crystalline and high strength properties.

☐ Used for machining highly abrasive products at high cutting rates.

mesh/mesh number The number of openings per linear inch of a mesh screen that determines the size of an abrasive grain.

mesh size The size of a superabrasive grain as defined by an internationally approved sieving method, hence the approximate diameter of particles below which they will pass through and above which they will be retained on the screen.

metal bond A form of impregnated bond in which sintered powdered metal is used to bond the superabrasive in the form of a grinding wheel.

microcrystalline Same as polycrystalline (*see* POLYCRYSTALLINE) except that individual crystallites are of micrometer or submicrometer size.

microfracture A type of superabrasive grain wear where extremely small particles break from the surface of the abrasive grain in the grinding process to resharpen themselves. *(See figure.)*

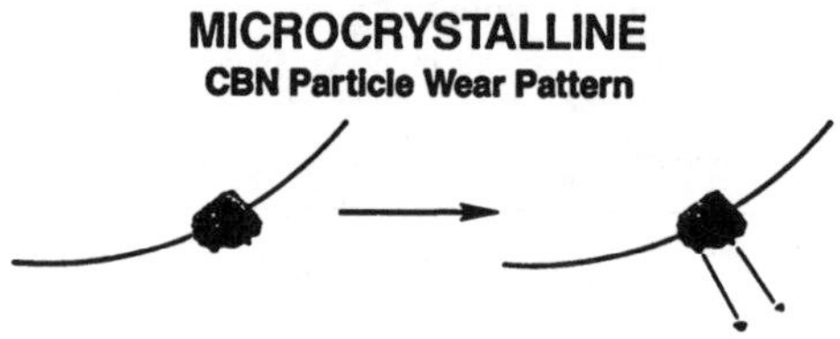

microfracture. *(Courtesy GE Superabrasives.)*

microinch One-millionth of an inch.

micron One millionth of a meter.

microstructure The structure of an object, organism, or material as revealed by a microscope at magnifications greater than 25×.

modulus of elasticity The ratio of stress to strain within the proportional limit of a material in tension or compression. The coefficient is a measure of the softness or stiffness of the material. The modulus is also known as *Young's modulus.*

monocrystalline A single crystal with a regular arrangement of atoms in all three dimensions. The first-generation CBN abrasives were monocrystals.

MRR *See* MATERIAL-REMOVAL RATE.

material-removal rate The time rate at which material is removed from a workpiece, measured in cubic inches or cubic centimeters per unit time.

octahedron A form of cubic system that is bounded by eight similar faces, each an equilateral triangle with plane forming 60° angles.

oscillating grinding *See* RECIPROCATING GRINDING.

out-of-round(ness) or out-of-truth A condition in which the rim of the grinding wheel is not concentric (true) with the spindle axis of rotation (peripheral wheels), or is not perpendicular to it (cup wheels).

outside diameter (OD) grinding The cylindrical grinding of an outside diameter. *See* EXTERNAL (CYLINDRICAL) GRINDING.

paradichlorobenzene A compound used to induce high porosity into vitrified grinding wheels.

PDA The DeBeers PremaDia Indicator system that assists toolmakers in the diamond selection process.

□ Abrasive is described on the basis of particle strength (room temperature), particle strength (high temperature), and particle structure.

pellets *See* SEGMENTS.

peripheral grinding The process of grinding using the outside diameter or periphery of a grinding wheel; at zero infeed the wheel rim and the workpiece touch each other over a line of contact parallel to the wheel's axis. *(See figure.)*

peripheral grinding. *(Courtesy DoAll Co.)*

phenolic resin Synthetic thermosetting materials obtained by the condensation reaction of phenol or substituted phenols and aldehydes. They produce very strong physical bonds with many materials at low concentrations, and, as such, they are also treated as binders.

plastic deformation The permanent (inelastic) distortion of materials under applied stresses that strain the material beyond its elastic limit.

plunge grinding Grinding in which the only relative motion of the wheel is radially toward the work, as in grinding a groove.

polycrystalline A solid body comprising many small crystals randomly or partially randomly oriented and intimately bonded together.

□ Individual crystals are large enough to be identified under a microscope at 20× magnification.

□ It may be homogeneous (one substance) or heterogeneous (two or more crystal types or combinations).

porosity The voids in a grinding wheel that make it appear almost spongelike to provide chip clearance for the material removed from the work by a grain.

productivity The efficiency or effectiveness with which tools, raw materials, and labor can be applied in the production process.

profile copy grinding A form grinding process in which the workpiece contour or shape is formed with a pointed profile wheel displaced by means of a template and a pantograph.

projection/protrusion When a grit extends noticeably above the general rim level, there is protrusion or projection of the grit away from the bond.

reciprocating grinding Grinding mode in which workpiece and wheel contact each other repeatedly at shallow infeeds, with a large number of entry and exit impacts.

regenerative chatter The most important type of self-excited vibration caused when a tool cuts a surface that has a roughness or disturbances from the previous cut.

replication A repeat run in a test sequence, performed to ascertain the repeatability of obtaining a given performance.

residual stress The internal stress existing in a body at rest, in equilibrium, at uniform temperature, and not subjected to external forces. This usually results from some prior treatment such as grinding, severe temperature changes during quenching, or chemical differences as in carburized surfaces.

resin bond/resinoid bond A form of impregnated bonding in which a hot-pressed phenolic resin is used to bond the superabrasive particles to the rim of the grinding wheel.

resinoid Any of the class of thermosetting synthetic polymers, either in their initial temporarily fusible state or in their final infusible state.

resinoid wheel A grinding wheel bonded with a synthetic resin.

rigidity The stiffness of a system which is measured by the deflection of a system under a given load. *(See figure.)*

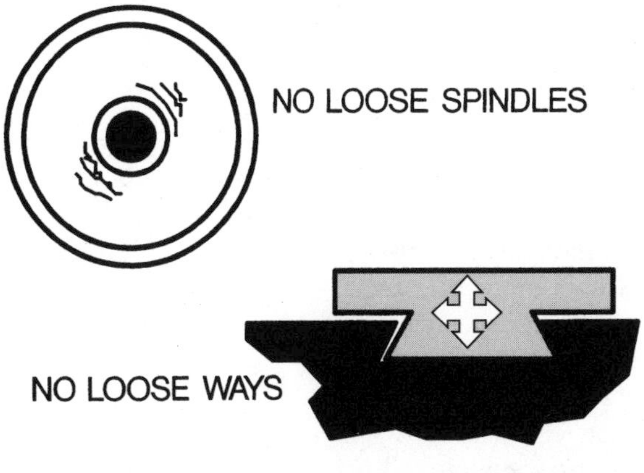

rigidity. *(Courtesy GE Superabrasives.)*

rotary-powered truing device A pneumatically, hydraulically, or electrically powered device using metal-bonded or electroplated diamond truing wheels.
- □ The axis of the truing wheel may be set at an angle of 30°, 45°, 60°, or 90°, or parallel to the axis of the grinding wheel.
- □ This device is most often used in production grinding operations for truing cylindrical, centerless, internal, double-disk, and large CBN grinding wheels.

roughing/rough grinding/machining Grinding and machining without much consideration to tolerances and surface finish, usually to be followed by a subsequent operation.

runout *See* OUT-OF-ROUND OR OUT-OF-TRUTH.

RVG A trademark of GE Superabrasives for an elongated friable diamond abrasive suited primarily for use in resinoid and vitrified grinding wheels for grinding tungsten carbide.

RVG-D A trademark of GE Superabrasives for a copper metal-coated diamond abrasive suited specifically for use in resinoid bonded grinding wheels for dry-grinding cemented carbides.
- □ Using a 50 weight percent coating of copper on the RVG grains achieves the same holding power in the bond and increases the overall thermal conductivity of the wheel.
- □ The RVG-D crystal is designed for dry grinding with resin bonds at higher material-removal rates without excessive loss of pulled-out grains.

RVG-W A trademark of GE Superabrasives for a nickel-coated diamond abrasive suited specifically for use in resinoid grinding wheels.
- □ The nickel coating is much less thermally conductive than RVG diamond itself and provides a cooler surface to the resin, less deterioration, and better retention of grains.
- □ The 56 weight percent nickel coating RVG-W is used primarily for wet grinding of cemented tungsten carbide, while the 30 weight percent RVG-W is used for carbide-steel composites.

segments Shaped bonded abrasive sections fitted together on a core and positioned to form an interrupted grinding rim. The segments or pellets, often circular in shape, are used frequently in surface grinding with cup wheels, to allow optimal coolant penetration and to reduce the contact surface.

self-sharpening A mechanism by which a grinding wheel continually develops sharp cutting points through progressive, controlled wear. This occurs via double self-sharpening, namely, self-sharpening through partial grit fragmentation (micro- or macrofracturing), and self-sharpening through bond erosion.

semifinishing Preliminary operations performed before finishing operations.

sf/min *See* SURFACE FEET PER MINUTE.

shear properties Characteristics of metal deformation where parallel planes within the metal are displaced by sliding but keep their parallel relation to each other.

silicon carbide (SiC) A manufactured abrasive usually green (more friable) or black (less friable) in color, used to grind nonferrous and nonmetallic materials. Also called *carborundum.*

sparkout A grinding pass taken over a surface with no downfeed to eliminate previous grinder spindle or wheel deflection.

specific energy The amount of energy used to remove a unit volume of material.

steadyrest A supporting device for cylindrical workpieces to prevent their deflection under the forces of grinding.

stiffness The ability of a material or shape to resist elastic deflection.

stock The workpiece material to be removed.

stock-removal rate The rate at which material is removed from a workpiece.

structure The relative size, proportion, and position of the abrasive grain, bond material, and porosity in a grinding wheel matrix. *(See figure.)*

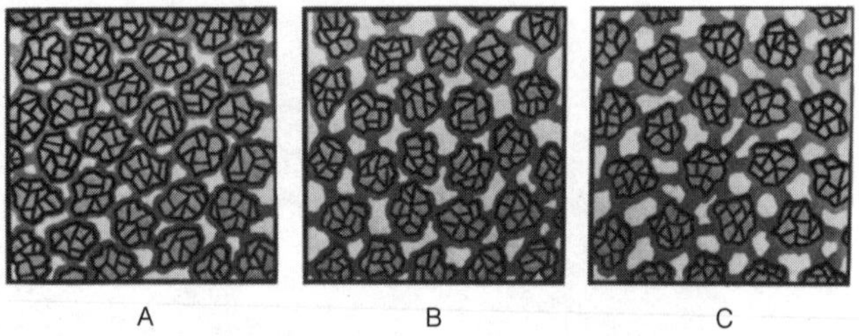

structure. *(Courtesy Carborundum Co.)*

- □ The structure of bonded abrasives (grinding wheels) ranges from dense to open.
- □ Wheels with open structures provide greater chip clearance than do those with dense structures and remove material faster than do dense wheels.

substrate The base of cemented carbide that provides support and strength for the layer of polycrystalline superabrasive bonded to its surface.

superabrasive Collectively refer to diamond and cubic boron nitride (CBN), both of which are produced by synthetic means. Because of their high hardness, abrasion resistance, and other unique properties, these materials find extensive use in a wide variety of abrasive or cutting applications.
- □ Diamond is used extensively to grind stone, concrete carbides, glass, ceramics, plastics, and composites.
- □ CBN is used for grinding and machining ferrous materials such as hardened steel, cast iron, nickel- and cobalt-based superalloys, and hardfacing materials.

superalloys Heat-resistant alloys based on nickel (Ni), iron-nickel (Fe-Ni), or cobalt (Co) that exhibit a combination of mechanical strength and oxidation-corrosion resistance that is unmatched by other metallic alloys.
- □ Superalloys are used primarily in gas turbines, coal-

conversion plants, and chemical process industries, and for other specialized applications requiring high heat and corrosion resistance.

surface feet per minute (*sf/min*) A unit for measuring rim or wheel speed.

surface finish The geometric irregularities in the surface of a solid material. It is measured by scratching a precision stylus across the surface and measuring the amplitude of the fluctuations of the stylus.

surface grinding The machining of a flat, angled, or contoured surface by passing a workpiece beneath a grinding wheel in a plane parallel to the grinding wheel spindle.

surface integrity Describes not only the topological (geometric) aspects of surfaces and their physical and chemical properties but also their mechanical and metallurgical properties and characteristics.

□ It is an important consideration in manufacturing operations because it influences the properties of the product, such as its fatigue strength, resistance to corrosion, and service life.

surface roughness Fine irregularities in the surface texture of a material, usually including those resulting from the inherent action of the production process. Surface roughness is usually reported as the arithmetic roughness average, R_a, and is given in micrometers or microinches.

swarf Intimate mixture of grinding chips and fine particles of abrasive and bond resulting from a grinding operation.

Syndite The DeBeers trade name for polycrystalline diamond (PCD) products that are used for machining hard or abrasive nonferrous and nonmetallic materials.

table The part of a grinding machine that supports the workpiece.

tailstock A part of a cylindrical grinding machine that provides center support to a workpiece. The tailstock is situated on the workslide opposite the headstock or workhead.

tensile strength In tensile testing, the capacity of a metal to withstand a pull or tension applied to it. It is obtained by dividing the maximum load by the specimen cross-sectional area before straining. Also called *ultimate strength*.

thermal conductivity Ability of a material to conduct heat.

thermal damage The metallurgical damage that occurs to a material when it is subjected to temperatures which will affect its metallurgical structure. Thermal damage often occurs before the visible burn marks. *(See figure.)*

thermal stability The stability of a system to remain within tolerance during a fluctuation in temperature.

tool-work interface The area where the edge of a cutting tool and the workpiece contact.

topography Configuration of the rim of a superabrasive wheel at the microscopic level.

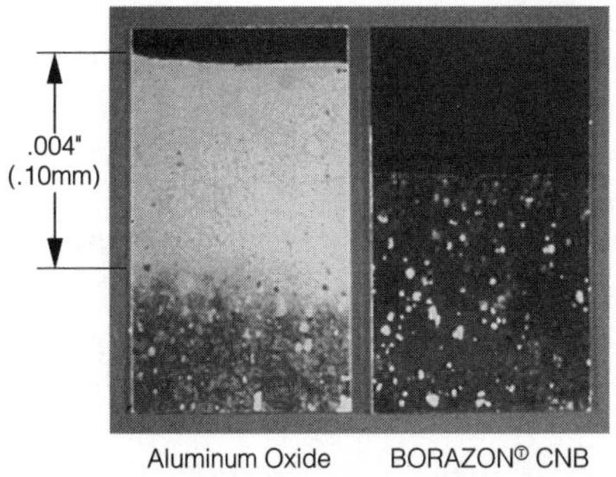

thermal damage. *(Courtesy GE Superabrasives.)*

□ This term refers to the number of grits still present on the rim's surface, the number of holes left by grits torn out, the condition of the bonding material, the presence or absence of grinding debris, the projection of the grits away from bond level, and the state of the cutting edges visible on the individual grits.

trilogy Refers to the necessity, in any machining operation, to consider the characteristics and the limitations of all the partners involved in the process, namely, workpiece, machine, and cutting tools.

truing The process of making a grinding wheel round and concentric with its spindle axis and to produce the required form or shape on the wheel. *(See figure.)*

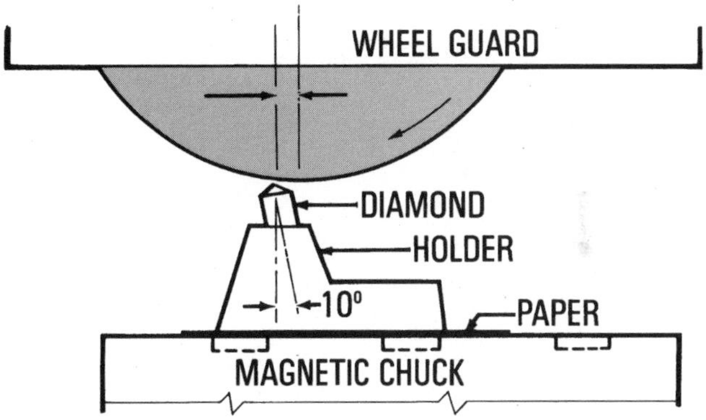

truing. *(Courtesy Kelmar Associates.)*

ultrahard abrasive *See* SUPERABRASIVE.

up-cutting A mode of grinding in which the grinding wheel is rotating in a direction that, at the arc of cut, opposes the motion of the workpiece.

velocity The time rate of motion in a stated direction, usually measured in feet or meters per minute.

vibrational stability The stability of a system to remain within tolerance across a range of vibrational frequencies.

vitrified bond A form of impregnated bond in which the superabrasive grains are attached to the rim of the grinding wheel by the sintering of a mixture of abrasives and feldspar or silica in a kiln.

wear Damage to a solid surface, generally involving progressive loss of material, due to a relative motion between that surface and a contacting surface or substance.

wear resistance The ability of a body to resist abrasion and wear.

weight percent (wt%) The percentage of the total weight of a metal-coated superabrasive made up by the coating itself.

wheelhead The part of a grinding machine that houses the grinding spindle.

wheel specification All the parameters required to specify a grinding wheel, namely, wheel type and diameter, up to five supplementary dimensions: bore, grit type and size, grit concentration, and bond.

wheel speed The peripheral speed of the grinding wheel.

wheel wear During grinding, a wheel is subjected to attrition phenomena affecting both bond material and abrasive grit.

- □ Progressive wear maintains the freeness of cut via partial grit fragmentation, pullout of dull grits with wear flats and bond erosion.
- □ A controlled, progressive wear is necessary for proper superabrasive wheel performance.

wire EDM A process that is used to remove metal through the action of a controlled electrical discharge of short duration and high-current density between the tool or wire electrode and the workpiece.

- □ Dielectric fluid is circulated under constant pressure between the tool and the workpiece to conduct the current and wash away the chips.

work hardening/strain hardening A unique property of many metals that allows the strength and hardness to actually increase as they are machined or formed.

workhead or headstock The part of a cylindrical grinding machine that holds and rotates the workpiece.

work speed The rotational or translational speed of the workpiece while being machined.

work-wheel interface *See* TOOL-WORK INTERFACE.

Young's modulus *See* MODULUS OF ELASTICITY.

SECTION 5

Dies, Jigs and Fixtures, Molds

Steve F. Krar
Consultant
Kelmar Associates
Welland, Ontario

In the manufacture of metal parts, many methods may be used to produce the final product. *Machining operations,* such as milling, turning, and drilling, are used to produce parts manually through conventional operations, or automatically by computer numerical control. *Jigs and fixtures* can be used to locate and hold parts for machining operations to ensure consistent accuracy. For metal stampings, *dies* can be used to draw sheet metal into three-dimensional shapes; form sheet metal by bending, coining, extruding, and other means; and by trimming, piercing, and so on to the required size and shape required. Other parts can be produced by *pouring liquid metal* into molds and allowing it to cool, or by heating the metal and *forging* it to shape while it is hot.

Dies

angular clearance The relieved angular space below the straight portion of the die that is used to relieve the internal pressure of the blank as it passes through the die and allows it to drop through the bottom of the die. *(See figure.)*

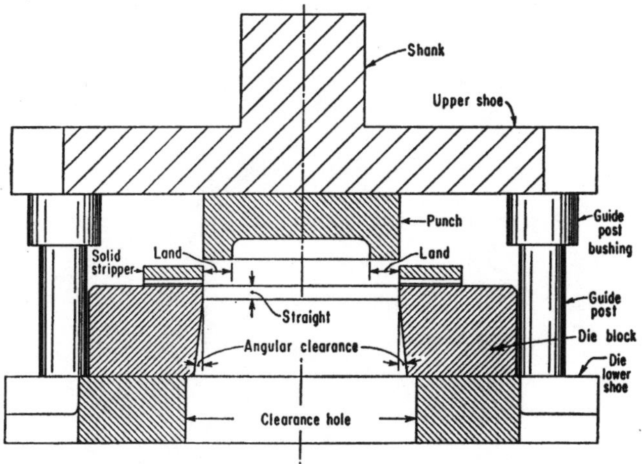

angular clearance.
(Courtesy Society of Manufacturing Engineers.)

backup blocks A section used to support notch, trim, and bend punches when shearing forces are not balanced.

ball cages Retainers used to hold ball bearings in a die set or stripper plate having a ball-bearing guidance system.

ball-lock punch A type of punch that is held vertically and radially by a spring-pressured steel ball in a punch retainer designed for rapid, in-press punch removal and replacement.

ball-lock punch retainer A type of quick-punch-release retainer, used to hold ball-lock pierce punches. *(See figure on next page.)*

bell-mouthed A radial entry opening for slugs that always lets slugs pass through ever-increasing openings.

bend allowance The amount of material in the flat that is consumed when the material is formed (bent) from a starting point tangency to an ending-form tangency.

bending The operation in which flat metal is forced around an inside radius.

blade punch Normally a thin and/or fragile shearing punch; usually inserted into the side of a holding

ball-lock punch retainer. *(Courtesy Dayton Progress Corp.)*

retainer which has a hardened backup plate that supports both the holder and the blade-punch insert.

blank The flat blank size required to produce the finished stamping.

□ In a progressive die, blank size is the coil width times (multiplied by) the coil advance.

blanking The shearing of material to a certain shape in preparation for finishing stamping to print size. *(See figure.)*

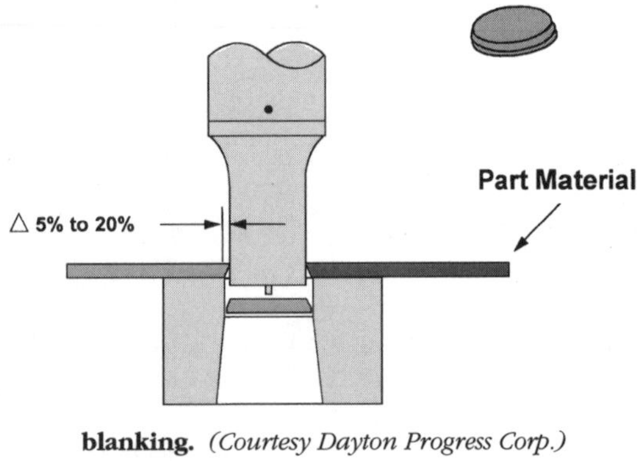

blanking. *(Courtesy Dayton Progress Corp.)*

break side The side of a die cut edge that fractures during a shear, such as the bottom side of a pierced hole.

buckling Severe upward or downward bowing of material during an automatic press feeding operation.

burnish A high-luster finish left on a part due to clearances less than the material thickness between the die section and/or the drawing-forming punch during a drawing or forming operation.

□ Many times this is required to burnish to size and/or control springback.

burr Small, tearaway edge of metal left on the break side of a stamping.

cam A mechanical die component used to transfer vertical force into a horizontal force or near-horizontal force.

camber The side-to-side curvature in a coil or blank sheet of steel.

□ It is produced when coils are slit with a slitting tool, duller on one edge than the other, causing a curvature in one direction that handicaps die feeding when the coil is unwound.

Catinay loop The loop produced by a coil of steel between the straightener and the feed mechanism.

clearance The space left between the punch and die to allow the metal to be cut or formed.

coining Operations that force metal to flow a relatively short distance in a self-contained manner. *(See figure.)*

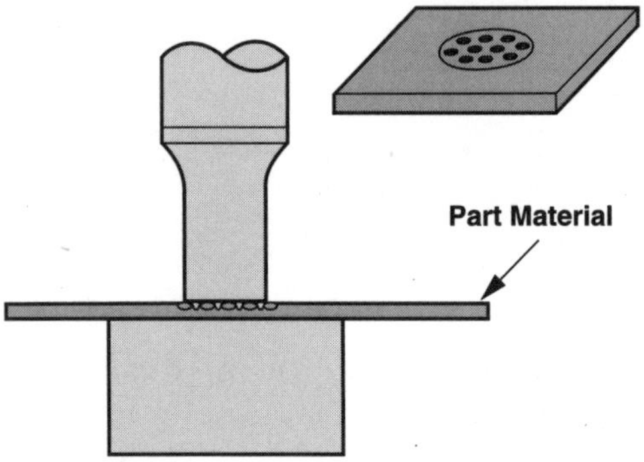

coining. *(Courtesy Dayton Progress Corp.)*

cushion (air) A built-into-press pressure source for form and draw dies.

cutting (shearing) The stressing of metal in a shear between two cutting edges to the point of fracture, and beyond its ultimate shear strength.

cutting clearance The clearance, usually a percentage of the material thickness, provided between the punch and the die to produce the least burr possible.

cutting edge The portion of the die section or punch that cuts the material.

cutting land The straight portion of the die section that allows a die to be resharpened a number of times without changing the die shape or size.

die **1.** A complete production tool that is used to produce a large number of sheet-metal parts to the required shape and accuracy *(see figure).* **2.** The female part of a die, generally fastened to the lower die shoe; it can also be on the upper shoe.

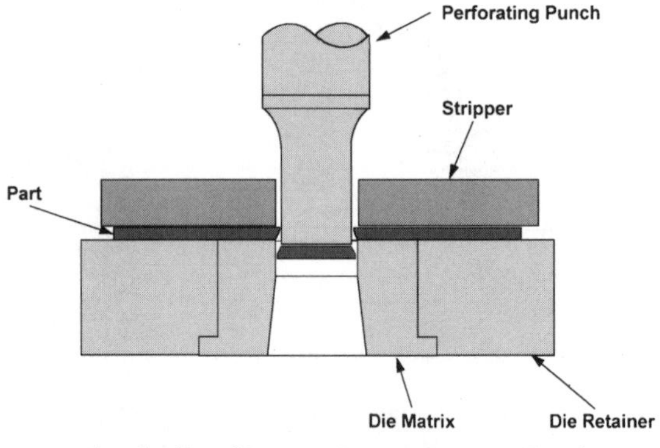

simple die. *(Courtesy Dayton Progress Corp.)*

die buttons A hardened, tool steel component that establishes shear clearance for piercing.

die components The parts that make up a complete die set, consisting of items such as pins, bushings, top (upper) shoe, and bottom (lower) shoe.

die life The straight (vertical) section of the lower die cutting edge before the relief starts.

die ring A simple die-block construction in which dowel pins are used to locate and position the die ring on the die pad, or directly on the lower die shoe.

die types

Acute-angle dies. These dies are used for 90° bends or less.

Beading dies. These dies are used to stiffen flat sheets and sometimes allow thinner material to be used as a result. *Open beads* extend from edge to edge; *closed beads* fade out in the sheet.

Bending dies. This term refers to dies that involve a forming operation.

- ▶ Pressure pad drawing dies are equipped with pressure pads actuated by springs, rubber cylinders, or nitrogen.

Blanking dies. These dies shear a flat part from a strip of sheet metal.

Box-forming dies. These dies require a high punch and a low die so that the box can be formed.

Brehm trimming (shimmy) dies. The design and action of these dies make it possible to trim boxes, cups, and shells that include tabs, notches, and angles.

Bulging dies. These dies are used to expand or bulge drawn shells to produce products such as artificial limbs, doorknobs, kettles, and teapots.

Channel-forming dies. These dies are used to form channels in gooseneck or single-stroke channel dies. These dies have a spring-pressure and release mechanism to eject the formed part out of the die.

Coining dies. These dies are used to change the metal thickness through a squeezing operation in order to leave an impression of the face of the punch, die, or both, on the part.

Combination dies. These are single-station dies that combine both cutting and noncutting operations in one press stroke.

Compound dies. These are single-station dies that usually combine two or more cutting operations (generally blanking and piercing) at every press stroke. *(See figure.)*

- ▶ A *compound blank and pierce die* is a blanking die with the addition of piercing punches.

Compression dies. These dies use compression or squeezing operations to form a slug or blank to shape by plastically deforming metal.

Corrugating dies. These dies are used to provide a variety of corrugations in one press stroke to strengthen the sheet, or for design purposes.

Curling dies. These dies are used to form a curl or coiled-up end on a sheet or part.

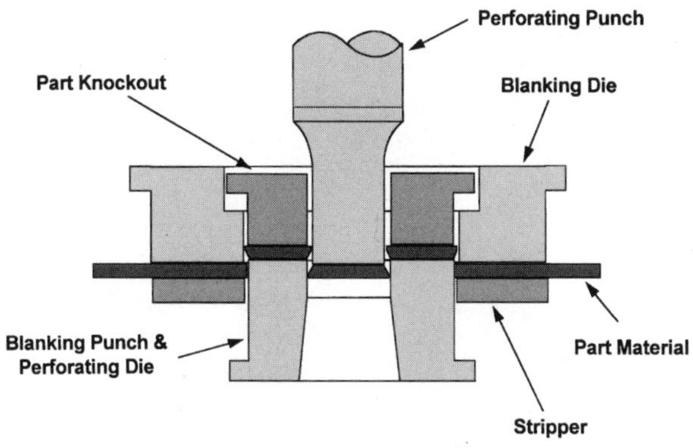

compound die. *(Courtesy Dayton Progress Corp.)*

Draw dies. These dies are used primarily to reduce flat, round, and rectangular blanks to a specific three-dimensional form by holding the blank with a pressure pad and forcing metal to flow between the punch and the die. *(See figure.)*

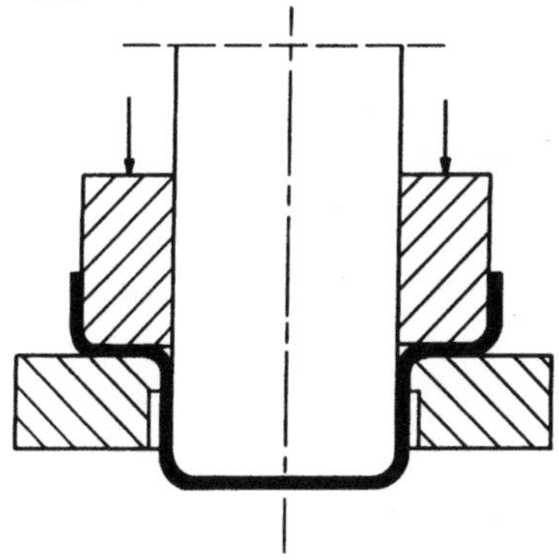

draw die. *(Courtesy Society of Manufacturing Engineers.)*

- ▶ The most common shells produced are cylindrical and rectangular containers.

Forming dies. These dies are used in conventional presses to form a part to the exact shape required. Along with the form on the die and punch, cams and slides are used to produce special forms or bends on the part.

- ▶ Common operations include drawing, bending and flanging, embossing, beading, bulging, compression, and ironing.

Forming dies, 90°. These are among the most common press-brake dies used to produce 90° angle bends on material.

Gooseneck dies. These dies are essentially V-shaped, provided with clearance for return flanges.

Hemming dies. These dies are used to turn over the edges of a sheet to provide a rigid, smooth edge.

Inverted dies. These dies, where the die components are mounted opposite to conventional dies, are used to produce parts too large to pass through the press bed or bolster plate. They are well suited for secondary operations such as shaving and trimming.

- ▶ The female (die block) part of the die is mounted on the punch shoe and moves with the press ram.
- ▶ The male part of the die is mounted on the bolster plate and/or the press bed.

Ironing dies. These dies are used to reduce the wall thickness and produce a smooth, uniform finish on shell walls by having the clearance space between the punch and die wall a little smaller than material thickness. *(See figure.)*

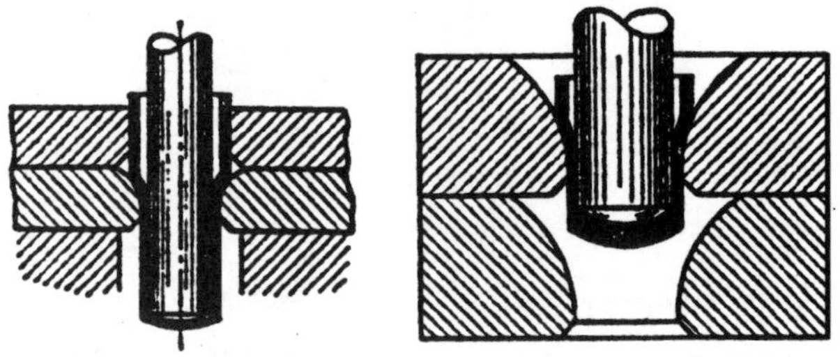

ironing die. *(Courtesy Society of Manufacturing Engineers.)*

Lamination dies. These are progressive dies that are specifically designed to produce high-volume parts for any stamping industry.

- ▶ Many lamination dies are built in clusters and use four-doweled pins to locate and lock each segment of the die.
- ▶ They use stock centralizers to align stock that is slightly misaligned before the entry of the pilot and punch.

Multibend dies. These dies are used in large production runs to perform a number of operations that would normally require two or more dies and operations.

Offset dies. These dies are used to form an offset by making two bends with a 90° or acute-angle die.

Piercing dies. These dies are used to produce holes in flat stock that is fed manually or automatically and kept on a straight path through the die by stock guides.

Pinch-trimming dies. These dies, which perform a primarily trimming operation, usually combine drawing, forming, and sizing to produce a uniform trimmed edge on the part.

Progressive dies. These dies generally consist of two or more stages where different operations are performed on a metal strip to produce a finished part. *(See figure.)*

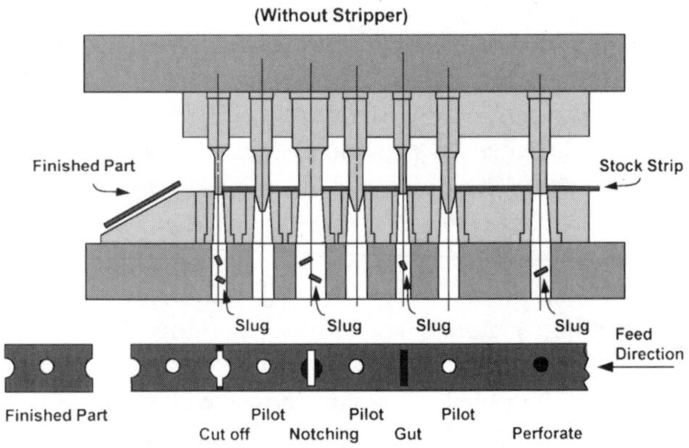

progressive die. *(Courtesy Dayton Progress Corp.)*

- ▶ At each press stroke, usually two operations are performed on a strip; therefore, a five-stage die would require four press strokes to produce the first finished part, and then a part is produced on each succeeding stroke.

Radius dies. These dies are used where the radius to be formed is more than 4 times larger than the material thickness. These bends are made with a V-die machined to less than 90° and a full radius punch.

Rocker edge dies. These dies can be used to form parts that would be difficult with a die that acts only in a vertical motion.

Shaving dies. These dies are used to produce finished edges or surfaces and sizes on light and heavy stampings, forgings, tubing, and other materials, for high or low production runs. *(See figure.)*

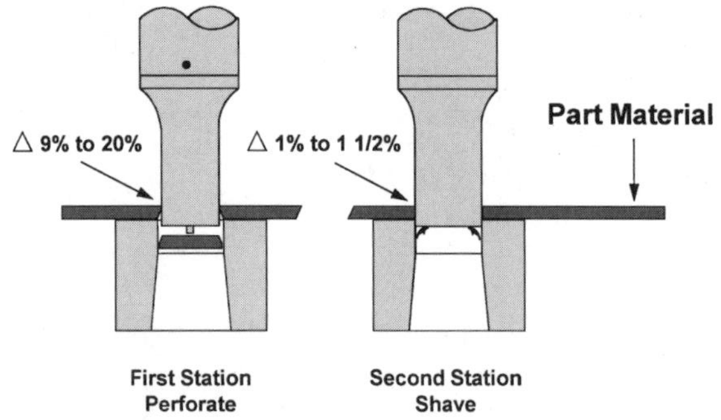

shaving die. *(Courtesy Dayton Progress Corp.)*

Seaming dies. These dies are used to make seams in sheets or tubes.

Side-action dies. These dies perform operations at an angle to the stroke of the press ram. They are useful for secondary cutting operations such as piercing, notching, and trimming, and forming operations such as bending and curling.

- ▶ The angular motion is provided through cams, toggles, bellcranks, pivots, and other components that are built as part of the die.

Sintered carbide dies. Sintered tungsten carbide is used as die material for most types of dies where high production runs and close tolerances are required.

- ▶ Carbide dies are widely used to manufacture electrical motor laminations at lower cost than with steel dies.

Tube (pipe)-forming dies. These dies are similar to curling dies.

U-bend dies. Operation of these dies is similar to the channel operation, but stock springback must be considered and provided for.

dowel pin A hardened and ground steel pin used to align and keep die parts in a positive location.

draw bead The rib on the draw surface of the draw pad that helps to control the flow of material.

draw-die radius The curved working surface of a draw die, or draw ring, where the material flows over during the drawing operation.

drawing The process of forming of material into a three-dimensional contour. *(See figure.)*

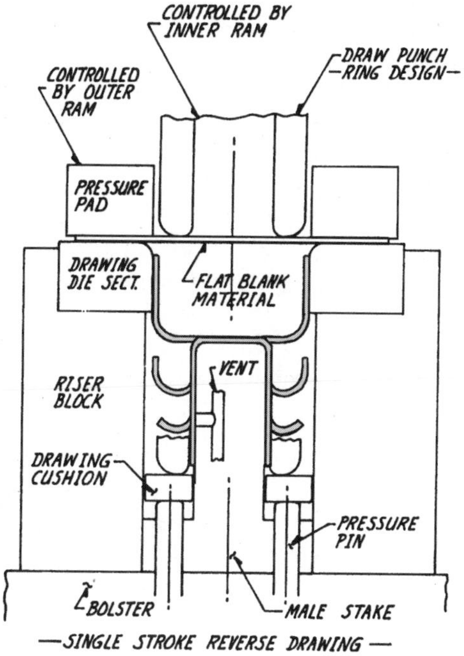

drawing.
(Courtesy Die Makers Handbook, Arntech Publishers.)

- ☐ A *shallow draw* is where only one drawing operation is required to produce a finished part.
- ☐ A *deep draw* is where a part requires one or more redrawing operations to produce a finished part.

draw pad (pressure pad) The part of the die used to hold the material tight against the draw punch during the drawing operation.

draw pin A steel pin used to transfer pressure from the press cushion to the die draw pad.

draw ring The part of die that is used to hold the material against the draw punch during the metal flow phenomena.

eccentric A hub connected to the crankshaft of the press that allows the distance from centerline to be adjusted to regulate the amount of movement in the feed system in a crank-driven slide feeder.

ejector A mechanism in a die that is used to remove the part, such as a spring-loaded plunger on the end of a punch to eject a slug.

extrude Forming of a short flange in one direction to make the walls of the form higher than metal thickness.

extruding Consists of operations that compress or force metal to flow plastically through a die orifice, usually in a continuous length of uniform cross section.

eyelet-machine tooling The tools and brass parts used in the production of brass-barrel eyelets. The die operations involved include blanking, drawing, flanging, coining, and piercing.

feed stops Items used on dies to stop the feeding of the stock strip at the correct location. Common types include the solid, plain, headed, pin, finger, pusher, trim, and pivoted auto stops. They serve two important functions: align the strip at the stop position and at the registry position. *(See figure.)*

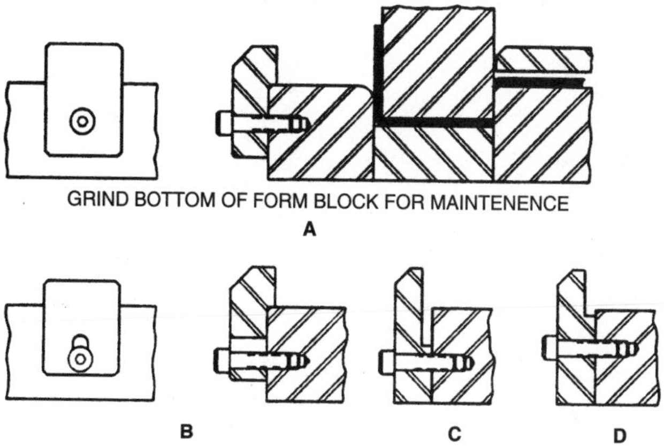

feed stops.
(Courtesy National Tooling & Machining Association.)

- ☐ The *stop position* is the location of the actual stopping point or surface against which the stock strip is stopped.
- ☐ The *registry position* is the exact location where the stop strip must be, so that the work will be dimensionally accurate in subsequent stations of the die.

forming A metalworking process in which the shape of the punch and die is reproduced in the metal.

form pressure The pressure required by the form punch to form a part; also the holding pressure required to hold the part flat during the forming operation, which must always be greater than the bending pressure.

form punch The punch that is used to form a part.

form radius The radius on a punch section where the material flows over during forming.

fracture A condition in which material is pulled apart during a draw operation, or a break at bend during forming.

French (progression) notch A cut on one or two edges of the material, the exact length of the die progression, that is made in the first station of a progressive die.

- ☐ The cut edge is then fed against a positive stop position in the die for the second station.

gall The material being formed that is fused by friction to die steel during the forming operation.

guides Used to make sure that the stock stays in the proper slot or groove while in the die.

heel block (heel) A block or plate, generally mounted on the lower die shoe, to prevent or minimize the deflection of punches and cams. A punch may have an extension or heel to guide it and reduce or eliminate its deflection.

heel of punch The section of the punch that is used for mounting in the punch pad.

holding pressure The amount of pressure required to hold a part flat during forming operations.

inch A press mode that allows the setup person to start and stop the press for setup purposes with short jogging strokes or a full revolution.

insert A small die section that is fitted into a larger section, such as a die pad.

jackscrew A series of threaded holes through a die section to aid in removal of components.

knockout A device or mechanism such as a pressurized pad, usually connected to and operated by the press slide, for pushing or lifting a part from the cavity of a form or draw die.

□ A common knockout is where a rod forces the knockout plate to strip the part from the die.

lancing The operation of cutting material without removing a slug. *(See figure.)*

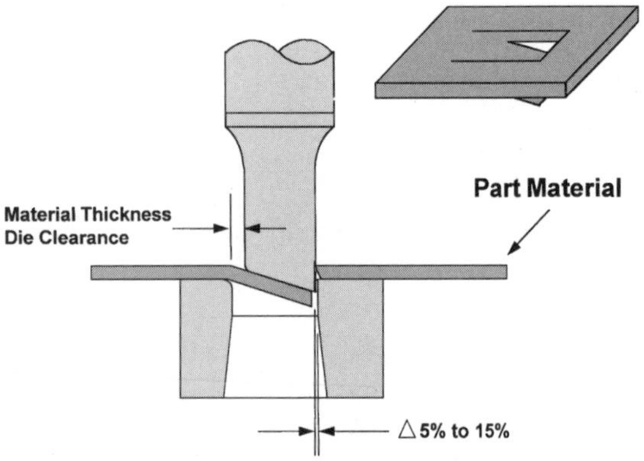

lancing. *(Courtesy Dayton Progress Corp.)*

land The flat (usually horizontal) surface at the cutting edge of a die that is ground and reground to keep the cutting edges sharp.

leader pin bushings The guiding bushings, usually located in the upper die shoe, that guide the leader pins in the lower shoe.

leader pins Precision-ground pins that are used to guide and maintain the location of die sets as they open and close.

lifter A spring-loaded insert that is used to raise the stock material above the die level.

lifter pins Spring-loaded pins, extending from working surfaces of dies, to stop the stamped part from sticking to the working surface of the die.

locator A type of extension, above the die level, that is used to locate a part in a certain place.

misfeed detectors Devices which generally incorporate a spring-loaded detector to activate a switch that stops the press in case of a stock misfeed.

□ They are generally used in progressive dies and are designed like a punch whose point is a little smaller than the pilot hole.

mismatch A condition in which two die cuts in different dies or stations do not blend together because of an offset.

multiple-slide press tooling This tooling, along with auxiliary and cutoff slides, is used to pierce, trim, and letter-stamp the part during the manufacturing process.

pad A high-pressure blank holder used to hold the part during a forming or drawing operation.

parallels Precision-ground blocks that are used to raise the die shoe above the press working surface for correct feed-line height.

parting scrap The piece of material left on the metal strip between each part that is made. It is generally .250 in and is cut out in the final station of a progressive die.

piercing The operation of producing holes or small slots through a part or blank. *(See figure.)*

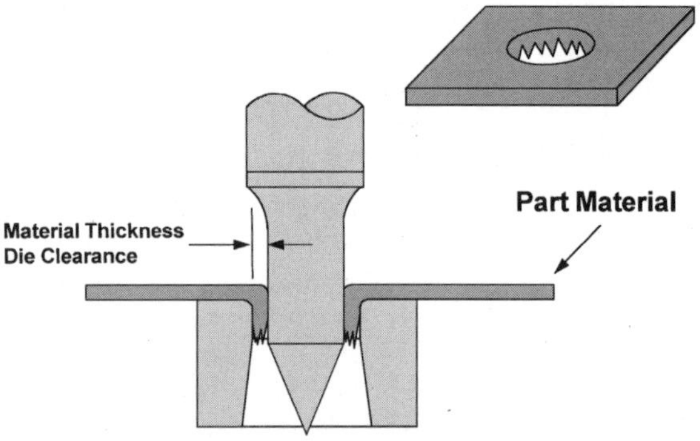

piercing. *(Courtesy Dayton Progress Corp.)*

pilot holders Usually consist of a block of steel into which punches are press-fitted; this assembly is fastened to the punch shoe.

pilot punch A punch in a die, with a lead or nose ground on the point extending below the stripper, that is used to accurately locate the part or strip just before the stripper contacts the material. *(See figure.)*

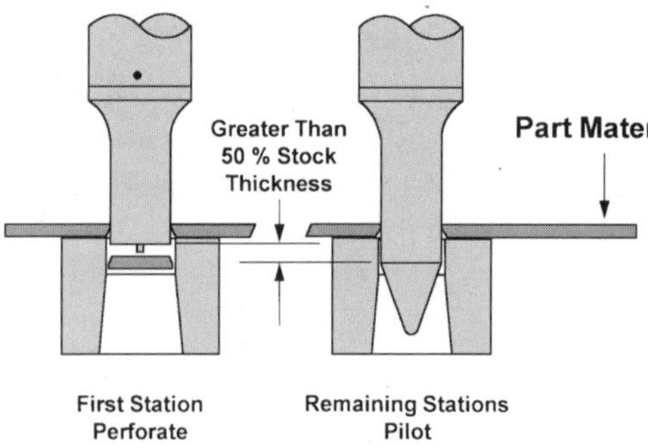

pilot punch. *(Courtesy Dayton Progress Corp.)*

□ The punch should be strong enough to align the stock repetitively without bending.

□ Pilot punches should be a good grade of tool steel hardened to Rockwell Rc 56 to 60.

pinch trim A die operation that cuts material and produces a knife edge by closing to near zero clearance at the end of the draw or form wipe.

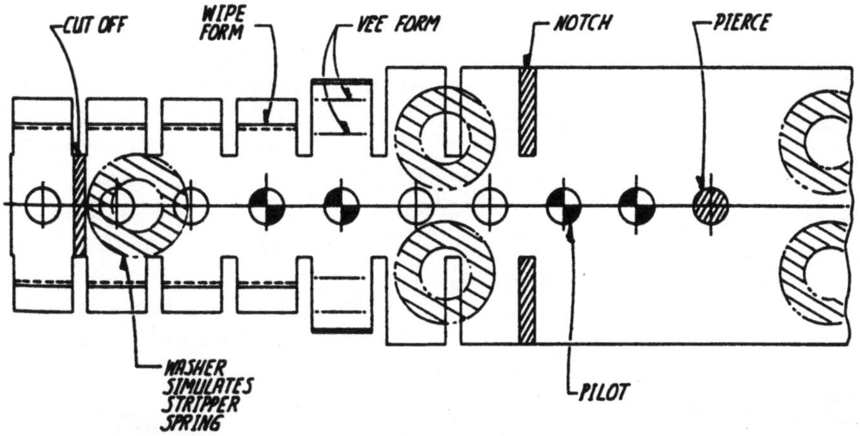

die strip design.
(Courtesy Die Makers Handbook, Arntech Publishers.)

pressure mark A mark or a shiny, burnished look on the material surface caused by high pressure.

pressure pad The part of the die that applies holding pressure to the blank or part. It holds the part during bending operations, serves as a bottoming block for setting up bends, and acts as a stripper to remove the part from the die.

progressive die designs The planning of a progressive die to determine the sequence of operations (piercing, forming, etc.), to reduce scrap on the strip to a minimum, and to remove the part and scrap strip from the die. *(See figure.)*

punch **1.** The male part of a die, fastened to the punch holder, which is moved directly or indirectly by the press slide. **2.** To cut a hole in a part or stock material.

□ Common punches include the plain, headless, bossed, flanged, heeled, pedestal, and offset pedestal types.

punch nose radii The radius at the end of a punch, which prevents excessive thinning at the bottom of the cup. In multiple draws, each radius should be smaller in proportion to the preceding shell.

punch plate (pad) A plate in which punches are mounted in the proper location and then this unit is fastened to the upper die shoe.

punch retainer A block or section of steel mounted on the upper die shoe which holds a punch, or punches.

punch shank The section of the punch that is fitted to the punch retainer.

punch shoe Usually the upper part of the die set which contains the guidepost bushings and holds the punches.

pushoff A spring-loaded insert that is used to push material or the slug away from the top of the die; also called the "kicker."

retainers Devices used to hold punches and dies when the spacing between these components is close.

riser A large, permanent shim that is used to raise or lower a group of sections to a desired height.

rubber-pad forming process (guerian process) A process generally used for forming shallow parts, such as aircraft parts, from lightweight materials.

□ The rubber pads are in a steel pad holder attached to the moving platen; the form block is attached to the bolster plate.

□ The most common dies of this type include laminated plastic, zinc, and wood dies.

section A piece of steel mounted to the lower half of the die, such as the pierce section, cutting section, and form section.

set blocks Spacers between the top and bottom shoe that are used to adjust the working height of the die; they also aid the die setter in adjusting the die to the desired shut height.

set edge The offset on the wipe side of the form section that is timed to hit the radius of the part at the bottom of the press stroke in order to form a 90° angle.

shear (shear angle) **1.** An inclination between the cutting edges of a punch and die created by grinding a relief on either part to reduce the shearing force. **2.** To cut with shearing dies or blades. **3.** A tool for cutting (shearing) metals and various materials by the closing action of the two cutting edges of the punch and die.

shearing action The stages that a piece of stock goes through as a result of the pressure applied by the blanking or piercing punch during a shearing operation. *(See figure on next page.)*

shedders Devices used to remove (knock out) a part from the die cavity after it has been formed.

□ Common shedders are the *spring-laminated* and the *flangeless* types.

shim A material of specific thickness that is used to raise a punch or die section to a specific height.

shoe Upper and lower parts of a die set used to mount the punches, die sections, and other components.

shut height The distance from the bottom of the die shoe to the top of the punch holder when the die is in its closed working position.

sizing The operation of flattening and smoothing areas of castings, forgings, and stampings, by squeezing the metal to the desired dimensions.

slugging Clogging and blocking of die slugs (scrap) that normally fall through the base.

spotting The operation of removing a small amount of material in a specific spot on a die to relieve the pressure.

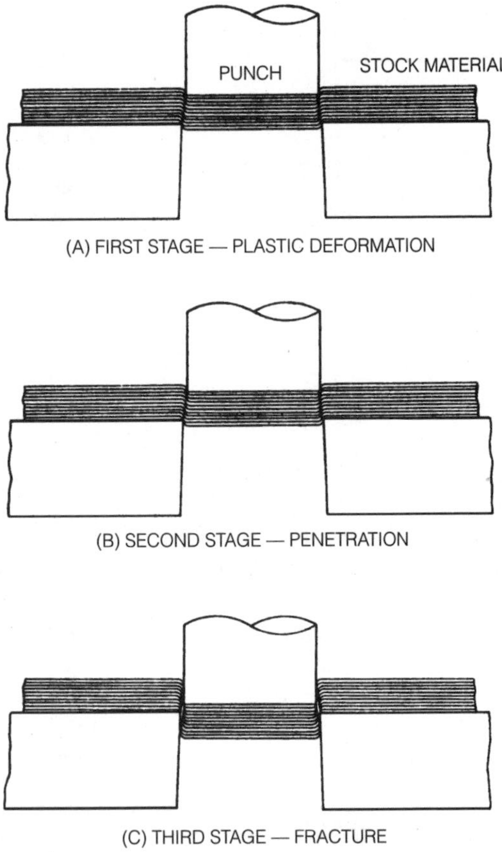

shearing action.
(Courtesy National Tooling & Machining Association.)

springback The amount that a metal springs back as it tries to return to its original shape after a bending or forming operation. At the bottom of a press stroke, the elastic limit of the metal is exceeded, but the ultimate strength is not.

□ The amount of springback varies with the size of the bend radius and with the thickness, type, and condition of the metal.

spring retainers Devices used to allow the stripper plate and pads to be removed to prevent the springs from falling out.

stake The shaped inner form of a draw die over which the part is formed.

standoff A die insert used to hold one part away from another, generally to control pressure.

station The area in a progressive die covering one progression or coil advance.

stock guides Guides that are mounted on a progressive die to control the centerline of the coil width.

stock pushers Devices used to make sure that the part of the metal strip is against the gaging surface of the die.

stops Most dies require one or more stops—a solid registering stop or a temporary stop—for starting the stock through a progressive die.

straight land The straight surface of a die between its cutting edge and the beginning of the angular clearance.

strip development Used on progressive dies to determine the sequence of piercing and forming operations required to produce parts that are accurate and to the required shape.

stripper **1.** The spring or pressure load part of a die that is used to push the part off punches after they have pierced the holes and the die has completed its work *(see figure)*. **2.** The bridge solid stripper (not spring-loaded), mounted to the bottom of the die with pierce or trim section and clearance for punches, pulls material off the punches when the holes have been pierced.

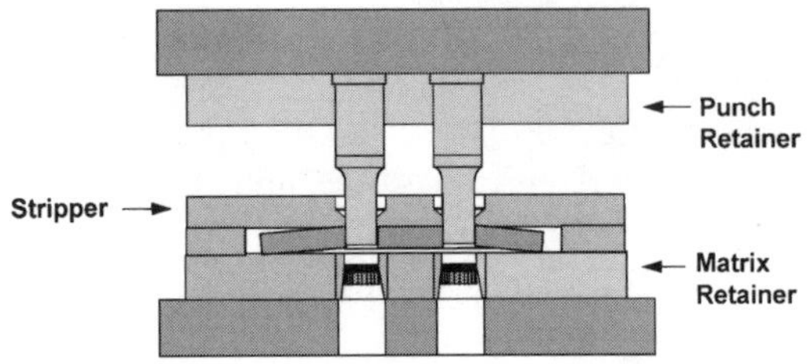

stripper. *(Courtesy Dayton Progress Corp.)*

□ Common types of strippers are the box, spring-actuated plunger, hook, positive knockoffs, and mushroom-type knockoffs.

stripper bolt A bolt with a special head that is used to hold a stripper at a specific height.

stripper plate A fixed or movable plate that removes the part or the metal strip from the punch or die; may also guide the feedstock.

stripper pressure The amount of pressure applied by the stripper to hold the part during the press cycle.

swaging A process in which the shape of the blank or slug is greatly changed as the metal is forced into the contours of a die; the remaining unconfined metal generally flows at an angle to the direction of the applied force.

thrust block A block of steel mounted on the top shoe to guide and support the die set.

trim The operation of cutting along the outside edge of a part to finish it to size and form.

vise jaws Specially designed interchangeable vise jaws are available or can be manufactured to securely hold a relatively small part in a vise of almost any shape.

□ Vises can be of the hand or air-operated type.

□ Multiple or equalizing clamping is possible with some types of vises.

wipe The area where a form punch contacts and moves over a part as it performs the forming operation.

wipe block The name of the steel part that does the forming.

wipe radius The radius on the punch that forms the part.

wrinkles The results from the uncontrolled flow of material because a drawing radius is too large, and the subsequent loss of the compressive flow phenomena.

Jigs and Fixtures

Jigs and fixtures provide a means of producing interchangeable parts because their function is to establish a direct relationship between the work and the cutting tool on all parts of similar shape and size.

angle brackets Angle brackets are often used when a right-angle alignment or reference is required. The common types are the 90° models; however, there are also styles that can be adjusted to suit the angle required on the workpiece.

bushings (jigs and fixtures) Mechanisms used to guide the tool and keep it in proper location in relation to the workpiece, so that operations such as drilling and reaming can be performed accurately on many similar parts.

bushing types *(See figure.)*

bushings. *(Courtesy Dayton Progress Corp.)*

Liner bushings. Sometimes called master bushings, these bushings are available with or without heads, and are permanently installed in a jig to receive renewable bushings.

Press-fit bushings. These bushings are press-fitted directly into a jig without the use of a liner bushing, and are available with or without heads. They are generally used for short production runs and will not require replacement.

Renewable bushings. These are used in liner bushings where there could be considerable wear, and where several bushings must be interchanged in the same hole.

- ► *Fixed renewable bushings* are installed in the liner bushing and are left in until they are worn out.
- ► *Slip renewable bushings* are interchangeable in a given-size liner and have a knurled head for easy removal.

Rotary or pilot bushings. These bushings are used where the location of the tool is very important; wear on the bushing is not permitted in order to maintain precise accuracy.

clamping devices The purpose of a clamp is to exert force to press a part (workpiece) in a jig or fixture against the locating surfaces and hold it in position in opposition to the forces of the cutting action.

- □ A clamp should be designed to produce the best clamping force when activated by the smallest force expected, and strong enough to withstand the largest force expected.

clamp locking devices Many types of clamping devices are used to lock a part in a jig or fixture. These may consist of simple wrench or cam types operated manually, or air or hydraulic devices that operate automatically.

Cam-operated clamps. These are available in a wide variety of styles to suit the clamping operation or the part. Common types are simple cam-operated, cam-operated with return bar, cam-operated with screw, and cam-operated by air or hydraulics.

Manually operated clamps. These include types such as double-acting screw, equalizing-pinch, leaf-type with rocker, quick-rising, screw-operated with hand knob, wedge-operated, and wedge-operated with screw.

clamp types The main categories of clamping devices are simple hand-operated clamps, quick-acting hand-operated clamps, and power-operated clamps.

Aluminum clamp straps. These clamps are used where weight is important, or where softer clamps are needed to avoid marring the work surface.

C clamps. These incorporate a screw clamp in a C-shaped frame and are the most widely used in manufacturing. They are available in a wide variety of types and sizes.

Cam-actuated clamps. The surface applying the clamping force is on an eccentric, and by turning the clamping lever approximately one-quarter of a turn, enough pressure is applied to hold the part securely.

Clamp straps. These are forged steel clamps used for general-purpose clamping applications where extra toughness is required. *(See figure.)* Common types include the plain, tapered-nose, wide-nose, U-shaped, gooseneck, and double-end straps.

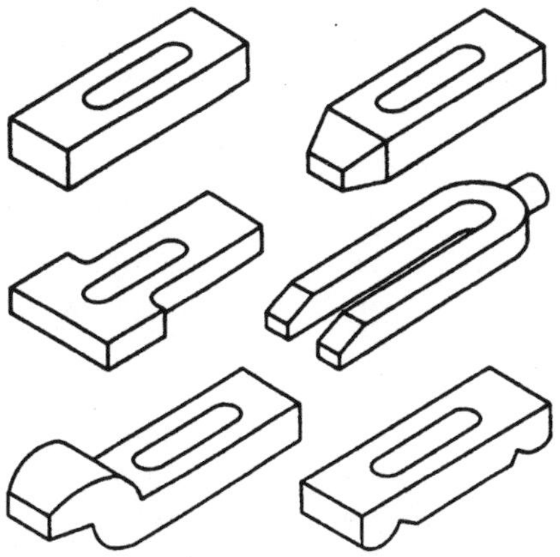

clamp straps. *(Courtesy Carr Lane Mfg. Co.)*

- ► The *slotted-heel clamp strap* has a slotted heel to locate the clamp strap on the clamp rest, and a slotted stud hole that allows the clamp to be moved clear of the workpiece for easier loading and unloading.

- The *tapped-heel clamp strap* has a tapped hole in the heel end of the strap that allows the heel rest to be adjusted to suit the height of the surface to be clamped.

Eccentric clamps. These apply the clamping force with the action of an eccentric circle. The off-center location of the mounting hole produces the radial movement of the cam through its clamping cycle.

Edge clamps. These are used for clamping on the side of a part to transfer the motion of an internal thread into a sliding motion that moves both forward and down.

- These clamps usually have gripping serrations on the clamping jaws that provide a better clamping action.

Floating clamps. These consist of a clamp and a work-holding plunger that operate together to hold the work in the proper location. When the clamp is tightened, it draws the stud downward, and at the same time, binds the wedge collar against the tapered seat on the plunger shank and locks it.

Forged-steel clamp straps. These are used for general-purpose clamping applications requiring extra toughness.

High-rise clamps. These consist of a number of riser elements, special contacts, and clamp straps that can be combined to make a variety of styles. The clamp design allows the workpiece to be clamped directly on the base element, or raised off the base with various accessories.

Hinged clamp. The simplest form of a hinged clamp is where the adjustment of the screw against the work pivots the clamp so that its heel grips the work with equal pressure.

Hydraulic clamps. These clamps use hydraulic liquids in a cylinder to apply the clamping pressure.

Latch-action clamps. These are designed for applications where a straight-pulling action is required. They are available in the adjustable U-bolt, adjustable hook, and fixed-length styles.

Screw clamps. These clamps must be strong enough to withstand the repeated stresses, and large enough to avoid failure fatigue from the stresses in the thread root.

Spiral clamps. These are similar to eccentric clamps, but have an involute curve on the clamping face that provides a wide range of clamping positions instead of just one.

Strap clamps. A typical strap clamp has a central slot about .06 in wider than the bolt or stud passing through it. The most common strap clamps used in the machine-tool industry are the finger, straight, and U-clamp.

Swing clamps. These devices use a swinging arm to speed up the loading and unloading of parts.

Toggle clamps. These operate on a pivot-and-lever principle and provide the clamping action through a series of fixed-length components connected by pivot pins. *(See figure.)*

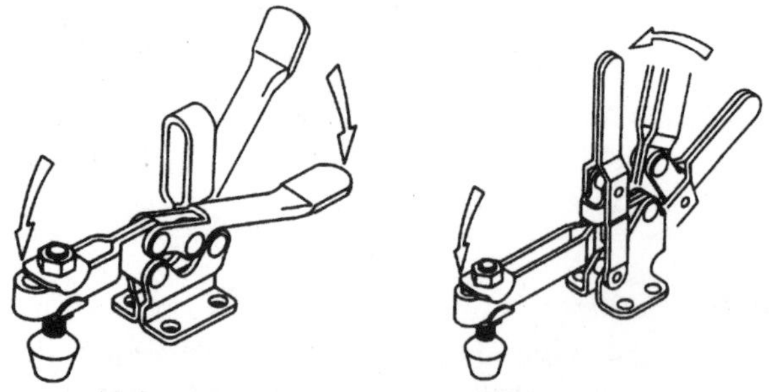

toggle clamps. *(Courtesy Carr Lane Mfg. Co.)*

- They are widely used in machine shops because they are so versatile and efficient.
- In the locked position, the levers are set slightly past the center point to provide a positive locking action.
- *Push/pull toggle clamps* are used for applications where a straight-line holding action is required.
- *Hold-down toggle clamps* are available with horizontal or vertical handles.

Up-thrust clamps. These have a clamp-type lever that pulls the lower jaw upward and allows the workpiece to be located on its top surface.

- The underside of the top jaw element is ground and acts as a precision locator for the workpiece.
- The jaw element rotates and has two clamp openings, one for thin parts and the other for thicker parts.

diamond pin A locating pin that has its sides ground away in the form of a diamond, leaving a small contact diameter width for locating at right angles to the axis of a hole center.

edge support A type of locator, having a precision-ground top surface and step, that is used for modular workholders. The step can locate a workpiece in both the horizontal and vertical axes. *(See figure.)*

edge support. *(Courtesy Carr Lane Mfg. Co.)*

ejectors These provide a means of easily removing a part, especially when it is located on a stud, plug, or ring. Effective ejectors include a pin ejector located under the work, hinged ejector, and the push-type ejector that can operate one or more pins.

fixture A tool or device used on machine tools to position and clamp a part for machining operations or inspection purposes.

fixture types

Angle-plate fixtures. These are variations of the basic plate fixture and are used when the locating surface is at an angle to the machine table.

- ▶ The two main types are the right-angle (90°) and the modified-angle fixtures that have an angle other than 90°.

Inspection fixtures. Inspection and gaging fixtures are used to check and compare workpiece features and accuracy against standards of known size.

- ▶ Plug gages can be used to check hole locations, hole sizes, and distances between holes.

Milling fixtures. These fixtures are designed to hold a workpiece firmly to allow milling operations to be performed.

Plate fixtures. The plate fixture, the most basic and common, is used to hold a workpiece parallel to the machine table *(See figure.)*

plate fixtures. *(Courtesy Carr Lane Mfg. Co.)*

- ▶ It is built with a mill fixture base, cast flat section, tooling plate, or similar material, and has all locators and clamps mounted on the plate.

Welding fixtures. These fixtures are specifically designed to hold parts where welding operations must be performed.

jacks (supports) The screw-type adjustable jack with a spring plunger is the most popular means of supporting specific or fragile areas of a workpiece.

- □ Spring jack assemblies can be made and used in fixture bodies whenever required.

jig A tool or device that positions and clamps a workpiece; also guides or controls the location of the cutting tool.

jig types

Box jigs. Sometimes called "tumble jigs," these contain locators inside the box to precisely position the workpiece. The top is hinged or removed to load or unload the part.

- ▶ Box jigs completely enclose the workpiece and allow machining operations to be performed from all six sides.

Indexing jigs. These are generally used where it is necessary to drill holes in a pattern around a center axis.

- ▶ A hand-retractable plunger is used to position the workpiece after the first hole has been drilled.

Leaf jigs. These are simplified box jigs which have a bushing plate that swings open for easy loading and unloading *(See figure.)*

leaf jigs. *(Courtesy Carr Lane Mfg. Co.)*

- ▶ Locators are usually mounted in the base, and the drill/ream bushings are mounted in the leaf or hinge plate.

Multistation jigs. A type of jig used for repetitive, simultaneous operations on many identical parts.

Plate jigs. These are similar to template jigs, but they have a self-contained clamping device.

Pump jigs. These jigs act in the same way as the leaf jig but have a positive rack-and-pinion hinge arrangement to move the top plate clear of the workpiece for easy unloading.

Sandwich jigs. These jigs are designed for thin or delicate workpieces and use two jig plates for performing machining operations from two sides.

- ▶ Locating pins and bushings maintain the proper relationship of the plates.

Table jigs. These jigs are variations of the plate and template jig but are raised up on legs. They are designed for applications where the surface to be machined also locates the workpiece.

Template jigs. These are generally used for layout or light machining operations, and are held in place by hand.

- ▶ When drilling multiple holes, a pin is placed in the first drilled hole, and it references the jig.

Trunnion jigs. A type of jig that can rotate heavy or odd-shaped workpieces on precision bearing mounts called *trunnions.* The trunnions, generally used in pairs, can be locked in the required place for operations to be performed on the workpiece.

locating pin Used to locate workpieces from either an internal or external surface *(See figure.)* Common types include the round, bullet-nose, diamond, floating, and locating pins.

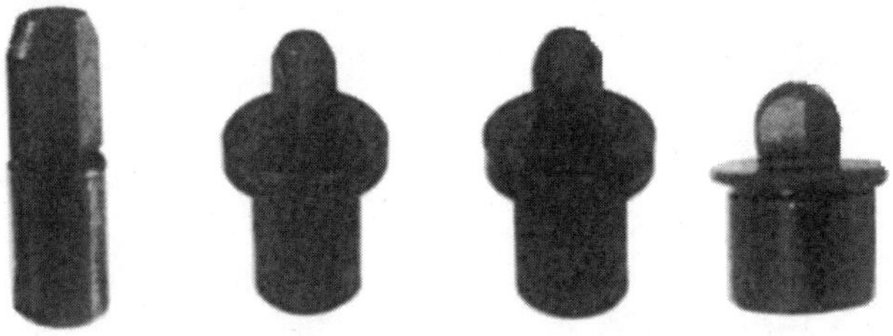

locating pins. *(Courtesy Carr Lane Mfg. Co.)*

- □ For internal location, the pin must match the size of the hole that is to be used as the location point.
- □ For external location, the size of the locating pin must be strong enough to withstand the machining forces.

locating screws These are precision screws used to locate and mount modular components in multipurpose holes. They are installed through the component and aligned in the precision bushing with the threaded end securely attached to the base element.

locators Devices used to establish an accurate relationship between the workpiece (parts) and the jig or fixture.

- □ In a drill jig, the cutting tool is guided by bushings in the jig body while the part is held in location in the jig by special clamping devices.
- □ Fixtures are usually fastened to the machine table, and the fixture or the cutting tool is aligned in the proper position for the machining operation.

locator types Many types of locators are used to accommodate the wide variety of workpieces held in jigs or fixtures *(see figure).*

Adjustable locators. These are generally used on sand castings, welded fabrications, and parts where a size variation can be expected.

- ▶ These locators can be adjusted to compensate for the differences in part size, yet provide a solid and accurate locating point.

Conical locators. These are cylindrical pins that use a spring cone to locate a part from a hole. The part rests on a flat surface and is centered by the spring cone.

- ▶ Fixed cones are used to locate from a hole without a flat surface.

Equalizing locators. These consist basically of a screw and nut, and are used to support a part having two or more rough or varying surfaces.

Expanding locators. These locators have fingers that are pivoted so that they move outward uniformly when the knurled hand wheel is turned. They are used for a rough bore of medium or large size.

Retractable locators. These locators may use a locating pin, or pins, that are retracted for a moment when loading or unloading so that the part can fall free on completion of the operation.

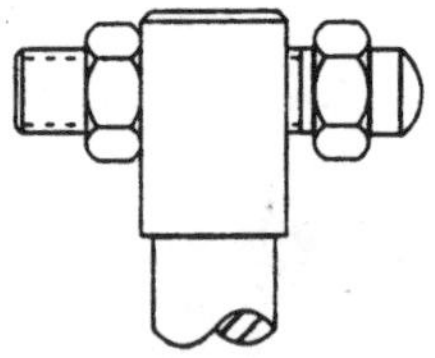

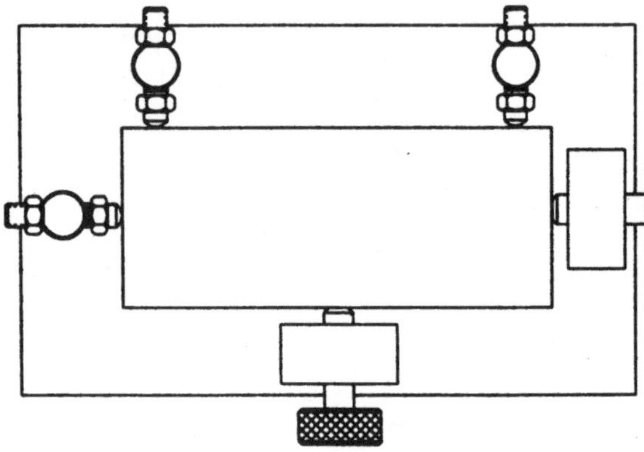

locator types. *(Courtesy Carr Lane Mfg. Co.)*

Spring-loaded locators. These locators are used for locating surfaces that may vary in size and position.

modular fixturing A workholding system that uses a series of standard components to build a specialized workholding device for many types of parts. *(See figure on next page.)*

- □ Standard off-the-shelf tooling plates, supports, locators, clamps, and similar components are assembled on a baseplate.

screw jacks Heavy-duty supports containing an adjustable screw that provides a means of adjusting their height. Most screw jacks are supplied with interchangeable tips to allow the contact area to suit the workpiece.

tool bodies Tool bodies provide the mounting area for all the locators, clamps, supports, and other devices that position and hold the workpiece.

- □ The common tool body forms are the cast, welded, and built-up designs.

tooling blocks Vertical tooling blocks allow parts to be located and clamped on their surfaces for operations on a horizontal machining center.

- □ The common types, two-sided and four-sided, can be mounted on an index plate so that parts can be clamped to all their vertical surfaces.

tooling plates Basic flat plates provided with methods to clamp the plate to a machine table and fasten the locating and clamping components to its surface.

- □ They are available in rectangular, blank, round, square, rectangular pallet, platform, and angle styles.

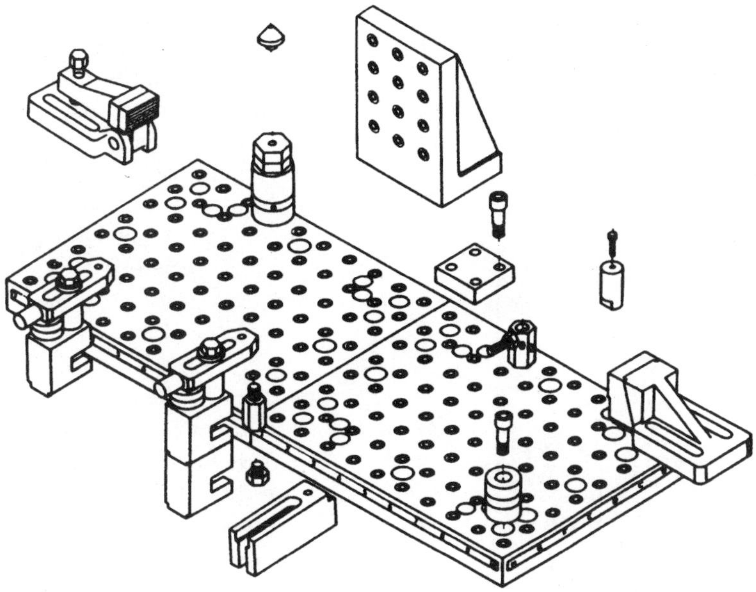

modular fixturing. *(Courtesy Carr Lane Mfg. Co.)*

vise jaws Specially designed interchangeable vise jaws that are used to hold almost any shape of relatively small part securely in a vise.

□ Vises can be of the hand- or air-operated type.

□ Multiple or equalizing clamping is possible with some types of vises.

Molds

casting A process in which molten metal is poured or injected and allowed to solidify in a shaped sand, metal, plaster, or ceramic mold cavity.

□ After cooling, the part is removed from the mold and then processed.

casting methods

Centrifugal casting. In this process molten metal is poured into a mold that is being rotated, or starts to rotate at a certain point in the pouring cycle.

▶ In *vertical centrifugal casting,* the molten metal is subject to centrifugal force in a horizontal direction.

▶ In *horizontal centrifugal casting,* the center hole remains uniform regardless of the length of the casting.

Cored casting. This method is used to form internal surfaces using cores made from mold material such as sand, which offers greater resistance to the harsh and chemically active reactions in the mold cavity before and during solidification. *(See figure.)*

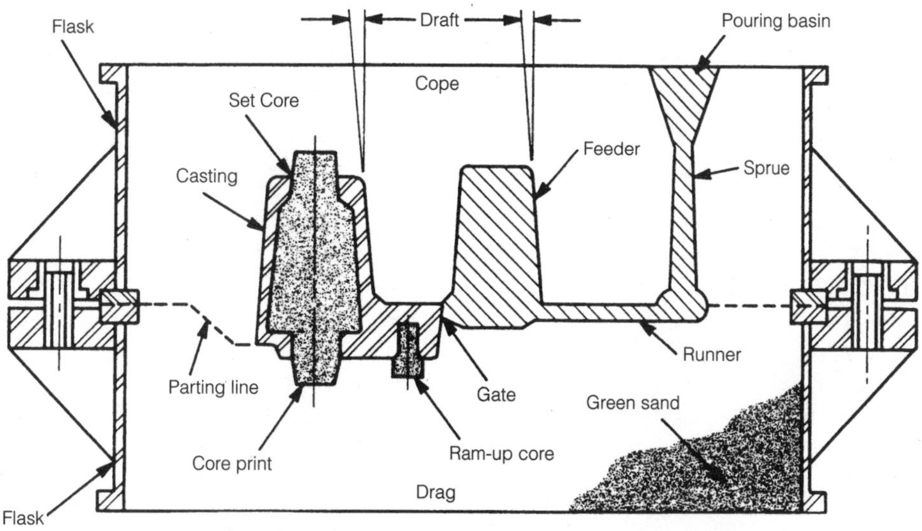

cored casting. *(Courtesy Society of Manufacturing Engineers.)*

Coreless casting. This is a process such as centrifugal and slush casting that is used to produce hollow parts.

Die casting. In this process molten metal is forced into a metallic die under pressure *(See figure.)* It is widely used for high-volume production of aluminum and zinc castings.

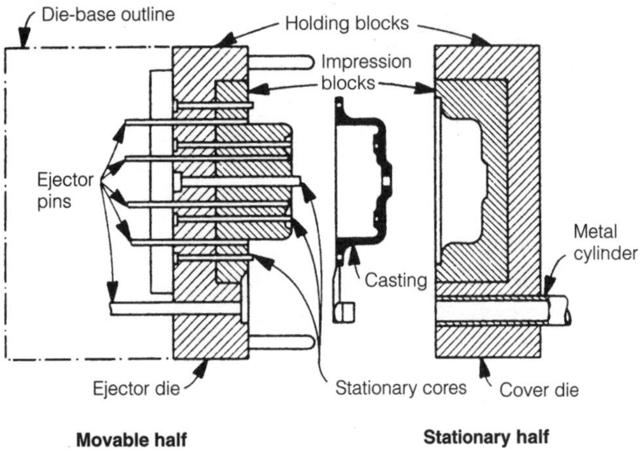

die casting. *(Courtesy Society of Manufacturing Engineers.)*

- ▶ In hot-chamber machines (for zinc), metal is pumped into the die.
- ▶ In cold-chamber machines (for aluminum), metal is ladled into the shot chamber.

Investment casting. A wax pattern is coated with a refractory shell by casting or dipping. The assembly is then heated to melt out the wax pattern, and molten metal is poured into the ceramic mold.

- ▶ This method is used for precision casting.

Metal mold casting. This process is used to produce metal parts with an excellent surface finish, close dimensional tolerance, and good physical characteristics.

- ▶ Common methods of producing castings in metal molds are high-pressure die casting, permanent and semipermanent mold casting, low-pressure casting, and centrifugal mold casting.

Near-net-shape casting. This is a casting process using unique naturally bonded, microfine sand that gives malleable iron castings their superior finish, dimensional accuracy, and intricate detail.

- ▶ The machining time is greatly reduced because the casting can be produced within .015 to .040 in of finished size.

Plaster and ceramic mold casting. Commonly called *precision casting,* this process is characterized by the use of nonmetallic ceramic and plaster mold material.

- ▶ The material is usually a fine powder suspended in the binder as a slurry.
- ▶ This method is used for parts that require extreme accuracy, tolerance, and the reproduction of very fine details.

Sand-mold casting. In this process a shape or pattern of the finished product is made by ramming sand that is bonded with clay or other chemicals, around a wood or metal pattern. Molten metal is poured into the mold and allowed to solidify.

- ▶ This method is widely used for small- and large-production-run parts.

Semipermanent mold castings. These are made in a mold similar to a permanent mold, but a sand core is used instead of a metal core.

- ▶ Sand cores are used where cored openings, undercuts, and recesses make it difficult to remove metal cores.

Slush molding. This procedure is when metal is poured into a mold and only the shell is allowed to solidify.

- ▶ The thickness of the shell depends on the amount of time the shell is allowed to solidify.
- ▶ This method is used for ornamental surfaces that require a good appearance and not high strength.

cold shut A short crack caused by metal flowing outward and then back on itself.

cope The upper part of a casting mold box.

cores Forms made of sand or metal that are placed in the mold cavity to form the inner surfaces of a casting, such as a hole or cavity.

die shift Misalignment between the top and bottom forging dies.

drag The lower part of a casting molding box or flask.

drop forging Process in which castings are made by shaping hot metal, through drop-hammer blows, within the upper and lower sections of a form die.

flakes Internal breaks or ruptures occurring in some grades of alloy steel as a result of cooling too rapidly.

gating system The channel system used to distribute the molten metal in the mold cavity and risers. It consists of the sprue, runner, gate, and riser or reservoir.

mistrimmed forging An effect caused by incorrectly placing the forging in the trimming die.

molding processes

Cement molding. This process uses cement molds made of washed and graded silica sand, plus 10 percent Portland cement mixed with 4 to 5 percent water. It is used to form large molds for ship parts and for ingots.

Dry-sand molding. A method of molding in a sand mold that has been dried after the pattern has been removed.

- ▶ The drying process increases the strength of the molding sand many times.
- ▶ This method is used primarily for large castings.

Green-sand molding. A process in which a mold is formed around a pattern with a sand-clay-water mixture and other additives. The molding mixture is compacted, and when the pattern is removed, the compacted mass retains a reverse of the pattern's shape.

Shell molding. A process used when high-dimensional accuracy and precise intricate shapes are the prime requirements.

- ▶ A metal pattern that is coated with a mixture of sand and resin, which is heated to allow the resin to melt and cure. The excess sand is dumped, the hardened mold is stripped, and the part duplicates the pattern in reverse.

Vacuum molding. A process (V-process) in which the sand is held in a mold by vacuum. Both halves of the mold are covered with a thin sheet of plastic to retain the vacuum and create an airtight molding process.

overheated metal Condition caused by improper heating, or by soaking the metal too long.

pattern A wood, metal, or plaster form, in one piece or in sections of an object to be made by a casting method. These are slightly larger than the finished casting, to allow for shrinkage during cooling and leave material for machining operations.

risers These are liquid reservoirs of metal that feed the casting as it shrinks and solidifies.

runner The vertical channel to the interior of a mold through which molten metal is poured.

ruptured-fiber structure A discontinuity in the flow lines of a forging caused by working some of the alloys too quickly, improper stock size, or improper die design.

scale pits Irregular shallow depressions caused by the scale on a die that is worked into the surface of a forging.

sprue The channel or channels through which the molten metal is led from the runner, down-gate, or pouring gate, to the mold cavity.

SECTION 6

Hand Tools and Operations

J. W. Oswald
Consultant
Kelmar Associates
Niagara Falls, Ontario

As hand tools developed over the centuries, it became easier for humans to satisfy their needs and improve their standard of living. The human hand and arm movements set the pattern for modern tool operations. Although the use of hand tools is declining because of CNC machine tools and automated processes, there will continue to be a need for hand tools. However, it is always better to perform an operation on a machine tool, if available, because it can generally be done faster and more accurately.

abrasive A natural or artificial material such as sandstone, emery, aluminum oxide, or silicon carbide.
□ Used in the manufacture of grinding wheels, abrasive cloth, and lapping compounds.

abrasive cloth Cloth on which an abrasive material has been bonded with glue or another adhesive.

accurate Without error; within the tolerances allowed; precise; correct; conforming exactly to a standard.

acute angle An angle that is less than a right angle (90°).

adjacent Near; adjoining; bordering. For example, two lines meeting at a common vertex in a triangle are called *adjacent sides*.

align To adjust or set to a line, as the tailstock center of a lathe is aligned with the headstock center.

allowance The intentional or desired difference between the dimensions of mating parts to provide a given class of fit.

angle The amount of opening or divergence between two straight lines that meet at a vertex or that intersect each other. This divergence is measured in degrees.

angle iron Structural iron made in the shape of a right angle.

angle plate A precision tool made from cast iron, tool steel, or granite. It is used as a fixture for holding work to be laid out, machined, or inspected. Two main faces are at right angles (90°) to each other and may contain holes or slots for holding the work. *(See figure.)*

apprentice One who is bound by a legal agreement, to learn a trade or business under the guidance of a skilled trade or craftsperson.

arbor A shaft or spindle used to hold cutting tools, especially on a milling machine.

arbor press A hand-operated machine using a rack-and-gear principle, capable of applying high pressure for the purpose of assembling or disassembling parts. *(See figure on next page.)*

arc A section of the circumference of a circle bounded by two equal radii.

assembling To put together, in correct relation, the parts of a unit or mechanism.

assembly An assembled unit (or units) required in the construction of a machine, such as the headstock or tailstock assemblies of a lathe. A machine may consist of several unit assemblies.

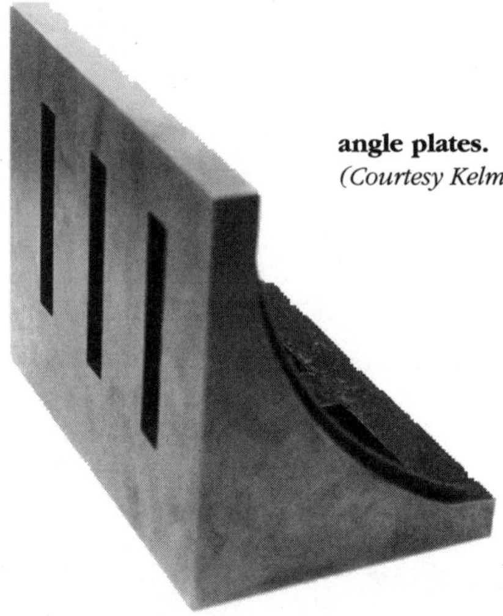

angle plates.
(Courtesy Kelmar Associates.)

arbor press. *(Courtesy Kelmar Associates.)*

assembly drawing A drawing that shows a completed product with all the parts in correct position and properly labeled or numbered.

axis The line, real or imaginary, that passes through the center of a body and about which the body would revolve.

ball bearing An antifriction bearing having an inner race that fits on a shaft, and an outer race that fits into a housing or support.

- □ Hardened steel balls are used between inner and outer races to reduce friction.
- □ The inner race is generally a press fit on the shaft, and the outer race is generally a hand push fit in the housing.

ball-peen hammer A machinist's hammer that has a ball end for riveting and forming metal. The other end is flat and used for striking a chisel or other such tools.

bar coding A data-collection system used to monitor quality control and collect information on assembly lines through multiple electronic reading devices. *(See figure.)*

bar stock Metal bars of various lengths made available in round, square, flat, hexagonal, and octagonal shapes from which parts are machined. They may be hot-rolled, cold-rolled, or extruded.

bastard file A rough-cut file having coarser teeth than a second-cut file. It is used to remove material quickly.

bearing A support or carrier in which a shaft rotates. It may be of plain, ball, or roller type.

bell-mouthed hole A hole that is larger at one end or at both ends and is not cylindrical throughout its length.

belt dressing An adhesive and softening material applied to leather belts to increase the pulling force and prevent slippage between the belt and pulley surface.

bevel **1.** Any surface not at right angles (90°) to the face of the workpiece. If a bevel is at a 45° angle, it is often called a *chamfer.* **2.** The name given to a hand tool used for measuring, laying out, and testing the accuracy of angles machined on a workpiece.

bill of material A list of the stock, sizes, part names, and quantity required, to produce a given project or component.

bisecting an angle Dividing an angle into two equal parts.

blind hole A hole that is machined to a certain depth on a workpiece but does not pass through it.

blue vitriol A chemical mixture of copper sulfate, water, and sulfuric acid. It is applied to polished steel to produce a copperlike finish for layout purposes.

bore **1.** The inside diameter of a cylinder or pipe. **2.** The operation of machining a circular hole in a workpiece.

boss A projection or raised section above the body of a casting or a forging, usually cylindrical in shape, through which a hole may be machined. The top of the boss is usually machined or spotfaced square with the axis of the hole to provide a seat for a bolthead or other fastening devices.

box wrench A type of closed-end wrench made in many styles for specific sizes and shapes of boltheads or nuts.

broach A long tool with a series of cutting teeth at progressive heights that is pushed or pulled through a hole in a metal part or across a surface to form the desired shape and size. *(See figure on next page.)*

buffing The operation of polishing or finishing a metal to a very smooth surface and a high luster. The

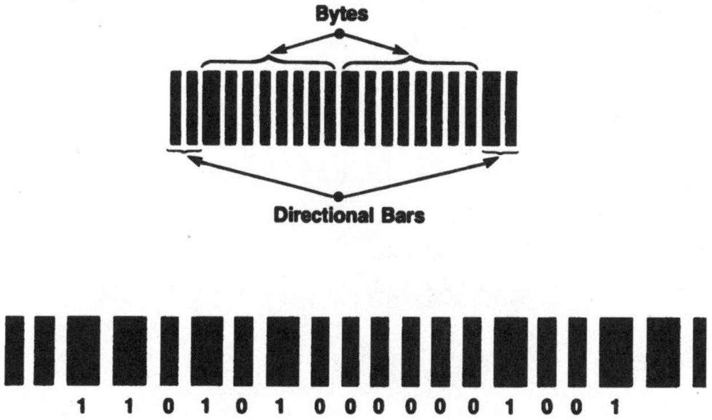

bar coding. *(Courtesy Hewlett-Packard.)*

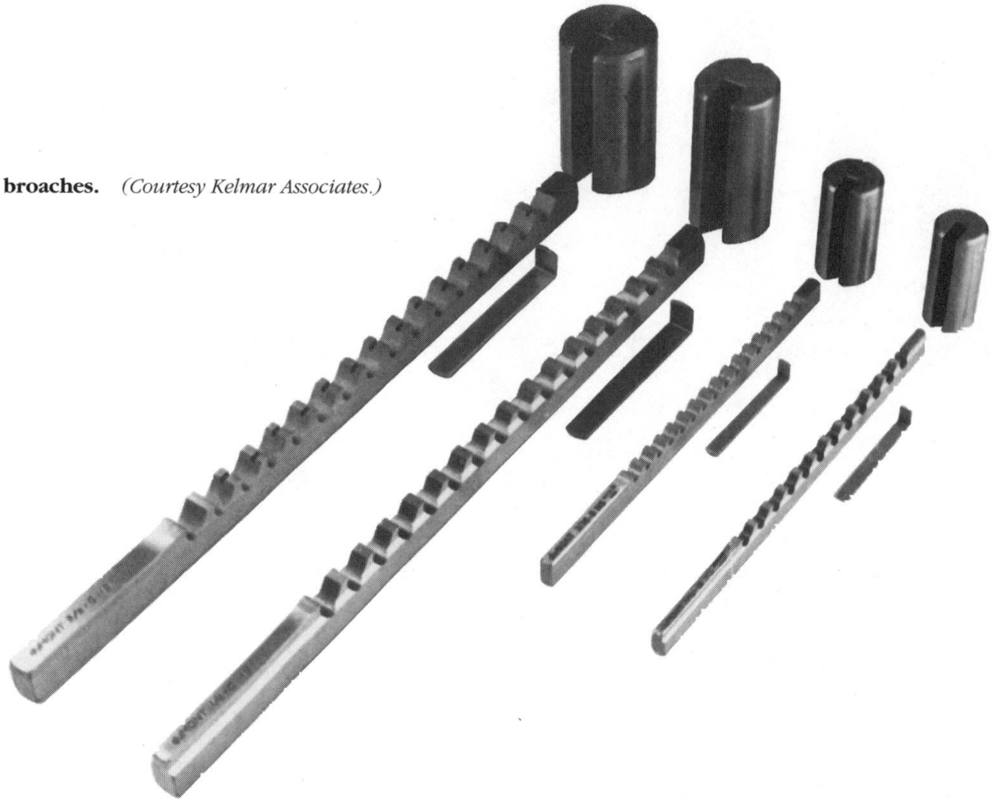

broaches. *(Courtesy Kelmar Associates.)*

work is held against a cloth, felt, or leather buffing wheel, to which a compound containing a very fine abrasive has been applied.

burnishing Finishing metal by contact with another, hardened, metallic surface that compresses the surface of the softer metal to produce a shiny surface.

burr The ragged, sharp edge on a metal workpiece as a result of cutting, machining, grinding, or punching.

bushing **1.** A removable sleeve or lining for a bearing or a drill. **2.** A hardened sleeve used to guide a drill to produce a hole in a specific location.

cap screw A finished screw, 1/4 in diameter or larger, used for fastening two pieces together.

□ The screw is passed through a clearance hole in one part and threaded into a tapped hole in the other. Heads may be of hexagon, round, flat, fillister, or socket types.

center A fixed point about which the radius of a circle or an arc moves.

□ The centers on a workpiece which are used to support the work for machining in a lathe.

center drill A combined drill (on the end) and a 60° countersink used for drilling the centers in a workpiece that is to be held between centers for machining or grinding. *(See figure.)*

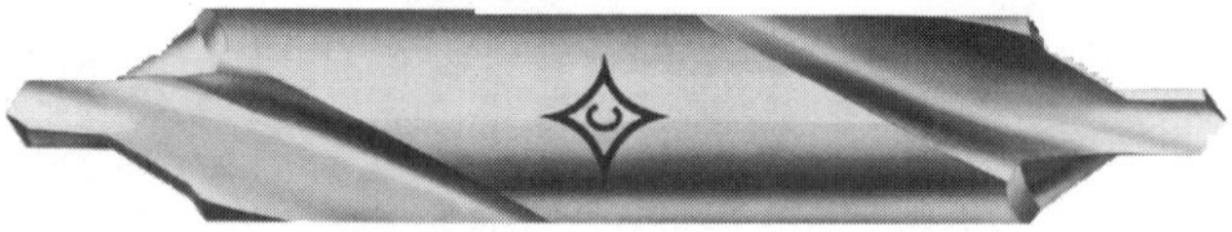

center drill. *(Courtesy Cleveland Twist Drill Co.)*

center gage A small, flat gage having 60° angles that is used for checking threading tool profiles and setting thread cutting tools in the lathe. Also used for checking the angle of lathe centers. *(See figure.)*

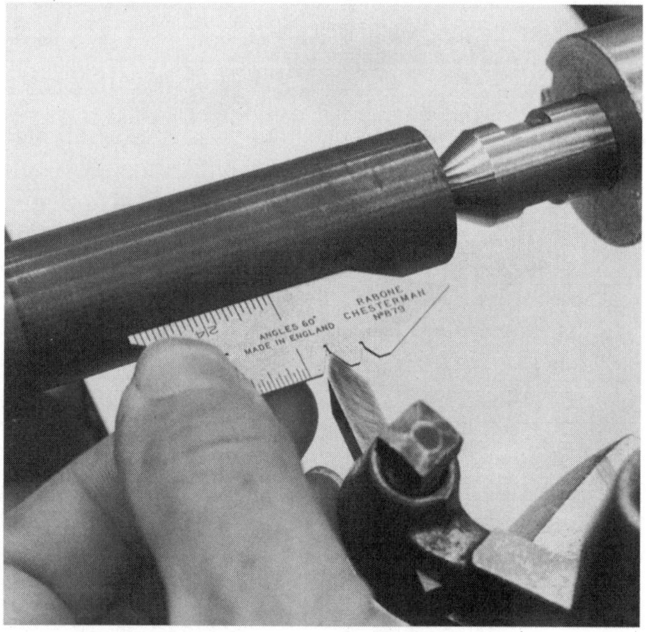

center gage. *(Courtesy Kelmar Associates.)*

center line A line used on drawings and layout work to show the centers of objects and holes. The centerline consists of alternate long and short dashes.

center punch A small, hardened steel tool tapered at one end and having a sharp conical point ground at approximately a 90° angle. The center punch is used to indent metal for starting a drill in the correct location.

center reamer A countersink having a 60° included angle for sizing and smoothing center holes in workpieces that are to be turned or ground between machine centers.

center square A tool consisting of a 90° V-shaped head with one edge of a steel rule passing through the center. It is used to locate the centers of round workpieces in layout work. *(See figure.)*

chamfer **1.** To bevel or remove the sharp edge or corner of a machined part. **2.** The beveled edge on adjoining surfaces of a workpiece.

chaser A handheld formed thread-cutting tool having a number of teeth rather than a single point; generally used on a lathe to "clean up" a thread form.

chasing threads Cutting threads on an engine lathe.

□ Producing threads by the use of self-opening dies and collapsible taps which have inserted cutting blades called *chasers*.

chatter Vibrations which develop during machining operations between the cutting tool and workpiece, producing a rough wavy surface on the workpiece.

chipping The operation of cutting or removing metal with a cold chisel.

circumference The length of a boundary line forming a circle or enclosing any area.

circumscribe To enclose within certain lines or boundaries.

□ To draw around the outside of another figure so that it touches at several points without intersecting, such as drawing a circle around a triangle.

clearance angle An angle ground on a cutting tool to allow it to cut metal and provide chip clearance.

concave A curved depression, such as the inner surface of a bowl.

concentric Two or more circles having the same center.

cone A solid figure that tapers uniformly from a circular base to a point (vertex).

contour A line, or outline, that shows the shape of an object.

convex An outside curve, such as the outer surface of a bowl.

corundum Extremely hard aluminum oxide used in the manufacture of grinding wheels, abrasive stones, and abrasive cloth.

counterclockwise From right to left in a circle, or in the opposite direction of the hands of a clock.

countersink **1.** To machine a conical recess in a hole for the head of a screw or rivet. **2.** A drill used to produce conical holes.

crankshaft A driving or driven shaft having one or several offset cylindrical bearing surfaces used for changing reciprocating motion to rotary motion or vice versa.

cross-peen hammer A hammer with a wedge-shaped peen at one end that is at right angles to the direction of the handle.

dead smooth file A Swiss-type file of the finest cut.

demagnetizing Removing the magnetism from the workpiece after it has been held on a magnetic chuck.

depth gage A measuring tool used to determine the depth of a hole, slot, or groove, or the length of a shoulder.

detail drawing A drawing that shows and describes a single part or several parts individually and gives all the necessary information needed for fabricating these parts. It may be part of an assembly drawing.

center square. *(Courtesy Kelmar Associates.)*

die **1.** A tool used to form and blank metal parts. It consists of a punch and a die usually mounted in a punch press to punch or stamp out duplicate pieces in quantity. **2.** Also the name of a tool used to cut screw threads on the outside of a rod or cylinder.

divider A compass-type instrument used for measuring and/or setting off distances and laying out circles and arcs.

dog A tool that is clamped on a workpiece to drive it while held between centers, such as a lathe dog.
- □ A projecting part on a machine tool that strikes and moves another part, such as the reversing dogs on a grinding machine.

draw chisel A pointed cold chisel, usually diamond- or V-shaped, used for moving the center of a hole being drilled to its correct location.

drawfiling A method of filing flat surfaces to a smooth finish by holding the file near each end and pushing and pulling it over the length of the workpiece.

driftpin A round tapered steel pin used to align rivet holes to accept the rivet.

drill drift A wedge-shaped piece of steel that is used to remove tapered shank tools from spindles, sockets, and sleeves.

drill gage A thin, flat steel gage having accurately machined holes for various sizes of drills, with the size of each hole marked on it. It is used for checking the diameters of drills.

drill grinding gage A tool for checking the angle and length of the cutting lips on a twist drill when it is being ground. *(See figure.)* Also called a *drill-point gage* or a *drill-angle gage.*

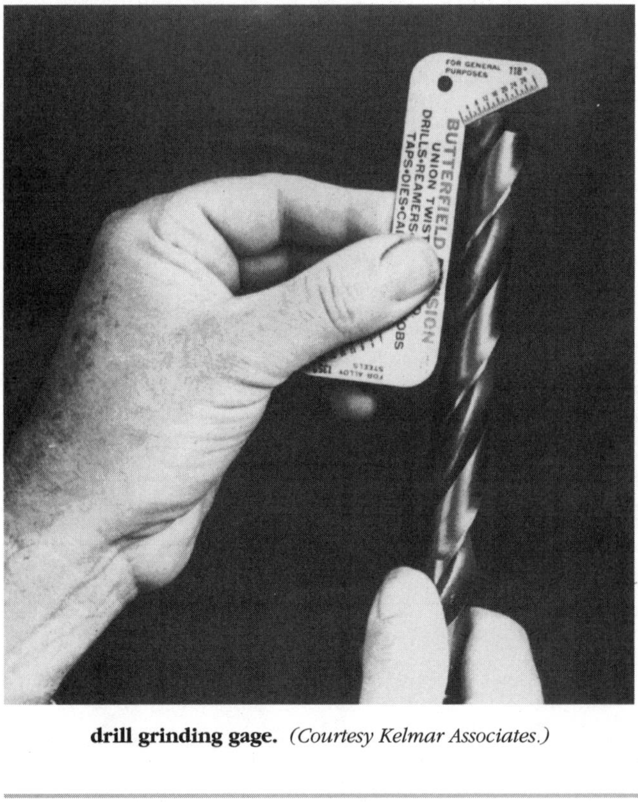

drill grinding gage. *(Courtesy Kelmar Associates.)*

drive fit One of the several classes of fits in which two parts are assembled or held together by forcing one part into another with hammer blows.

eccentric A device used to convert rotary motion to reciprocating motion on an engine or machine.

emery A natural abrasive material, dark in color, and very hard and brittle. Used for grinding and polishing.
- □ It has been replaced by manufactured abrasives such as aluminum oxide and silicon carbide.

emery cloth Cloth coated with emery grit or powder that is used for polishing and removing very small amounts of metal.
- □ Although artificial abrasives have generally replaced emery, the term *emery cloth* is often incorrectly used for the term *abrasive cloth.*

emery wheel A grinding wheel made from emery or natural abrasive. The term *emery wheel* is still widely and incorrectly used to indicate the wheel of an off-hand tool grinder even though such grinding wheels are now generally made of artificial abrasives.

expansion reamer A type of hand or machine reamer in which the diameter may be slightly increased by an expanding screw or wedge. *(See figure.)*

expansion reamer. *(Courtesy Cleveland Twist Drill Co.)*

external The outside surface of any workpiece.

Ezy-Out A tapered spiral plug-type tool, having left-hand spiral flutes; used for removing a broken stud or bolt from a hole.
- □ A hole is drilled into or through the stud remaining in the hole. The Ezy-Out is then screwed into the hole in a counterclockwise direction and the broken portion is removed from the hole.

FAO An abbreviation used on detail drawings to indicate that the part must be "finished all over."

feather A sliding key or spline which is used to prevent a pulley or gear from turning on a shaft but allowing it to move along the shaft. The feather is usually fastened to the sliding part.

feeler (thickness) gage A tool having several blades of different thicknesses varying from .0015 in to as much as .200 in.
- □ Generally used to check the clearance between mating parts and for gapping spark plugs.
- □ Paper strips are often used in shop work as feelers to establish clearance between parts.

file test for hardness An approximate method of testing the hardness of a heat-treated part by rubbing and pressing a file edge or corner against the hardened surface to see if it nicks the metal.

filing The shaping and sizing of workpieces by removing metal with a hand file. Files are manufactured in many shapes, sizes, and degrees of coarseness.

fillet A convex surface connecting two surfaces that meet at an angle.

fit The amount of clearance or interference between two mating parts when assembled. There are two main classes of fits: clearance and interference.

flush When the surfaces of adjacent parts are even with each other, they are said to be *flush*.

force fit A class of fit in which one part is forced into another part to become one unit. Kinds of forced fits are light-, medium-, and heavy-drive fits, and press fits.

- □ The difference between each depends on the standard limits required between the mating parts to produce the correct fit.

forge **1.** To form or shape hot metal by hammering or squeezing. **2.** A device for heating metal, such as a blacksmith's forge.

free fit A class of fit used where accuracy is not essential, or where high temperature variations are likely to occur in running fits of machine parts.

gage A hardened master tool that can be used to check the accuracy of a hole, dimension, or threads.

gib An angular and tapered strip of metal placed between two sliding bearing surfaces to ensure a proper fit and provide adjustment for wear.

graduate **1.** To divide into equal parts by engraving or cutting lines or graduations into the metal, such as the graduations on a steel rule. **2.** A vessel marked in units for measuring liquids.

hacksaw **1.** A light frame that holds a metal-cutting saw blade for sawing metals by hand. **2.** A reciprocating-type machine which uses a larger, heavier saw blade to cut metal.

hand tap A hardened and tempered steel tool for cutting internal threads. *(See figure.)* It has a thread cut on the circumference and has three or four flutes that produce the cutting edges.

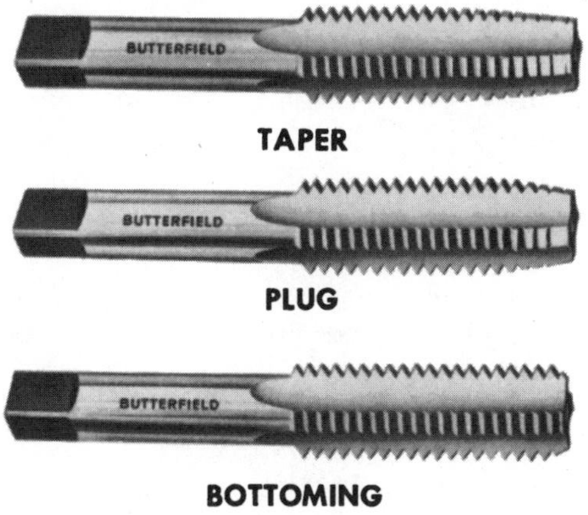

hand taps. *(Courtesy Union Butterfield Corp.)*

- □ A set of hand taps consists of the taper, plug, and bottoming taps.
- □ A square at the end of the shank provides a means for turning the tap by hand.

hand tools A general term which applies to a wide variety of small tools that are hand- rather than power-driven. Examples are layout tools, wrenches, and hammers.

hexagon A regular polygon having six equal sides; the length of a side is equal to the radius of the circumscribed circle.

hex nut A nut having six sides of equal length and shaped like a hexagon.

honing A process of finishing internal or external cylindrical surfaces to a high degree of accuracy and finish, not possible by any other method.

- □ Bonded abrasive sticks are applied to the surface under controlled pressure using a combination of rotary and reciprocating motions.

inspection The process of measuring, testing, or gaging workpieces to make certain that each dimension is within the specified size and limits shown on the print.

internal threads Screw threads cut inside a hole as in a nut. *(See figure.)*

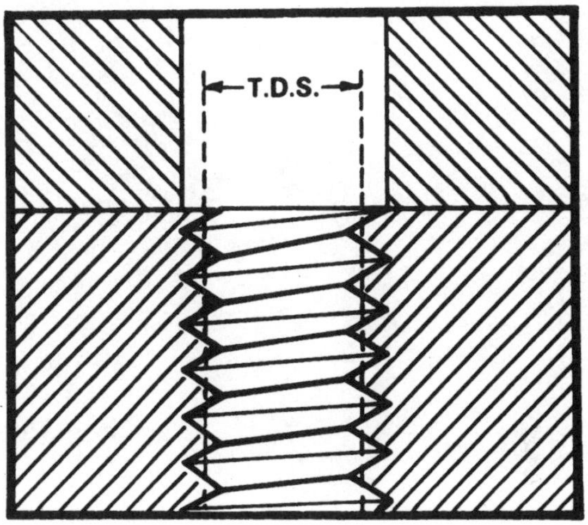

internal thread. *(Courtesy Kelmar Associates.)*

jack A mechanical device used for lifting heavy loads through short distances with a minimum expenditure of manual power. They may be operated by levers, a screw, or a hydraulic system.

jackscrew A small jack which uses a screw thread to level and support workpieces for machining purposes.

Jarno taper A standard taper having .600 in of taper per foot. The formulas used for Jarno calculations make it easy calculate the length and large and small diameters.

journal The part of a shaft or axle that has been machined and finished to fit into a bearing.

journal box A housing and support for a bearing that supports the journal of the shaft.

key A piece of metal, usually rectangular, that fits into a keyseat in a shaft and projects above the shaft. This fits into a mating slot in the center hole of a gear or pulley to provide a positive drive between the shaft and the gear or pulley.

keyseat The slot or recessed groove cut in a shaft or gear hub to receive the key. Also called a *keyway*.

lap A tool used for finishing internal and external surfaces of workpieces. The contacting surfaces of the lap, which are charged (loaded) with a fine abrasive powder, are made from a material softer than the surface being lapped.

lapping A process of precision finishing surfaces with fine abrasives such as diamond powder or abrasive flours.

□ This may be done by hand or machine to improve the size, shape, and finish of workpieces.

layout The operation of scribing or drawing lines to show the exact shape, size, and location of holes, circles, angles, and other surfaces that require machining, bending, or forming. *(See figure.)*

layout. *(Courtesy Kelmar Associates.)*

layout plate A smooth, accurately finished flat steel plate which may be used for layout operations on certain workpieces.

lead hammer A soft hammer made of lead for tapping and adjusting workpieces to prevent damage to the finished surface, and to accurately seat the part prior to machining.

left-hand screw A thread that advances into a mating part when turned to the left or counterclockwise.

lever A simple device, consisting of a rigid bar which pivots on a fixed support (fulcrum), that is used to obtain mechanical advantage when lifting or moving heavy objects.

limits or tolerance The size or variation from the nominal size that a part may be machined to and still be acceptable.

locknut A type of nut that is prevented from loosening under vibration.

□ A second nut, threaded onto a bolt or stud and jammed against the main nut to prevent it from loosening.

□ A type of nut that uses a tight-fitting fiber ring on the upper part of the thread that prevents the nut from backing off.

longitudinal movement A lengthwise movement, such as the feed of a lathe carriage along the ways.

machinist A person trained during apprenticeship to set up, adjust, and operate all types of machine tools skillfully.

□ The machinist must also be able to read and understand prints and plan the required procedures for doing a job.

□ This person should also have a knowledge of metallurgy and heat-treating processes.

male part The external part of any workpiece that fits into a hole, slot, or groove of the mating part.

mandrel A hardened and ground shaft that may be tapered, slightly tapered, and/or threaded, which is used to support and rotate a workpiece between the centers of a lathe. *(See figure.)*

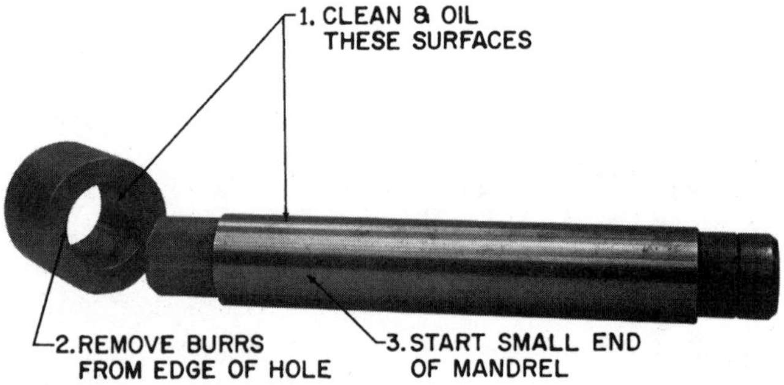
1. CLEAN & OIL THESE SURFACES
2. REMOVE BURRS FROM EDGE OF HOLE
3. START SMALL END OF MANDREL

mandrel. *(Courtesy Kelmar Associates.)*

□ The part to be machined may be held on the mandrel by a press fit, or possibly by the nut.

□ A device fitted into the lathe headstock taper and threaded on the outer end to hold small parts for facing.

mechanical advantage In a machine, this is the ratio of the energy required to produce work (output) compared to the applied force (input).

morse taper A self-holding, standard taper generally used on drilling tools, drilling machine spindles, and lathes.

music wire A high-carbon select steel wire used for making mechanical springs.

nut An internally threaded metal fastener of hexagonal, square, or other shape, that screws onto a bolt, stud, or arbor.

object lines The heavy, full lines on a drawing that indicate the shape of the object.

off-center Not on the true centerline or axis; offset, eccentric, or inaccurate.

oil stones Abrasive materials such as silicon carbide and aluminum oxide combined with a bonding material and molded into sticks or blocks of various shapes for sharpening and honing cutting tools.

orthographic view A method used in drafting that

uses several drawings, generally three, to show the profile and dimensions of the front (face), top, and right-side view of an object. *(See figure.)*

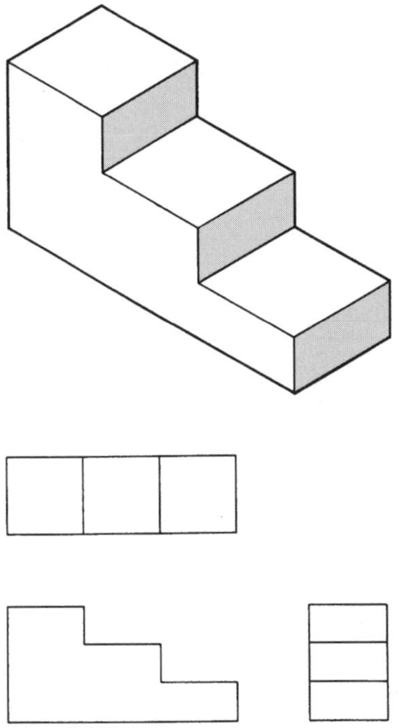

orthographic view. *(Courtesy Kelmar Associates.)*

□ A three-dimensional (pictorial) view of an object from which the orthographic views are developed.

out-of-true Eccentric, inaccurate, or out-of-balance.

overall dimensions Dimensions that give the entire length or width of an object as contrasted with detailed dimensions such as the location of a hole, or thread size.

parallel Two lines in the same plane equidistant from each other at all points over the entire length.

parallels Flat metal pieces of suitable length usually made in pairs, hardened and precision-ground, and having precisely the same sizes over all corresponding sides and with all sides square to each adjacent surface. *(See figure.)*

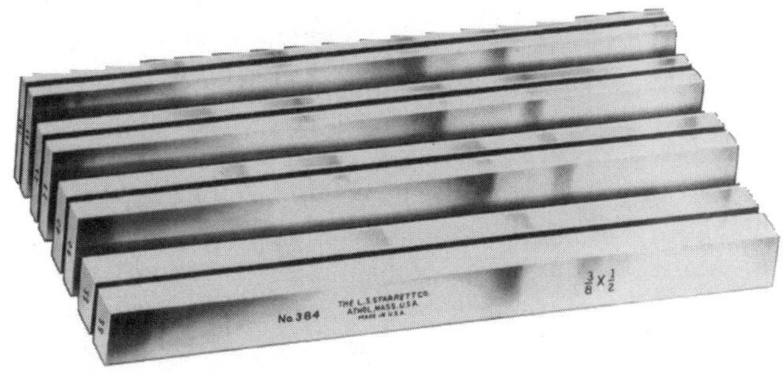

parallels. *(Courtesy L.S. Starrett Co.)*

□ Used for supporting and seating work in a vise or on a machine table, or for inspection work on a surface plate.

peen The end of a machinist's hammer opposite the face, such as ball, straight, or cross-peen, that is used for peening or riveting.

periphery The circumference of a circle, ellipse, or similar figure.

perpendicular Referring to a line or surface that meets another line or surface at right angles (90°).

press fit The operation in which one part is pressed into another to a degree of tightness that produces a more or less permanent assembly.

prick punch A hardened metal punch ground to an angle of 30° to 60° that is used to permanently mark the location of layout lines.

production The entire process of producing a manufactured product which involves planning, materials, tools, equipment, and the sequence of operation.

□ To produce many parts as opposed to several single parts.

pulley A wheel having a plain or V-groove rim over which a belt runs for the transmission of power from one shaft to another.

punch **1.** The male part of a punch and die that pierces, blanks, or stamps out various-shaped pieces of sheet metal in a punch press. **2.** Various hand tools such as center punch, pin punch, or prick punch.

push fit The type of fit that permits one part to be pushed into another part by hand without interference or looseness.

radial Arranged outward from the center, as the spokes of a wheel.

radius The distance from the center of a circle to the circumference, which is equal to one-half the diameter.

reamer A precision cutting tool for finishing and sizing a drilled hole smoothly and accurately by removing a small amount of metal.

recess A groove cut below the normal surface of a cylindrical workpiece.

rivet A one-piece fastener which has a head and a body and is used for fastening two or more pieces together. The body is passed through a hole in each piece, and the protruding end of the body is peened to form a second head which holds the parts together. *(See figure.)*

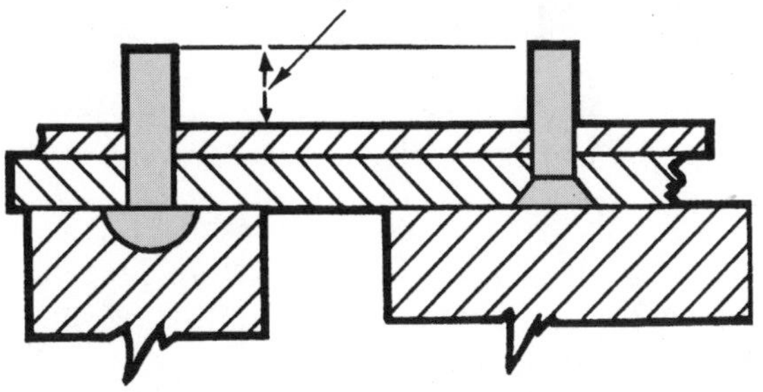

riveting. *(Courtesy Kelmar Associates.)*

round-nose chisel A cold chisel having a rounded cutting edge for chipping oil grooves in bushings or internal bearing surfaces.

running fits A series of standard fits that provide for clearance between the mating parts and allowance for lubrication.

sandblast A process in which sand is blown by compressed air against the surfaces of the workpiece in order to remove dirt, rust, and scale to permit further operations on the workpiece.

scale **1.** The outside rough, sandy coating of a casting. **2.** A measuring instrument used by drafters. **3.** An incorrect shop term for a steel rule. **4.** The information on a print that indicates size of the part shown, such as full-scale, half-scale (1:2), or quarter-scale (1:4), etc.

scraping A hand-tool- or power-driven operation controlled by hand for removing the high spots from the surfaces of round or flat bearings, precision surface plates, and machine-tool slides and ways.

screw A helix formed or cut on the outside of a cylinder that advances along the axis to the right or left.

scriber A slender, hardened steel rod, with or without a handle, having a sharp point on one or both ends, used for drawing lines on metal surfaces when laying out work.

scribing The operation of drawing or marking lines on a metal surface for layout work.

sectional view Used on a print to show interior details that are too complicated to be shown by outside views or hidden lines. *(See figure.)*

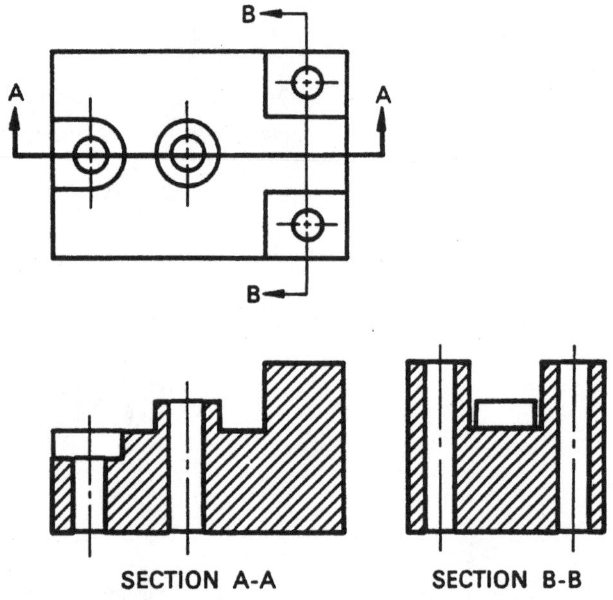

sectional view. *(Courtesy Kelmar Associates.)*

sector A device mounted on the front of the index plate of a dividing head for indicating the number of holes to be included for each advance of the index crank when dividing circles.

□ The sector arms can be adjusted and set to form any part of a complete turn required to cover the desired number of holes for each movement of the crank.

selective assembly A procedure used when conditions affecting the sizes of mating parts become so exacting that they are difficult to maintain. The completed parts are sorted or mated as to size for assembly purposes.

setscrew Usually a hardened steel screw having either a square head or a recess on the top end to receive an Allen or Torx type of wrench, that is used to tighten the setscrew and secure the machine part in place.

setup The procedure for mounting the work and setting the cutting tool when preparing a machine tool for a particular operation or operations.

shaft A length of cylindrical steel of suitable diameter on which pulleys, gears, clutches, or other parts are mounted to transmit power by rotation.

shims Thin pieces of sheet metal of various thicknesses that are placed between parts to obtain the desired clearance.

shoulder screw A screw having two or more diameters or shoulders that is used for supporting levers and other machine parts that must operate freely.

□ The screw body (unthreaded section) is slightly longer than the thickness of the piece which pivots on the screw to allow the piece to move freely.

shrink fit A process in which a part is bored to slightly less than the shaft diameter, then heated and slid onto the shaft to the proper location and cooled. This causes the part to shrink and grip tightly on the shaft.

slide fit A fit that provides enough clearance between mating parts to allow free movement and accurate location.

soft hammer A hammer—the head of which is made of copper, lead rawhide, or plastic—used to prevent marring or damaging finished surfaces of workpieces or machine parts.

soldering A method of joining metals together by means of a nonferrous alloy (solder) whose melting point is below 800°F. A flux is used to dissolve oxides and prevent oxidation of the work surfaces during the operation.

spanner wrench A type of wrench having C-shaped ends with a pin or a hook, used for tightening or loosening threaded circular collars that have either slots or holes to receive the hook or the pin on the wrench.

spring A device made of hardened steel wire or thin flat strips in various shapes or forms that have elastic characteristics *(see figure on next page)*. There are several types of spring:

Compression spring. A helical cylindrical spring that tends to shorten in action and produce a pushing effect.

Leaf spring. A built-up spring made of several layers of flat stock clamped together in which each adjacent leaf is a little shorter than the next, as in an auto spring.

Spiral spring. A flat spring wound in a continuous spiral with one coil around the other, as in a clock spring.

Tension spring. A helical spring that tends to lengthen in use, thus producing a pulling action.

Torsion spring. A helical spring that operates with a coiling or uncoiling action, as in the hinges used for light self-closing doors.

springs. *(Courtesy Dayton Progress Corp.)*

square A tool (solid or adjustable), used for laying out, inspecting, and testing the squareness of workpieces.

standard tapers Any of the numerous tapers specified in the American Standard System of Tapers that include the self-holding such as the Morse, Brown & Sharpe, Jarno, and steep tapers, or the self-releasing tapers used on lathe spindles.

step block A block of steel or cast iron having a series of steps that is used for supporting the ends of machine clamps when clamping work to the table. *(See figure.)*

step blocks. *(Courtesy Northwestern Tools, Inc.)*

stock A term used to indicate bars of metal used to make workpieces.

stock and die A threading die, mounted in a hand holder or stock, that is used for turning the die.

straightedge A hardened and ground flat piece of steel that is used for testing the flatness of surfaces.

strain A force or pressure that tends to cause a part to resist distortion.

stress The force acting on a part that causes it to distort or deform.

stud A plain cylindrical piece having a threaded portion of suitable length at each end or a continuous thread for the entire length. One end is screwed into a machine or workpiece, and a second part is placed over the stud and held in place by a nut.

superfinish A surface-improving process, similar to honing, that removes the relatively rough particles on the surface of the workpiece to create a highly polished surface finish which is more wear-resistant.

surface gage An adjustable hand tool consisting of a base, spindle, and hardened scriber that is used to lay out parallel lines, set up workpieces, check heights, and inspect parts. *(See figure.)*

surface gage. *(Courtesy Kelmar Associates.)*

sweating The process of soldering parts together without a soldering iron. The parts to be joined are "tinned" with a coating of solder, then brought together and heated with a torch until the solder melts and flows together to join the parts.

tangent A straight line that touches the circumference of a circle at only one point.

tap A hardened, threaded, and fluted steel tool for cutting internal threads.

- ☐ The flutes, running lengthwise, provide cutting edges for the threads, and a square at the end of the shank provides a means of turning the tap into the work with a tap wrench.

tap extractor A device or tool for removing broken taps from holes. Tempered steel fingers which extend down into the flutes of the broken portion are held in place by a sliding collar. If a wrench is attached to the extractor, the broken part of the tap in the hole may be loosened and backed out. *(See figure on next page.)*

tap wrench A hand tool that is adjustable to fit a number of tap sizes. It has handles and a means of holding the square shank end of taps for turning them into the hole to be tapped.

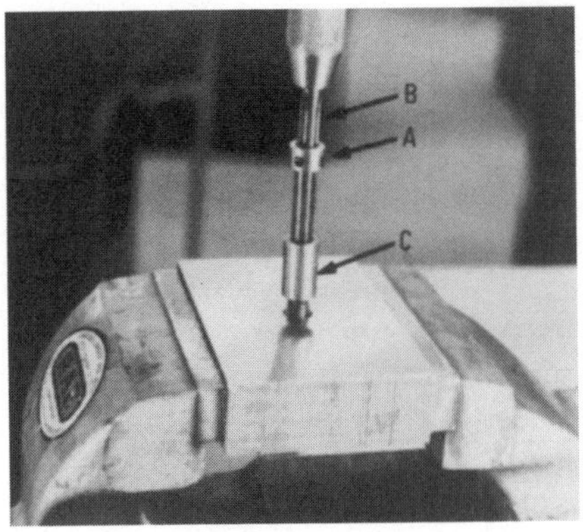

tap extractor. *(Courtesy Kelmar Associates.)*

taper A piece of work that increases or decreases uniformly in diameter or size along its length and assumes a conical or wedge shape.

taper gage A hardened and ground precision gage for checking the accuracy of a standard taper.

□ The *taper plug gage* is used to check the accuracy and fit of an internal taper.

□ The *taper ring gage* is used to check external tapers.

taper pins Standardized steel pins used for locating and holding machine parts in position on a shaft. They have a taper of 1/4 in per foot of length and vary in size from 6/0 to 10.

taper tap One of the three taps in a set of hand taps. The end threads are tapered or chamfered back for a length of six to eight threads for easy starting.

tapping The operation of producing internal threads by means of a tap.

T-bolt A threaded bolt having a square or rectangular end that fits into the T-slot of a machine table for clamping workpieces to the table.

technician A person working at a level between an engineer and a journeyperson or tradesperson, who has specialized training in one area of technology and a working knowledge of several other technologies.

□ The technician may assist the engineer in making cost estimates, preparing technical reports, or programming numerically controlled machines.

technologist A person who works at a level between a graduate engineer and a technician.

□ The technologist may perform such duties as design studies, production planning, laboratory experiments, and supervision of technicians.

template A flat pattern or master plate, usually made from thin steel or sheet metal, that is used when laying out a number of workpieces for drilling or machining operations.

tension A pulling force that tends to elongate a body to which it is applied.

□ For example, a coil spring will elongate under tension, and a test sample in a tensile tester will elongate under tension.

thumbscrew A type of screw having a winged or knurled head for turning with the thumb and forefinger when quick and light clamping is required.

tolerance The permissible variation from the basic dimension of a part.

tool-and-diemaker A skilled craftsperson who has above-average mechanical ability and can operate all machine tools.

□ This person must also have a broad knowledge of shop mathematics, print reading, machine drafting, design principles, machining operations, metallurgy, heat treating, computers, and process planning.

torque The turning or twisting effort which a shaft sustains when transmitting power.

try square A small square with a fixed blade used for laying out lines and testing right angles (90°). *(See figure.)*

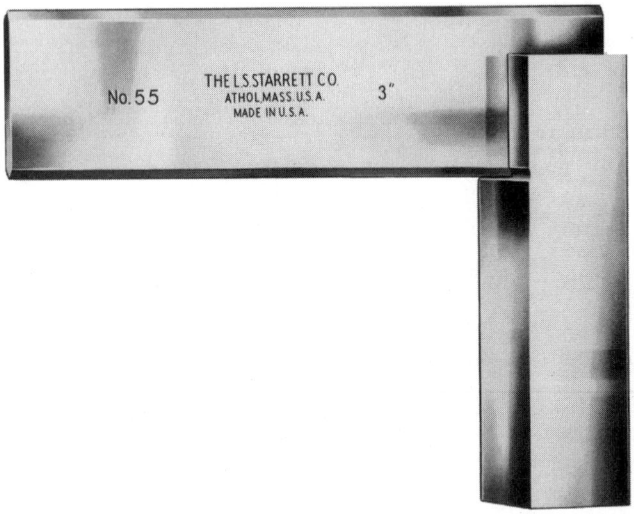

try square. *(Courtesy L. S. Starrett Co.)*

U-bolt A threaded rod which is bent into the shape of the letter *U,* with both ends threaded.

U-clamp A square piece of metal bent into a U shape. Used for clamping workpieces onto a machine table.

undercut A groove or recess cut adjacent to the shoulder of a workpiece to provide clearance for a grinding wheel when finish grinding a part. Also used to permit a mating part to fit squarely against a surface.

unified threads A system of screw-thread size and pitch combinations adopted jointly by the United States, Canada, and Great Britain, in which the angle of the thread is 60°, the crest of external threads may be either flat or rounded, and the root is rounded in both external and internal threads. *(See figure on next page.)*

□ This thread is a combination of the British Standard Whitworth and the American National Thread forms. It was adopted to achieve interchangeability of parts during World War II.

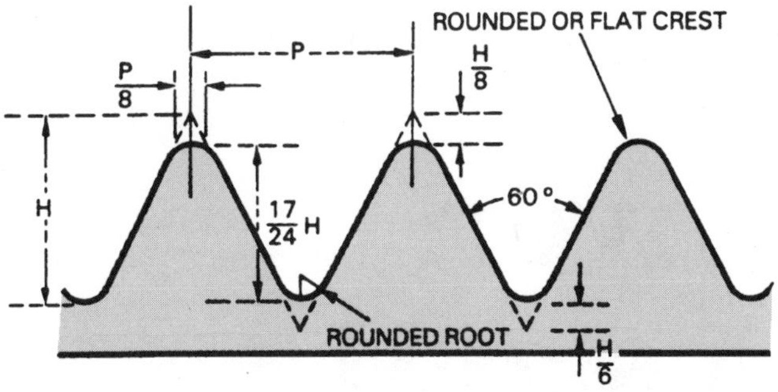

unified thread. *(Courtesy Kelmar Associates.)*

V-blocks Rectangle-shaped blocks of steel that are usually hardened and accurately ground. They have a 90° V-groove through the center and slots in the sides to accommodate clamps for holding round workpieces. *(See figure.)*

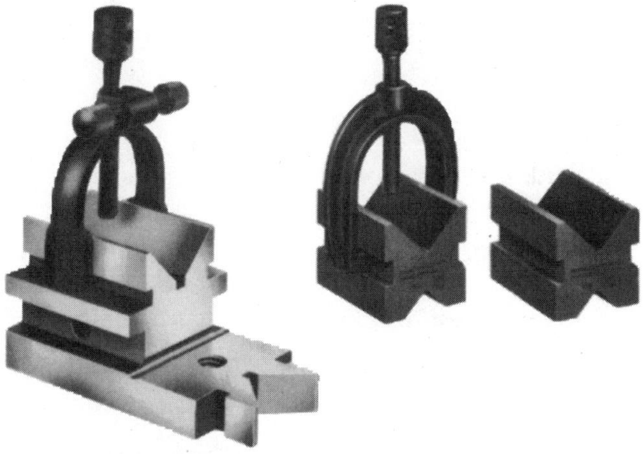

V-blocks. *(Courtesy Brown & Sharpe Manufacturing Co.)*

□ Used for laying out and holding round work for various layout and machining operations.

vise A mechanical device used to clamp work for hand or machining operations.

□ Vises are generally made of malleable iron or steel and have two jaws: one fixed and one movable.

□ The movable jaw may be moved inward to clamp the work by means of a screw, air pressure, or hydraulic pressure.

vixen file A flat file with curved-cut teeth that is generally used for filing soft metals and round work in a lathe.

washers There are two basic types of washers.

□ *Flat washers* are small flat disks with a hole and are used to provide a flat surface between the bolt head or the nut, and the workpiece. They distribute the fastening force over a larger area, and also prevent damage to the work surface.

□ *Spring-lock* or *tooth-type* lock washers are used to prevent the nut or machine screw head from loosening.

wedge A device, usually made of wood or metal, having two faces which form an acute angle (less than 90°). May be used to create great pressure for locking purposes, for splitting material, or for leveling workpieces.

wire gage A flat, circular metal disk having a series of notched slots used to measure the diameter of wire or the thickness (gage) of sheet metal.

Woodruff key A flat, semicircular piece of metal used as a key in a mating keyseat slot cut in a shaft to drive a gear, pulley, or other part.

work A mechanical principle which indicates the overcoming of resistance. Work is measured in foot-pounds (force×distance through which it acts).

working drawings Drawings that give complete information concerning the shape, size, and specifications of a single part or of a complete assembly.

wrenches Hand tools used for turning bolts and nuts and other fasteners used to hold machine parts together.

wringing fit A class of fits that require a twisting motion when the parts are being assembled.

wrist pin A steel pin connecting the rod to the piston of a gas-powered engine.

SECTION 7

Fluid Power: Hydraulics and Pneumatics

Dr. George C. Ku
Professor, Technology Education
Central Connecticut University
New Britain, Connecticut

The two main divisions of fluid power are hydraulics and pneumatics.

- *Hydraulics* is the branch of science that stores a fluid, moves it from one point to another, and controls it to transmit force and motion. Hydraulics is best suited for situations that require precise control of speed or position.
- *Pneumatic systems* use air or some other gas such as natural gas or carbon dioxide as the power-transmitting medium. Pneumatics are best suited for clamping, feeding, ejecting workpieces, and air motors that run at high speeds.

absolute Free or independent of any positive or negative pressure (vacuum). A measure using its zero point as the base—a complete absence of the entity being measured.

absolute pressure The air above us creates atmospheric pressure [14.7 psi (lb/in^2)] on the earth's surface. *(See figure.)*

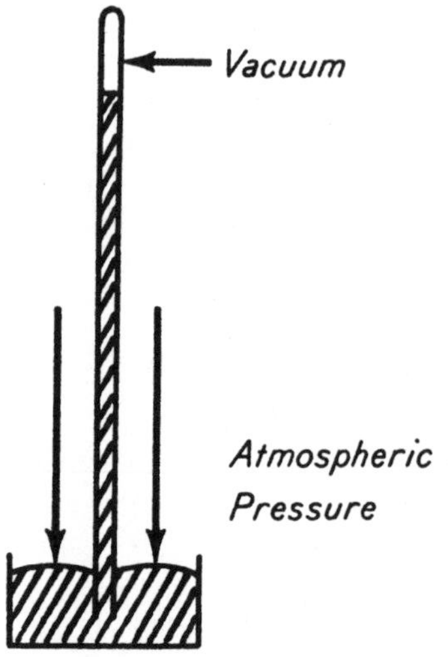

absolute pressure. *(Courtesy General Motors Corp.)*

□ A column of air one square inch in cross section supports a tube of mercury at 14.7 psi at sea level.

□ Any condition less than atmospheric pressure is a vacuum or partial vacuum.

accumulator A container in which fluid is stored under pressure and used as a source of energy in fluid power. *(See figure at top next page.)*

□ It contains a piston or a bladder-diaphragm and can be loaded by spring, weight, or compressed gas to supplement pump delivery during the peak demand or to be a source of emergency power.

actuator A linear or rotary device that converts hydraulic or pneumatic energy into mechanical energy. *(See figure on next page.)*

□ A linear actuator is a cylinder or ram and can supply force and/or motion outputs in a straight line.

□ A rotary actuator or motor produces torque and rotating motion.

aeration Aeration is air in the hydraulic system that will cause noise and damage a pump because it lacks lubrication.

□ Excessive aeration causes the fluid to appear milky and components to operate erratically as a result of the compressibility of the air.

amplifier A relatively simple circuit used to raise the level (or increase the amplitude) of an electronic signal. The amplified output signal is proportional to the input signal.

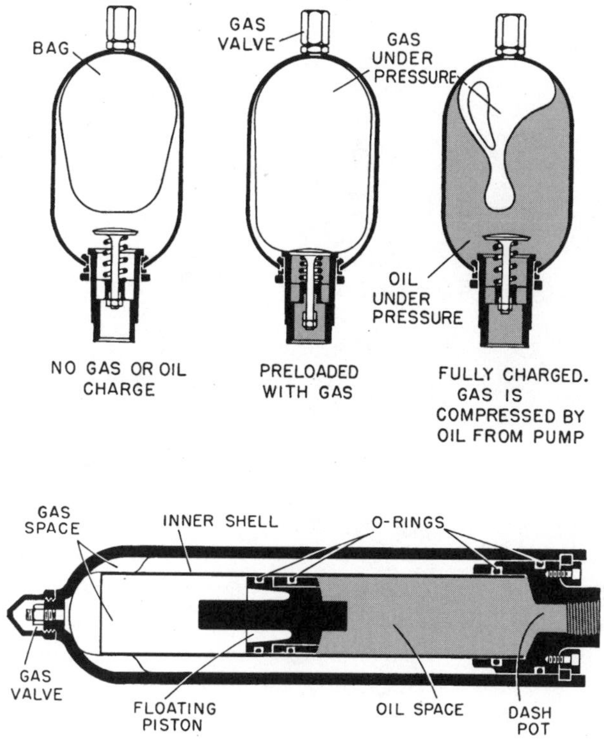
BAG
GAS VALVE
GAS UNDER PRESSURE
OIL UNDER PRESSURE
NO GAS OR OIL CHARGE
PRELOADED WITH GAS
FULLY CHARGED. GAS IS COMPRESSED BY OIL FROM PUMP
GAS SPACE
INNER SHELL
O-RINGS
GAS VALVE
FLOATING PISTON
OIL SPACE
DASH POT

accumulator.
(Courtesy Mobil Oil Corp.)

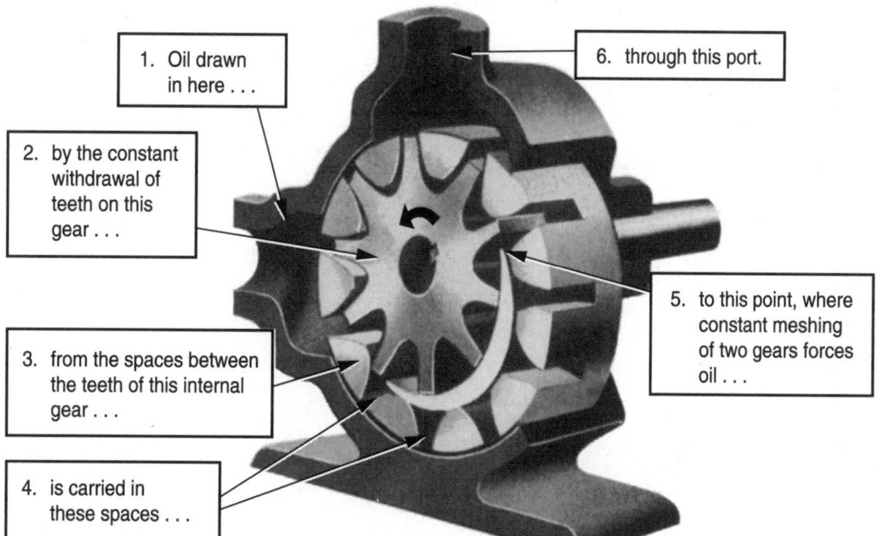
1. Oil drawn in here . . .
2. by the constant withdrawal of teeth on this gear . . .
3. from the spaces between the teeth of this internal gear . . .
4. is carried in these spaces . . .
5. to this point, where constant meshing of two gears forces oil . . .
6. through this port.

actuator. *(Courtesy Sun Oil Co.)*

amplitude of sound The loudness of a sound.

annular area A ring-shaped area that often refers to the net effective area of the rod side of a cylinder piston; an example is the cylinder area minus the cross-sectional area of the piston rod.

atmospheric A pressure of 14.7 psi produced by the atmosphere.

atmospheric pressure Pressure of the air in our atmosphere due to its weight. Atmospheric pressure at sea level is approximately 14.7 pounds per square inch absolute (psia).

□ At high altitudes, atmospheric pressure is less. Below sea level the pressure is more than 14.7 psia.

back-connected A condition in which pipe connections are on normally unexposed surfaces of hydraulic equipment. Gasket-mounted units are back-connected.

back pressure A pressure in series with the main line that usually exists on the discharge side of a load.

□ The purpose of creating a back pressure is to control the rate of a runaway load.

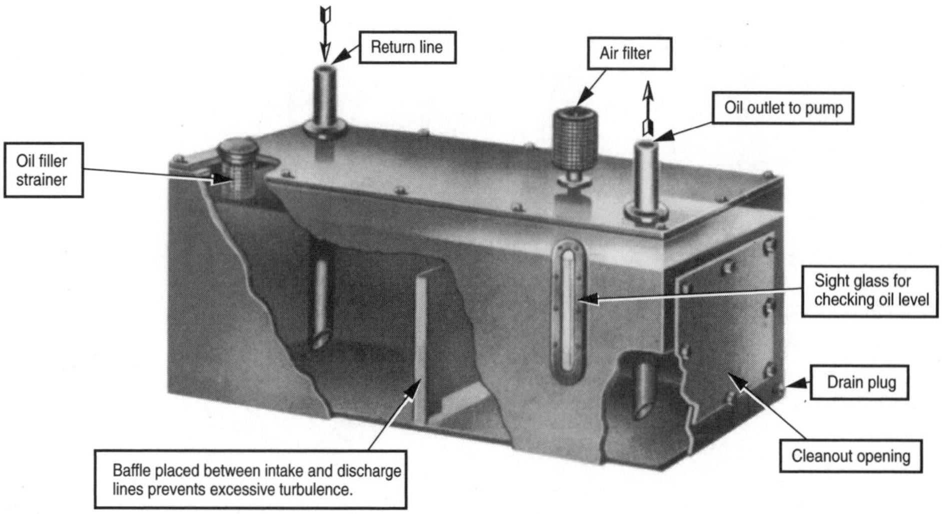

baffle. *(Courtesy Mobil Oil Corp.)*

baffle A baffle plate in the reservoir is used to separate the suction line from the return line that will cause the return oil to circulate around the outer wall for cooling before it can get to the pump again. *(See figure.)*

bleed-off Used as a method of flow control; it controls the cylinder speed by diverting the main-line oil into reservoir prior to the directional control valve. *(See figure.)*

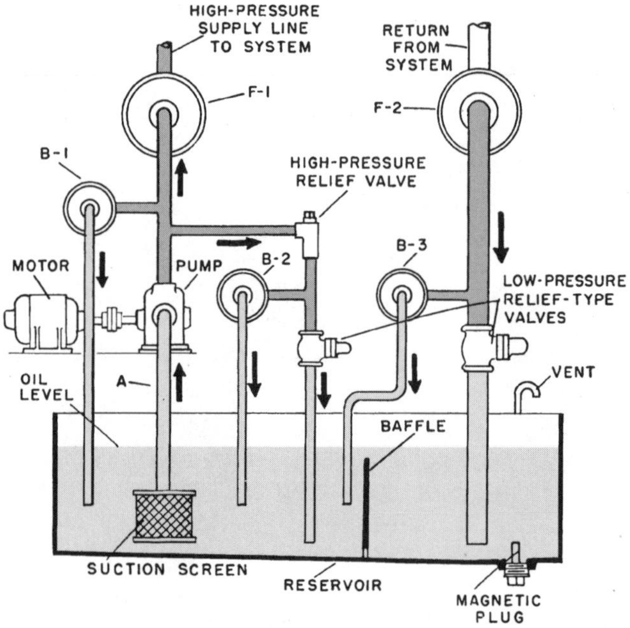

bleed-off. *(Courtesy Mobil Oil Corp.)*

□ There is no excess flow going over the relief valve, and the pump operates at only the pressure that is needed to move the workload on the cylinder, which saves energy but does not prevent a load from running away.

breather A device that permits air to move in and out of a reservoir or tank to accommodate the air exchange that results from the constant change of pressure and temperature within the tank.

□ A breather must be capable of handling the airflow required to maintain atmospheric pressure.

bypass A channel or passage to conduct fluid around the main line as the alternate route.

cartridge This term is used to denote a self-contained element that can be readily inserted in a fluid filter case, a vane pump frame, and a valve housing.

1. In a fluid filter, it is used to replace the element.
2. In a vane pump, a cartridge can alter displacement and change horsepower output.
3. A cartridge valve can be inserted into a standardized cavity in a manifold block.

cavitation A localized gaseous condition within a liquid stream that occurs where the pressure is reduced to the vapor pressure.

□ Cavitation occurs when there are air cavities and the fluid does not entirely fill the space provided for it in the pump.

□ Cavitation creates an action similar to an implosion which causes the metal parts of the pump to disintegrate, produces excessive noise, and causes pump vibration.

chamber A compartment within a hydraulic unit that may contain elements to aid in operation or control of a unit. Examples are the spring chamber and the drain chamber.

channel A fluid passage that has a greater longitudinal dimension than cross-sectional dimension.

charge To load or fill a component to the usual capacity. Examples are

1. To replenish an inlet of a hydraulic system with atmospheric pressure.
2. To refill an accumulator with fluid or gas under pressure to its capacity.

charge pressure The pressure at which replenishing fluids are forced into the hydraulic or pneumatic system (above atmospheric pressure).

check valve A simple one-way valve. The valve opens to allow flow in one direction, but closes to prevent flow in the opposite direction. *(See figure.)*

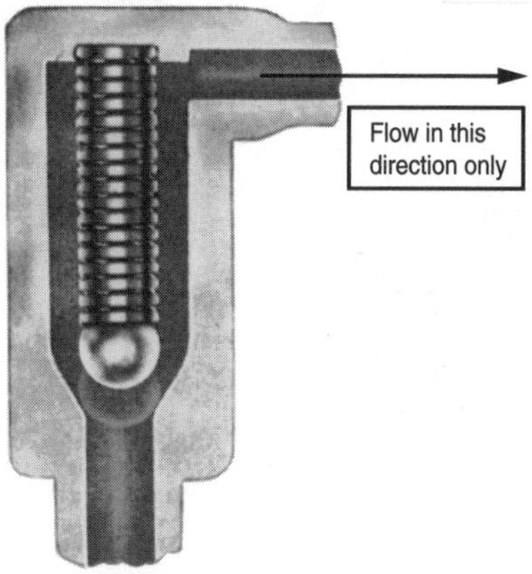

check valve. *(Courtesy Sun Oil Co.)*

□ The valve is opened by system pressure; the valve closes when inlet pressure drops.

□ Sometimes, a pilot-operated check valve allows reverse flow when the cylinder must be stroked. A pilot piston can force the check valve open during these cylinder strokes.

choke A restriction—which is greater in length than in cross-sectional dimension—that is used to slow down the movement of an actuator.

circuit An arrangement of interconnected components that function as part of a system to perform a specific task.

□ A circuit is capable of performing one or more specific tasks, but not a complete work cycle.

closed-center circuit In a closed-center circuit, all four parts are blocked in a neutral position. Trapped oil holds a cylinder piston in place, while oil stays at full system pressure. *(See figure.)*

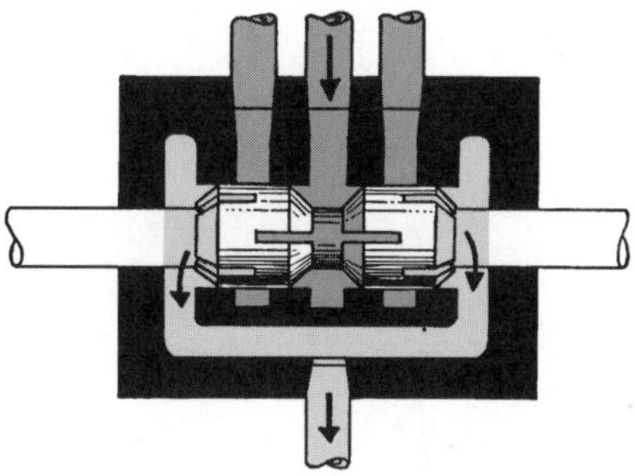

closed-center circuit. *(Courtesy Mobil Oil Corp.)*

closed-center valve In a four-way directional control valve, all four ways are blocked in the center or neutral position.

closed loop A system in which the output of one or more elements is compared to other signals to provide feedback dictating the output of the loop. *(See figure.)*

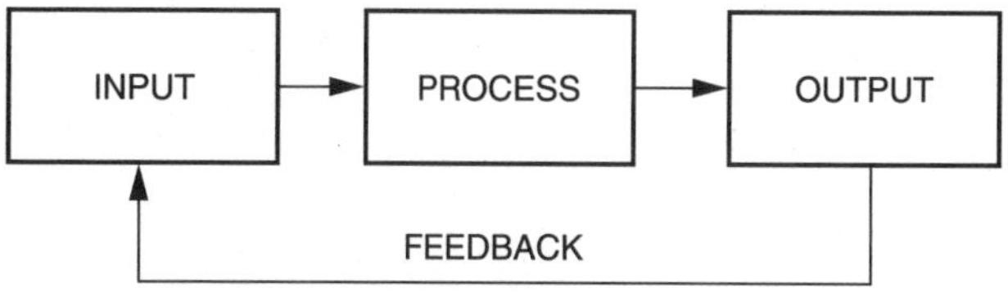

closed-loop circuit. *(Courtesy Kelmar Associates.)*

□ Hydrostatic drive and the pressure-reducing valve application illustrate application of a closed loop.

command signal (or input signal) An external signal that detects the movement of the servo.

compensator control A displacement control for variable pumps and motors that changes displacement in response to pressure changes in the system as related to its adjusted pressure setting.

component A single part of a hydraulic or pneumatic unit.

compressibility The change in volume of a unit of fluid when it is subjected to a change in pressure.

controls Devices such as multimedia valves, hydraulic control, and mechanical and manual compensators that are used to regulate the function of a unit.

□ Pneumatic control signals are often used to regulate the flow of liquids.

□ Electrically controlled valves, both electrohydraulic and electropneumatic, are becoming popular in microprocessor-based controllers.

coolers Heat exchangers, such as air-to-oil and water-to-oil, are used to remove heat from the hydraulic fluid.

1. *Air-to-oil.* The cooler has fins that direct the air over long coils of oil tubes to transfer heat from the oil to the air. *(See figure.)*

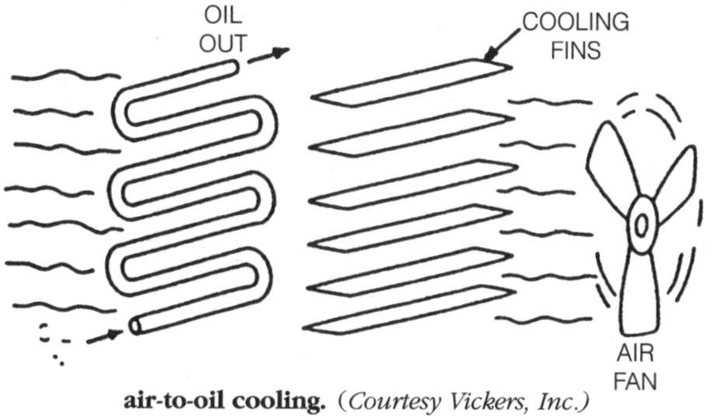

air-to-oil cooling. *(Courtesy Vickers, Inc.)*

2. *Water-to-oil.* Water flows through tubes and oil circulates around cooling tubes. *(See figure.)*

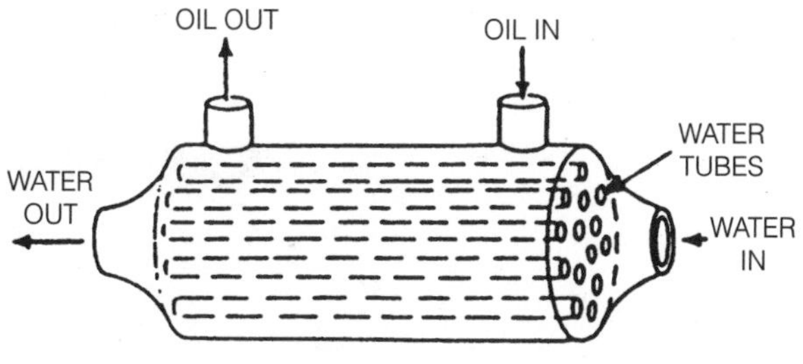

water-to-oil cooling. *(Courtesy Vickers, Inc.)*

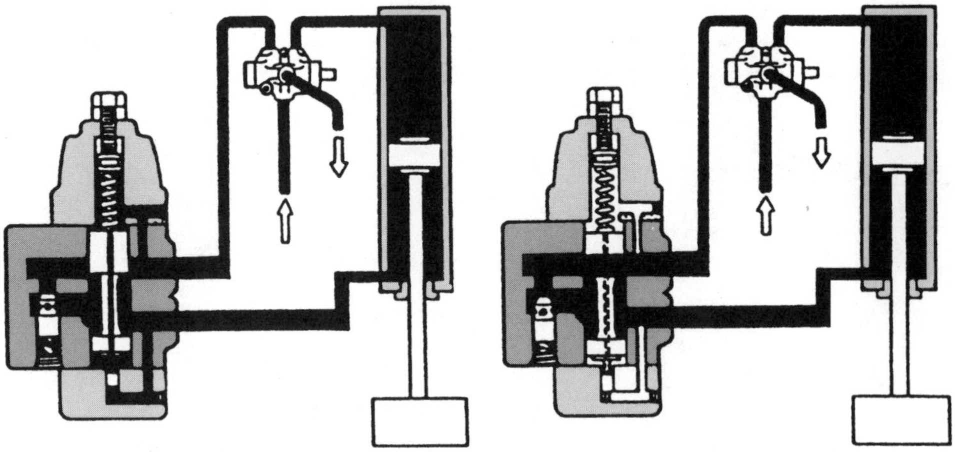

counterbalance valve. *(Courtesy Vickers, Inc.)*

counterbalance valve A pressure control valve used to maintain control over a vertical cylinder to prevent the load from falling because of gravity. *(See figure.)*

- □ The primary part of the valve is connected to the lower cylinder part and the secondary part to the directional valve. The pressure setting is slightly higher to prevent the load from falling.
- □ When the cylinder piston is forced down, pressure at the primary part increases and raises the spool, opening a discharge path through the secondary part to the directional valve.
- □ When the cylinder is raised, the integral check valve opens to permit free flow for returning to the cylinder.

coupling Sometimes called a *fluid clutch,* this mechanism joins a motor (engine) at the power-input end with some mechanism at the output end hydraulically. *(See figure.)*

- □ It produces a very smooth flow of power and protects mechanisms from the damage of vibration and shock loads.

cracking pressure The pressure at which the valve first begins to divert flow.

cushion A device built into a hydraulic or pneumatic cylinder that restricts the flow of fluid at the outlet port, thereby stopping the motion of the piston rod.

- □ Cylinder cushions are often installed at either or both ends of a cylinder to slow down the movement of the piston near the end of its stroke to prevent the piston from hammering against the end cap.

cylinder A device usually consisting of a movable element such as a piston and piston rod, plunger rod, plunger, or ram, operating within a cylinder bore, in which fluid power is converted into linear mechanical force and motion. *(See figure on next page.)*

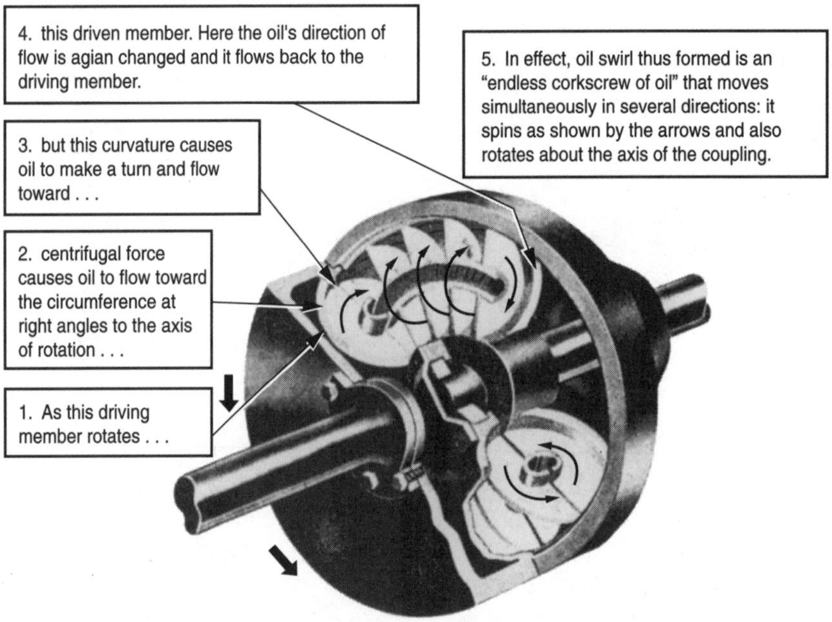

4. this driven member. Here the oil's direction of flow is agian changed and it flows back to the driving member.
5. In effect, oil swirl thus formed is an "endless corkscrew of oil" that moves simultaneously in several directions: it spins as shown by the arrows and also rotates about the axis of the coupling.
3. but this curvature causes oil to make a turn and flow toward . . .
2. centrifugal force causes oil to flow toward the circumference at right angles to the axis of rotation . . .
1. As this driving member rotates . . .

coupling. *(Courtesy Sun Oil Co.)*

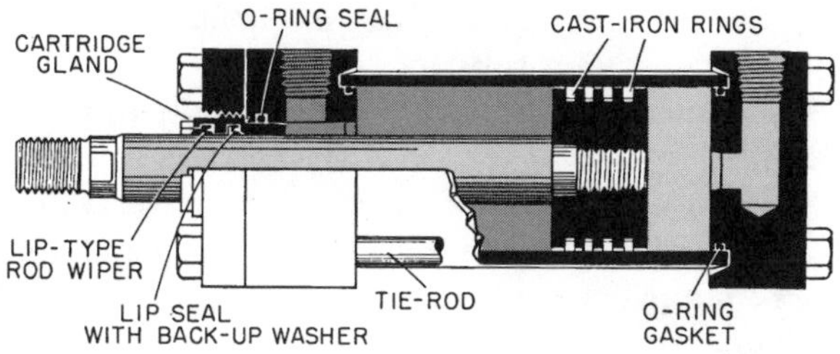

cylinder. *(Courtesy Mobil Oil Corp.)*

deadband The region or band of no response when an error signal will not cause a corresponding actuation of the controlled variable.

decompression The slow release of confined fluid to gradually reduce pressure on the fluid.

delivery The volume of fluid in gallons per minute at a specific speed, discharged by a pump in a given time.
- ☐ The pump's delivery is proportional to drive shaft speed.

devent To close the vent connection of a pressure control valve, permitting the valve to function at its adjusted pressure setting.

differential current The algebraic summation of the current in the torque motor; measured in milliamperes (mA).

differential cylinder Any cylinder in which the two opposed piston areas are not equal.
- ☐ Differential cylinder has unequal areas exposed to pressure during the extend and retract moments.
- ☐ The difference is caused by the cross-sectional area of the rod, reducing the area under pressure during retraction.
- ☐ During the extension stroke, the cylinder working area is larger than that of the piston side; therefore, the force is greater than the retraction stroke.

directional valve A valve that selectively directs or prevents fluid flow to desired channels.

displacement The amount of fluid, in cubic inches per revolution or cubic inches per stroke, that passes through a pump, motor, or cylinder in one revolution or stroke.
- ☐ Displacement of a cylinder is area × length of the cylinder: area $= 0.7854 \times D^2$.

dither A low-amplitude, relatively high-frequency periodic electrical signal, sometimes superimposed on the servo valve input to improve system resolution.
- ☐ This signal keeps the valve spool continually in motion, to reduce hysteresis.
- ☐ Dither is expressed by the dither frequency [in hertz (Hz)] and the peak-to-peak dither current amplitude (mA).

double-acting cylinder The most common industrial cylinder in which hydraulic force is applied to either port, giving powered motion when extending or retracting.

drain A passage in, or a line from, a hydraulic component that returns leakage fluid internally to the reservoir or to a vented manifold.

efficiency The ratio of output to input of a pump.
- ☐ The overall efficiency of a hydraulic system is the output power divided by the input power. Efficiency is usually expressed as a percentage:
  Efficiency = actual output\theoretical output × 100
- ☐ If a pump was designed to deliver 10 gallons per minute (gpm) but actually output only 9 gpm at 500 r/min and 1000 psi, the efficiency of the pump at that speed and pressure is 90 percent:
  $$\text{Efficiency} = 9/10 = 90\%$$

electrohydraulic servo valve An electrohydraulic servo valve is a force amplifier that uses a variable or electrical signal to control the spool movement of a directional valve in metering the amount of oil and detecting the directional flow of oil.
- ☐ It couples with the proper feedback sensing devices to provide accurate control of position, velocity, or acceleration of an actuator.

enclosure A rectangle drawn around a graphical component or components to indicate the function(s) of a component or an assembly.

energy The ability or capacity to do work measured in units of work (ft/lb).

error (signal) The signal that is the algebraic summation of an input signal and a feedback.

feedback (or feedback signal) In certain hydraulic operations, the results of output send a signal back to the input to dictate the movement of the machine or component.

feedback loop The feedback loop system monitors output behavior and feeds appropriate command signals back to various control elements to maintain system performance.
- ☐ Such feedback loop systems provide automatic control of many manufacturing and processing operations.
- ☐ Typical feedback loops are found in servo operations and proportional valves.

filter A device whose primary function is the retention by a porous medium of insoluble contaminants below 50 μm in size from a fluid. *(See figure.)*
- ☐ Many filters are capable of removing undesirable liquids from a system's fluid, such as eliminating water from hydraulic oil.

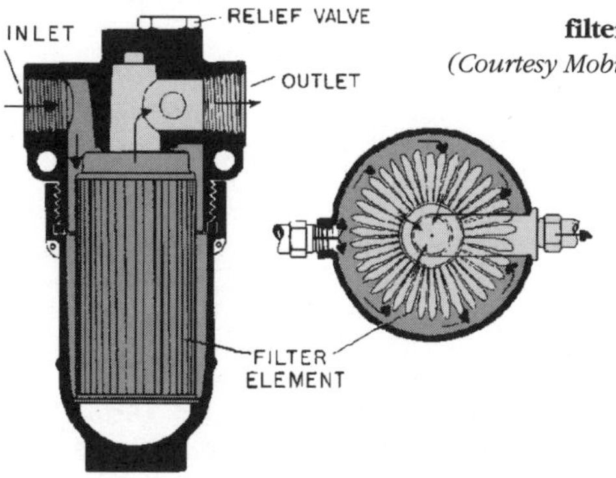

filter. *(Courtesy Mobil Oil Corp.)*

flooded A condition where the pump inlet is charged by placing the reservoir oil level above the pump inlet port.

flow-control valve A valve that controls the rate of oil flow.

flow rate The volume, mass, or weight of a fluid passing through any conductor per unit of time.

fluid **1.** A liquid or a gas. **2.** A liquid that is specially compounded for use as a power-transmitting medium in a hydraulic system.

follow valve A control valve that ports oil to an actuator so that the resulting output motion is proportional to the input motion to the valve.

force A pushing or pulling effort measured in units of weight.

□ Force (ounces, pounds, tons, etc.) = pressure (pounds per square inch) × area in square inches.

four-way valve A directional valve with four flow paths. *(See figure.)*

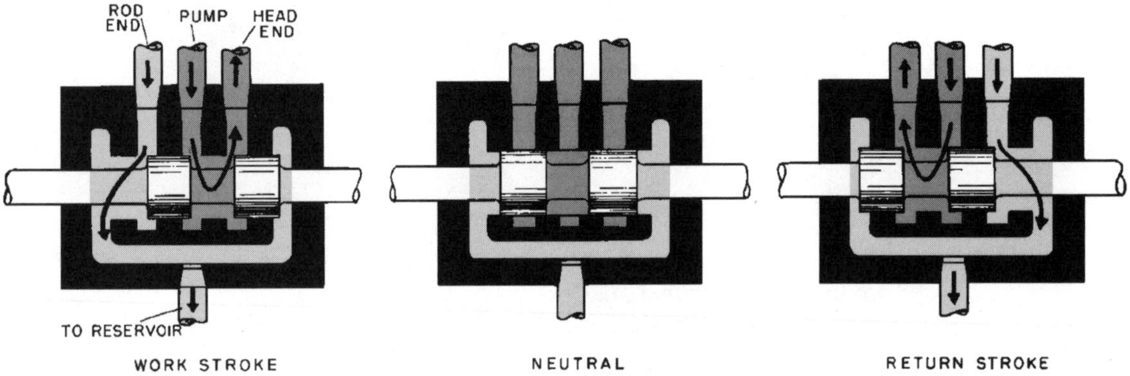

four-way valve. *(Courtesy Mobil Oil Corp.)*

□ The function is to direct flow from the pressure port to either of the two outlet ports; it can be used to move an actuator in either direction.

frequency The number of times an action occurs in a unit of time. A pump or motor's basic frequency is equal to its speed in revolutions per second multiplied by the number of chambers.

front-connected A condition in which piping connections are normally on exposed surfaces of hydraulics components.

full flow In reference to a filter, this means that all the flow into the filter inlet port passes through the filtering element.

gage pressure A pressure scale that ignores atmospheric pressure. Its zero point is 14.7 psi absolute.

gear pump The constant-discharge gear pumps use spur, helical, or herringbone gears to carry oil from the suction side of the pump to the discharge side.

□ They are used mainly for pressures less than 1500 psi, but are available for pressures as high as 3000 psi.

head The pressure due to the height of a column or body of fluid that is often used to indicate gage pressure.

heat A form of energy that has the ability to increase the temperature of a substance.

□ Heat is measured in calories or British thermal units (Btu).

□ One Btu is capable of raising the temperature of one pound of water one degree Fahrenheit (1°F).

heat exchange A device designed to dissipate heat in a hydraulic system.

□ The three types are heaters, air coolers, and water coolers.

horsepower (hp) The power required to lift 550 pounds (lb) 1 foot (ft) in one second (s) or 33,000 lb 1 ft in 1 minute (min).

□ One horsepower is equal to 746 watts (W) or to 42.4 Btu per minute. Hydraulic hp = gpm × psi × .000583.

hydraulic balance A condition of equal opposed hydraulic forces acting on a part in a hydraulic component.

hydraulic control A control of movements which is actuated by hydraulically induced force.

hydraulics Engineering science pertaining to liquid pressure and flow. It may be used to multiply force or modify motion.

hydraulic system The principal elements of a hydraulic system are an oil-storage tank, a pump to move the oil under pressure, valves to control oil flow, a motor (piston and cylinder) to convert hydraulic power to mechanical power, and piping or tubing to connect the various parts. *(See figure.)*

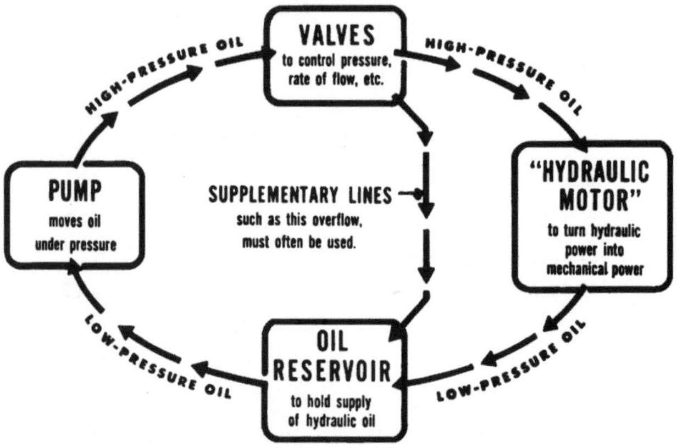

hydraulic system. *(Courtesy Sun Oil Co.)*

☐ A hydraulic system is a collection of mechanical elements joined by *fluid-connecting rods,* and in many cases, *fluid levers.*

hydrodynamics Engineering science that governs the movement of liquid and the forces opposing that movement.

☐ Impact by a moving stream (kinetic energy) to produce motion is one example of hydrodynamics in action.

☐ Hydraulic couplings and torque converters are examples of hydrodynamics application.

hydrostatics Engineering science pertaining to the energy of liquids at rest.

☐ Most industrial hydraulic systems have pumps that provide a constant pressure and direction of flow.

☐ The hydraulic jack is an example of how hydrostatics is used to increase force.

kinetic energy Science that deals with changes in movements of matter produced by forces. Falling water is a form of kinetic energy.

laminar flow A condition in which the fluid particles move in a continuous path in parallel at low velocity. With laminar flow, friction is minimized.

leverage A gain in output force over input force by sacrificing the distance moved. Force multiplication results in leverage.

lift The height that a column of fluid is raised, for instance, from the reservoir to the pump inlet.

☐ Frequently, lift is referred to vacuum.

line A tube, pipe, or hose for conducting fluid to various parts of a hydraulic system.

linear actuator A device that converts hydraulic energy into linear mechanical motion, such as a cylinder or ram.

manifold A conductor that provides multiple connection ports.

manual control A control actuated by the operator, such as a lever control for a directional valve.

manual override A means of manually actuating an automatically controlled device, for example, a solenoid manual override.

maximum pressure valve A valve that ensures the flow rate through any orifice will remain constant. *See* RELIEF VALVE.

mechanical control Any control actuated by linkage, gears, screws, cams, wheels, plunger, or other mechanical elements.

meter A control that regulates the amount or rate of fluid flow.

meter in To regulate the amount of fluid flow into an actuator by placing a flow-control valve between the pump and the actuator. *(See figure below left.)*

☐ Pump delivery in excess of the metered amount is diverted to a tank over the relief valve.

☐ Meter-in circuits can be used only with opposing loads.

meter out To regulate the flow of discharge fluid from an actuator by placing a control valve on the outlet side of the actuator. *(See figure below right.)*

☐ Meter-out circuits are designed for runaway loads.

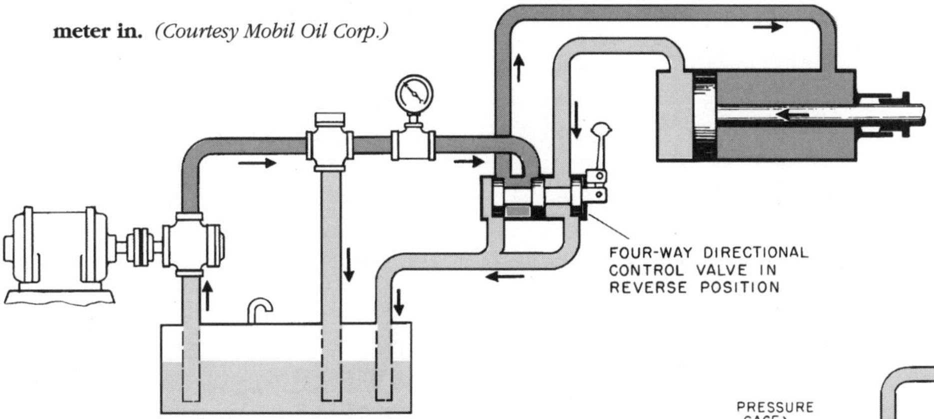

meter in. *(Courtesy Mobil Oil Corp.)*

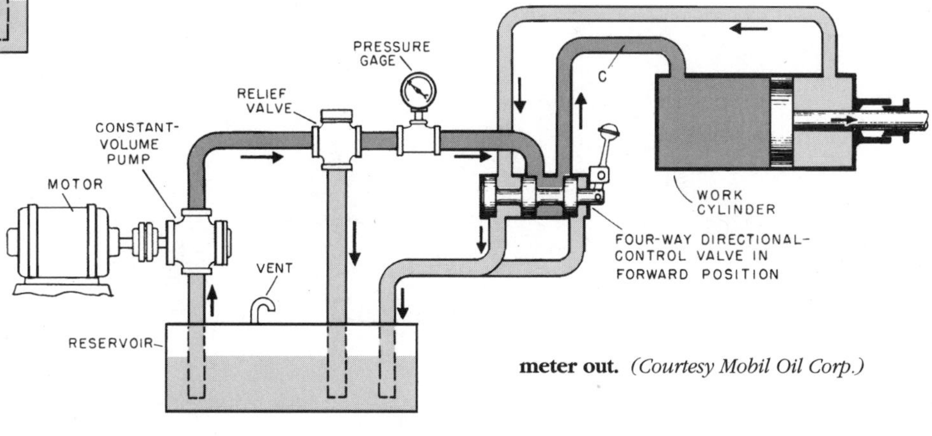

meter out. *(Courtesy Mobil Oil Corp.)*

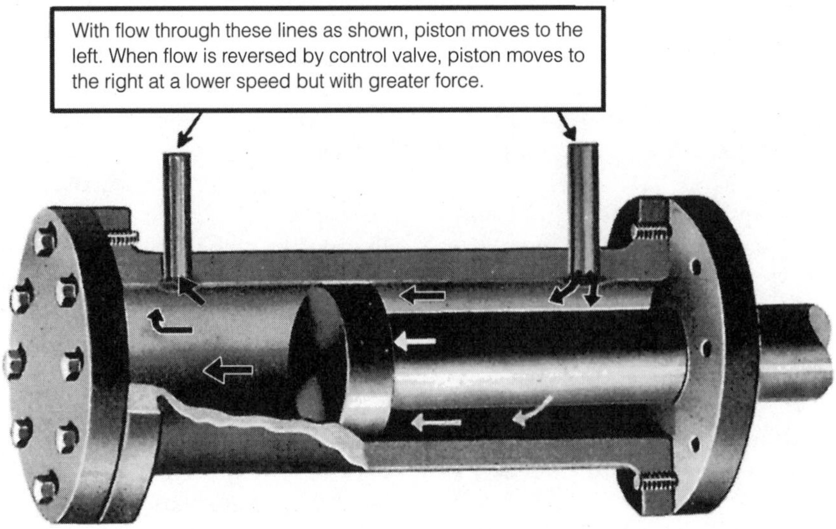
With flow through these lines as shown, piston moves to the left. When flow is reversed by control valve, piston moves to the right at a lower speed but with greater force.

motor. *(Courtesy Sun Oil Co.)*

micrometer A unit of length equal to one millionth of a meter or about .00004 in.

micrometer rating The size of the particles in a fluid that a filter will remove.

motor A device that converts fluid power into mechanical force and motion. *(See figure.)*

- □ When oil flows into the cylinder from the right, the piston moves a given distance to the left in a given time.
- □ When oil flow is reversed and fed to the left end, the piston moves to the right at a slower speed.

open-center circuit A circuit in which pump delivery flows freely through the system and back to the reservoir in neutral.

open-center valve A directional control valve in which all ports are interconnected and open to each other in the center or neutral position.

orifice A restricted passage used to control flow or create a pressure difference (pressure drop).

Pascal's law A pressure applied to a confined fluid at rest is transmitted with equal intensity throughout the fluid. *(See figure.)*

- □ Pascal's law is the foundation of fluid power application.

passage A machined or cored fluid-conducting path that lies within or passes through a component.

pilot pressure Auxiliary pressure used to operate a valve or other control.

pilot valve An auxiliary valve used to control the operation of another valve.

- □ The pilot valve can be used on a two-stage valve, check valve, directional control valve, or pressure and reducing valve.

piston A solid cylinder that fits snugly into a larger cylinder and moves back and forth by fluid pressure.

- □ A reciprocating engine, a pump, a motor, and a compressor incorporate pistons.

plunger A cylindrically shaped part that operates with repeated thrusting or plunging movement such as a piston or ram.

poppet The part of certain valves that prevents flow when the valve closes against a seat.

port A terminus or passage in a component to which conductors can be connected.

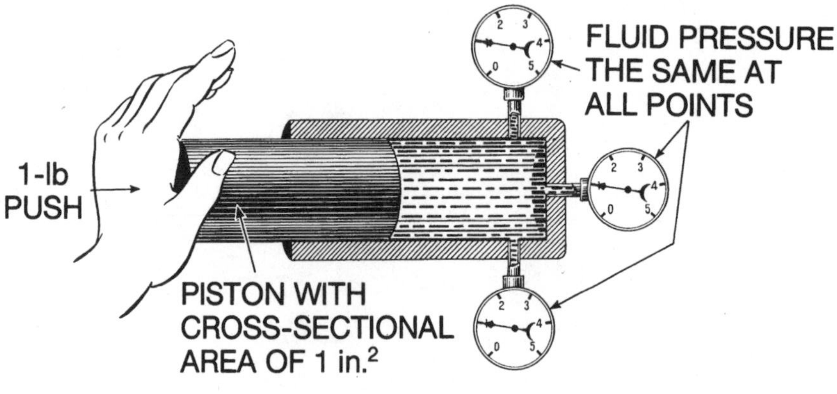
FLUID PRESSURE THE SAME AT ALL POINTS
1-lb PUSH
PISTON WITH CROSS-SECTIONAL AREA OF 1 in.²

Pascal's law. *(Courtesy Sun Oil Co.)*

positive displacement A characteristic of a pump or motor that has the inlet positively sealed from the outlet so that fluid cannot recirculate in the component.
□ Most pumps used in hydraulic systems are positive displacement.

potentiometer A control element in the servo system that measures and controls electrical potential.

power Power is the rate of doing work:
Power = work / time = (force × distance) / time

power pack An integrated power supply unit usually containing a pump, a reservoir, a pressure-relief valve (also referred to as *relief valve*), and a directional valve.

precharge pressure To fill the gas chamber to a desired pressure with an inert gas in an accumulator prior to the admission of hydraulic fluid.

pressure Pressure is the result of resistance to flow.
□ In hydraulic application, pressure equals the force to the load divided by the piston area.
□ Pressure is expressed in pounds per square inch.

pressure drop The difference in pressure between any two points of a system or a component. Also pressure differential ΔP.
□ To effect oil flow through an orifice, there must be a pressure difference or drop through the orifice.

pressure line The line carrying the fluid from the pump outlet to the pressurized part of the actuator.

pressure override The difference between the cracking pressure of a relief valve and the pressure reached when the valve is passing full flow.

pressure plate A side plate in a valve pump or motor cartridge on the pressure port side.

pressure-reducing valve A valve which limits the maximum pressure at its outlet regardless of the inlet pressure. *(See figure.)* Both direct-acting and pilot-operated versions are in use.

pressure switch An electric switch operated by fluid pressure.

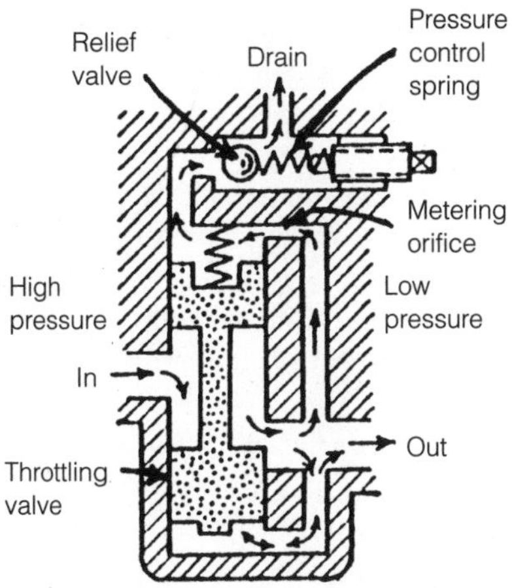

pressure-reducing valve. *(Courtesy Vickers, Inc.)*

proportional flow In a filter, the condition where part of the flow passes through the filter element in proportion to pressure drop.

pump A device that converts mechanical energy into hydraulic fluid power by pushing fluid into the system.

ram A single-acting cylinder with a single-diameter plunger rather than a piston-and-rod component. The plunger is a ram-type cylinder.

reciprocation Back-and-forth straight-line motion or oscillation.

regenerative circuit A piping arrangement for a differential-type cylinder in which oil discharged from the rod end joins the pump delivery to be directed into the head end to increase speed during piston advancing.

relief valve A pressure-operated safety valve that bypasses pump delivery to the reservoir to release the excess fluid when pressure exceeds the preset maximum.
□ A check valve is a simple relief valve.

replenish To add fluid to maintain a full hydraulic system.

reservoir A container for storage of liquid in a fluid power system. *(See figure.)*

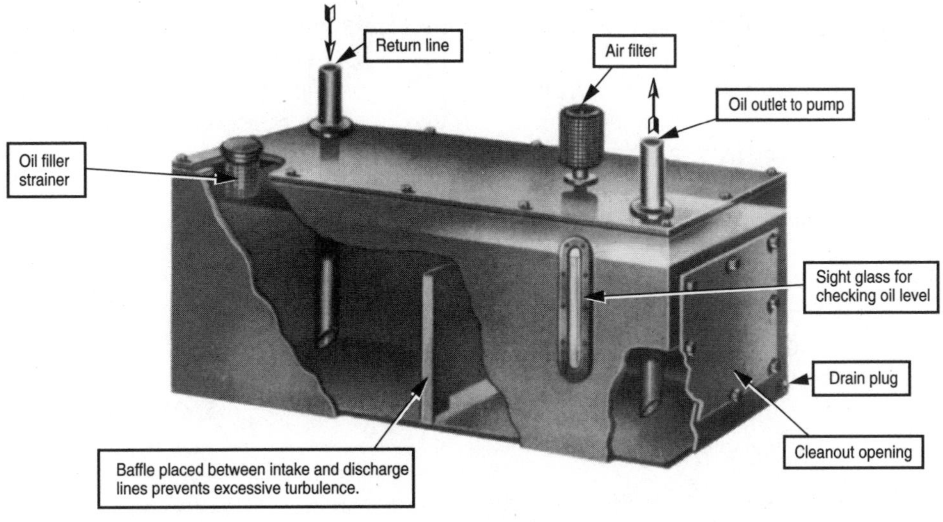

reservoir. *(Courtesy Sun Oil Co.)*

restriction A reduced cross-sectional area in a line or passage which produces a pressure drop

return line A line designed to carry exhaust fluid from the actuator back to the reservoir.

reversing valve A four-way directional valve designed to reverse a double-acting cylinder or reversible.

rotary actuator A device for converting hydraulic energy into rotary motion: a hydraulic motor. There are three types of hydraulic actuators: linear, rotary, or oscillatory.

sequence The order that is followed to perform a series of operations or cause movements.

□ To divert flow to accomplish a subsequent operation or movement.

sequence valve A pressure-operated valve which, at its setting, diverts flow to a secondary line while holding a predetermined minimum pressure in the primary line. *(See figure.)*

streamline flow *See* LAMINAR FLOW.

stroke **1.** The length of travel of a piston or plunger in a cylinder. **2.** To change the displacement of a variable-displacement pump or motor.

subplate A mounting style widely used for industrial application that provides a means of connecting piping to the component.

suction line The hydraulic line connecting the pump inlet port to the reservoir or sump.

sump A reservoir or tank.

supercharge *See* CHARGE.

surge A momentary or sudden rise of pressure in a hydraulic system.

swash plate A stationary-canted plate in an inline axial piston pump. The cylinder block is turned by the drive shaft. As the block turns, the piston shoes follow the swash plate, causing the piston to reciprocate.

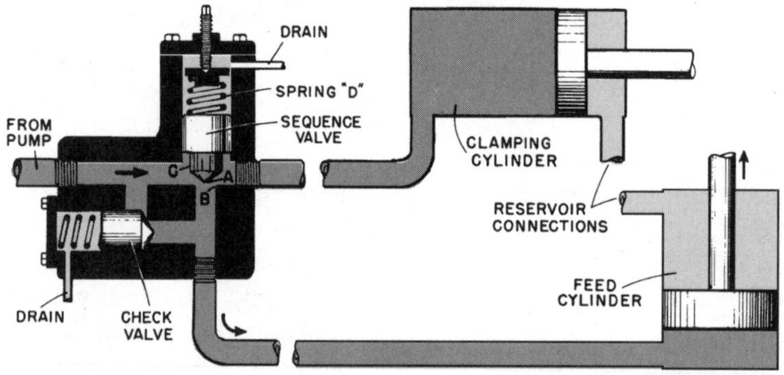

sequence valve. *(Courtesy Mobil Oil Corp.)*

□ It is used where one machine movement must be completed before another begins.

servo mechanism (servo) A mechanism subjected to the action of a controlling device that will operate as if it were directly actuated by the controlling device, but capable of supplying power output many times that of the controlling device.

servo valve A force amplifier for positioning control that modulates output as a function of an input command.

signal A command or indication of a desired position or velocity.

single-acting cylinder A cylinder in which hydraulic energy can produce thrust or motion in only one direction. Gravity or a spring may return the cylinder to its starting point.

slip Internal leakage of fluid that occurs in a hydraulic system.

spool A spool sliding within a cylinder bore is often used in directional control valves and other hydraulic components. The spool is usually hardened and ground to produce a smooth, accurate, and durable surface.

strainer A coarse filter that removes large particles from a hydraulic fluid.

1. Maximum swash-plate angle for maximum displacement
2. Decreased swash-plate angle for partial displacement
3. Zero swash-plate angle for zero displacement

synchro A rotary electromagnetic device generally used as an AC feedback signal generator which indicates position. It can also be used as a reference signal generator.

tachometer—(AC) (DC) A device that generates an AC or DC signal proportional to the speed at which it is rotated and the polarity of which is dependent on the direction of rotation of the rotor.

tank The reservoir or sump that holds the oil supply for the hydraulic system.

throttle To permit passing of a restricted flow. This may control flow rate or create a deliberate pressure drop.

torque A rotary thrust. The turning effort of a fluid motor is usually expressed in inch-pounds. The general torque-power formula for any rotating equipment is

$$\text{Torque} = \frac{63{,}025 \times \text{hp}}{\text{r/min}} = \frac{\text{torque} \times \text{r/min}}{63{,}025}$$

torque converter A rotary-fluid coupling that is capable of varying the ratio of the input torque to the output torque *(See figure on next page.)*

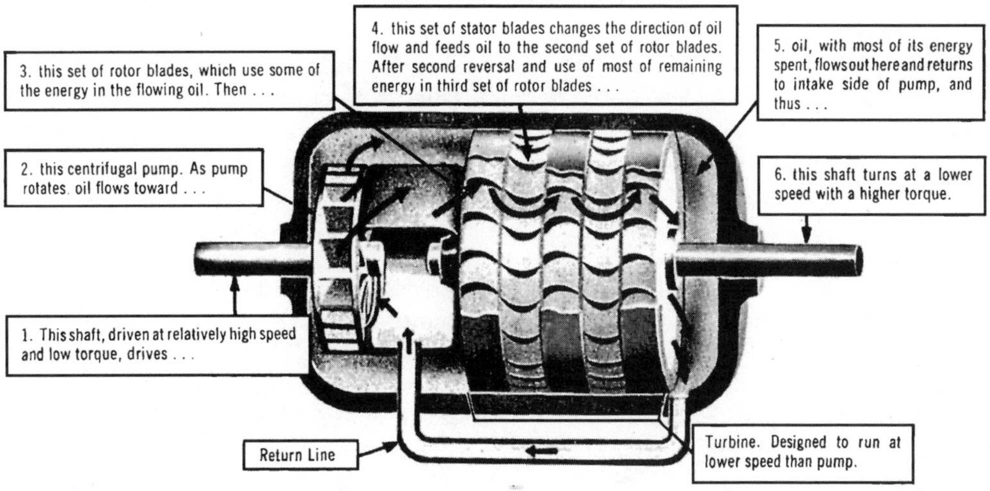
4. this set of stator blades changes the direction of oil flow and feeds oil to the second set of rotor blades. After second reversal and use of most of remaining energy in third set of rotor blades . . .
3. this set of rotor blades, which use some of the energy in the flowing oil. Then . . .
5. oil, with most of its energy spent, flows out here and returns to intake side of pump, and thus . . .
2. this centrifugal pump. As pump rotates. oil flows toward . . .
6. this shaft turns at a lower speed with a higher torque.
1. This shaft, driven at relatively high speed and low torque, drives . . .
Return Line
Turbine. Designed to run at lower speed than pump.

torque converter. *(Courtesy Sun Oil Co.)*

- □ It is a hydraulic gearbox, acting as a clutch and transmission, that is capable of a smooth flow of power.
- □ Any combination of high torque at low speed, low torque at high speed, or any of the two within the limits of the equipment is possible.

torque motor A type of electromechanical transducer having rotary motion that is used to actuate servo valves.

transducer (or **feedback transducer)** A device that converts one type of energy to another. An example would be the transducer-sensing pressure and generation of an electrical signal in proportion to the pressure.

transformer A device that transfers AC energy from one circuit to another without *electrical* contact between the two circuits.

turbine A rotary device that is actuated by the impact of a moving fluid against blades or vanes.

turbulent flow (turbulence) A condition in which the fluid particles move in random paths rather than in continuous parallel paths.

two-way valve A directional control valve with two flow paths.

unload To release flow, usually directly to the reservoir, to prevent pressure being imposed on the system or portion of the system.

unloading valve A valve that bypasses flow to a tank when a set pressure is maintained on its pilot port.

- □ The unloading valve directs pump-output oil back to the reservoir at low pressure after system pressure has been reached.
- □ In some hydraulic systems, pump flow may not be needed during part of the cycle. This is where an unloading valve works best; it saves energy.

vacuum Pressure less than atmospheric pressure. It is usually expressed in inches of mercury (inHg), referred to as the *existing atmospheric pressure.*

valve A device which controls fluid flow direction, pressure, or flow rate.

vane pump A constant-volume pump having vanes that are forced by the contact with a stationary cam ring to slide in and out of slots in a rotating hub. *(See figure.)*

- □ Oil is drawn through the suction port and expelled through the discharge port.

velocity The speed of flow through a hydraulic line

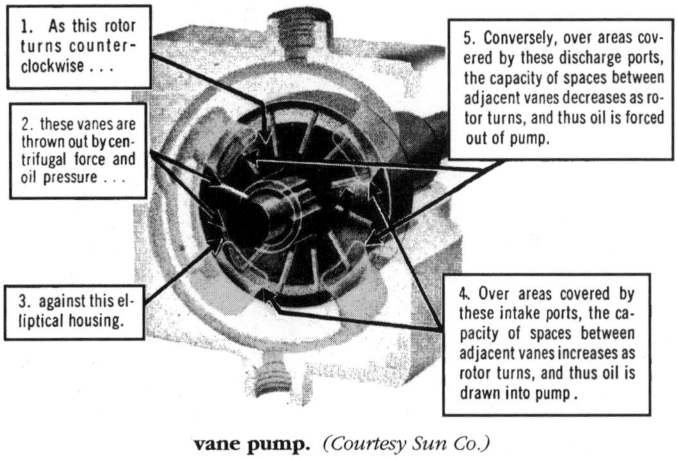
1. As this rotor turns counter-clockwise . . .
2. these vanes are thrown out by centrifugal force and oil pressure . . .
3. against this elliptical housing.
5. Conversely, over areas covered by these discharge ports, the capacity of spaces between adjacent vanes decreases as rotor turns, and thus oil is forced out of pump.
4. Over areas covered by these intake ports, the capacity of spaces between adjacent vanes increases as rotor turns, and thus oil is drawn into pump.

vane pump. *(Courtesy Sun Co.)*

expressed in feet per second (fps or ft/s) or inches per second (ips or in/s), or the speed of a rotating component measured in revolutions per minute (r/min).

□ Speed of a cylinder piston is dependent on its size (piston area) and the rate of flow into it:

$$\text{Velocity, in/min} = \frac{\text{flow, in}^3\text{/min}}{\text{area, in}^2}$$

or

$$\text{flow} = \text{velocity} \times \text{area}$$

vent The opening of a pressure-control valve by opening its pilot port (vent connection) to atmospheric pressure.

□ An air-breathing device on a fluid reservoir used to remove trapped air from a component.

viscosity The measure of the fluid's resistance to flow; or an inverse measure of fluidity.

□ A thin fluid has a low viscosity.

□ A fluid that flows with difficulty has a high viscosity.

viscosity index An arbitrary measure of a fluid's resistance to viscosity change with temperature changes.

□ A fluid that has a stable viscosity at temperature extremes has a high viscosity index.

□ A fluid that is very thick when cold and very thin when hot has a low viscosity.

volume The size of a space or chamber in cubic units; loosely applied to the output of a pump in gallons per minute (gpm).

wobble plate A rotating canted plate in an axial-type piston pump that pushes the pistons into their bores as it "wobbles."

work Work is force acting through a distance:

$$\text{Work} = \text{force} \times \text{distance}$$

Example. Work, in • lb = force (lb) × distance, in.

SECTION 8

Manufacturing Processes

Steve F. Krar
Consultant
Kelmar Associates
Welland, Ontario

Manufacturing is a broad term referring to the planning, tooling, manufacturing, and marketing of a product. The terms included in this section are primarily those which involve the design and planning, materials, manufacturing, and process control of a product. Processes involving a single machine tool, a group of machines (flexible manufacturing cell), or an entire factory (flexible manufacturing systems) are also included. There is an emphasis on the use of computers in manufacturing and the terms coined to cover computer-assisted and computer-controlled processes.

abrasive machining Various grinding, honing, lapping, and polishing operations that use abrasive particles to create new shapes, improve surface finishes, and cut by removing metal or material.

abrasive waterjet *See* HYDRODYNAMIC MACHINING *and* WATERJET MACHINING.

abrasive-wire bandsawing A variation of bandsawing that uses a small-diameter wire with diamond, cubic boron nitride, or aluminum oxide abrasives, bonded to the surface as the cutting blade.

age of automation The result of increased innovation and invention that is recognized by the automatic machines and processes used on the production line. *(See figure.)*

age of information The remarkable increase in knowledge and information since the 1960s when the computer was first developed.

agile manufacturing Combines the state-of-the-art fabrication and product delivery techniques to custom-made products to suit each customer's taste,

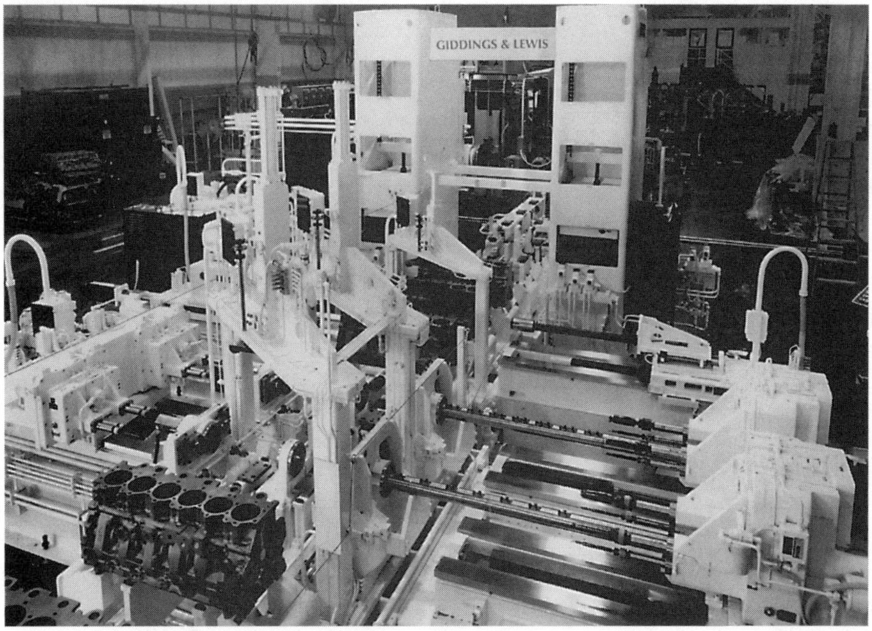

assembly automation.
(Courtesy Giddings & Lewis Inc.)

specifications, and budget.

alternative energy sources Sources of energy that are seldom used to a great extent, such as water, wind, and solar power.

artificial intelligence (AI) A new manufacturing tool which combines the use of artificial vision, expert systems, robotics, natural-language comprehension, and voice recognition.

assembly The joining together of two or more parts to complete a unit or structure.

□ The ease with which a part can be assembled should be considered when a part is designed for automation.

assembly line A production line in which a conveyor system moves the product, and workers along each side perform specific tasks in the assembly of the product. *(See figure.)*

assembly line. *(Courtesy Giddings & Lewis Inc.)*

asynchronous (nonsynchronous) This can refer to data-communications messages, assembly processes, or materials movement, where two or more events occur independently.

automated guided vehicle (AGV) A computer-controlled robotic handling system used to move tools and materials from a storage-retrieval area to a machine station, or vice versa.

automatic identification system A system in which data is collected automatically and produced by advanced technology such as laser-read bar coding, machine vision, and radio frequency identification.

automation A manufacturing approach in which all or a part of a machining or manufacturing process is accomplished by setting in motion a sequence of events that completes the process without further human intervention.

□ The system can be controlled by a combination of mechanical (stops, cams, etc.), electrical (relays, contact switches, etc.), or electronic (computer- or microprocessor-controlled) devices.

band polishing A variation of bandsawing which uses an abrasive band to smooth or polish parts that have been sawed or filed.

bandsawing Power bandsawing, often called *band machining,* uses a long endless band with many small teeth traveling over two or more wheels in one direction. The saw band produces a continuous and uniform cutting action with a series of evenly distributed low, individual tooth loads.

barrel finishing A finishing process that involves low-pressure abrasion resulting from tumbling workpieces in a barrel together with an abrasive slurry.

batch manufacturing A type of manufacturing process in which products are produced in small quantities.

bill of materials A complete record of every product part, with part specifications, part costs, and the total cost for each product.

bonding The process of joining solid materials by cohesively mixing their molecules, or by adhesively gluing or cementing them.

boring The operation of truing and enlarging a previously drilled or cored hole with a single-point, lathe-type tool.

brake forming The use of a manual or powered pan brake to bend sheet metal into any preferred angle.

broaching An operation that uses a hardened multitooth form tool to produce a similar form on the inside or outside of a surface.

brushing The use of rapidly spinning wires or fibers to economically remove burrs, scratches, and imperfections from the surface of components.

□ It is widely used in the manufacture of bearing races and gears to remove sharp edges.

buffer inventory An extra inventory, sometimes kept to cover unforeseen problems and shortages.

buffing The operation of smoothing and shining a surface by using a soft wheel coated with an abrasive compound against the workpiece.

CAD *See* COMPUTER-AIDED DESIGN.

CAD/CAM *See* COMPUTER-AIDED DRAFTING/COMPUTER-AIDED MANUFACTURING.

CADD *See* COMPUTER-AIDED DRAWING AND DESIGN.

CAE *See* COMPUTER-AIDED ENGINEERING.

CAM *See* COMPUTER-AIDED MANUFACTURING.

capital-intensive Manufacturing that is based on tools, machines, and other equipment, especially computerized and automated tools.

casting and molding Materials-forming processes in which industrial material is either dissolved, melted, or compounded into a liquid, and allowed to flow by gravity into a hollow mold or cavity of the desired shape. *(See figure on next page.)*

catalyst A special material mixed with thermosetting-plastic resins to produce a chemical action that solidifies the resins into permanent shapes.

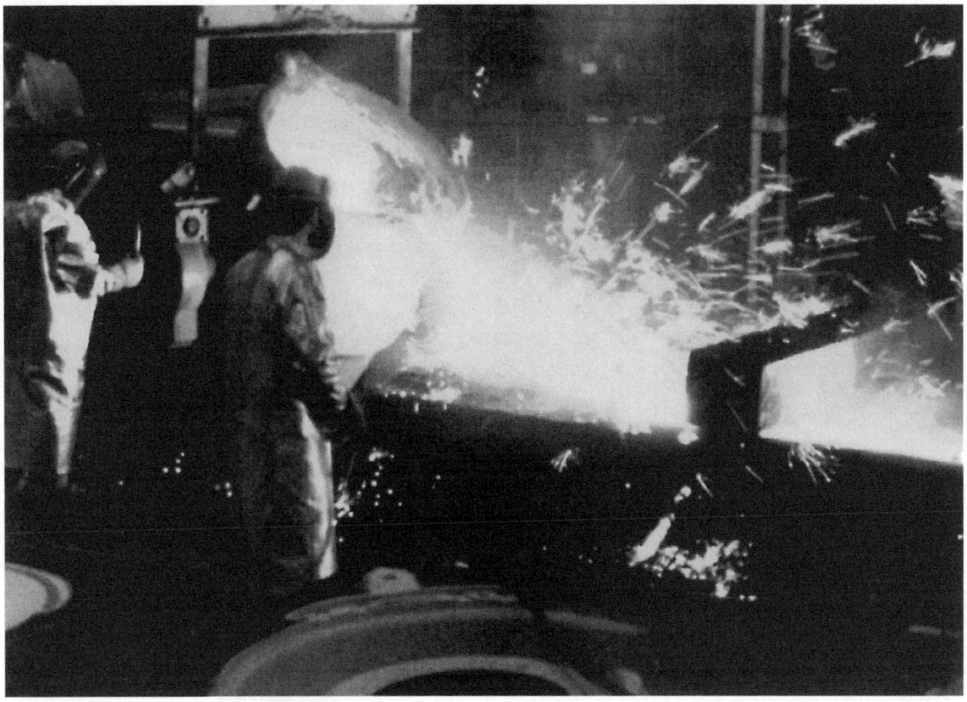

casting and molding. *(Courtesy Giddings & Lewis Inc.)*

cellular manufacturing The grouping of equipment, people, and processes together to manufacture a specific family of parts (parts that have similar characteristics).

center drilling Drilling tapered holes in the ends of a cylindrical workpiece for mounting work between a machine tool's centers.

☐ Center-drilled holes are also used as starter holes for drilling larger holes in the same location.

centering The process of locating the center holes of a workpiece which is to be mounted on lathe centers.

centerless grinding A grinding operation for long, thin parts or shafts where the workpiece rests on a knife-edge support, rotates through contact with a regulating or feed wheel, and is ground by a grinding wheel.

centrifugal feeder (rotary feeder) Equipment that uses centrifugal force to orient and deliver parts, at a specified feedrate, to an assembly machine.

ceramics A group of manufacturing materials that includes china, clays, glass, porcelain, and pottery.

chamfering Machining a bevel on a workpiece or tool to improve a tool's entrance into the cut, improve its appearance, and remove sharp edges.

chip-producing machines Machines such as lathes, milling machines, and grinders that form metal to size and shape by cutting away the unwanted sections. *(See figure.)*

chip removing A materials-separation process that cuts material in the form of chips, and reduces the size of standard stock using special single- and multiple-point cutting tools.

chucking/turning center A CNC machine tool used for producing round forms on a workpiece. *(See figure on next page.)*

☐ The *chucking center* is used mainly for machining work held in a chuck.

☐ The *turning center* is used for machining work between centers, or long work held by a chuck and supported on the end by a center.

☐ The *combination turning/milling center* is used for round turning, plus operations such as milling, drilling, and tapping, on round parts.

chip-producing machine. *(Courtesy DoAll Co.)*

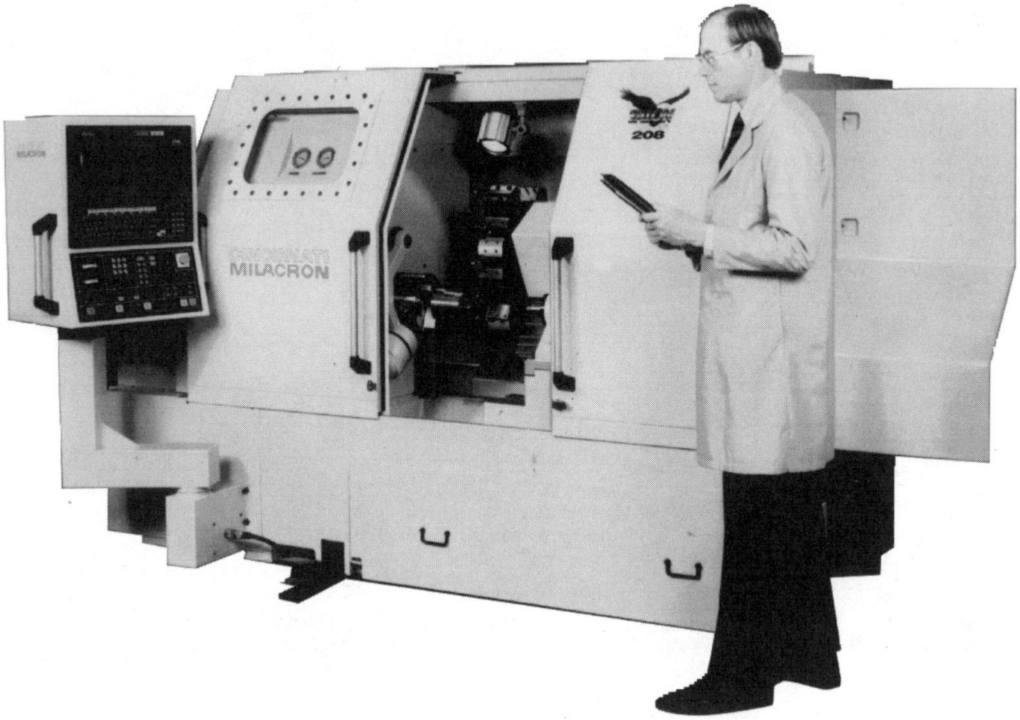

chucking center. *(Courtesy Cincinnati Milacron, Inc.)*

CIM *See* COMPUTER-INTEGRATED MANUFACTURING.

cohesive A bonding process in which chemicals, heat, pressure, or combinations of these are used to cause molecules of materials to permanently mix or cohere.

combining Includes the mixing, bonding, coating, and fastening processes that are used to finish one-piece or parts of multiple-piece products together in subassemblies and final assemblies.

composite A classification of manufacturing materials which are combinations of materials, such as plywood and fiberglass or metals and plastics.

computer-aided design (CAD) Computer systems and software programs used by drafters to design a product and create technical drawings of each part. *(See figure.)*

computer-aided design/computer-aided manufacturing (CAD/CAM) CAD/CAM results when CAD systems in engineering and CAM systems in manufacturing are joined together to produce a product.

computer-aided drawing and design (CADD) Computer and software systems used by engineers to design, draw, and test a product on a computer before manufacturing begins.

computer-aided engineering (CAE) The use of computers and special software to perform various engineering functions such as determining a material's ability to withstand stresses and strains.

computer-aided manufacturing (CAM) The use of computers to control all phases of manufacturing, including the tools, equipment, and machines that are used. A *CAM-actuated assembly* is a rotary or linear assembly machine that uses cam mechanisms to sort and/or assemble parts.

computer-integrated manufacturing (CIM) A central computer system that joins all the different phases of manufacturing, such as management, engineering, production, finances, marketing, and human resources, into one cohesive unit.

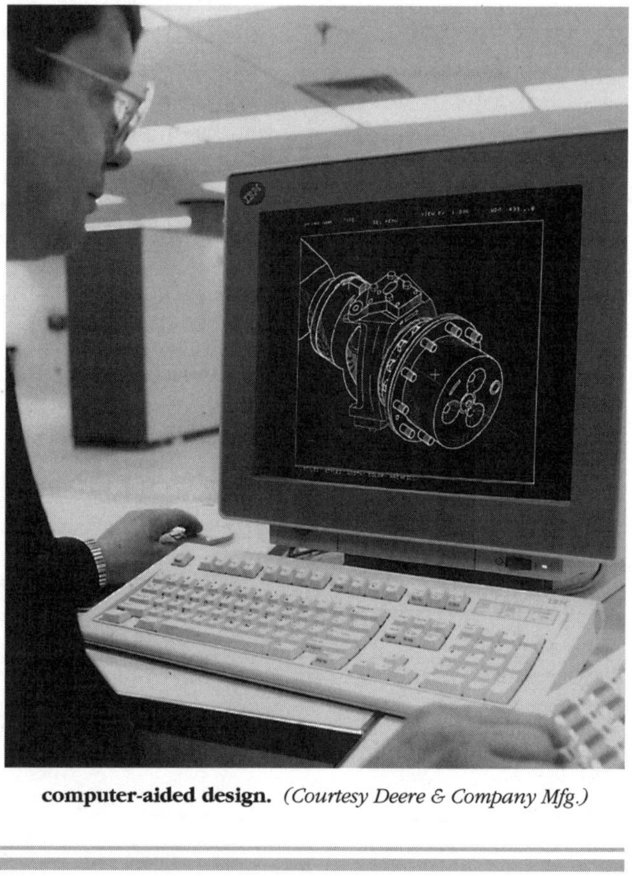

computer-aided design. *(Courtesy Deere & Company Mfg.)*

concurrent engineering Sometimes called *simultaneous engineering*; a concept based on time—spending more time up front to save even more time downstream.

□ It is used to develop a new product from technology, taking the concept through production and continuous improvement, and also for research and development.

conditioning Any materials-forming process that uses chemicals, heat, or mechanical pressure to change the internal molecular structures and improve the properties of a material, such as ductility, hardness, toughness, and moisture content.

continuous corporate renewal A management of change and vision which includes continuous change, rapid response, and evolving quality standards.

continuous flow The use of a conveyor system to provide uninterrupted movement of materials and parts at a constant rate of speed along an assembly line.

conversion coating A type of process in which the surface of a material is chemically changed by dyeing, oxide coating, or phosphate coating.

conveyor system A machine used to move parts and materials along an assembly line at a constant rate of speed.

coordinate measuring machine (CMM) A computer-controlled inspection machine, capable of making many very accurate measurements of three-dimensional objects, in a short period of time. *(See figure.)*

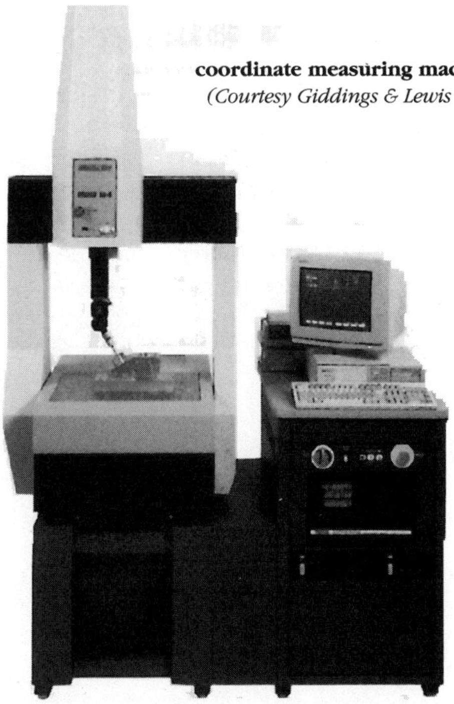

coordinate measuring machine. *(Courtesy Giddings & Lewis Inc.)*

cost-efficient A product or process that is profitable for a company to make because it provides a good return on the time and money invested.

counterboring The process of enlarging one end of a drilled hole so that bolt heads and nuts can be set below the surface of the part.

countersinking Cutting a beveled edge at the entrance of a hole so that the head of a fillister screw can be flush with or slightly below the workpiece surface.

creep-feed grinding An operation in which the grinding wheel is set to the full depth of the form required, and then slowly fed into the work at a slow speed to finish the form in one pass.

custom manufacturing Producing a made-to-order product that is generally manufactured by one person or a few people who work on the product from beginning to end.

cutoff The operation of removing a slug, blank, or other piece of material from the original bar stock so that it can be machined on conventional or CNC machine tools.

cycle time The total time it takes to complete a product, starting from the raw materials to the finished product.

debugging The process of making corrections to a new system to obtain proper and cost-efficient operation.

design criteria The factors of a product that a designer must consider, such as function, form, ergonomics, aesthetics, ease of manufacture, durability, and cost.

design engineering The process of creating ideas for products by drawing sketches, creating models and prototypes, and preparing working drawings. *(See figure.)*

design engineering. *(Courtesy Giddings & Lewis Inc.)*

design for assembly (DFA) An engineering practice in which the product designer works with knowledgeable assembly people to ensure that as few parts as possible are used and that they can be assembled quickly and easily.

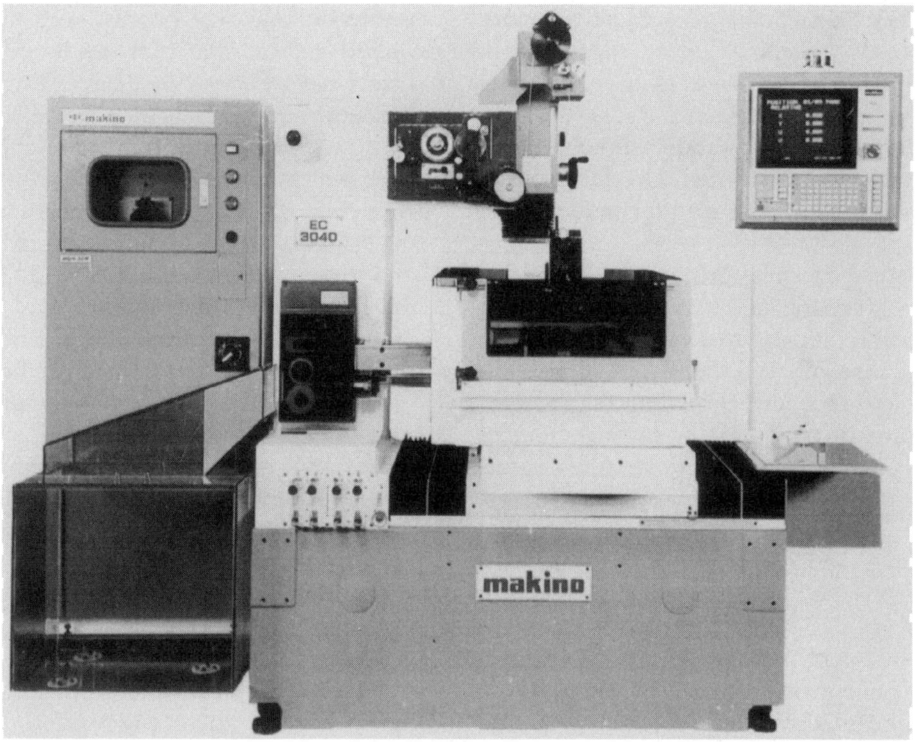

electrical discharge machining. *(Courtesy Makino, Inc.)*

diamond bandsawing An operation in which a band coated with diamond abrasives is used to machine carbides, ceramics, and other extremely hard materials.

direct steelmaking A steelmaking process designed to bypass the blast furnace and coking ovens, to produce steel directly from iron ore.

disk grinding An operation in which the workpiece is held and ground against the side of a wheel, rather than on the wheel's periphery.

distribution The process of getting a product to the consumer in the most economical and efficient way.

diverse product line The variety of related or unrelated products manufactured by a company.

downtime Any time, such as setup, maintenance, or for other reasons, that a machine is not operating when it should be.

drawing The operation of using a punch to force a metal blank into a die cavity to produce a variety of steel parts, such as automobile body panels, fenders, and doors, to the desired shape.

drilling An operation that uses a hardened rotating tool to produce a round hole in a workpiece.

EBM *See* ELECTRON-BEAM MACHINING.

economies of scale The goal of achieving low per-unit costs by producing in the largest possible volume, so that costs can be spread over a large number of products.

economies of scope The goal of achieving low per-unit costs by computerized production and flexible manufacturing systems that allow goods to be manufactured economically in small lot sizes.

electrical discharge machining (EDM) A metal-removal process that uses an electric spark for the removal of metal *(see figure)*. The most commonly used EDM machines are the *plunge*, or *ram*, and the *wire-cut*.

electroforming A process for making thin parts where an electrode deposits metal on a mandrel or mold. This process, although similar to electroplating, produces deposits that are much more than that of an electroplated deposit.

electrohydraulic forming A method of shaping hollow tubes or preforms by discharging electrical energy, created by a spark discharge or exploding bridge wires, that produces shock waves inside the workpiece to do the forming.

electron-beam machining (EBM) The operation of using a pulsating stream of high-speed electrons produced by a generator to generate thermal energy for removing material. The beam is focused by electrostatic and electromagnetic fields to concentrate energy on a very small area of work.

electroplating An electrical process, for depositing metal on a material with a conductive surface, that uses a cathode in an electrolytic bath containing dissolved salts of the metal being deposited.

end milling A vertical or horizontal milling machine operation in which an end-milling cutter is mounted in the machine's spindle, rather than on an arbor.

□ Common end-mill cutters have two, three, or four flutes that cut on the end and also on the periphery.

energy converter A device or apparatus that is used to change energy from one form to another more usable form, such as generators, turbines, motors, and engines.

environmentally soft manufacturing Manufacturing

that considers its effect on the environment, and creates healthy working conditions.

ergonomics Designing manufacturing systems where the goal is to improve the efficiency, safety, and well-being of all workers.

expert systems A form of AI that uses a knowledge base to solve problems which formerly required some form of human expertise.

face milling A form of milling that produces a flat surface, generally at right angles to the rotating axis of a cutter. *(See figure.)*

face milling. *(Courtesy Association for Manufacturing Technology.)*

□ The cutter has teeth or inserts on its periphery and on its end face.

facing A type of cleanup operation that provides a true, flat reference surface for use as a reference or base point for succeeding operations.

factory system An organizational system for production that uses special machines and buildings for manufacturing.

family of parts Parts grouped by their shape, size, and machining operation for most efficient manufacturing.

filing Usually a hand operation in which a cutting tool with numerous small teeth is used to round sharp corners and shoulders, and remove burrs and nicks.

□ Filing can also be performed under electric power on a contour bandsaw with a special filing attachment.

finishing Any of many different processes used for surface, edge, and corner preparation, as well as conditioning, cleaning, and coating.

fixed costs Manufacturing costs that stay the same, or fixed, no matter how many products the company makes or sells. These costs, which are affected by the cost of raw materials and labor, may remain fixed for a limited time period.

fixed-position layout Plant or static layout is used when manufacturing large structures, such as airplanes or electric-power generators.

flame cleaning A method of cleaning the surface of metal parts using a flame from an oxyacetylene torch to burn off oils, dry water residues, or heat scale and cause it to flake off.

flexible assembly system A system that uses robots which can be programmed to assemble a family of parts or a group of similar assemblies.

flexible manufacturing system (FMS) An automated manufacturing system consisting of a number of CNC machine tools, serviced by a materials-handling system, under the control of one or more dedicated computers. It is designed to produce parts with a minimum of production changeover time. *(See figure.)*

flexible manufacturing system. *(Courtesy Cincinnati Milacron, Inc.)*

flow process chart A chart showing the sequence of manufacturing operations, inspections, transportation, delays, and storage performed on each part.

fluting The operation of cutting straight or spiral grooves in drills, end mills, reamers, and taps, to provide their cutting edges and aid in chip removal.

FMS *See* FLEXIBLE MANUFACTURING SYSTEM.

forging The plastic deformation of metals, generally at elevated temperatures, into desired shapes by compressive forces exerted through a die.

□ The simplest forging operation is by upsetting the compression of metal between two flat, parallel platens.

forming Any operation which may involve casting, molding, compressing, stretching, and conditioning processes that bring about external and internal form changes in metallic, ceramic, and polymeric materials.

friction sawing A high-speed sawing operation, using a special band machine capable of achieving band velocities of 15,000 sf/min or more, to cut hard alloy steels.

□ Frictional heat softens the metal at the point where the band contacts the metal surface; then the teeth remove the molten material.

gang cutting, milling, slitting Milling operations that use several cutters mounted on a single arbor, generally for simultaneous cutting of a form or contour.

global competition A term referring to worldwide competition for manufactured goods, regardless of where they are made.

grinding Any operation involving removal of material from the workpiece by a powered abrasive wheel, stone, belt, paste, sheet, compound, slurry, or other agent. *(See figure.)* The most common forms of grinding are

□ *Surface grinding,* which is used to produce flat and/or squared surfaces.

□ *Cylindrical grinding,* which is used to grind internal and external parts.

grinding. *(Courtesy Bridgeport Machines, Inc.)*

grooving A machining operation that is used to cut grooves and shallow channels in parts such as ball-bearing raceways.

group technology A system where a large number of different parts can be identified by shape, configuration, holes, threads, size, and other properties before creating families of parts. *(See figure.)*

□ This can determine the type of machines required in a cell for the efficient flow of parts between machines and operations.

gun drilling A drilling process that uses a single-lip, self-guiding tool to produce deep, precise holes. High-pressure coolant is fed to the point of the drill through holes in the drill's shank and body.

hard technology Also called *high technology*; a technology based on improving productivity and profits without considering human, environmental, and social consequences.

histogram A quality-control tool used as a final inspection for lot-size parts, to see if a process is holding tolerance and where it lies in relation to the minimum and maximum limits.

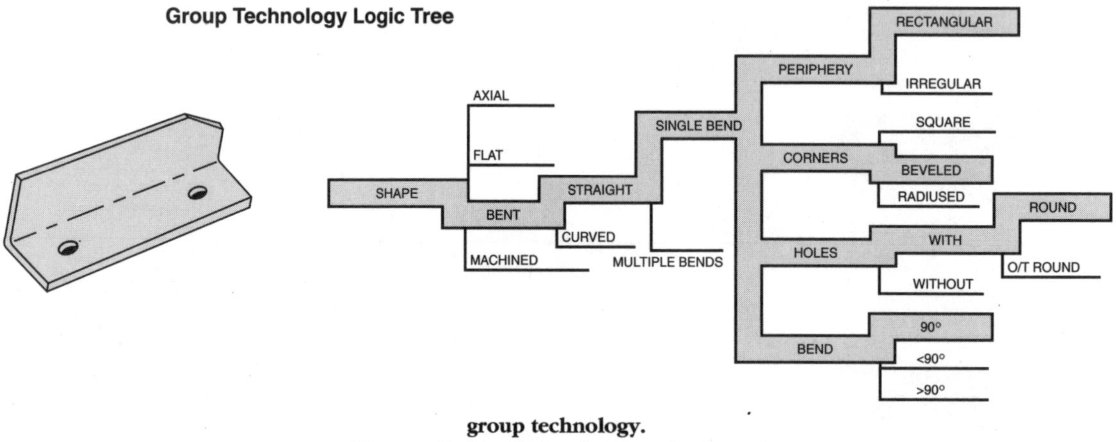

group technology.
(Courtesy Society of Manufacturing Engineers.)

hobbing A gear-tooth-generating process consisting of rotating and advancing a hardened fluted-steel worm cutter past a revolving blank.

holemaking The operation of using a tool such as a drill, punch, or electrode, to produce holes in a workpiece.

honing A low-velocity abrading process which uses abrasive sticks (aluminum oxide and silicon carbide) mounted on a metal mandrel, to size and improve the finish of internal cylindrical surfaces.

hydrodynamic machining A process that produces a narrow kerf to remove workpiece material by the cutting action of a fine, high-pressure (up to 60 ksi, 414 MPa), high-velocity stream of water or water-based fluids with additives.

ideation The design stage of an engineering process in which the designer combines certain bits of information learned during the preparation stage into new ideas.

indexing machine An assembly machine with intermittent rotary motion, designed to assemble parts sequentially where each station consists of specialized tooling for the operation to be performed.

industrial materials Raw materials that have been refined and changed into standard sizes, shapes, or weights, by primary manufacturing companies.

Industrial Revolution A period of tremendous industrial change that started in England around the mid-1750s when power-driven machine tools began to replace hand tools in the manufacture of products. This created a tremendous increase in the production of many kinds of goods.

innovators People who find ways to make an existing product better by combining other ideas to create a new, better, and different product.

in-process gaging A system using probes, lasers, optical devices, or similar instruments to measure or inspect work while it is being machined. *(See figure.)*

input This includes all the resources required to make a product, such as tools, machines, materials, people, and energy.

interchangeable parts Parts and components produced to specific tolerances to allow them to be substituted for one another, resulting in high-volume output and lower manufacturing costs.

inventory control The steps taken to keep the proper levels of raw materials and finished goods in relation to consumer demands.

jig boring A high-precision operation which was initially used to produce holes to accurate locations for jig and fixture manufacturing. Machining centers now are capable of greater accuracy and the need for jig boring processes has declined.

just-in-time (JIT) manufacturing A system in which materials are available in the amount and at the time they are needed for production. Its purpose is to improve productivity, reduce costs, reduce scrap and rework, overcome the shortage of machines, reduce inventory and work in progress, and use manufacturing space efficiently.

Kanban A Japanese inventory replenishment system, similar to JIT, that was developed by the Toyota Company.

keyseating Milling or grinding an internal keyway.

knowledge-base building blocks The base of an expert system which consists of a large body of information that has been assembled, widely shared, and accepted by experts, as well as a diagnostic computer software package *(See figure on next page.)*

knurling The operation of producing diamond or straight-shaped impressions on the surface of a handle, or other round part, to provide a better gripping surface and improve the appearance.

□ In the automotive industry, knurling is used to enhance clearances and help pistons and valve guides retain oil.

in-process gaging. *(Courtesy Giddings & Lewis Inc.)*

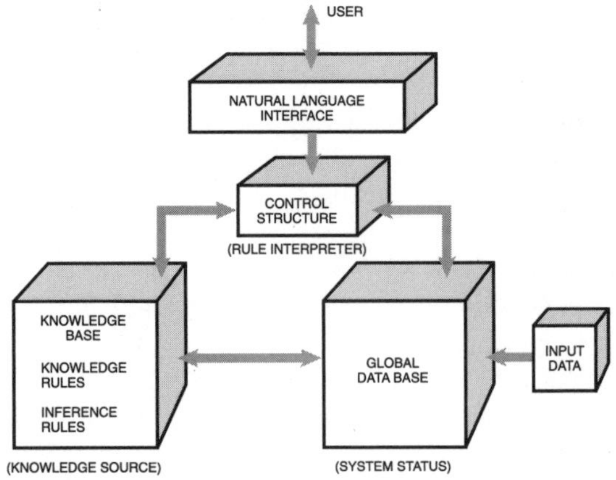

knowledge-base building blocks. *(Courtesy Society of Manufacturing Engineers.)*

lapping A finishing operation using a loose, fine-grained abrasive in a liquid medium, to correct minor shape imperfections, improve surface finishes, and produce a close fit between mating surfaces.

laser *L*ight *a*mplification by *s*timulated *e*mission of *r*adiation. A tool which amplifies and intensifies light into a narrow beam that can used to read bar codes, cut, melt, and vaporize materials, and perform numerous other applications in science and medicine.

lasercaving A process, similar to milling, that uses lasers to produce cavities in material too hard to mill.

lathe turning A machining operation in which a round workpiece is rotated against a stationary cutting tool to remove material and produce external or internal forms.

layout The operation of creating a part outline on the surface of a workpiece, using layout tools such as scribers, dividers, and prick punches, to show a machinist the part shape that acts as a guide during machining.

limited (nonrenewable) energy sources Sources of energy that are not replaceable and will eventually be completely used up, such as coal, petroleum, natural gas, and uranium.

linear motor A high-force motor which eliminates the need for ball screws and any other mechanical means of driving the machine axes *(See figure.)*

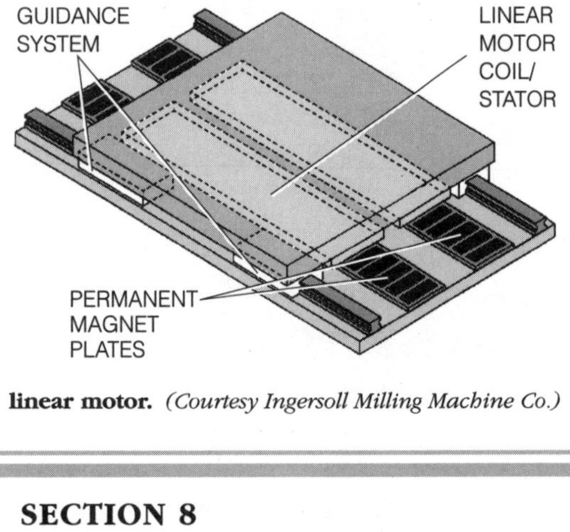

linear motor. *(Courtesy Ingersoll Milling Machine Co.)*

□ Magnetic force alone drives the machine axes faster and more accurately than do conventional machines.

live tooling The cutting tools used on combination turning-milling centers are held in a special turret; each tool has its own individual drive.

load balancing A means of distributing the workload so that manufacturing takes place efficiently on the available machine tools.

machining The process of producing a new shape on a workpiece and cutting or shaping it on conventional or CNC machine tools.

machining center A machine tool, similar to a milling machine, that is CNC-controlled to perform a wide variety of machining operations automatically under the control of the part program.

□ Common types are the *vertical, horizontal,* and *universal* machines.

manufacturing cell A group of machines combined to perform all operations on a part before it leaves the cell. *(See figure.)*

manufacturing cell. *(Courtesy Cincinnati Milacron, Inc.)*

□ The cell and part movement are computer-controlled for automatic operation.

□ A cell can consist of one or more of the following: CNC machining center, CNC turning center, parts-handling system, and automatic part inspection.

manufacturing engineering Planning the layout of tools, machines, materials, and people on a production line, to produce the most efficient and cost-effective manufacturing operation.

market research Research conducted to determine what products are required, and who is likely to buy these products.

mass distribution Products that are widely marketed in retail and wholesale stores in any one country, or even worldwide.

mass media Any medium, such as newspapers, radio, magazines, and television, that is used for advertising to reach as many people as possible.

mass production Another name for continuous manufacturing in which the product is moved along the production line continuously.

material-handling system The transportation systems that include accessories such as a bar feeder, loader/unloader, and robot which are used to move materials, supplies, and work in progress, through a plant to increase productivity. *(See figure.)*

automated materials-handling system.
(Courtesy Giddings & Lewis Inc.)

materials resource planning (MRP) A planning technique used for ordering raw materials, parts, and subassemblies.

mechanical energy The energy found in moving objects, such as cutting tools, twist drills, and rotating saw blades.

mechanical fastening Material-combining processes that use threaded devices (screws, nuts, bolts, clamps) or nonthreaded fasteners (nails, straps, fabric thread, pins) to hold materials together.

metalcutting, material cutting Any machining process that uses some form of cutting tool to cut or shape metal or material into a new form, or give a workpiece a new configuration.

metalforming Any manufacturing process in which products are given new shapes by casting, or by some form of mechanical deformation such as forging, stamping, or bending.

metalworking Any manufacturing process, such as design and layout, heat treating, material handling, or inspection in which metal is processed or machined so that the workpiece is given a new shape.

methods engineering The sequence-planning processes required to make parts and assemble finished products.

microslicing A process used for cutting very small or thin parts from expensive materials such as silicon, germanium, and other computer-chip materials.

milling A machining operation in which metal or other material is held in a fixture or vise and a rotating cutter takes a straight or formed cut from its side or surface.

- ☐ In *vertical milling,* the cutting tool is usually an end mill that is mounted vertically on the spindle.
- ☐ In *horizontal milling,* the cutting tool is mounted horizontally, either directly on the spindle or on an arbor.

modular design, construction The manufacturing of a product in subassemblies that allows for fast and simple replacement of defective assemblies and the use of the product for different purposes.

modular tooling A complete tooling system which combines the flexibility and versatility to build a series of tools necessary to produce a part. Modular systems combine accuracy and quick-change capabilities to increase productivity. *(See figure.)*

modular tooling.
(Courtesy KPT Kaiser Precision Tooling, Inc.)

MRP *See* MATERIALS RESOURCE PLANNING.

new-generation machines Machines that can perform operations that would be very difficult, or impossible, to perform on chip- or nonchip-producing machines.

non-chip-producing machines Machines that form metal to size and shape by a pressing, drawing, bending, extruding, or shearing action. Examples of these machines are the punch press, forming press, hobbing press, and similar.

nondurable Manufactured goods that are destroyed during use, or shortly after.

nontraditional machining A variety of metal-removal operations that use chemical, electrical, mechanical, and thermal processes for machining workpieces.

octahedral hexapod A relatively new innovation in machine tool design consisting of a computer-controlled machining head, supported by six computer-controlled struts containing ball screws driven by servo motors, and a spindle that floats nearly free in space. *(See figure.)*

octahedral hexapod. *(Courtesy Giddings & Lewis Inc.).*

□ The machine can mill in six axes and is very rigid, fast, and more accurate than a CNC machining center.

offhand grinding The operation of holding a workpiece by hand and bringing it into contact with a revolving grinding wheel on a bench or pedestal grinder.

□ This is one of the most common methods used for sharpening drills and cutting tools in most shops.

operation process chart A graphic display of each part of a product that shows operations and inspections performed.

part families An approach of classifying standard parts, based on their physical characteristics, so that they can be grouped together with similar parts that can be produced using the same manufacturing process.

parting **1.** A lathe operation that cuts off a completed part from a longer bar stock held in a chuck or collet using a narrow, flat-end parting tool. **2.** A stamping operation that performs two cutoff operations in the same stroke, leaving a small amount of scrap between successive blanks cut from a strip of metal.

part orientation The operation of designing the assembly machine, feeding mechanism, and the parts to be assembled, so that they are all aligned before and during the assembly operation.

parts feeder A feeding mechanism that delivers parts to an assembly machine at a regulated rate and oriented properly.

peripheral milling A type of milling operation that machines a finished surface parallel to the rotating axis of a multitooth cutter.

physical coating A type of coating process, such as brushing, rolling, dipping, spraying, plating, or printing, in which a thin layer of a coating material is applied to the surface of another material.

pilot run A practice session, or dry run, in which a system is tested to make sure that all tools and systems operate correctly as planned.

planing A machining operation that creates flat surfaces when the workpiece is reciprocated in a linear motion against one or more stationary single-point tool.

pneumatics A fluid-power system that uses the pressure of air to do work.

polishing An abrasive finishing process that uses abrasive particles attached to a flexible backing to improve the surface finish of a part.

powder metallurgy A widely used procedure for forming precision metal parts and shapes from metal powders that are mixed with other powders, blended, compacted, and sintered into their final form. *(See figure on next page.)*

power brushing A process that uses a power-driven, rotary industrial brush made of metal wires, synthetics such as nylon and polypropylene, or natural animal hairs to deburr, clean, or finish a metal part.

power hacksawing A cutoff process that uses the reciprocating (back-and-forth) motion of a straight saw blade to cut the workpiece.

press forming Any forming or bending operation performed in a hydraulic or mechanical press using special tooling.

primary manufacturing Manufacturing processes that use raw materials, refine these raw materials, and convert them into usable forms of industrial materials.

problem solving A technique that uses the skills of a production team to gain an in-depth understanding of a problem and then applies these ideas to solve the problem.

process layout A plant layout in which machines and tools used to perform similar operations, or processes, are grouped together to reduce product handling time and provide for a cost-efficient manufacturing operation.

production control A manufacturing control system that makes the best use of workers, materials, and

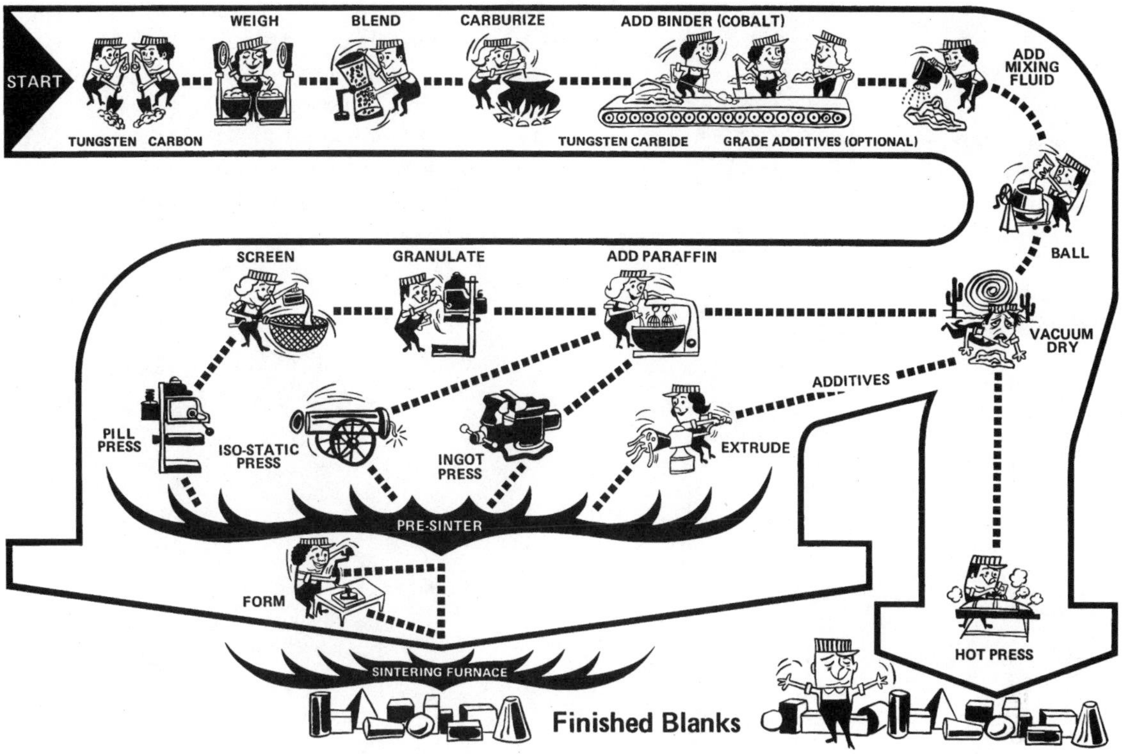

powder metallurgy. *(Courtesy Carboloy, Inc.)*

machines, to meet customer orders and delivery schedules.

productivity A measure of the efficiency of manufacturing, determined by dividing the number of products manufactured by the time required to manufacture them.

product layout A plant layout that provides a continuous and efficient flow from raw materials to finished product.

product planning The planning process, based on market research, that considers what products are required, old ones that should be discontinued, and what changes should be made to existing ones.

profiling The operation of machining contour forms on workpieces by hand, using an end mill in the spindle of a vertical milling machine, automatically using a template, or through a CNC program.

prototype A full-size model of the product that can be tested under conditions as close as possible to those encountered in actual use to make sure that it will function properly.

punching The operation of producing holes on a punch press using special punches and dies. *(See figure.)*
□ Operations such as nibbling, notching, piercing, perforating, slotting, pointing, and marking are very similar to punching.

quality control A list of procedures, programs, and activities, designed to gather and analyze data so that the manufacturing process does not produce defective products.

quality-control (QC) circles Groups of workers, with similar skills or experience, who meet on a regular basis to discuss and solve problems related to the quality of the manufactured goods.

radiant energy Also called *electromagnetic energy*. Radiant energy is both visible and invisible light energy.

rapid protoyping and manufacturing (RP&M) Same as *stereolithography*.

rapid response The ability of a company to respond quickly to any manufacturing challenge to suit customer needs or changes in the market conditions.

raw materials Any material such as metal, ceramic, or polymer (plant and animal) materials obtained from air, earth, or water, before it is converted into industrial materials by primary manufacturing companies. *(See figure on next page.)*

reaming A process that uses a multiedge, fluted cutting tool to smooth, enlarge, and accurately size a drilled or cored hole.

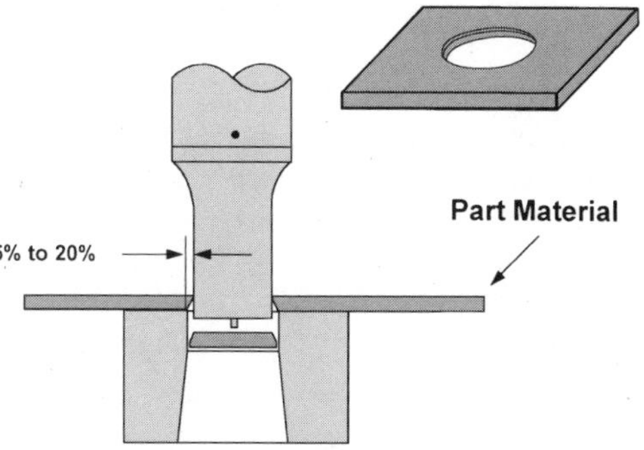

punching. *(Courtesy Dayton Progress Corp.)*

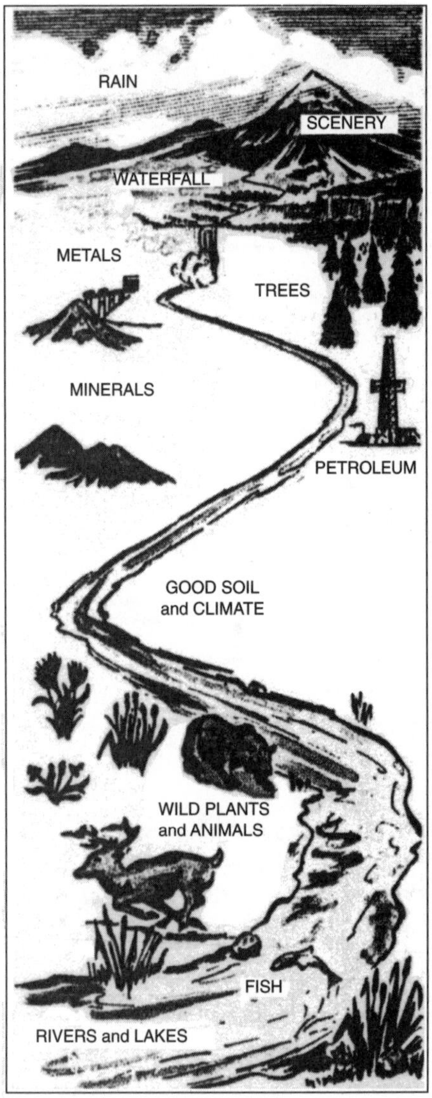

raw materials. *(Courtesy DoAll Co.)*

- □ Reaming can be performed by hand, using a hand reamer, or on machines such as the drill press, lathe, or milling machine using machine reamers.

research and development (R&D) A process performed by design and engineering departments in which prototypes for new products are conceived and developed.

resources planning Management method, normally computer-aided, for cost-effective control of manufacturing support functions, such as inventory, production equipment, and personnel.

- □ MRP was the initial, somewhat limited method; MRP-II implies a more sophisticated system.

roll bending A process for forming round or semiround shapes from sheet or plate steel by feeding the work edgewise into the gap of at least three small-diameter straight rolls.

- □ The size of the gap (space) between rolls and the sheet thickness will determine the size of the bend radius formed.

roll forming A continuous process of feeding metal sheet, strip, or coiled stock, between successive pairs of rolls that progressively shape it until the desired cross section is produced.

sawing An operation in which a machine using a cutting blade with hardened teeth is used to cut material to length and/or a new shape. The common methods of sawing are

Contour bandsaw. *(Courtesy DoAll Co.)*

- □ *Hacksawing.* In a power or manual operation, the blade is moved back and forth through the work; the cutting generally occurs on the forward stroke.
- □ *Cold or circular sawing.* A rotating, circular, toothed blade, similar to a table or radial-arm saw blade, is used as the cutting tool.
- □ *Bandsawing* A continuous flexible-toothed blade rides on driving and driven pulleys (wheels) under tension and is guided through the work. *(See figure.)*
- □ *Abrasive sawing.* Abrasive grains attached to a fiber- or metal-backing wheel cut material similar to a grinding cutoff operation.

scrap The product of a manufacturing process that is not suitable as intended, and cannot be reused easily or economically.

secondary manufacturing Manufacturing processes that use forming, separating, and combining operations to convert standard industrial materials into finished products.

setup The operations required to prepare and organize the machines, tools, and materials necessary for a manufacturing operation.

shear forming A lathe operation similar to spinning, also known as *flow turning*. During the flow process the metal is also thinned by shear forces.

shearing A material-separating process that removes excess material, or reduces the size of standard stock, using blades, punches, dies, or rotary cutters.

shuttle mechanism A reciprocating mechanism, used on an assembly machine, where a part is inserted at the end of each forward stroke.

simulation A term that describes a variety of forecasting techniques, such as modeling, role playing, and simulation.

slicing, slitting An operation that uses a number (gang) of hardened, thin circular blades to cut strips from a piece of sheet steel.

slotting **1.** A *milling operation* that produces slots and grooves, and forms such as T-slots and dovetails, in workpieces. **2.** A *punching operation* used to produce elongated and rectangular holes.

soft technology The technology that considers how it affects the people, environment, and society, as it considers productivity and profits.

spade drilling A drilling operation that uses flat, interchangeable end-cutting blades mounted on the end of a bar or holder, for drilling long or large-diameter holes. *(See figure.)*

spade drill. *(Courtesy DoAll Co.)*

specialization A stage in the development of manufacturing during which workers or manufacturers have gained experience and are considered experts in a particular product or process.

spindle finishing A finishing process in which a number of parts are individually mounted on spindles, and then lowered into a rotating tub containing the finishing media.

- □ Generally the spindles rotate between 10 to 3000 r/min, depending on the application, and in some cases the spindles also oscillate up and down.

spinning A chipless deformation process of forming metal cylindrical blanks into various shapes.

- □ The preform or blank is plastically deformed when it is brought into contact with a rotating mandrel (part form) by axial or axial-radial motions of a tool or rollers.
- □ Cones, hemispheres, tubes, cylinders, and other radial forms can be produced in a wide variety of sizes and contours.

spiral milling Also called *helical milling*; an operation in which the table of a milling machine is set at an angle and the workpiece is rotated and fed under a revolving formed cutter to create a spiral form; commonly used for cutting helical flutes on drills, taps, reamers, end mills, and so on.

spotfacing The operation of enlarging the surface around the top of a hole to provide a flat, recessed area for a nut or washer.

standardization A characteristic of mass production that produces standardized parts; often related to precision measurement.

standard stock The common sizes, shapes, and weights in which industrial materials are produced for consumers and industrial use.

- □ Examples include steel coils, sheets, bars, pipes, and other common materials used in manufacturing.

statistical process control (SPC) A quality-assurance method of using performance data to identify product and process errors that lead to the production of faulty goods. Correct analysis of this data should lead to correcting the errors and producing only acceptable goods.

stereolithography (SL) The process which takes a CAD design and makes a solid three-dimensional prototype (model) using a combination of laser, photochemistry, optical scanning, and computer software technology. *(See figure.)*

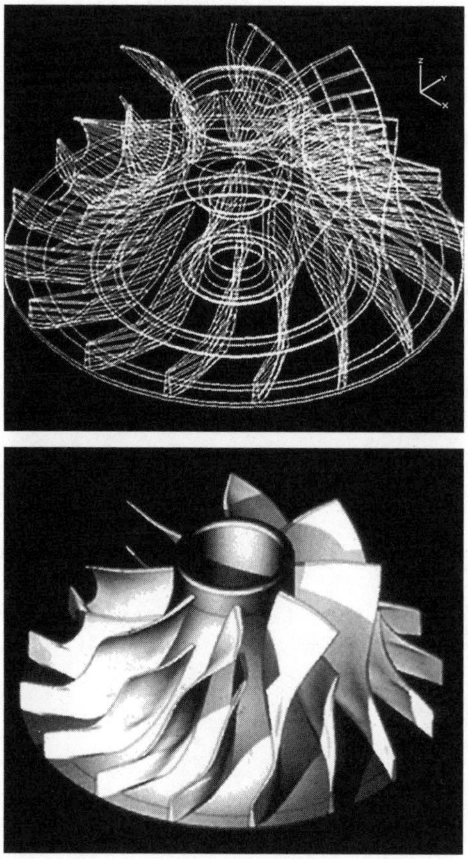

stereolithography.
(Courtesy Society of Manufacturing Engineers.)

strategic planning A listing of existing skills, resources, and capabilities that can be compared to the manufacturing needs of the future.

stretch flanging A concave bending operation that occurs when the material undergoes tension as the flange is being formed. It is good practice to keep the width of stretch flanges to a minimum to reduce the possibility of tension tears.

stretch forming A forming process widely used in the aircraft industry to produce parts with large radius of curvature by primarily applying tensile forces to stretch the sheet metal over a tool or form block.

stretching A material-forming process that uses tensile force to pull materials into desired shapes.

subsystem A system that works in conjunction with other subsystems to make a larger system function more efficiently. *(See figure.)*

technology transfer The transfer of knowledge about machines, tools, materials, and processes used in communication, transportation, manufacturing, and energy systems.

threading An operation of producing a helical ridge of uniform cross section (thread) around an external or internal form by cutting, turning, and rolling of threads into some type of material.

- □ Common thread forms include the American Standard, Unified, ISO Metric, Whitworth, Acme, Square, and International metric.
- □ All threads (external and internal) are made to standardized specifications so that threads are uniform throughout the world.

tooling The tools and equipment that help industrial personnel make product parts of consistent size, shape, and quality.

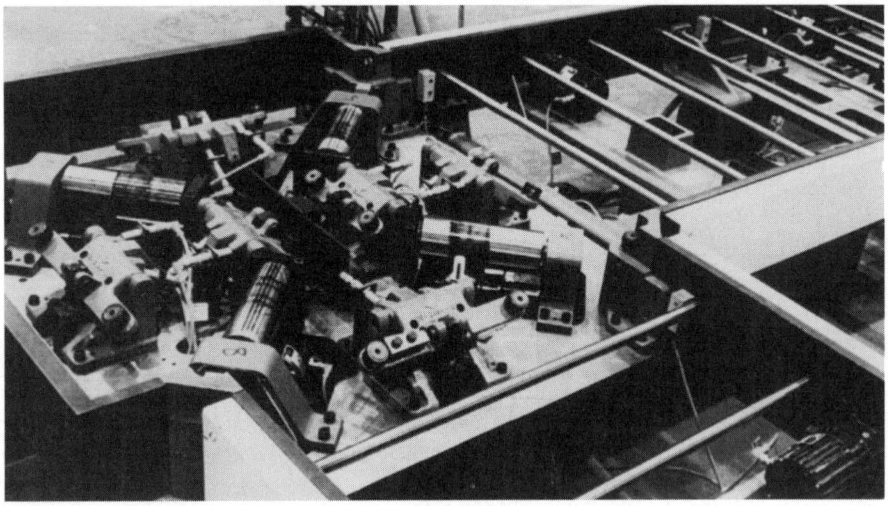

subsystem. *(Courtesy Giddings & Lewis Inc.)*

synchronization A characteristic of mass production that plans for the machines, tools, materials, workers, and processes to be in the right place at the right time.

synchronous transfer A part-transfer system in which all parts are progressively moved to the next work or tooling station at the same time.

tapping A hand or machining operation where a tap, with teeth on its periphery, cuts internal threads in a predrilled hole having a smaller diameter than the tap diameter.

- □ Hand taps have a square on the driving end for a tap wrench; machine taps are fitted into a driving chuck and have spiral flutes to assist in removing chips from a hole.

technological literacy The ability to have a basic understanding of how technology works, to be able to adapt to changes created by new technologies, and to understand how new technologies will affect people, society, their work, and the economy of a country.

tool-monitoring system The most common tool-monitoring system measures the force or load on the machine spindle that is required for a machining operation. Once the load is exceeded, which indicates a dull tool, the feed and speed are generally reduced automatically. *(See figure on next page.)*

trepanning The operation of drilling deep holes that are too large to be drilled by gun drills or high-pressure coolant drills. The trepanning tool consists of a heat-treated alloy-steel body with carbide-tipped replaceable inserts and two carbide-tipped replaceable wear pads.

turbine An energy converter that is used to change the mechanical energy of water, wind, or steam into rotary mechanical energy.

turning The operation of revolving a piece of round stock held in a chuck, between centers, or mounted on a faceplate against a stationary cutting tool, to create a cylindrical shape. Turning operations can be performed on lathes, machining centers, or other machine tools where the rotating work is fed past a stationary cutting tool.

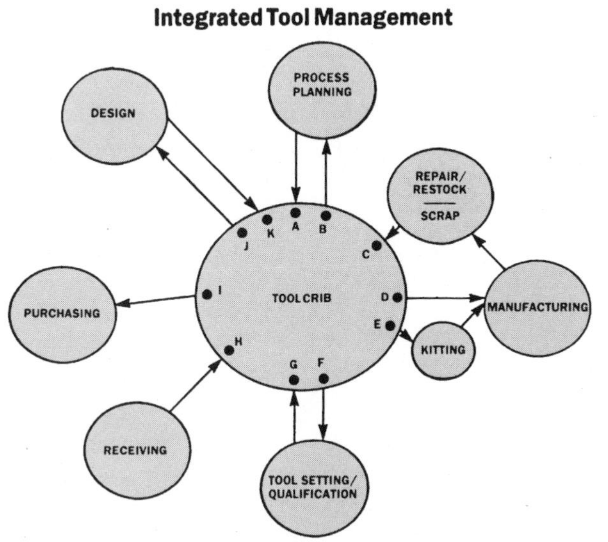

tool-monitoring system.
(Courtesy Society of Manufacturing Engineers.)

□ Common turning operations include rough and finish turning, taper turning, step turning, chamfering, facing, and cutting threads.

turning center A machine tool, similar to a lathe, that is CNC-controlled to perform a wide variety of machining operations automatically under the control of the part program. *(See figure.)*

undercut **1.** A section of a diameter, by a shoulder or at the end of a thread, that is cut below the diameter to provide accurate seating on assembly for a mating part. **2.** In CNC applications, a cut shorter than the programmed cut resulting after a command change in direction.

vacuum metallizing A process in which the surfaces of a part are thinly coated with metal by exposing them to the vapor of metal that has been evaporated under vacuum.

variable costs The manufacturing costs that increase or decrease with the number of parts produced.

variax A type of machining center which has a lower platform to hold the workpiece, and an upper platform containing a free-floating spindle to drive the cutting tool. The two platforms are connected by six legs which form triangles. *(See figure.)*

variax. *(Courtesy Giddings & Lewis Inc.)*

verification The stage in the design engineering process that uses prototypes to make sure that a product actually works as it is supposed to.

turning center. *(Courtesy Cincinnati Milacron, Inc.)*

vibratory bowl feeder A hopper-conveyor mechanism that uses a vibratory motion along with a spring suspension system, to deliver properly oriented parts to an assembly machine at a specified rate.

waterjet machining A process that uses a low-pressure (about 250 psi, 1.7 MPa) stream of water or water-based fluids primarily to remove burrs or for a finishing process.

work in process The components that are in the process of being manufactured before the final product is assembled.

zero-defect manufacturing A quality-improvement program whose goal is to continually improve manufacturing until no defects are produced during manufacturing operations.

SECTION 9

Measurement and Inspection and Quality Control

Paul J. Wanner
Manufacturing Technology Department
Advanced Technology Center
Clackamas Community College
Wilsonville, Oregon

Ken "Kun" Li
Manufacturing Engineer
GEC Precision Corporation
Wellington, Kansas

Measurement and inspection refers to the processes of making a part to the correct size and the verification of that size. Accurate measurement-and-inspection systems are very important to develop manufacturing operations that produce high-quality products. New measuring tools and instruments, capable of measuring in millionths of an inch, can check parts while they are being manufactured and feed this information back to the machine control unit to take corrective action, if necessary. Digital and noncontact laser tools reduce many reading errors common with conventional tools. The terms included in this section are involved primarily in common measurement systems, practical applications of inspection, and the tools of inspection and inspection procedures. There is an emphasis on terms of quality control and quality improvement.

Measurement and Inspection

accuracy (measurement) The deviation of a part or a measuring system from a known value. Also defined as an *unbiased true value.* The difference between the average of several measurements and the true value.

acting size The distance between the working gaging surface or point and the gaging surface in contact with the stack of gage blocks.

actual size The measured size of a part after its has been produced. Also known as *produced size.*

acute angle Any angle that is less than a right angle (90°).

addendum The height of a gear tooth above the pitch circle or the radial distance between the pitch circle and the top of the tooth.

adjustable snap gage A caliper-type gage used to determine if the external dimension (size) of a part is within specified limits. It consists of a C-frame with a fixed anvil and has an adjustable upper jaw. *(See figure.)*

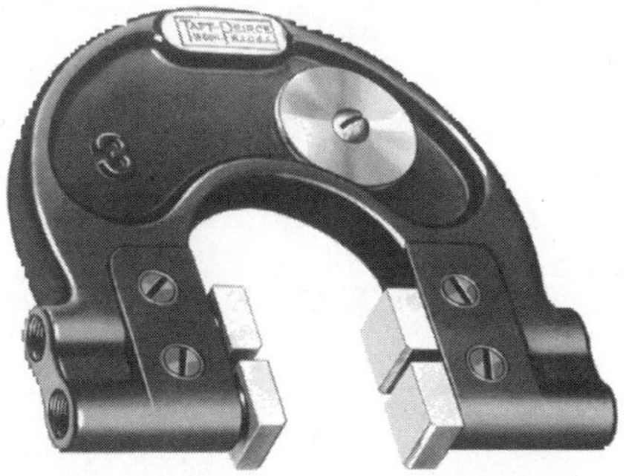

adjustable snap gage. *(Courtesy Taft-Peirce Mfg. Co.)*

adjusting slots Slots provided in ring gages for the expansion or contraction of the gage size. They are generally radial and end in an adjusting slot hole.

air bearings An accurate form of support for the moving axis of coordinate measuring machines and other precision measuring instruments where pressurized air is used to create a cushion between machineways and bearing pads.

air gages A form of comparison measurement used to compare workpiece dimensions with those of a master gage by means of air pressure or flow.

alignment A means of aligning the coordinate system of a machine to the coordinate system of the part.

allowance The difference in the dimensions between two mating parts, for example, the difference between the maximum diameter of a shaft and the minimum diameter of the mating bore.

□ *Formula.* Allowance = maximum material condition (MMC) of internal feature − MMC of external feature.

American Gage Design Standard Designation that gages are made to the design specifications of the American Gage Committee.

amplification The movement of a measuring device's contact points in relation to the amount of the needle movement or readout on the scale.

angle An angle is the difference between two rays with endpoints intersecting at the vertex. The degree of difference between two features which eventually meet in a point.

□ An angle can be measured in units of degrees, minutes, and seconds.

□ The symbols are degree (°), minutes (″), and seconds (′).

□ There are 360° (degrees) in a complete circle, 60″ (minutes) in a degree, and 60′ (seconds) in a minute.

□ *See* ACUTE ANGLE, OBTUSE ANGLE, SUPPLEMENTARY ANGLE, COMPLEMENTARY ANGLES, RIGHT ANGLE, *and* REFLEX ANGLE.

angle of thread The angle included between the sides of the thread measured perpendicular (90°) to the axis of the thread.

angular measurement The part of a circle included between two radii that is generally expressed in degrees of a complete circle divided into 360 equal parts.

□ Angles can be measured with a bevel protractor and calculated using trigonometry.

angularity The condition of a surface, axis, or center plane that is at a specified angle (other than 90° or 0°) from a datum plane or axis.

annular plug gage A plug gage in which the gaging surface is in the form of a ring supported by two or more arms and with the handles fastened to the center where the arms cross.

anvil The fixed nonadjustable block of a snap or other gage such as micrometers, the face of which is used as the datum (reference plane) from which the dimension is taken.

arithmetic average (AA) The average distance between peaks (high points) and valleys (low points) of surface roughness.

attribute gages Gages that measure on a GO/NOGO basis. An example is a plug gage for a hole. *(See figure.)*

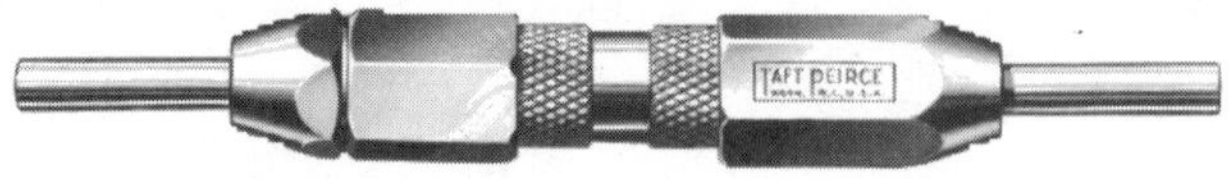

attribute gage. *(Courtesy Taft-Peirce Mfg. Co.)*

attributes Characteristics that must (or are chosen to) be studied in terms of whether they are good or bad, on or off, GO or NOGO.

axis of a screw The centerline running through the screw, about which linear motion occurs during rotation.

axis of measurement The line of measurement about which linear motion occurs. A micrometer spindle creates an axis of measurement.

B1.2 The ANSI/ASME standard for gages and gaging for unified inch screw threads.

B4.4 The ANSI standard for inspection of workpieces.

B5.54 Methods for the performance evaluation of CNC machining centers.

B46.1 The ANSI/ASME standard for surface texture.

B89.1.12 The American Standard (ANSI/ASME) for determining coordinate measuring machine accuracy and performance.

B89.3.1 The ANSI standard for the measurement of out-of-roundness.

B89.3.4 The American Standard (ANSI/ASME) for axis of rotation.

B89.6.2 The American Standard (ANSI/ASME) for temperature and humidity environment for dimensional measurement.

ball bar A three-dimensional gage consisting of two precision balls separated by a bar; used to determine volumetric accuracy.

base channel The outer channel of a channel-type end standard holder that is equipped with the clamping mechanism.

base of thread The bottom portion of the thread closest to the centerline of the screw; also called the *root* on external threads.

basic dimension Provides a basis or dimension from which deviations in size and form are measured; a theoretically perfect dimension that has no tolerance. *(See figure on next page.)*

□ Used in conjunction with geometric dimensioning and tolerancing (GDT) to show the exact size, location, form, or orientation of a part or feature.

□ Represented on a drawing by placing a rectangular box around the dimension and not stating any tolerance.

basic size The exact theoretical size from which all limiting variations (tolerances) are measured.

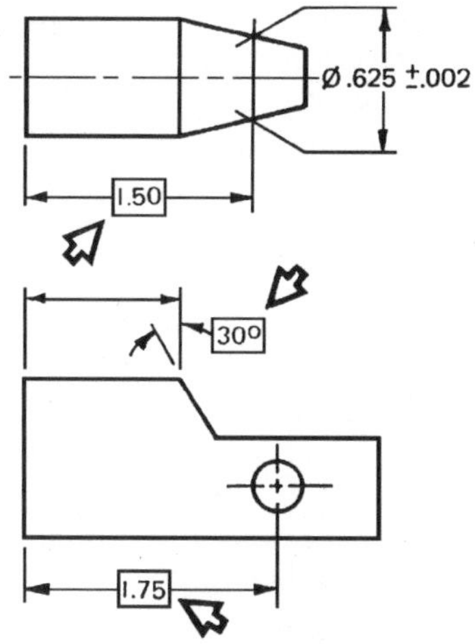

basic dimension.
(Courtesy Engineering Drawing and Design, Glencoe/McGraw-Hill.)

bench micrometer An instrument that has a spring-loaded, constant-correct-pressure anvil. Its indicator comes to the zero position to signal that the correct pressure has been applied to the workpiece and the reading can be taken.

bend test A test used in determining relative ductility and toughness of metal that is to be formed.

bevel An angled edge other than 90°.

bevel vernier protractor A precision angle-measuring device that contains a vernier scale enabling accuracy, discrimination, and precision to 5 minutes (5″) or 1/12th of a degree.

bias in measurement This bias occurs when one uses a measuring instrument incorrectly. An example would be when using a caliper to measure a part and squeezing the jaws to get the desired reading, not the actual reading.

bilateral tolerance A tolerance that is allowed to vary in two directions from the specified dimension. *(See figure.)*

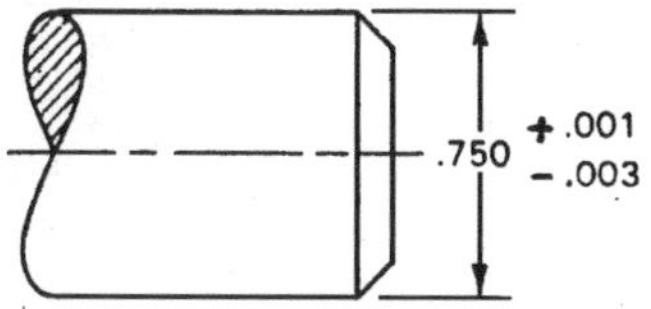

bilateral tolerance. *(Courtesy Kelmar Associates.)*

bolt circle A group of holes on a circular centerline sharing a common center axis.

bore An inside diameter (hole).

bore gage A precision measuring device used to accurately measure inside diameters. The reading is taken from an indicator head.

brale A penetrator used with a Rockwell hardness tester; usually spheroconical in shape and made of diamond.

Brinnell hardness test Performed by pressing a 10-mm-diameter hardened steel or carbide ball under a load into a flat surface. The diameter left by the impression is measured with a microscope.

- □ Once the diameter of the impression and load are known, the Brinell hardness number (BHN) can be obtained from Brinell hardness tables.

calibration The process of comparing one standard against a higher-order standard of greater accuracy, or comparing an instrument of known accuracy with another instrument.

calibration interval A specific amount of time between calibrations where the accuracy of gages (test equipment) is considered valid.

calibration labels Identification stickers used to track
When the tool was calibrated.
Who calibrated it.
When it is next due for calibration.

caliper A comparison instrument used to test the dimensions of a workpiece to an accuracy of .015 in; commonly used with steel rules. These indirect or transfer measuring tools are available in four types characterized by the mechanical joint that connects the two sides of the unit: spring, firm, lock, and hermaphrodite.

caliper gage for external members Similar to a snap gage and for internal members, it is similar to a plug gage.

caliper-type end standard holder A variation of a channel-type end standard holder that operates in a manner similar to a machinist's caliper or dividers.

cartesian coordinate system A grid system consisting of three mutually perpendicular number lines that is used to describe points in space in terms of *X, Y,* and *Z* axes. *(See figure.)*

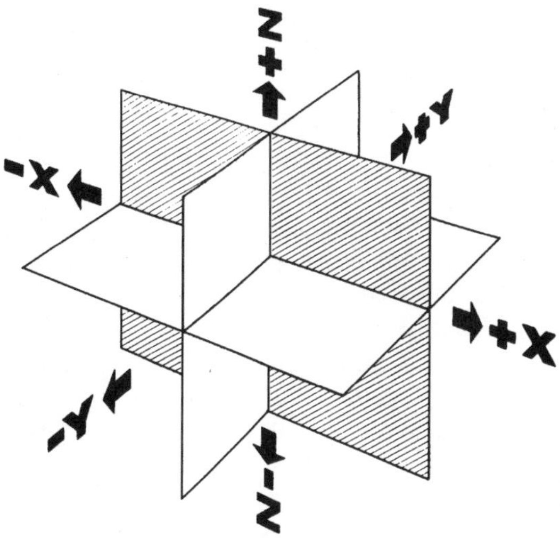

cartesian coordinate system. *(Courtesy Superior Electric Co.)*

Charpy test A test for impact strength or notch toughness that uses a notched specimen supported at both ends as a simple beam.

- □ The specimen is broken by a single blow from a falling pendulum.
- □ The energy absorbed is measured by the subsequent rise of the pendulum

□ *See* IZOD TEST *and* IMPACT TEST.

clearance The difference in the dimensions of mating parts or the difference between the smallest diameter of a shaft and the largest diameter of the mating bore.

□ *Formula.* Clearance = LMC of shaft − LMC of hole.

comparator A mechanism for amplifying and measuring a small displacement. The amount of magnification must be sufficient to permit a visual reading.

□ It is used to compare the unknown dimensions of a part with some standard, such as gage blocks.

complementary angles When the sum of two angles equals 90°, they are called *complementary angles.*

coordinate measuring machine (CMM) A precision measuring device consisting of a granite or ceramic surface plate, a bridge for the *X* axis, ways for the *Y* axis, and a *Z* rail onto which a touch tip probe or hard probe is mounted.

□ This machine is used for checking part features for size, location, shape, and distance from other features or from a part zero.

□ It may also may be used to scan or reverse-engineer a component for purposes of duplication.

core diameter The same as minor diameter. It is the smallest diameter of a screw or nut.

cosine error A linear measurement error that occurs when the indicator travel is at an angle to the surface of the workpiece being measured.

crest The top surface joining the two sides of a thread.

crest clearance The space between the crest of a thread and the root of its mating thread.

cutoff The electrical response characteristic of the instrument that is selected to limit the spacing of the surface irregularities to be included in the roughness measurement (.03 if none is specified).

datum A reference plane from which measurements are taken. It is considered to be a theoretically exact surface, plane, axis, centerplane, or point from which dimensions for related features are established.

□ Represented by processing equipment such as surface plates, angle plates, and machine tables.

datum feature The actual feature of the part that is used to establish a datum.

datum-feature simulator A surface of adequately precise form, such as a surface plate, a gage surface, or a mandrel, that contacts the datum feature and establishes a simulated datum.

datum reference frame Three mutually perpendicular planes that intersect at what is called an *origin* from which measurements are taken.

dedendum The depth of tooth space below the pitch circle or the radial dimension between the pitch circle and the bottom of the tooth space.

defect An undesirable flaw in a part such as scratches, dents, or any other flaw that would cause the part to be rejected. It could be a single characteristic on a part, or each part can have several defects.

defective A part that has one or more defects is still called a *defective part.*

degree A division of a circle (1/360th of a circle) that represents angular measurement. Represented by a symbol such as 45°.

depth of engagement The length of contact of two mating parts.

depth of thread The depth, at right angles to the axis of the screw, between the crest and the root of the thread.

deviation The difference between a measurement and its stated value.

dial bore gage Consists of three spring-loaded centralizing plungers in the head, one of which actuates the dial indicator. *(See figure.)*

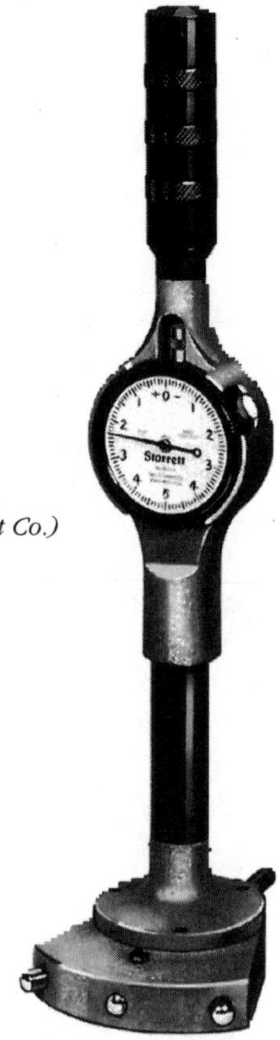

dial boar gage.
(Courtesy L. S. Starrett Co.)

□ This provides a quick and accurate method of checking hole diameters and bore sizes for out-of-round, taper, bell-mouth, hourglass, or barrel shapes.

□ It must be set to size with a master gage, and then the hole size is compared to the master.

dial/dial test indicators Dial indicators are used to compare sizes and measurement to a known standard and to check alignment of machine tools, fixtures, and workpieces prior to machining.

□ It operates on a rack-and-pinion gear principle, and any movement of the spindle is magnified and transmitted to a hand or pointer over a graduated scale.

□ The graduated dial may be adjusted and locked in any position.

differential measurement The use of a device that transforms actual movement into a known value such as a dial indicator.

dimension A numerical value representing the number of units of length of any object that describes size, shape, location, geometric characteristics, or surface texture.

direct measurement Where the standard is directly used on a part and a reading can be taken, such as a steel rule.

discrimination The direct distance between two lines on a scale or the fineness of the instrument's divisions of unit of measurement.

divider A divider consists of an adjusting screw, a spring, and two sharp-pointed legs that are used for transferring measurements from steel rules to the part or vice versa.

draft Tapered shapes in parts allowing them to be easily taken out of a mold or die.

drift hole/slot A hole or slot through which a drift (tapered piece of metal) may be inserted for ejecting a part or tool.

eddy-current test A nondestructive electromagnetic test in which eddy-current flow is induced into the test object, detecting changes in flow caused by variations in the object.

end standard The functional gage block that is designed for use in the end standard holder.

end standard holder A device designed to securely hold a buildup of gage blocks and end standards in such a manner that the gaging surfaces at each end are completely exposed and unobstructed.

environment All the conditions surrounding and affecting manufacture or inspection and measurement of a part or product such as temperature, humidity, and vibration.

Erichsen test A ductility test consisting of a restrained (except the center) piece of sheet metal that is deformed by a cone-shaped spherical-end plunger until fracture occurs. *See also* OLSEN TEST.

extension channel The inner channel of a channel-type end standard holder.

external thread A thread on the outside of a part, such as a bolt.

feature The general term given to any physical portion of a part such as holes, slots, surfaces, and pins.

feature of size The term given to any feature that has size associated with it.

fit The dimensional relationship between mating parts. It can be classified as running and sliding, or location and force fits.

fixture A device for locating and holding the workpiece for machining operations.

flatness The maximum peak-to-valley deviation of a surface.

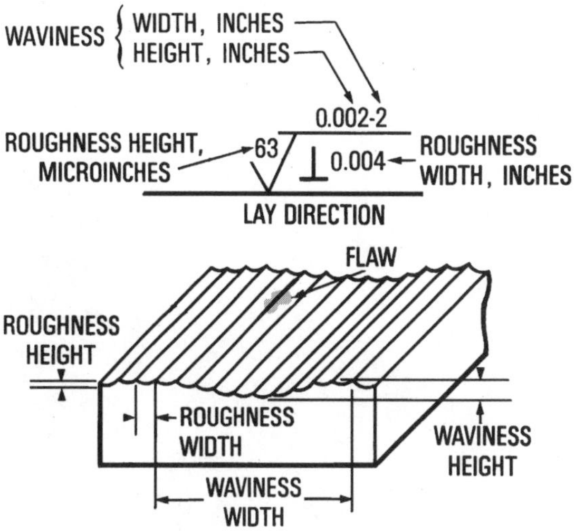

flaw. *(Courtesy Kelmar Associates.)*

flaw A defect in the surface that is not part of the surface roughness or lay. *(See figure.)*

□ Flaws may be deep tool marks or a void in a surface.

□ The effect of flaws should not be included in roughness average measurements.

fluorescent penetrant inspection Using a fluorescent liquid penetrant on a surface to locate flaws with the use of an ultraviolet light.

form A category of geometric dimensioning and tolerancing symbols that consist of straightness, flatness, circularity, or cylindricity of a feature.

frame The body of a gage to which anvils, pins, buttons, and other items are attached.

fringe One of a number of light or dark bands produced by the interference of lightwaves such as those seen through an optical flat when the flat is placed on the surface of a gage block.

full indicator movement (FIM) The total movement of an indicator's needle when applied to a surface to measure its variation.

functional gage A GO/NOGO gage, made for one size or part, that represents the function of the part to be measured.

gage An instrument for determining whether the dimensions of a part are within specified limits.

gage block dimension The dimension of gage blocks required between any two end standards to achieve the desired working dimension.

gage blocks Rectangular blocks of hardened and ground alloy steel that have been stabilized through alternate cycles of extreme heat and cold until the crystalline structure of the metal is left without strain. *(See figure on next page.)* They are the accepted industrial standard of accuracy that provide a means of maintaining sizes to specific standards or tolerances.

□ The two measuring surfaces are lapped and polished

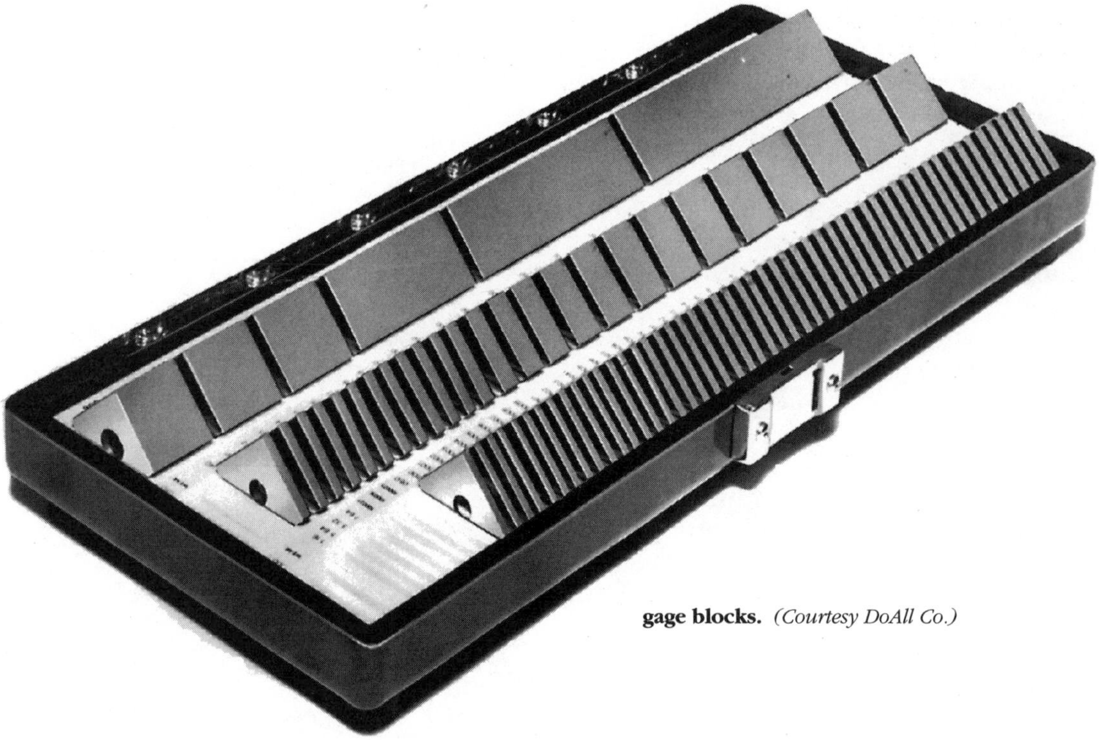

gage blocks. *(Courtesy DoAll Co.)*

to an optically flat surface.

- □ Available in three grades accurate to specific size within millionths of an inch.
  1. *Master blocks.* Grade AA: .000002-in accuracy.
  2. *Inspection blocks.* Grade A: .000005-in accuracy.
  3. *Working blocks.* Grade B: .000010-in accuracy.
- □ They enable high rates of production and have made interchangeable manufacture possible.

gage pin A pin used to determine hole sizes. Gage pins are available in plus and minus sets and produced within .0002-in accuracy.

- □ They are available in different grades that determine accuracy levels.

gaging The process of determining if the dimensions of a part conform to the sizes specified.

gaging button The adjustable gaging part of a gage.

gaging member The part of a gage that is accurately finished and comes in contact with the piece being measured.

gaging pin A pin used as the adjustable member of an adjustable snap gage.

gaging section The portion of a gage that comes into contact with the part being measured.

geometric dimensioning and tolerancing (GDT) A term associated with the use of the ANSI/ASME Y14.5M-1994 standard that describes a design philosophy and a standard set of symbols for dimensioning and tolerancing prints or drawings.

geometric tolerance The general term applied to the variation allowed in form, profile, orientation, location, and runout.

GO/NOGO gage A gage constructed so that the GO portion allows the part being measured to pass over the gaging surface but prevents the part from passing over the NOGO section. The dimensions of the GO/NOGO surfaces are within the limits specified.

graduations Accurate divisions on a scale, dial face, micrometer, or other tool.

granite A dense, wear-resistant mineral that is capable of being finished to excellent flatness and is used in the construction of surface plates.

hardness The ability to resist penetration, usually measured by a Rockwell or Brinell hardness tester. *(See figure.)*

hardness. *(Courtesy Praxair, Inc.)*

hard probes A mechanical solid probe consisting of a precision ball or tapered shape being mounted to a shaft.

height gage Consists of a precision finished base, a beam that is at a right angle to the base, and a scriber or indicator attachment. These gages are used for direct or comparison measurements and range from

10 to 72 in in height.

helix angle (lead) The angle made by a thread or screw at the pitch diameter with the plane perpendicular to the screw axis.

impact test A single-blow impact test, like the Charpy or Izod test, used to determine the behavior of materials that are subjected to high rates of loading, such as bending, tension, or torsion.

included angle The combined angles of two smaller symmetrical angles.

indexing heads Also called *dividing heads*; used to divide the circumference of a workpiece into equal divisions or angular spacing.

indicating gage A measuring instrument that shows visually the variations in the dimensions of the part being measured. It usually has a pointer or shadow that travels over the face of the graduation scale.

inner gaging surface The lapped surface of an end standard that contacts the gage block stack when mounted in a holder.

in-process gaging When inspection occurs at the point of production with a contact or noncontact measuring tool such as a laser scanner system.

inside micrometer A tool used to measure inside dimensions from 1.50 to over 100 in. It consists of a micrometer head and extension rods of various lengths.

inspection The process of examining or comparing a product with various tools and techniques to determine its conformance to the required specifications.

inspection gages Gages used by inspectors to determine if a part being inspected comes within the specifications.

inspection record Recorded data concerning the results of inspection actions.

interference bands The dark bands observed when using an optical flat to check a flat surface. *(See figure.)*

interferometer An optical instrument using the principles of interferometry for the measuring and calibration of precise dimensions.

interferometry The science of measuring by the use of light.

INTERMIK A hole-measuring device consisting of a head with three contact points spaced 120° apart, attached to a micrometer-type body.

internal thread A thread on the inside of a part, such as on a nut.

Izod test A test for impact strength or notch toughness.
- □ A notched specimen is supported at both ends as a simple beam and is broken by a single blow from a falling pendulum.
- □ The energy absorbed is measured by the subsequent rise of the pendulum.
- □ *See* CHARPY TEST *and* IMPACT TEST.

Knoop hardness A test for microhardness using a pyramid-shaped diamond indentor.

laboratory gages Gage blocks of a high degree of accuracy (2 μin/in) used in a laboratory for masters to check fine measurements and other gage blocks. Generally used in a temperature- and humidity-controlled room.

laser A system or a device used to generate a very intense, spectrally pure, and spatially coherent beam of light, making use of stimulated emission from excited atoms in an optical resonator.

LaserMike An optical micrometer that is highly accurate because of a helium-neon laser beam that is projected in a straight line, with almost no diffusion, to mirrors mounted on the shaft of a precision electronic motor. *(See figure on next page.)*

lay The direction of predominant surface pattern that may be caused by tool marks or grain.
- □ Measurements should be taken in a direction perpendicular to the lay of the surface pattern.

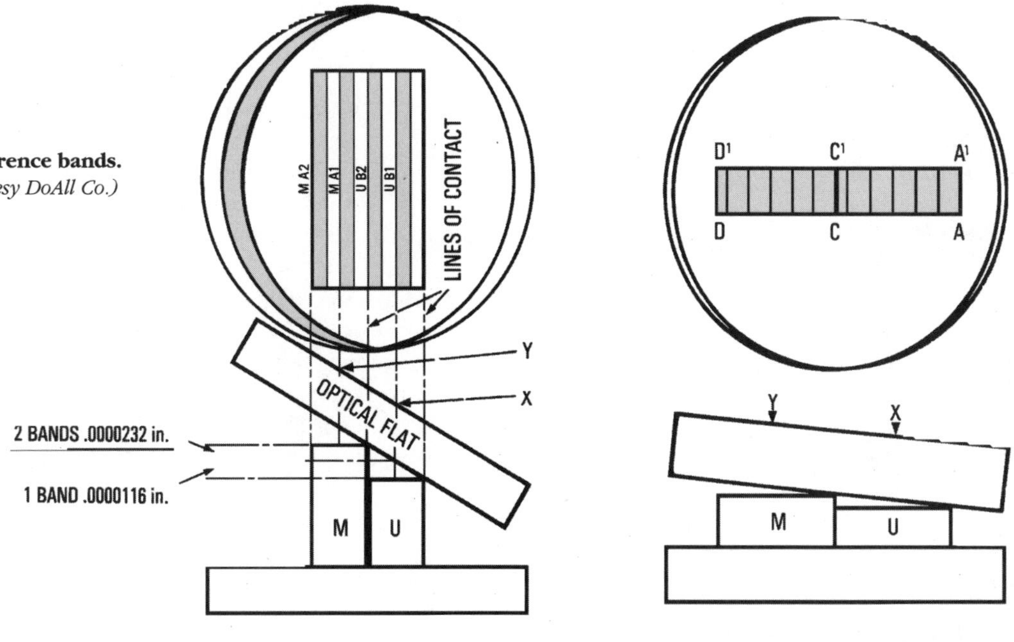

interference bands.
(Courtesy DoAll Co.)

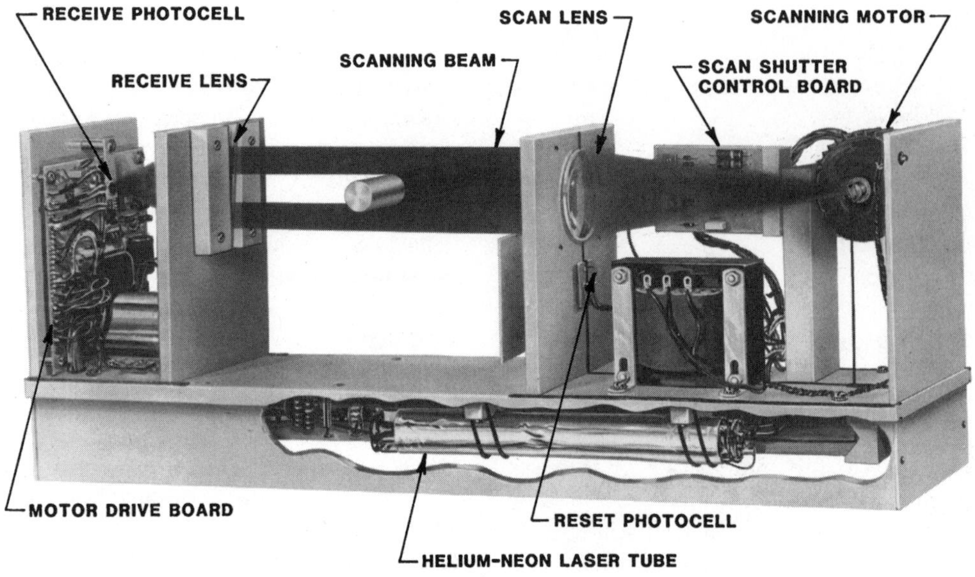

LaserMike. *(Courtesy LaserMike Division of Techmet Co.)*

lead The distance through which a screw will advance parallel to the axis during one complete revolution.

least material condition (LMC) The condition in which a part has the least material or weight within the given limits of a dimension. The opposite of maximum material condition.

□ *Example.* The smallest external dimension or the largest internal dimension.

lightening holes Holes drilled or cored in gages to reduce the weight.

limits Maximum and minimum dimensions specified for a part.

linear accuracy A nonspecific term sometimes used in reference to positional accuracy or to axial length accuracy.

location A category of GDT symbols that control the locations of features to other features of reference datums. The characteristics of position, concentricity, and symmetry.

locking slot A slot that passes through the wall of a thread ring gage, and when used with the gage locking device, it permits the changing of a gage size.

lower control limit *See* CONTROL LIMITS in the next (*Quality Control*) section.

magnetic-particle inspection A nondestructive test which uses fine magnetic particles that are applied to a magnetized part to determine the existence of surface cracks and similar imperfections.

major defect A defect other than critical that may cause the product to fail, cause poor performance, shorten life, or prevent interchangeability.

major diameter The outside or largest diameter of a screw or internal thread.

marking disk A plate attached to a gage on which the size or identification for the gage can be marked.

master gages Gages used for checking other gages.

maximum material condition (MMC) The condition in which a part contains the most material or weight within the given limits of a dimension. The largest possible external (shaft) dimension. The smallest possible internal (hole) dimension.

measured surface The surface of a movable measuring tool from which the measurement is taken, such as a micrometer spindle.

measurement error The difference between the measured value and the actual value.

measurement pressure The pressure used to measure the part should be the same as the pressure used to calibrate the tool.

measurement standard A standard of measurement that is a true value and is recognized by all as a basis for comparison.

measuring and test equipment All devices used to measure, gage, test, inspect, diagnose, or examine products to determine compliance with the required specifications.

mechanical optical comparator Also called a *reed-type comparator*; combines a reed-type mechanism with a light beam to cast a shadow on a magnified scale to indicate the dimensional variation of the part.

metalcutting dynometer A device used to measure the cutting forces incurred during the machining process.

metrology The science of measurement.

microinch One-millionth (.000001) of the U.S. standard inch, abbreviated μin.

micrometer (Formerly *micron.*) One-millionth (0.000001) of a meter, abbreviated μm.

micrometer A measuring tool used to take measurements accurate to within .001 in or 0.02 mm on various shapes of material. The most common micrometers are *outside, internal,* and *depth* micrometers. *(See figure on next page.)*

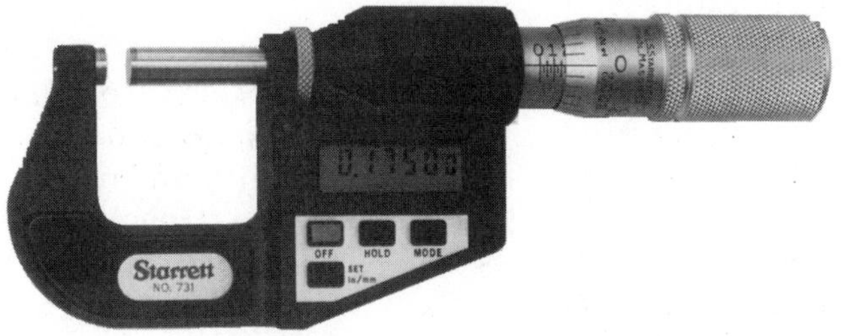

micrometer. *(Courtesy L. S. Starrett Co.)*

- □ All are based on the relation of a screw's circular movement to its axial movement.
- □ Micrometers are available in standard, direct-reading, and digital.
- □ A laser micrometer can measure to an accuracy of millionths of an inch.

micrometer depth gage A gage consisting of a flat base attached to a micrometer sleeve with various length rods. It is used for measuring depths of holes, slot recesses, or height of projections.

minor defect A defect that is not likely to reduce materially the usability of the unit, such as a small scratch in a car fender.

minor diameter The dimension from root to root through the axis on an external thread and measured across the crests through the axis on an internal thread.

minus tolerance The variation allowed under the specified dimension.

minute (of angle) One-sixtieth of an angle of 1 degree (°). The symbol is ″.

modulus of elasticity A measure of the rigidity of metal.

moiré fringe principle A superimposed ripple effect that occurs when one family of curves having a regular pattern is superimposed on another so that the curves cross at an angle less than 45°. A new family of curves called moiré appear which pass through the intersections of the original curves. *(See figure.)*

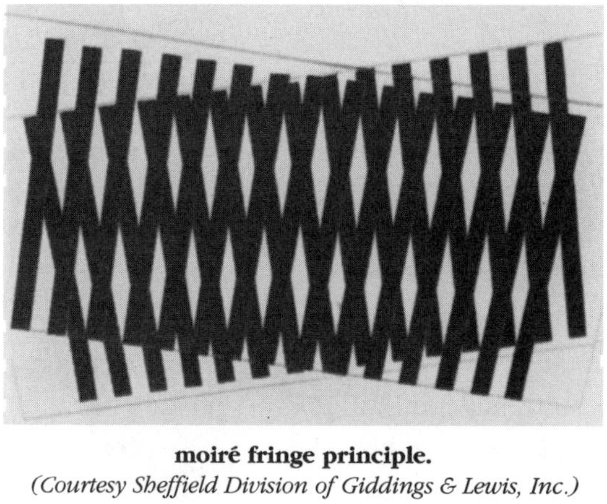

moiré fringe principle.
(Courtesy Sheffield Division of Giddings & Lewis, Inc.)

monochromatic light A light that gives off only one color and is used with an optical flat for purposes of surface flatness inspection.

neutral zone A positive allowance between two mating parts.

nominal size A dimension used for general identification such as stock size or thread diameter.

noncontact probe A coordinate measuring machine (CMM) whose probe is a vision or laser-scanning system that does not contact the part during measurement.

nondestructive test, inspection, and examination (NDT) The term given to a test that does not destroy the part or its performance.

number of threads The number of threads in one inch of the length measured parallel to the axis of a screw.

obtuse angle An angle between 90° and 180°.

Olsen test A ductility test comprised of a restrained (except at the center) piece of sheet metal that is deformed by a cone-shaped spherical-end plunger until fracture occurs. *See also* ERICHSEN TEST.

100 percent inspection The inspection of every part produced is costly because of the time involvement. In certain situations, such as destructive tests, 100 percent inspection cannot be done.

open-set up inspection Use of a surface plate to make setups with other reference planes, such as angle plates, and measure part features with indicating and referencing instruments.

optical comparator (shadowgraph) A device that projects an enlarged shadow of an object onto a screen where it may be compared to lines on a grid or to a master form that indicates the limits of the dimensions or the contour of the part being checked.

optical flats Highly polished pieces of transparent material such as plate glass, optical glass, Pyrex, or fused quartz. They are used for checking the measuring faces of gage blocks and anvil spindle faces of a micrometer. *(See figure.)*

optical flats. *(Courtesy DoAll Co.)*

- □ These are cylinders varying in size from 3/8- to 3/4-inch thickness and from 2- to 4-in diameter, with at least one surface polished so perfectly flat that surface waviness, warp, or irregularity is virtually immeasurable.
- □ A source of monochrome light is recommended for best use.
- □ The light and dark bands (interference bands) are read to determine the flatness of the part being inspected.

orientation The variation in the attitude of one feature relative to a datum feature reference. It consists of angularity, parallelism, and perpendicularity.

origin A zero point or datum.

original inspection The first inspection of a lot of parts prior to any decision of nonacceptance.

parallax error An error in reading an instrument at an angle so that the readings seem higher or lower than they actually are.

parallelism The condition of a surface, axis, or center plane that is equidistant at all points from a datum plane or axis.

perpendicularity The condition of a surface, axis, or center plane that is 90° from a datum plane or axis.

pitch diameter The diameter of an imaginary cylinder that passes through the thread at a point where the groove and thread widths are equal.

pitch of measurement Vertical deviation from a level plane as applied to the travel of a component along a given axis.

pitch of threads The distance from one point on one thread to the same point on the next thread.

plug gage A precision-ground steel cylinder of a specific size used to check the inside diameter of a straight hole.
- □ They are attribute gages that are used to measure diameters on a GO/NOGO basis.
- □ The NOGO plug is usually shorter than a GO plug for easy identification.

plus tolerance Variation allowed over the specified dimension.

polar coordinates Points in space that are described in terms of radius and angle. *(See figure.)*

precision The closeness of agreement between randomly selected individual measurements or test results.

precision measurement The practice of measuring and meeting exacting standards.

precision square A hardened and ground square, fixed or adjustable, that is used for inspection and setup purposes.

profilometer An electronic device that incorporates a stylus to determine surface finish in microinches.

protractor An instrument generally used to measure angles in 1° increments.
- □ Some bevel protractors have a vernier scale that can measure angles to within 5″ (angle minutes) of a degree.

qualification A procedure for establishing true size.

qualification sphere A precision sphere mounted on a post that can be fastened to a machine table and used to test the machine and probes to be used.

radius gages Thin steel blades with forms used to check the size and accuracy of radii of concave and convex corners and shoulders.

reference dimension A dimension without tolerance that is used for information purposes only.
1. It may be a repeat of another dimension.
2. It may be established from other values shown on a drawing.
3. It is represented on the drawing with parentheses around the dimension.

reference gages Also called *master gages*; gages used for testing or checking inspection gages and instruments.

reference surface The surface of a measuring device that is fixed, such as a micrometer anvil.

reflex angle An angle greater than 180°.

repeatability The ability of an observer to obtain consistent results measuring the same part (or set of parts) using the same measuring instrument.

reproducibility The overall ability of two or more observers to obtain consistent results after repeatedly measuring the same part (or set of parts) while using the same measuring equipment.

resolution The smallest change in a measured value that the readout will display.

rework The action taken on a nonconforming product to ensure that it will satisfy the specified requirements.

right angle An angle formed by the perpendicular intersection of two straight lines; an angle of 90°.

ring gage A hardened-steel inspection tool with a very accurate inside diameter. They are attribute gages used to measure outside diameters on a GO/NOGO basis. *(See figure on next page.)*

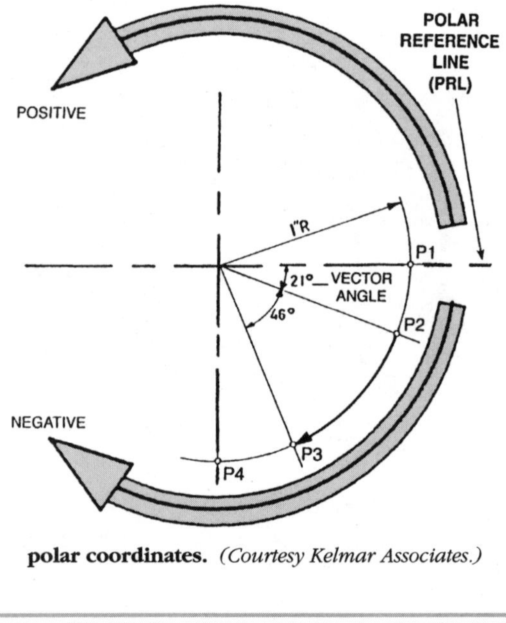

polar coordinates. *(Courtesy Kelmar Associates.)*

ring gage.
(Courtesy Taft-Peirce Mfg. Co.)

□ The NOGO ring usually has a groove on the outside for easy identification.

□ It can be used to reference or set other instruments such as air gages or indicating bore gages.

Rockwell hardness test An instrument that uses a hardened-steel ball or a diamond penetrator under a known pressure to test the hardness of a piece of steel; the hardness is read directly from a dial.

□ It has a wide range of applications from aluminum to carbides.

□ A 120° diamond penetrator, under a 300-lb (150-kg) load, is used to test hardened materials; the reading is taken from the C (outside) scale.

□ A 1/16-in (1.5-mm) hardened-steel ball penetrator, under a 220-lb (100-kg) load, is used to test soft metals, and the reading is taken from the B (red inner) scale.

roll of measurement The twist of an axis about a centerline.

root The bottom of the thread.

root mean square (RMS) The average height of crests and depths of troughs of the profile of the surface with reference to the centerline; it is the square root of the average of the sum of the squares of the heights being measured.

rotary table Also called a *circular milling attachment*; provides a precise method of indexing a workpiece for cutting radii, circular grooves, circular sections, and bolt-hole circles. *(See figure.)*

□ Most models are equipped with a vernier scale on the handwheel collar, while others may also have an indexing attachment for accurate indexing.

roughness Relatively finely spaced surface irregularities superimposed on the waviness pattern.

roughness height The arithmetic average deviation measured normal to the mean plane or centerline of a surface. Generally measured in microinches or micrometers.

roughness spacing In surface finish measurement, the distance parallel to the nominal surface between successive peaks.

roughness width cutoff A dimension that distinguishes surface roughness from surface waviness for a particular distance. *(See figure.)*

□ The cutoff length must be preset prior to using a profilometer.

rotary table. *(Courtesy Kelmar Associates.)*

□ When not specified on the drawing, use the cutoff length of .030 in.

rounding-off error An operator error that occurs when eliminating a decimal place. If 1.0024 were rounded to 1.002, and if the upper limit were 1.002, a bad part would be accepted.

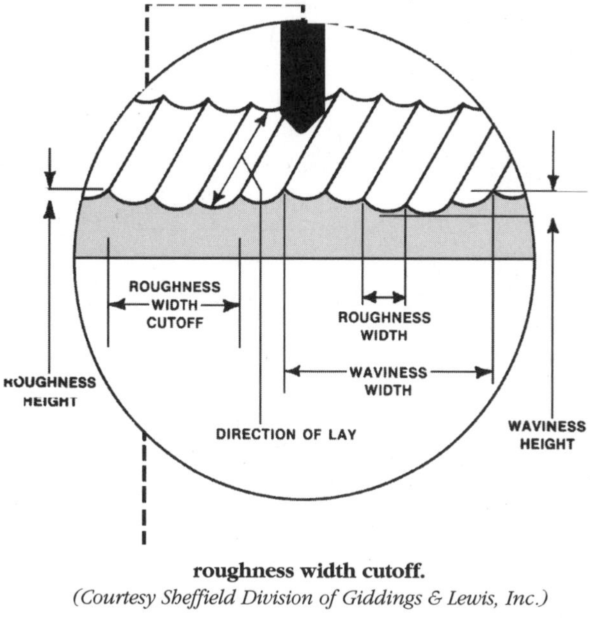

roughness width cutoff.
(Courtesy Sheffield Division of Giddings & Lewis, Inc.)

salt-spray test A fast corrosion test in which samples are exposed to a fine mist of sodium chloride solution to determine the resistance to, and rate of, corrosion shown by various materials.

screw pitch gages Thin steel blades whose profile checks the pitch error in screw threads by comparison.

screw thread A helical ridge of uniform section formed on the inside or outside of a cylinder or cone.

second A second (′) of arc is $1/3600$th of a degree (°) or $1/60$th of a minute (″).

secondary reference standards Standards used to perform calibration on equipment and lower-level standards.

- □ They are a lower-level standard to the primary reference standard.
- □ They must be calibrated to the primary reference standard.

sensitivity The ratio of the instrument's response to the measured variable or the ability to change mechanical contact with the workpiece into movement on the mechanical dial.

sine bar A precision-hardened and precision-ground steel bar with two hardened-steel lapped cylinders of equal diameter fastened near the ends. It is used to check or set up work to angles less than 5″ (minutes). *(See figure.)*

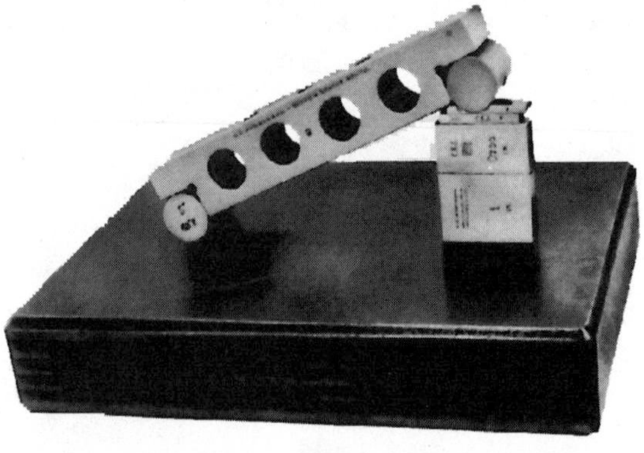

sine bar. *(Courtesy Brown & Sharpe Mfg. Co.)*

- □ The centers of the cylinders are on a line exactly 90° to the edge of the bar.
- □ The distance between the centers of the lapped cylinders is usually between 5 and 10 in.

sine error This error is caused by misalignment with a flat contact and the surface being measured.

small hole gages Small steel shafts with expanding rounded ends that are used to transfer the hole sizes from .125 to .500 in to a micrometer or other precision instrument.

snap gage A gage used to sort parts, on a GO/NOGO basis, within certain limits by comparing the part size to the preset dimension of the snap gage.

specified dimension The stated or listed part of a dimension from which limits are calculated.

spirit level A glass tube filled with a liquid solution that is used to level machinery and measuring equipment.

square A hardened precision instrument with two straight edges ground at right angles; used primarily for inspection and setup purposes.

- □ *Solid*
  1. Beveled-edge precision square that makes line contact with the part being checked.
  2. Cylindrical square, whose circumference is etched with several series of dots that form elliptical curved lines which indicate the amount that a part is out of square.
  3. Toolmaker's surface plate square used for checking work on a surface plate.
- □ *Adjustable*
  1. The combination square used for layout work, checking 45° and 90° angles, and a depth gage.
  2. The diemaker's square used to check the clearance angles on dies.

stability The dimensional stability of the gage is the characteristic that ensures the permanency of the dimension over a long period of time.

steel rule Narrow strips of steel with one or more fractional, decimal, or metric graduated scales used for making linear measurements that do not require great accuracy.

- □ Common rules are the spring-tempered, flexible, narrow, and hook types.

step gage A known, traceable reference bar having incremental steps.

straight edge A rectangular bar of hardened, precision-ground steel with both edges flat and parallel that is used for checking straightness.

straightness The maximum peak-to-valley deviation of the measured profile from the reference line.

stylus The portion of a probe, usually a synthetic ruby, that is mounted on a shank and makes contact with the part during inspection. *(See figure.)*

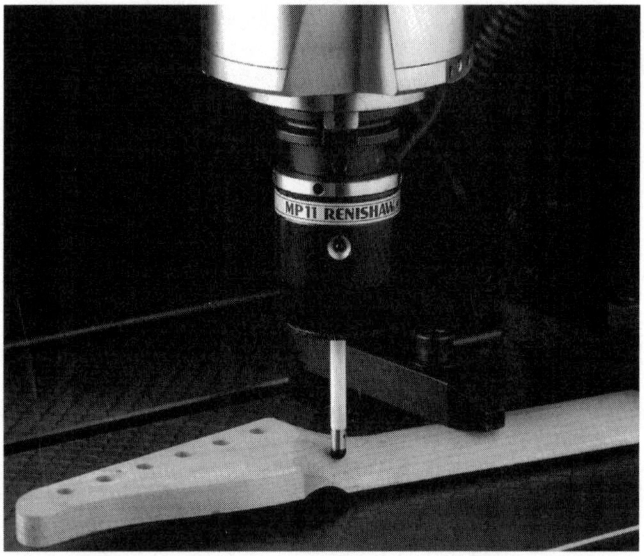

stylus.
(Courtesy Fadal Engineering, Division of Giddings & Lewis, Inc.)

superficial Rockwell hardness test A hardness test used to determine surface hardness or hardness of thin sections or small parts.

supplementary angle When the sum of two angles equals 180°.

surface deviation Any departure from the nominal surface pattern in the form of waviness, roughness, flaws, lay, and profile.

surface finish The smoothness of a machined, ground, or lapped surface as measured by a profilometer or other surface measuring instruments, in microinches.

surface gage Consists of a base, rocker, spindle, adjusting screw, and scriber that is used on a surface plate as a height-transfer tool for layout or inspection.

surface plate A cast-iron, steel, granite, or ceramic plate that provides an accurate reference plane from which measurements for inspection may be made.

□ Surface plates are available in the following grades:

1. Laboratory grade AA: accurate to ±25 millionths of an inch
2. Inspection grade A: accurate to ±50 millionths of an inch
3. Shop grade B: accurate to ±100 millionths of an inch

surface roughness The finely spaced irregularities on the surface of a part created by the cutting tool or process.

tangent Making contact at a single point along an arc; touching but not intersecting.

taper Gradual increase or decrease in the diameter, thickness, or width of an object measured along its length.

taper plug gage A hardened precision-ground tapered steel plug used for inspecting the size and taper of a hole. *(See figure.)*

taper plug gage. *(Courtesy Kelmar Associates.)*

□ Used to measure inside tapers on a GO/NOGO basis.

□ Usually has rings or steps to determine the correct diameter.

taper ring gage A hardened precision steel ring that is ground for a specific taper and used to check the size and taper of shafts.

□ Used to measure outside tapers on a GO/NOGO basis.

□ Usually has steps to determine the correct diameter.

telescoping gage A T-shaped instrument consisting of a pair of spring-loaded telescoping tubes or plungers connected to a handle; used to measure holes.

□ It is expanded inside a diameter at the exact center of a hole and located in this position.

□ The measurement over the ends of the plungers is taken with an outside micrometer.

testing A means of checking whether an item meets specified requirements by subjecting it to a set of physical, chemical, environmental, or operating actions and conditions.

thread micrometer A micrometer instrument that has V-shaped anvils that measure the pitch diameter of screw threads.

thread plug gage An accurately ground steel plug with a specific thread that is used to check the size and accuracy of internal threads.

thread ring gage An accurately ground steel shaft with a specific thread that is used to check the size and accuracy of external threads.

three-wire method A thread-pitch-diameter measurement that consists of placing three equal-diameter wires in the grooves of a thread and then taking a micrometer reading over the wires to determine the thread accuracy. *(See figure.)*

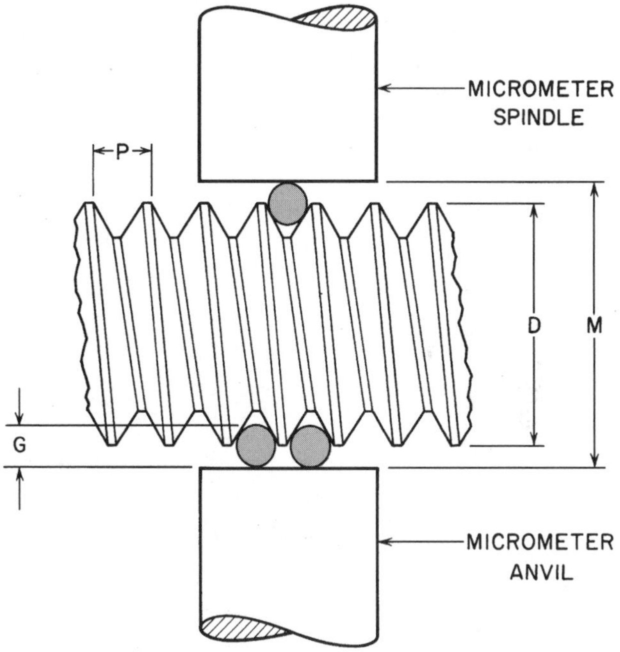

three-wire method. *(Courtesy Kelmar Associates.)*

□ The best-size wire (.57735P) should be used for greatest accuracy because it will contact the thread at the pitch diameter.

tolerance The amount of permissible variation in a dimension; also described as the difference between the lower limit and the upper limit.

□ Bilateral tolerance is both plus and minus (±).

□ Unilateral tolerance is only plus or minus.

total indicator reading (TIR) A term relating to the total movement of an indicator needle while moving the indicator along the part or while rotating the part.

touch tip probe A precision switching device that holds a stylus and makes contact with a surface, causing a mechanical change in the probe to be converted into a change of electrical voltage.

□ The change in voltage triggers a recording of the *X*, *Y*, and *Z* positions at the moment of contact.

trammel A trammel set consists of a beam with two sliding or adjustable heads with scriber points.

□ It is used for scribing large arcs and circles, and to check the distances between lines or centers.

TRI-ROLL A measuring device, manufactured by Brown and Sharpe, that incorporates an indicator and three round disks to measure the pitch diameter of threads.

ultimate strength The maximum stress of tension, compression, or shear that a material can withstand.

ultrasonic testing A nonconductive test that is applied to sound-conductive materials to test for structural discontinuities within a material by the application of an ultrasonic beam.

unilateral tolerance A tolerance that has variation in only one direction from the specified dimension.

unit of length A defined length used as a measure in space is known as the unit of length, generally compared to standard.

universal bevel protractor A precision instrument capable of measuring angles to within 5′ (minutes) (0.083°). *(See figure.)*

□ It consists of a base with a vernier scale, a protractor dial, and a sliding blade.

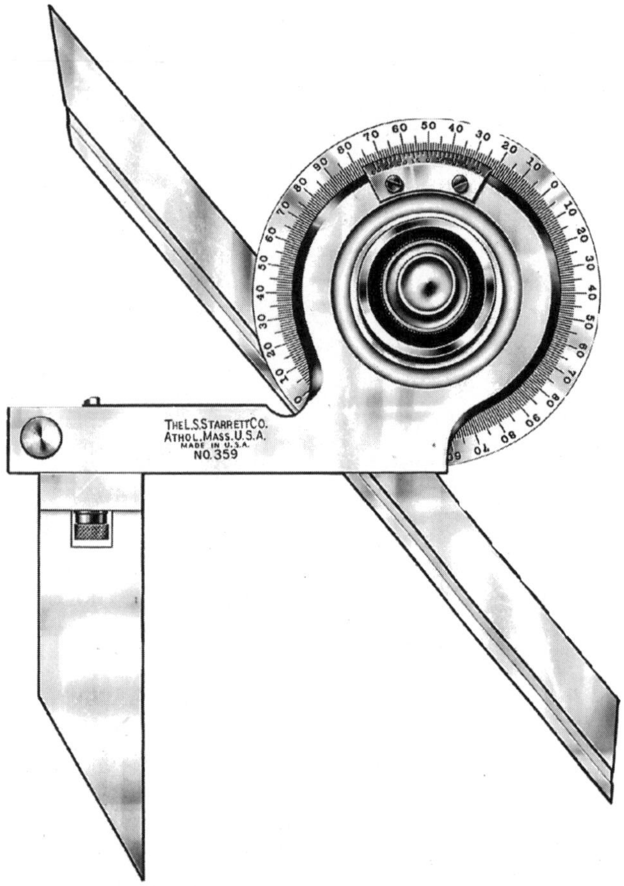

universal bevel protractor. *(Courtesy L. S. Starrett Co.)*

□ The protractor dial has graduations of 1°, and the vernier scale on the base is graduated into 5′ divisions.

variable Measurable characteristic of a part, such as inches of length, weight in pounds, or diameters in millimeters.

variable gages Gages that are capable of measuring the actual size of a part, such as a dial-indicating gage or a micrometer.

vernier instruments Precision measuring tools that use a vernier scale to make accurate measurements to within .001 in or 0.02 mm.

□ Common vernier tools include micrometers, calipers, protractors, and digital electronic calipers.

Vickers hardness test A hardness test that uses a 136° diamond indenter and variable loads so that one hardness test can be used for all ranges of hardness from soft lead to tungsten carbide.

volumetric accuracy Deviation among measurements taken with a ball, bar, or length standard.

waviness Widely spaced irregularities on the surface of a part.

working gages Gages that are used in production at the point of production.

working standard A lower-level standard that is calibrated to a secondary reference standard to perform equipment calibration.

wringing The adherence which takes place when two surfaces of clean gage blocks are slid together with a slight amount of pressure. *(See figure.)*

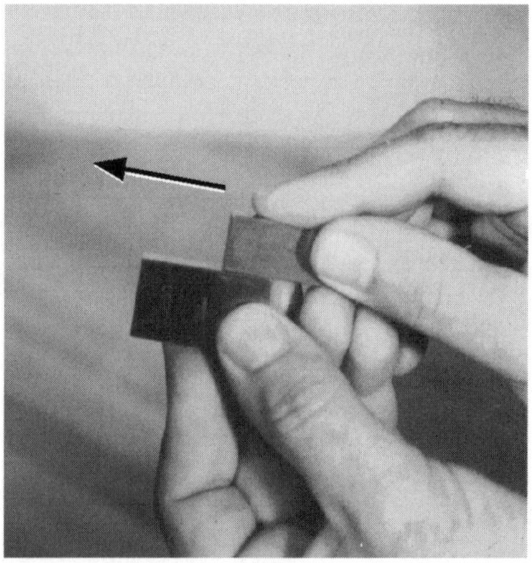

wringing gage blocks. *(Courtesy Kelmar Associates.)*

Y14.5 The ANSI/ASME standard for dimensioning and tolerancing.

Y14.5.1 Mathematical definition of dimensioning and tolerancing principles.

Y14.5.2 A certification program on ASME Y14.5 dimensioning and tolerancing.

yaw Side-to-side deviation from a straight line, as applied to the travel of a component along an axis.

Quality Control

acceptable quality level (AQL) This is usually defined as the worst quality level that is still considered satisfactory.

- □ MIL-STD-105 defines AQL as the maximum percent defective that, for the purpose of sampling inspection, can be considered satisfactory as a process average.
- □ AQL is an indicator of the expected quality level of the vendor's manufacturing process.

acceptance number The maximum number of defects or defective units in the sample that will permit acceptance of the inspection lot or batch.

acceptance sampling A technique for deciding whether to accept or reject a lot on the basis of inspection of samples from the lot.

- □ Acceptance sampling is most useful when testing is destructive or the cost of 100 percent inspection is prohibitively high.
- □ Sampling risks can be classified in two categories, the producer's or the consumer's risk.

acceptance sampling plan Indicates sampling sizes and the acceptance or rejection criteria used for determining the outcome of individual lots.

- □ There are single, double, multiple, sequential, chain, and skip-lot sampling plans.
- □ MIL-STD-414 and MIL-STD-105 are sampling plans.

American Quality Foundation An independent, nonprofit organization, established under the auspices of the American Society for Quality, to foster the application of quality practices by business leaders and public policy analysts.

American Society for Quality (ASQ) The world's leading professional, nonprofit association that develops, promotes, and applies quality-related information and technology in both the public and private sectors.

analysis of variance (ANOVA) A statistical technique used for analyzing data where the total variance in the data is subdivided into components associated with specific sources of variation. It tests for statistical significance.

assignable cause Any cause of variation in a process that is feasible to detect and identify. It may refer to both special and common causes, but is usually associated with special causes.

assignable variation/variability Manufacturing variation that indicates an exception from normal operating conditions and can be traced to a specific, controllable cause.

average outgoing quality (AOQ) The expected quality of outgoing product following the use of an acceptance sampling plan for a given value of incoming product quality.

average outgoing quality limit (AOQL) The maximum average outgoing quality (AOQ) over all possible levels of incoming quality for a given acceptance sampling plan.

bandwidth The total bidirectional deviation from a nominal value.

batch A definite quantity of some product or material produced under conditions that are considered uniform.

benchmark data The results of an investigation to determine how competitors or products compare as to their performance.

brainstorming A simple technique used to define problem areas, probable causes, probable solutions, and other factors.

C-chart A count or control chart for nonconformities (defects). The number or defect counts in the sample are plotted as a function of time.

capability The total range inherent variation of a stable process. The process must be in statistical control with the control limits well inside the tolerance limits.

cause-and-effect, fishbone, or Ishikawa diagram A fishbonelike diagram developed by Dr. Kaoru Ishikawa, a leader of Quality Improvement in Japan, to identify, explore, and display the possible causes of a specific problem or condition. *(See figure on next page.)*

- □ The effect or problem is stated on the right side of the chart, and the major influences or causes are listed on the left.
- □ It is useful in focusing the attention of the operators, manufacturing engineers, and managers on quality problems.

characteristic A property or dimension that helps differentiate between (separate) items of a given sample or population.

- □ The differentiation may be either quantitative (by variables) or qualitative (by attributes).

consumer's risk The probability of a bad lot being accepted; this risk corresponds to the beta (β) risk.

control chart A control chart distinguishes between random and assignable causes of variation through its choice of control limits.

- □ Control charts are one of the principal tools of statistical process control (SPC) and may be classified by the characteristic being tested.
- □ The sample average is known as an *X bar chart*. Averages are used because they are more sensitive to change than individual values and they measure the aim or centering of a process.
- □ The range of the measurements in the sample is known as an *R chart*; it measures variability about the aim of the process.

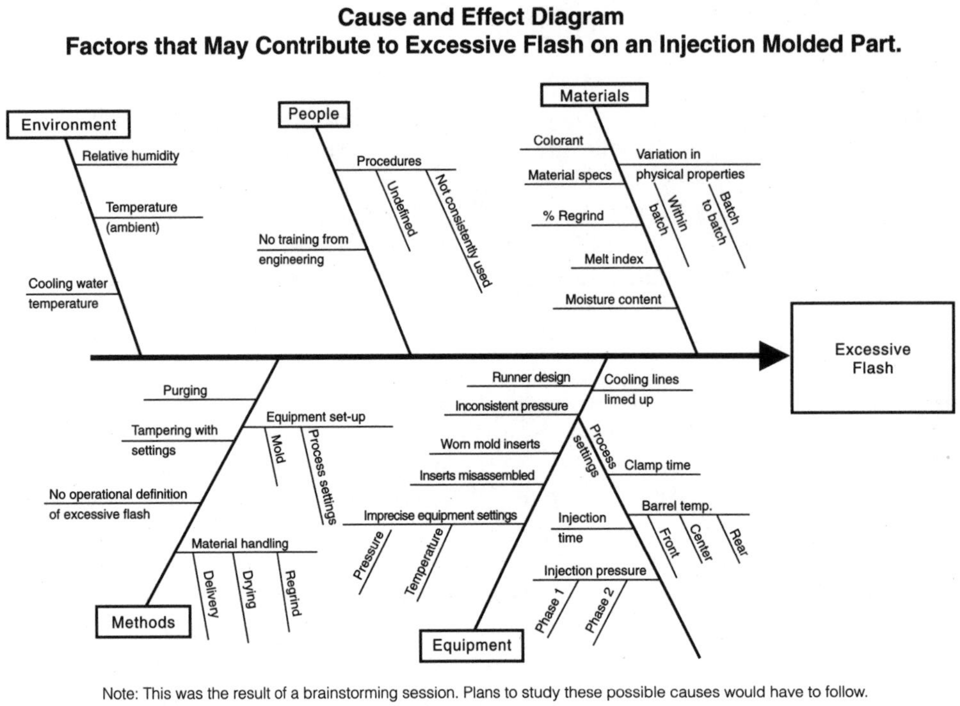

cause-and-effect diagram.
(Courtesy Powertrain Division of General Motors Corp.)

- ☐ The percent defective in the sample is known as a *p chart*.
- ☐ The number of defects in the sample is known as a *C-chart*.

control limits A pair of lines an equal distance above [upper control limit (UCL)] and below [lower control limit (LCL)] the centerline on a chart. *(See figure.)*

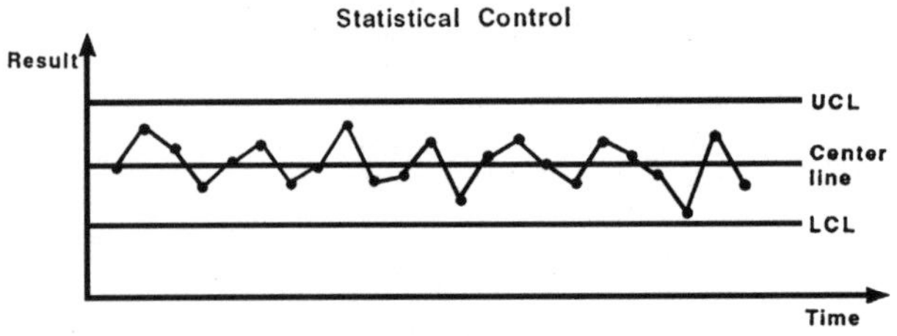

control limits.
(Courtesy Powertrain Division of General Motors Corp.)

- ☐ The upper and lower limits indicate the extreme statistical values of the process and are based on what the process is capable of doing.
- ☐ Their position is calculated by measuring outputs over time and shows the expected amount of variation that can be expected in a process.
- ☐ These limits are usually set at three standard deviations above and below the mean, and as long as outputs remain within the control limits, the process is in statistical control.

corrective-action plan A plan for correcting a process or part quality issue.

critical defect This classification of defect is one that experience or judgment indicates is likely to cause unsafe conditions for those who use, maintain, or depend on the product, or a defect that possibly prevents performance of the function of a major end item.

Deming's 14 points for management Dr. W. Edwards Deming's 14 management practices to help any organization or business increase their quality and productivity.

1. Create constancy of purpose for improving products and services.
2. Adopt the new philosophy.
3. Cease dependence on inspection to achieve quality.
4. End the practice of awarding business on price alone; instead, minimize total cost by working with a single supplier.
5. Improve constantly, and forever, every process for planning, production, and service.
6. Institute training on the job.
7. Adopt and institute leadership.
8. Drive out fear.
9. Break down barriers between staff areas.
10. Eliminate slogans, exhortations, and targets for the workforce.
11. Eliminate numerical quotas for the workforce and numerical goals for management.
12. Remove barriers that rob people of pride of workmanship and eliminate the annual rating or merit system.
13. Institute a vigorous program of education and self-improvement for everyone.
14. Put everybody in the company to work to accomplish the transformation.

design of experiments (DOE) An approach to systematically varying the controllable input factors and observing the effect they have on the output product parameters to find the most important variables affecting a quality characteristic.

Dodge-Romig sampling plans A system for lot-by-lot acceptance sampling by attributes developed by Harold F. Dodge and Harry G. Romig.

- □ These tables are used with the understanding that rejected lots will be screened or 100 percent inspected.
- □ When used properly, these sampling plans reduce the amount of inspection without increasing the risk of poor outgoing quality.
- □ This reduced inspection saves time and money and allows lots to be inspected quicker and still be effective in stopping poor quality.

failure-mode and effects analysis An analytic technique used to assure, to the fullest extent possible, that potential failure modes and their associated causes have been considered and addressed.

frequency distribution Graphs or charts of tally marks (or Xs) that display the frequency of data in column format. It is a tool used to help the user understand

The location of the process center.
The spread or variation of a process center.
The distribution of the process (e.g., bell curve).

goal of statistics The use of statistics from samples as estimates of parameters from populations.

histogram A vertical bar chart that shows the frequency (the number of times of occurrence) of raw data from a process. *(See figure.)*

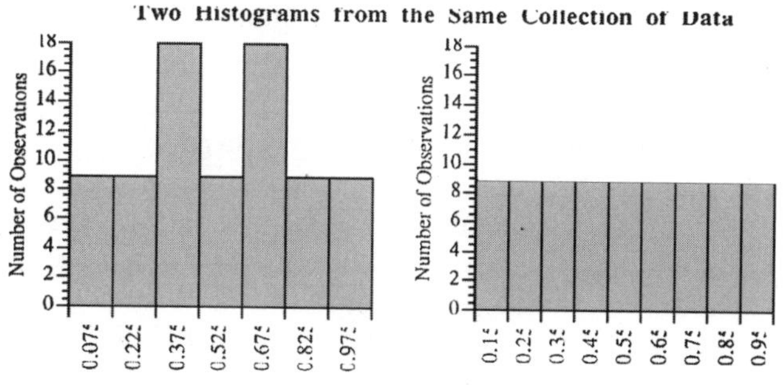

histograms.
(Courtesy Powertrain Division of General Motors Corp.)

ISO 9000 series standards A series of five individual but related standards (ISO 9000, 9001, 9002, 9003, and 9004) on quality management and quality assurance.

- □ It was developed by the International Standards Organization (ISO) in order to provide
  1. Guidance on the selection of an appropriate quality management program for a supplier's operations.
  2. Assistance to effectively document the quality system elements to be implemented to maintain an effective quality system.
  3. A good tool for companies to achieve a total quality management system with a worldwide scope.

kurtosis A measurement of the concentration of spikes near the mean as compared with a normal distribution.

lot A defined quantity of a product or material that is accumulated under conditions considered uniform for sampling purposes.

lot size The number of units in the lot.

Malcolm Baldrige National Quality Award The most prestigious quality award, established by the U.S. Congress in 1987 to raise the awareness of quality management and to recognize quality achievements by American companies.

mean The arithmetic average of a group of numbers. Sum the individual values X_i, and divide by the number of values n.

median A point at which one-half of the data is above the middle of the data and the other half below it when all data has been arranged in order. *(See figure.)*

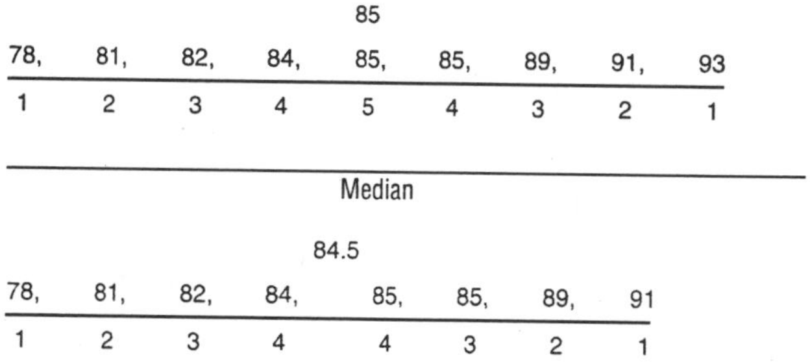

median. *(Courtesy Quality Assurance, Glencoe/McGraw-Hill.)*

- □ The middle value of an odd number of values when the series of numbers are arranged in order from smallest to largest.
  - ► *Example.* 2, 5, 7, 8, and 9, where 7 is the median.
- □ When there is an even number of values, take the average of the two middle values.
  - ► *Example.* 1, 3, 5, 7, 9, and 12, where 6 is the median.

mode The one single number that occurs the most often in a group of numbers.

- □ It can be found by placing the numbers in order and identifying the one that occurs the most.
  - ► *Example.* 2, 2, 2, 3, 5, 5, 6, 7, 8, and 9, where 2 is the mode.
- □ It may not exist, or there may be more than one.

natural (process) tolerance The measurability of the process itself according to its own inherent variation; stated as ±3 standard deviations of a process in statistical control.

- □ A process will produce parts according to its own variation, regardless of what the specifications are.
- □ The important thing is to make the process do what it can do consistently (in control), and then make sure that what it can do matches print requirements (process capability).

nonconforming unit A unit of product or service with at least one nonconformity.

nonconformity A departure of a quality characteristic from its intended level or state that causes an associated product or service not to meet a specification requirement.

nonrandom variations Variations that are not built into the process; they are called *special causes.*

- ☐ Special causes of variation are those that can be directly identified and corrected (particularly when their occurrence is known exactly).

normal distribution A normal distribution is assumed when a line drawn around a frequency distribution or histogram resembles a symmetrical bell-shaped curve. *(See figure.)*

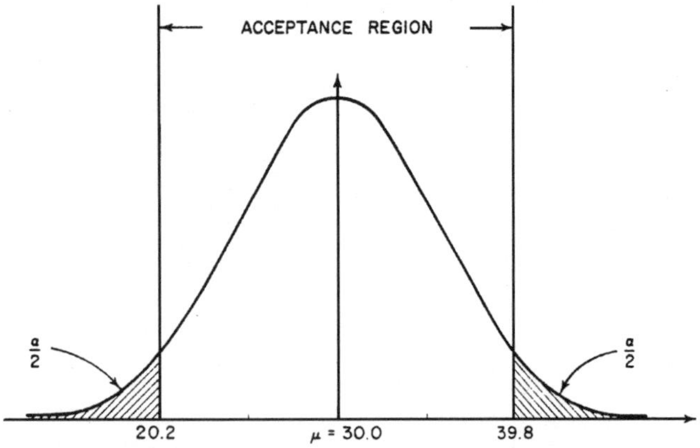

normal distribution. *(Courtesy Juran Institute.)*

- ☐ It is used to show the variation in a process and is also known as a *gaussian distribution.*
- ☐ A normal distribution in control (a stable process) can accurately predict the output of the process.

normal inspection Inspection, under a sampling plan, used when there is no evidence that the quality of the product being submitted is better or worse than the specified quality level.

***np* chart** An attribute control chart for the number of defectives when it is necessary to evaluate the stability of a process and it is desired to look at the number of units with particular defects from a series of lots.

- ☐ n is the sample size which remains constant from period to period.
- ☐ p is the fraction defective (rejected) as nonconforming to specifications, which is the ratio of the number of defects divided by the number of parts inspected.
  - ▶ *Example.* 20 parts were inspected and 3 parts were found to be defective: $p=3/20=.15$.
- ☐ The *np* chart is used when the number of defects is easier or more meaningful to report.

operating characteristic (OC) curve An OC curve indicates the percentage of lots or batches that may be expected to be acceptable under various sampling plans for a given process quality.

- ☐ OC curves are widely used by quality engineers or managers to
  - ▶ Assist in the selection of a sampling plan that is effective in reducing the risks of sampling (accepting poor quality or rejecting good quality).
  - ▶ Provide the ability to discriminate between good and bad lots when the OC curve is steeper (nearly vertical).
  - ▶ Keep down the high cost of inspection.

parameter A numerical value from the population (total aggregate), such as standard deviation, mean, mode, or range.

Pareto analysis (chart) A bar chart that illustrates in descending order the frequency of particular events or process outputs. *(See figure.)*

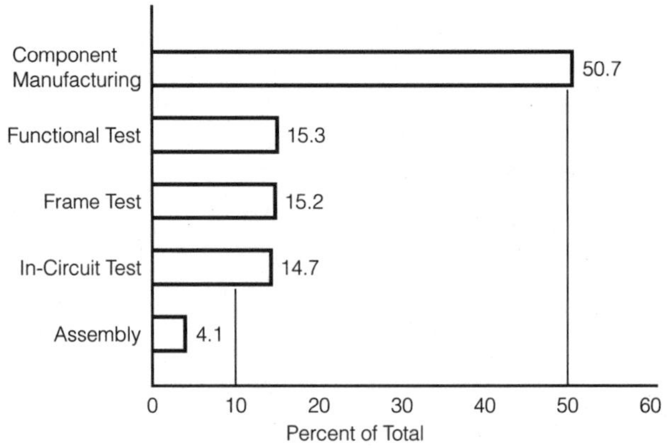

Pareto analysis. *(Courtesy Juran Institute.)*

- ☐ Used to reveal which causes among the many are responsible for the greatest effect (20 percent of the causes produce 80 percent of the effect)
- ☐ Used to identify the problems, causes, sources, or defects that should be addressed first in the problem-solving process

***p* chart** An attribute control chart for fraction defective (rejected) as nonconforming to specifications.

- ☐ p means fraction defective, not percent; it is the ratio of the number of defectives found, divided by the number of parts inspected.
- ☐ *Example.* If 20 parts were inspected and 3 parts were found to be defective, then $p=3/20$ or .15.

percent defective The percent defective of any given quantity of units of product is 100 times the number of defective units, divided by the total number of units inspected.

population A complete set of objects of interest; total aggregate.

primary reference standard A recognized true value. An agreed-on fundamental measurement unit defined in terms of its natural occurrence that is unchangeable.

probability A number that measures the likelihood that an event will occur when an experiment is performed.

probability of acceptance (P_a) The probability that a lot will be accepted under a given sampling plan.

process aim (target) The value(s) in which the adjustable process controls are set and directly controlled by the operator as adjustments are made.

process average The average value of some product characteristic, averaged over all the product produced.

process capability The statistical measure of the main process variability for a given quality characteristic.

- ☐ It is important to achieve statistical control before studying the capability, and there must be a normal distribution.
- ☐ It shows that the process can produce parts within certain limits of precision.
- ☐ An in-control process may or may not be able to produce parts that meet specifications.

process capability index A measure of the capability of a manufacturing process to meet established (customer) specifications given its natural variation. The most widely used indices are C_p and C_{pk}.

- ☐ C_p is a measure of process potential for two-sided specification limits. It is the difference between the upper and lower specification limit divided by the variation in the process known as sigma σ s (the estimated process standard deviation).
- ☐ C_{pk} measures the process variation and the location of the process average in relation to the allowable specification.

process control The use of statistical data gathered while monitoring the process during production to make a decision for adjustment or correction.

producer's risk The risk that a good lot could be rejected; this corresponds to the alpha (α) risk.

quality The number of features and characteristics of a product or service that affect its ability to satisfy given needs.

quality assurance All the planned or systematic actions necessary to provide adequate confidence that a product or service will satisfy given needs.

quality audit A systematic, independent examination and review to determine whether a company properly applies its quality plans and if these plans are suitable for achieving the desired results.

quality circle A group or team of people that practice quality improvement and brainstorming techniques that will improve the performance.

quality control The operation and use of techniques and activities that produce a quality of product or service that will satisfy given needs.

quality costs The cost of producing, identifying, avoiding, or repairing products that do not meet requirements *(See figure.)*. There are four categories of quality cost:

- ☐ *Prevention costs.* Costs related to efforts in design and manufacturing to prevent nonconformance.
- ☐ *Appraisal costs.* Costs of checking the degree of conformance to quality requirements.

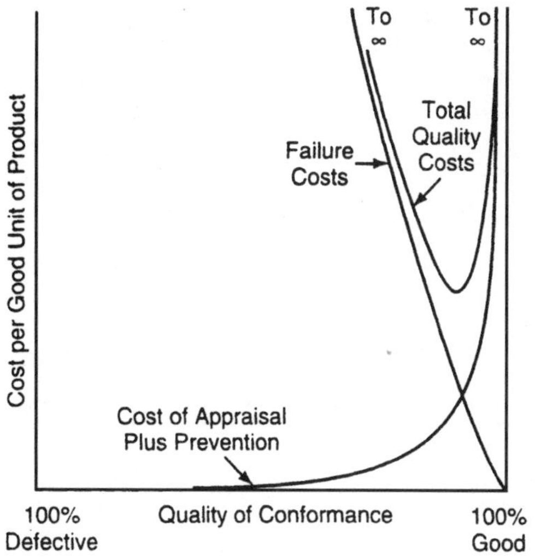

quality costs. *(Courtesy Juran Institute.)*

- ☐ *Internal-failure costs.* Costs related to defects found before the customer receives the product or service.
- ☐ *External-failure costs.* Costs that occur when the product does not perform satisfactorily after it is supplied to the customer.

quality engineering A technology discipline that establishes quality systems in the design and implementation of manufacturing processes at all stages to maximize the quality of the process itself and the product it produces.

quality management All the functions involved in the determination and achievement of quality.

quality manual A document that describes the elements of the quality system, such as

Assuring that customer requirements, needs, and expectations are met.

Responsibilities and authorities for each element of the quality system.

quality plan A document identifying the specific quality practices, resources, and sequence of activities related to a particular product, project, or contract.

quality planning A structured process for defining the methods of measurement, testing, and so on that will be used in the production of a specific product.

quality policy The overall goals and direction of an organization with regard to quality.

random sampling Taking a sample from a population or lot where each item has an equal chance of being included in the sample.

random variation The variation that has several common causes.

- ☐ Common causes of variation are things such as vibration, heat, and humidity, that either individually or collectively cause variation.
- ☐ These common causes are very difficult to identify and correct.

range An amount or extent of variation or the difference between the largest and smallest of a set of numbers.

□ In SPC it is a measure of the spread or dispersion of the sample data.
□ In measurement, it relates to the distance between the reference point and the measured point, such as
1. A 1- to 2-in micrometer has a one-in range.
2. A yardstick has a range of 3 ft.

***R* chart** Range chart, a control chart in which the subgroup range, *R,* is used to evaluate the stability of the variability within a process *(see figure).* It is a companion to the *X*-bar (average) chart and helps to identify an increase in the variability of sample averages.

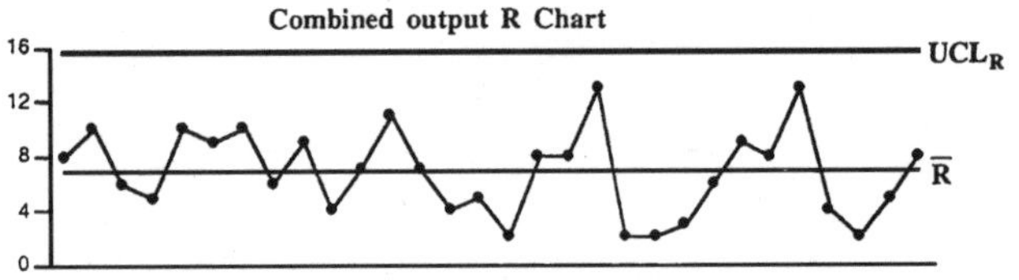

***R* chart.**
(Courtesy Powertrain Division of General Motors Corp.)

reduced inspection Inspection under a sampling plan using the same quality level as for normal inspection but requiring a smaller number of samples.

rejection number The minimum number of variants or variant units in the sample that will cause the lot or batch to be rejected.

reliability The probability that an item will perform its intended function for a specified period of time under certain conditions.

resubmitted lot A lot previously designated as unacceptable but resubmitted for acceptance inspection after having been further tested, sorted, reprocessed, and so on.

sample A subset of the population or representative samples.

sample percent defective Divide the number of defectives found in a sample by the sample size and then multiply by 100.

sample size The number of units selected at random from a lot, to be inspected in a sample.

sample standard deviation The square root of sample variance.
□ The main advantage of the sample standard deviation is that it is expressed in the original units of measurement.
□ The standard deviation does not consider the magnitude of the sample data, only the scatter about the average.

sampling plan A statement of the sample size or sizes to be used and the associated acceptance and rejection criteria.

sample variance The sum of the squared deviations of each observation from the sample average, divided by the sample size minus 1.

scatter diagram Scatter diagrams are used to test the strength of an assumed relationship between two variables (*X* and *Y*).
□ There are two variables of concern: the independent variable *X* and the dependent variable *Y.*
□ The *X* variable can be measured; the *Y* variable may be an important characteristic, but it should not be measured.

s chart Sample standard deviation chart; a control chart in which the subgroup standard deviation *s* is used to evaluate the stability of the variability within a process.

screening inspection When each item of product is inspected for designated characteristics and all defective items are removed.

***S-N* diagram** A chart used to show the relationship between stress and the number of cycles before fracture during fatigue testing.

special causes A variation that is not predictable and not due to causes as the system normally operates.
□ *See also* NONRANDOM VARIATIONS *and* VARIATION.

specification A document listing the engineering requirements for judging the acceptability of a product in terms of appearance, durability, performance, dimension, and other parameters.

specification limits The largest and smallest possible sizes that define the conformance boundaries for a manufactured feature.

standard A recognized true value. Most of the fundamental units are defined in terms of natural phenomena that are unchangeable.

standard deviation The average distance from the mean point that any data point is located. This average helps to find the spread of all data and is generally called a *bell curve.*
□ Symbols are *s* for samples and *s* for population.
► The formula for samples is $s=s^2=S(x-x)^2/n-1$
► The formula for population is $S=S(x-\mu)^2/n$.

statistic A numerical value from the sample, such as standard deviation, mean, mode, and range.

statistical process control (SPC) The application of statistical techniques to measure and monitor the performance of manufacturing processes to reduce variation and improve quality.

statistical quality control (SQC) Application of the laws of probability and statistical techniques to measure and improve product quality.

statistics The mathematics of collecting, summarizing, and analyzing numerical data so that generalizations can be made along with predictions of the future.

Taguchi methods Methods of quality engineering to achieve high quality and low costs by combining engineering and statistical methods to optimize product design and manufacturing processes. They were developed by the Japanese quality expert G. Taguchi.
□ Used to enhance cross-functional team interaction and implement experimental design.

ten-to-one rule Rule stating that the measuring or test equipment must be capable of dividing the range of the tolerance into ten parts.

total quality management (TQM) A set of management practices designed to improve quality and profitability, and increase customer satisfaction.

- □ Objectives are to develop processes that are capable of consistently designing, producing, and delivering quality products.
- □ Common features include customer-driven quality, strong leadership of management, continuous improvement, employee participation and empowerment, and action based on facts, data, and analysis.

traceability The ability to relate measurement results back to the primary reference standard. The ability to verify each higher step of calibration until the level of the National Institute of Standards and Technology (NIST) is reached.

***u* chart** An attribute control chart for the number of nonconformities in a unit.

unit A quantity of product, material, or service considered as a single entity from which a measurement or observation can be made.

variance Square of the standard deviation.

- □ Symbols are s^2 for samples and $(s)^2$ for population.
- □ The formula for sample variance is $s^2 = S(x-x)^2/(n-1)$.
- □ The formula for population variance is $(s)^2 = S(x-\mu)^2/n$.

variation The inevitable difference among individual outputs of a manufacturing process. There are two kinds of variation with regard to process control: random and nonrandom variation.

***X* bar chart** Average chart; a control chart used to plot the average measured value of a certain quality characteristic for each of a series of samples taken from the production process. It indicates how the process mean varies over a period of time. *(See figure.)*

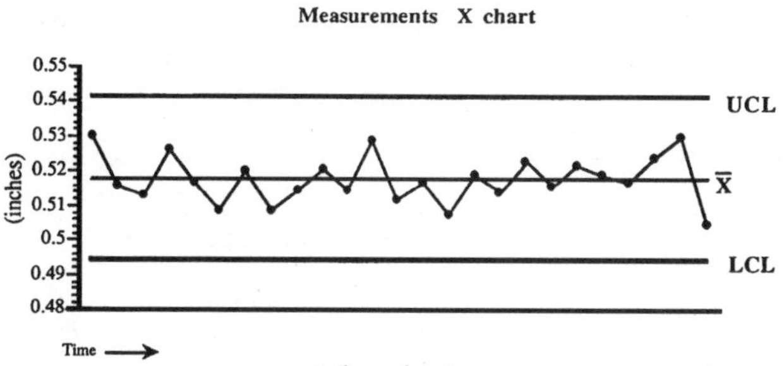

***X* bar chart.**
(Courtesy Powertrain Division of General Motors Corp.)

- □ The mean of the averages is the centerline of the chart.
- □ The upper and lower control limits can be calculated to determine whether the process is in or out of control and to determine process capability.

zero defects A quality assurance program with the basic philosophy that if a defect is identified, efforts will be made to isolate the source of the defect.

- □ Appropriate remedial action will be taken to ensure that the particular defect does not recur.

SECTION 10

Metallurgy, Heat Treating, and Testing

J. W. Oswald
Consultant
Kelmar Associates
Niagara Falls, Ontario

The need to reduce pollution and preserve the environment has lead to the development of direct ironmaking and direct steelmaking processes which are more efficient and create less pollution than did previous methods. Minimills can produce 180 to 200 tons of steel per hour at a lower cost, and avoid the coke and ironmaking steps used in larger steel mills. New steels are being constantly developed to meet new strength and temperature requirements. Under research are *smart structures* which can repair microscopic cracks and *shape memory alloys* that will return to their undeformed shape.

acicular ferrite A highly substructed ferrite formed on continuous cooling by a mixed diffusion and shear mode of transformation that begins at a temperature slightly higher than the transformation temperature range for upper bainite.

acid embrittlement A form of hydrogen embrittlement that can be induced in some metals by acid treatment.

affinity A strong chemical attraction of one element for another which, on contact, will unite to form a new compound.

age hardening Hardening by aging, usually after rapid cooling or cold working.

age softening A decrease in strength and hardness that takes place at room temperature in some strain-hardened alloys, especially those of aluminum.

aging Maintaining metals or alloys at room temperature for a prolonged period of time, or heating for a shorter time at suitable temperatures in a furnace.
□ It is generally used to increase the stability of the metal, often after the metal has been cold worked.

air hardening steel An alloy steel that will form martensite and develop a high hardness when cooled from its proper hardening temperature in an air blast or still air.

allotropic forms Different crystalline structures of the same substance. For example, coal, graphite, and diamonds are allotropic forms of carbon.

alloy A substance having metallic properties and composed of two or more chemical elements to produce a desired characteristic in the metal.

alloying element An element which is added to a metal to change the metal's properties.

alloy steel Carbon steel to which other elements have been added to produce the desired quality in the steel.
□ Alloying elements may be added to steel to increase

1. *Tensile strength*—chromium, manganese, or nickel
2. *Hardness*—chromium
3. *Red hardness*—cobalt
4. *Corrosion resistance*—chromium or nickel
5. *Toughness*—molybdenum, nickel, or vanadium

alpha iron The body-centered cubic form of pure iron, stable below 910°C.

alumel A nickel-base alloy containing about 94 percent Ni, 3 percent Mn, 2 percent Al, and 1 percent Si used chiefly as a component of pyrometric thermocouples.

alumina Refined aluminum oxide ore prepared for smelting.

aluminizing Forming a corrosion- and oxidation-resistant coating on a metal by coating with aluminum and fusing it with the base metal to form an aluminum-rich alloy.

angstrom A unit of length used for extremely fine measurements such as a wavelength of light. One unit (angstrom, Å)=10^{-8} cm.

anisotropic A crystalline structure having different properties in different directions.

anneal To accurately heat a metal to the proper temperature followed by slow cooling, usually in a furnace, through the transformation temperature range. *(See figure.)*

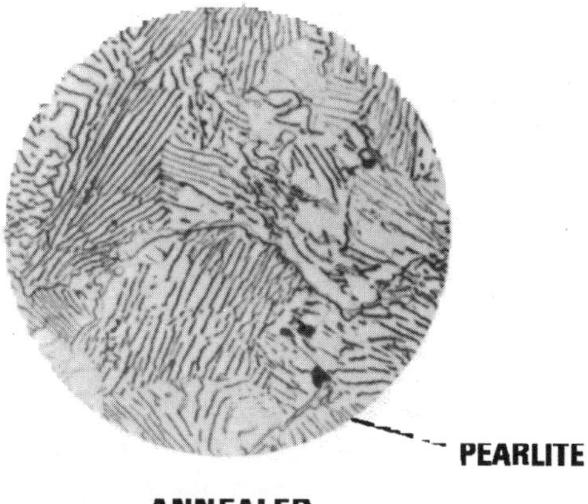

annealed tool steel. *(Courtesy Kelmar Associates.)*

annealing The heating of a metal to just above the upper critical temperature for a given time and then cooling at a suitable rate to produce the desired qualities in the metal.

□ Annealing may be done to relieve stresses, induce softness, improve physical properties, improve machinability, refine the crystalline structure, remove gases, and produce a specific microstructure.

argon-oxygen decarburization A refining process in which a combination of argon and oxygen gases is blown through the steel bath to remove carbon and control the chemistry of stainless steels. *(See figure.)*

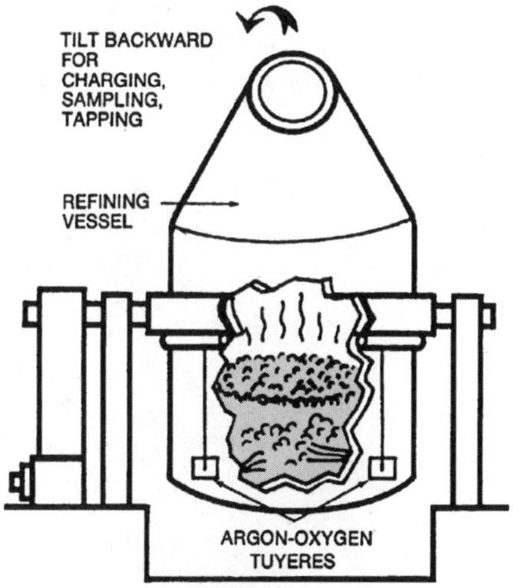

argon-oxygen decarburization.
(Courtesy American Iron & Steel Institute.)

□ Used to process steels containing large amounts of chromium and in producing extra-low-carbon grades.

atmosphere A gaseous environment in which the metal is heated for processing.

□ Atmospheric environments are used to protect the metal from chemical change or to alter the surface chemistry of steel through the addition or removal of carbon, nitrogen, ammonia, hydrogen, and oxygen.

atmospheric corrosion The gradual degradation or alteration of a material by contact with substances present in the atmosphere, such as oxygen, carbon dioxide, water vapor, and sulfur and chlorine compounds.

atom The smallest particle of an element that retains chemical and physical properties of that element after any chemical reaction.

austempering A heat-treating operation in which austenite is quenched to and held at a constant temperature (usually between 400 and 800°F) in a salt bath until transformation to 100 percent bainite is complete.

austenite A solid solution of iron carbide in gamma iron that forms at temperatures between the lower and upper critical temperatures of the steel.

□ Austenite is the structure from which all quenching heat treatments must be carried out.

□ Steel is nonmagnetic in the austenitic stage.

□ It has a face-centered cubic crystalline structure normally stable at room temperatures.

austenitizing temperature The temperature at which the steel structure changes to austenite.

bainite A needle-like structure of ferrite and cementite formed when austenite transforms between 450 and 900°F. *(See figure.)*

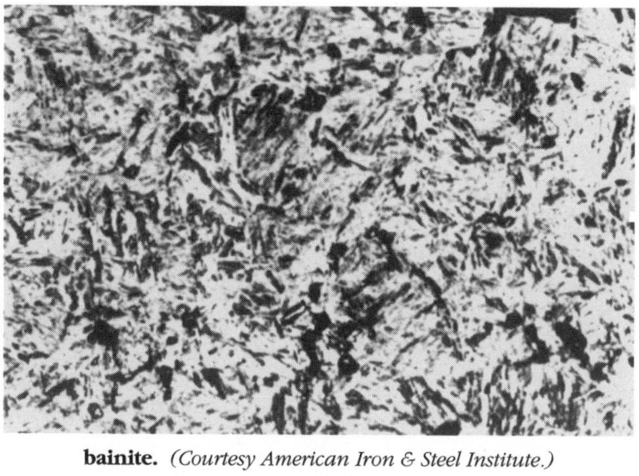

bainite. *(Courtesy American Iron & Steel Institute.)*

□ It varies in hardness between Rc 30 and 55.

banded structure A layering effect that is sometimes developed during the hot rolling of steel where alternating bands of different structures are formed in the steel in the direction of the hot working.

bar A length of steel of uniform cross section (round, square, rectangular, hexagonal, etc.) with sizes ranging from 1/4 to 8 in.

bark An old term used to describe the decarburized skin that develops on steel bars heated in a nonprotective atmosphere.

basic oxygen process A steelmaking process in which a high-pressure stream of oxygen is directed onto the top of a batch of molten pig iron and scrap steel to cause a rise in the temperature of the molten metal. *(See figure.)*

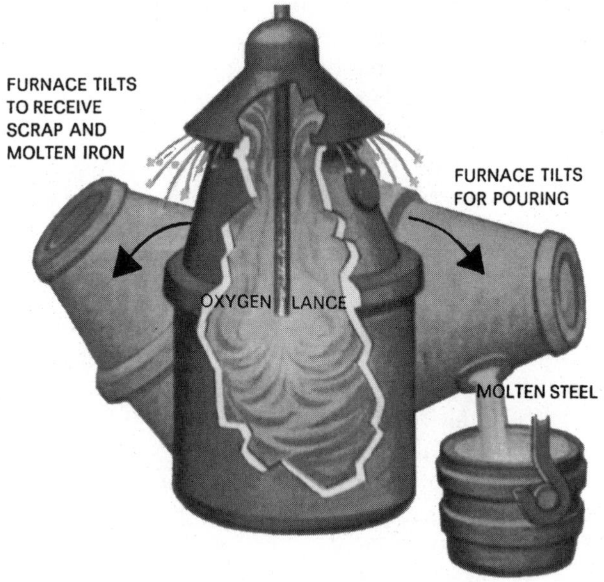

basic oxygen process. *(Courtesy U.S. Steel Corp.)*

□ A flux consisting of lime and fluorspar is added to remove the impurities.

□ This furnace can produce a 200-ton heat per hour.

beryllium (Be) A light, hard, noncorrosive metal used considerably in the aircraft industry. It is often alloyed with aluminum or copper for making window frames and other structural parts.

Bessemer process A former process of making steel by blowing air up through molten pig iron and thus oxidizing the carbon, manganese, silicon, and phosphorus.

billet A semifinished product in the rolling of steel, made in a blooming mill or a billet mill, and intermediate between a bloom and a bar.

□ A billet may be rectangular or square, and is less than 36 in^2 in cross-sectional area.

binary alloy An alloy containing only two component elements.

blast furnace A tall, brick and steel furnace, commonly 100 ft or more in height, in which hot air (blast) is blown to reduce the iron ore to pig iron. *(See figure.)*

bloom A semifinished product in the rolling of steel made in a blooming mill.

□ It is the first stage in the reduction of the ingot, square or rectangular in shape, and measures more than 36 in^2 in cross-sectional area.

brass An alloy containing varying amounts of copper and zinc. Brass fittings are used in water and gasoline-line tubing and fittings.

bright annealing Annealing work in a protective nonoxidizing atmosphere to avoid discoloration as the result of heating.

Brinell hardness test A test for determining the hardness of a material by forcing a hard steel or carbide ball of specified diameter into the material under a specified load.

□ The Brinell hardness number is the value obtained by dividing the applied load in kilograms by the surface area of the resulting impression in square millimeters.

brittle fracture Complete separation of a solid accompanied by little or no macroscopic plastic deformation.

□ Brittle fracture occurs by rapid cracking and requires less force than needed for ductile material fracture.

brittle tempering range Some hardened steels show an increase in brittleness when tempered in the range of about 1000 to 1250°F even though some tempering procedures may cause some softening.

bronze Copper alloys of tin or most other nonferrous metals other than zinc.

calcining A process of roasting metallic ores at about 600°C to remove water and foreign matter.

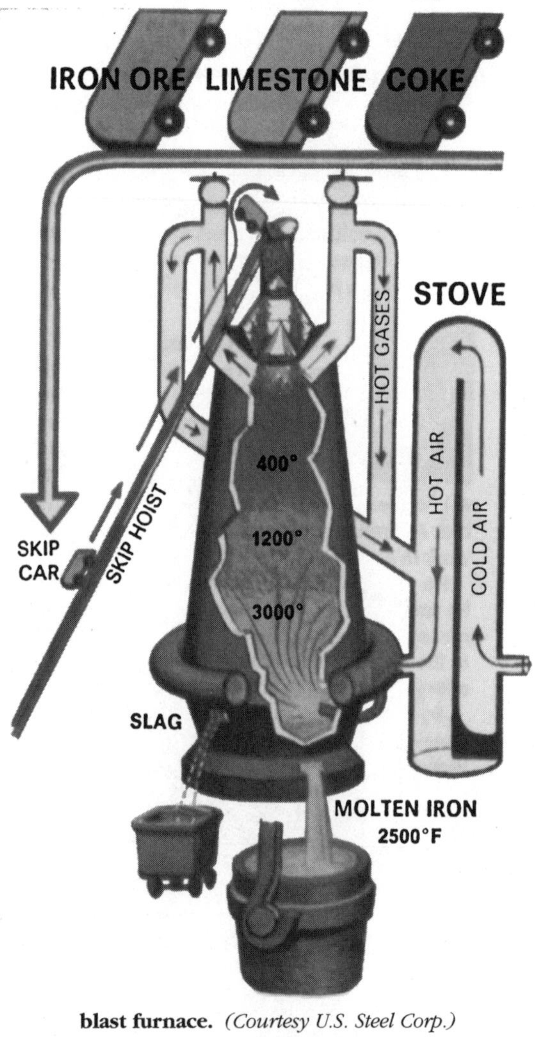

blast furnace. *(Courtesy U.S. Steel Corp.)*

carbide A compound of carbon and one or more metallic elements, which, when combined with steel, imparts hardness to the steel.

- □ The addition of tungsten, titanium, tantalum carbide, or a combination of these in a cobalt or nickel matrix provides hardness, wear resistance, and heat resistance to steel used for cutting tools.
- □ Other elements added to carbide include vanadium, niobium, and columbium.

carbonitriding A case-hardening process in which both carbon and nitrogen are absorbed by the surface of a steel workpiece when it is heated to the critical temperature.

- □ Carbonitriding may be done in a liquid or gas medium.

carbon steel Steel that is composed of iron plus carbon with no intentionally added alloy. Also known as *plain carbon steel* or *straight carbon steel.*

carburizing Adding carbon to the surface of steel by heating it in contact with carbon-rich solids, liquids, or gases.

case The surface layer of a steel whose composition has been changed by the addition of carbon, nitrogen, chromium, or other material at high temperature by a carburizing process.

case hardening A heat treatment in which the surface layer of a steel is made substantially harder than the interior by altering its composition by carburizing, cyaniding, or nitriding. *(See figure.)*

case hardening. *(Courtesy Kelmar Associates.)*

cassiterite The ore from which tin is made, SnO_2.

cast alloy Alloy cast from the molten state.

- □ Most high-speed steel is melted in an electric-arc furnace and cast into ingots.
- □ Many aluminum alloys are suitable for sand casting, die casting, or permanent mold casting.

cast cobalt-chromium-based alloy Used to make turning tools more heat-resistant and tougher than ordinary high-speed steel tools, which usually contain only tungsten and/or molybdenum.

castings Metal parts formed by pouring the molten metals into sand or metal molds in which they solidify into their final shape.

cast iron The large group of cast ferrous alloys in which the carbon content is more than can be dissolved in austenite at the eutectic temperature.

- □ Most cast irons contain at least 2 percent carbon, plus silicon and sulfur, and may or may not contain other alloying elements.
- □ The various forms of cast iron are gray iron, white iron, malleable iron, and ductile iron.

cemented carbide A hard, brittle, cutting-tool material consisting of tungsten and carbon powders in a metal-powder matrix (usually cobalt), capable of very high cutting speeds. *(See figure.)*

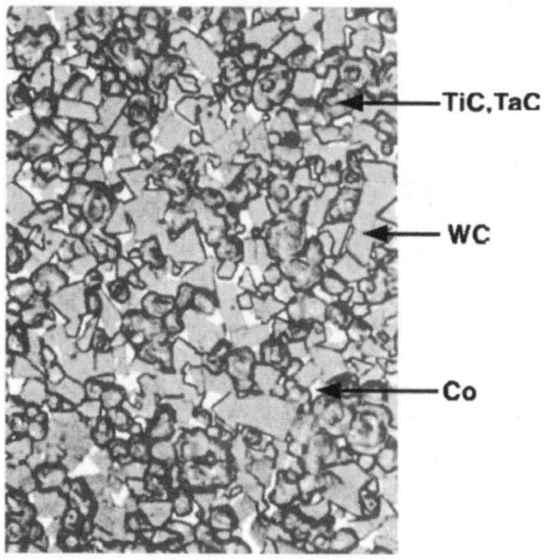

cemented carbide. *(Courtesy GE Superabrasives.)*

- □ Carbides of titanium or tantalum produce cutting tools that are more crater-resistant than tungsten carbide.

cementite A compound of iron and carbon, known chemically as iron carbide, Fe_3C. It is hard and brittle, has low tensile strength, and is the hardener in steel.

chalcocite A black-gray copper sulfide, Cu_2S. Also called *glance ore.*

chalcopyrite A yellow sulfide of copper and iron, $CuFeS_2$.

Charpy impact test A pendulum-type single-blow impact test in which the specimen, usually notched, is supported at both ends as a simple beam and broken by a swinging pendulum. *(See figure.)*

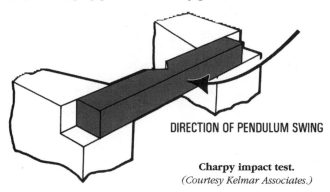

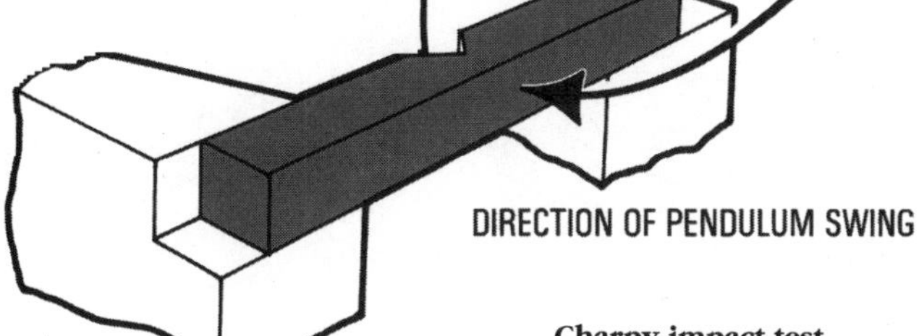

Charpy impact test. *(Courtesy Kelmar Associates.)*

□ The energy produced when the pendulum is released from a given height is a measure of impact strength or toughness of the metal.

chemically combined Two or more elements combined to form a chemical compound.

chlorinating Chemically combining a material with chlorine.

coated sheets Steel sheets coated with a layer of some other material such as galvanized steel or tinplate.

□ Steel may also be coated with paint, porcelain enamel, plastics, or aluminum.

coking coal The higher-carbon grades of coal used in coking ovens to produce coke in large, firm lumps.

cold drawing The process of pulling a hot-rolled bar through a die of similar shape but smaller dimensions, without preheating to improve machinability, and bring the material to accurate size.

cold reduction The process of reducing the thickness of steel by cold rolling or cold drawing, without preheating the steel. This adds strength, toughness, and a smooth, bright finish.

cold shut A discontinuity that appears on the surface of cast metal when two streams of liquid metal become chilled and will not fuse together when being poured into a mold.

□ A rolling defect on the surface of a forging or billet that was closed without fusion during deformation.

□ Cooling of the top surface of an ingot before the mold is full.

cold treatment Cooling or "deep freezing" tool steel to obtain total hardness, generally followed by low-temperature tempering.

cold working The permanent deformation of a metal and its grain structure when worked below the recrystallization temperature.

commercial-grade tool steel Low-grade tool steel; not controlled for hardenability.

compressive strength The maximum stress in compression that a material will withstand before rupturing or breaking.

□ The amount of force and the cross-sectional area are used to calculate the compressive strength.

conductivity The ability of various materials to conduct heat or electricity.

contact fatigue The failure of a metal as the result of fluctuating stresses (and/or loads) produced by rolling or sliding contact such as in a gear train used to transmit power.

continuous casting A casting technique in which molten steel is poured into a tundish (reservoir) and, in turn, into a mold for casting a slab, billet, or bloom. *(See figure.)*

□ As it solidifies at the bottom of the mold, the bar is withdrawn and may be cut to any desired length.

continuous wide-strip mill A mill in which slabs are reduced to sheets or strips by passing through a series of progressively finer-spaced rolls in a rapid, continuous, automatic process.

converters Bessemer, open-hearth, electric, or the

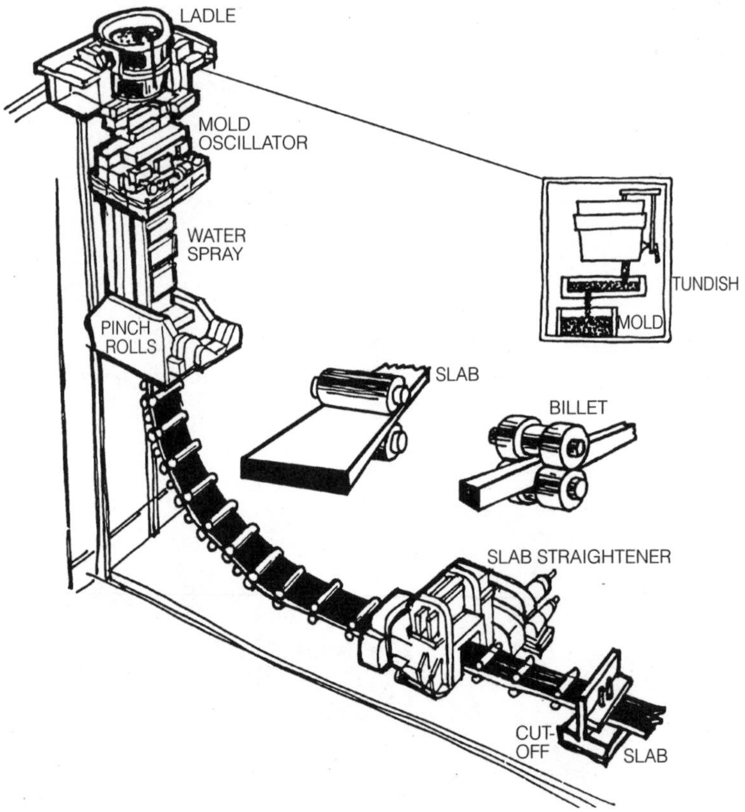

continuous casting. *(Courtesy American Iron & Steel Institute.)*

basic oxygen furnace in which blast furnace iron is converted into steel.

core **1.** The interior part of a steel whose composition has not been changed in a case-hardening operation. **2.** A body of green or baked sand which is mounted in a mold to permit a cored hole to be formed in the casting.

coring **1.** During the solidification of metal alloys, the first crystals to form (on the outside) are richer in the metal with the higher melting point, gradually reducing until the last crystals to solidify (in the center) contain richer amounts of the metal having the lower melting point. **2.** The use of a sand core, in metal castings, to produce large holes to save metal and machining costs.

corrosion The chemical or electrochemical reaction between a material, usually a metal, and its environment, or between two unlike adjacent metals that produces a deterioration of the material and its properties.

corrosion embrittlement The severe loss of ductility of a metal resulting from the corrosion of the metal.

corrosion fatigue The process in which a metal fails prematurely as a result of corrosion and fluctuating stresses on the part during the operating cycle.

corrosion resistance Ability of an alloy or material to withstand rust and corrosion.

□ The addition of nickel and chromium inhibits corrosion in alloys such as stainless steel.

creep A slow, permanent deformation in a metal part produced by a small, steady force below the elastic limit during a long period of time.

critical point A temperature point at which a chemical and/or structure change takes place when a material is being heated or cooled. *(See figure.)*

critical range The temperature range between the upper and lower critical points for a given material.

crystallization **1.** The transformation from a liquid state on cooling to a solid crystalline state. **2.** Often erroneously used to explain fracturing that actually has been caused by fatigue.

crystal structure types There are six basic types: cubic, hexagonal, tetragonal, orthorhombic, monoclinic, and triclinic.

cycle annealing An annealing process in which a predetermined and closely controlled time-temperature cycle is used to produce specific properties or microstructures.

decarburizing The loss of surface carbon when carbon steel is heated to above the lower critical temperature in an oxidizing atmosphere. This produces a soft outer skin on the metal.

dendrite A crystal that has a treelike branching pattern that usually forms in cast metals slowly cooled through the solidification range.

deoxidizing The removal of oxygen and undesirable oxides from molten metals by use of suitable deoxidizers such as manganese and silicon.

□ This term sometimes refers to the removal of other undesirable elements by the addition of elements or compounds that readily react with them.

□ In metal finishing, dexoidizing involves the removal of oxide films from metal surfaces by chemical or electrochemical reaction.

die casting A casting process in which molten metal is forced under high pressure into the cavity of a metal mold or die.

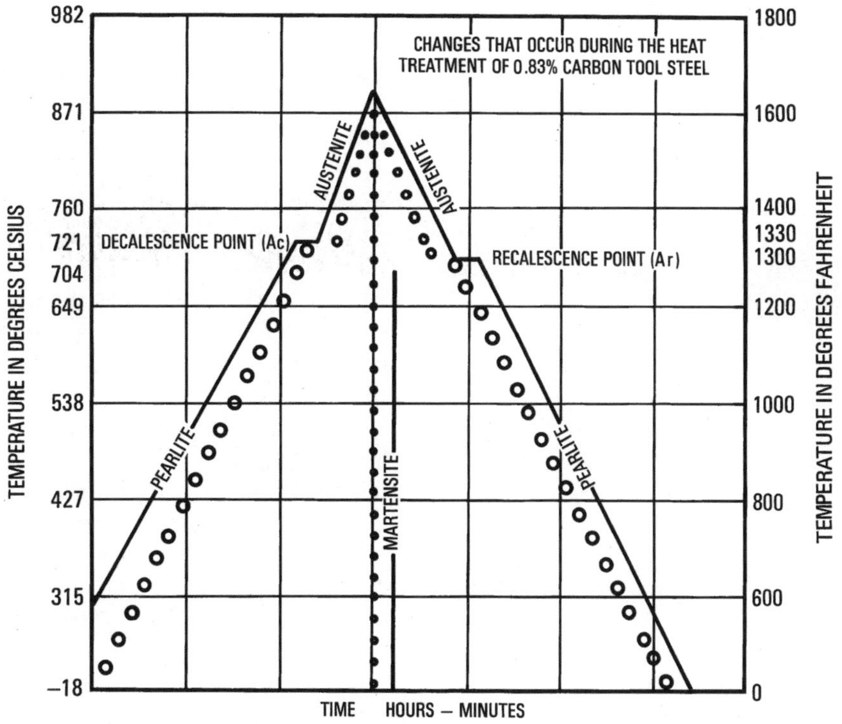

critical point. *(Courtesy Kelmar Associates.)*

direct hot charging A steelmaking process that links melting, casting, and hot rolling into a continuous sequence.

direct ironmaking A process based on the smelting of iron ore with coal injection, fluxes, and oxygen in a liquid bath to eliminate the coke process and its pollution. *(See figure.)*

ductile cast iron A cast iron, also known as *modular cast iron,* that has been treated while molten with certain alloying elements such as magnesium or cerium to induce the formation of spheroidal graphite nodules.

☐ These elements also lower the sulfur content in the cast iron and make it ductile.

AISI direct ironmaking flowsheet

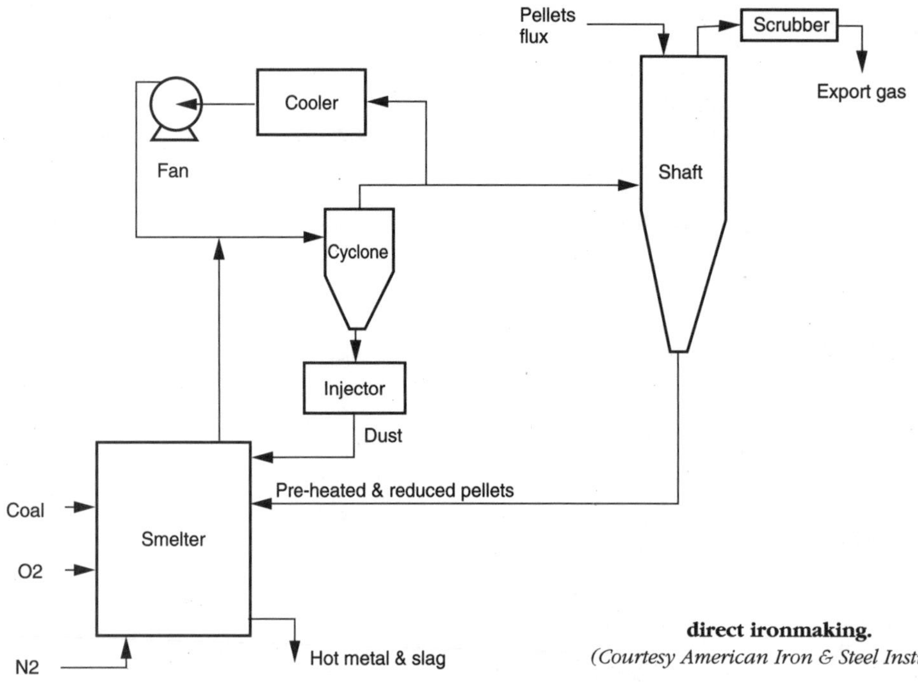

direct ironmaking.
(Courtesy American Iron & Steel Institute.)

direct-reduced iron An ironmaking supplement produced by re-forming natural gas and using the resulting hydrogen and carbon monoxide to strip oxygen from the ore to obtain a sponge iron.

☐ A pelletized iron ore (iron oxide and sulfide) that is refined to a highly metallized product without melting. It is an economically attractive supplement for steel scrap.

direct steelmaking A process which makes steel directly from iron ore that combines the operations of smelting, prereduction, offgas cleaning and handling, and refining.

☐ The process is almost pollution-free, and the heated gases are used to heat and partially reduce the iron ore pellets before they are charged in the smelter.

dissociation The chemical breakdown of a compound into simpler compounds or elements. A common example is the dissociation of ammonia (NH_3) into nitrogen and hydrogen.

dolomite A brittle marble-like rock containing calcium and magnesium carbonates.

☐ Used to make refractory bricks for high-temperature furnaces and also in the production of basic cast iron.

draw The common heat-treating term used interchangeably with tempering.

ductile fracture A separation in a metal caused by a tearing action and accompanied by mass plastic deformation.

ductility The ability of a material to be permanently deformed without fracturing.

☐ It may be measured by the elongation or reduction of area during a tensile test.

dye-penetrant testing Nondestructive inspection using a dye that penetrates a crack or defect in a material's surface.

elasticity The ability of a metal to return to its original shape after any force acting on it has been removed.

elastic limit The maximum stress to which a material may be subjected without any permanent deformation remaining, after the complete release of the stress. *See* TENSILE TEST.

electric furnace A type of furnace where the heat, amount of oxygen, and atmospheric conditions can be controlled to produce fine alloy and tool steels. *(See figure on next page.)*

☐ Selected steel scrap is loaded into the furnace, and the heat is produced by electric arcs between the carbon electrodes and the steel scrap.

electrolysis The application of a direct current to an electrolyte during which the positive ions are attracted

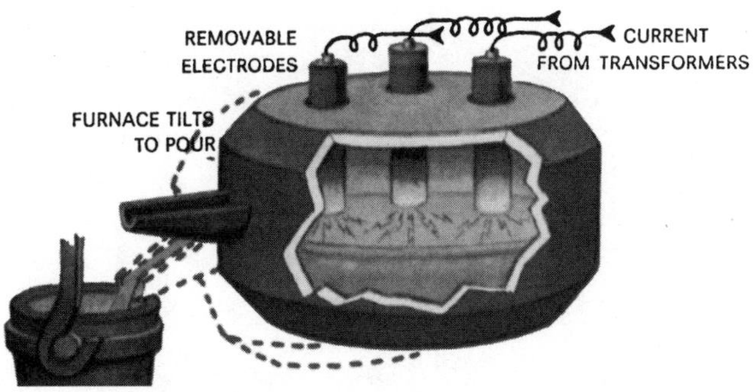

electric furnace. *(Courtesy U.S. Steel Corp.)*

to the cathode and the negative ions, to the anode.

electron An atomic particle that carries a negative charge.
- □ Electrons revolve in orbits around the nucleus of an atom.

electrostatic collector Tanklike vessel with electrically charged rods that collect solid dust from gases blown through the collector.

elements Fundamental substances that make up all matter.

elongation The amount of permanent extension provided on a metal during a tensile test to determine the strength of the metal.

embrittlement The loss of ductility in a metal due to a physical or chemical change.

endurance limit The maximum stress to which a material may be subjected without failure due to fatigue.

equilibrium diagram A curve drawn on a diagram or graph to show the various structural changes in an alloy when subjected to temperature changes, when in its most stable and permanent condition.

eutectic An alloy or solid solution that has a melting or freezing point below any of its components.

eutectic point The point of least solubility of one metal in another.

eutectoid steel Steel that contains just enough carbon to dissolve completely in the iron when the steel is heated to the critical range.
- □ This may be likened to forming a saturated solution of salt in water.
- □ Eutectoid steel contains from 0.80 to 0.85 percent carbon.

exfoliation A type of corrosion that forms parallel to the outer surface of the metal, causing layers of the corroded metal to be elevated.

experiment A scientific procedure which includes
- □ Statement of a theory.
- □ An attempt to record and analyze data.
- □ Discussion of results.
- □ Conclusions.

extrusion A process used to force metal, plastic, or other materials, by compression, through a die.
- □ Used to convert an ingot or billet into lengths of uniform cross section by forcing the plastic metal through a die orifice.

fabricated steel The fitting and assembly of individual parts by welding, riveting, or other joining processes, to produce a larger more complex object. The fabrication process is used in building ships, bridges, and other structures.

fatigue The phenomenon leading to failure under repeated stresses.
- □ Fatigue fractions are progressive, beginning as minute cracks that grow under continued fluctuating stress.

fatigue life The number of cycles of stress that can be sustained prior to failure under a given test condition.

fatigue resistance Ability of a tool or component to be flexed repeatedly without cracking or breaking; an example is leaf springs used in automotive suspension systems.

fatigue strength The maximum stress that can be sustained for a specified number of cycles without failure.

ferrite Alpha iron-carbon alloy, which exists below the lower critical temperature. It is magnetic, has face-centered cubic crystal structure, and contains a small amount of dissolved carbon.

ferroalloys Alloys of iron and some other element or elements (carbon excepted) which are added to molten steel to impart desired properties to the end product.
- □ The following ferroalloys are added to the molten steel:
  - ▶ *Molybdenum*—increases tensile strength and toughness
  - ▶ *Vanadium*—increases toughness and resistance to shock and fatigue
  - ▶ *Tungsten*—increases wear resistance; imparts red hardness
  - ▶ *Silicon*—acts as a deoxidizer; creates soundness in castings, forgings, and rolled products

flame hardening A process consisting of heating a localized area with a high-temperature flame followed by quenching to produce a hardened surface.

flash A thin web or film of metal on a casting that occurs at die partings, around air vents, and around movable cores. This excess metal is due to the necessary working and operating clearances in a die.
- □ The excess material squeezed out of the cavity as a compression mold closes, or as pressure is applied to the cavity.

flotation A process used to separate pulverized ores by placing them in a solution, causing the lighter particles to float to the surface and the heavier ones to sink.

flow stress The stress induced at the onset of plastic deformation in a metal.

flue dust Dust that is "airborne" in the exit gas stream of a blast furnace. It is collected in the dust catcher and the gas washer system.
- □ It is usually mixed and sintered with the ore and recharged to the blast furnace.

flux Limestone or similar materials added to a blast furnace to promote fusion of the ore.

forging The operation of shaping hot steel by means of hammers or presses.

formability The ability of a metal to be shaped through plastic deformation.

fracture stress **1.** The maximum stress applied when fracture occurs with unnotched tensile specimens. **2.** The (hypothetical) true stress that will cause fracture without further deformation at any given strain.

free carbon The amount of the total carbon in steel or cast iron that is present in the form of graphite or temper carbon.

free ferrite Ferrite that is formed in hypoeutectoid austenite (below 0.85 percent carbon content) that has been slowly cooled from above its upper critical temperature.

□ Because of the lower carbon content, pearlite and ferrite are formed as opposed to the formation of pearlite and cementite when hypereutectoid austenite is cooled.

free-machining steels Carbon and alloy steels that contain lead, sulfur, or other elements that improve machinability.

fretting A type of wear that occurs between close-fitting surfaces subjected to periods of relative motion of extremely small distances, often accompanied by corrosion of the wear debris.

full annealing Heating iron-base alloys to above the upper critical point for a required period of time, followed by slow cooling in the furnace, lime, or sand.

□ Annealing will soften the metal, relieve the internal stresses and strains, and improve the machinability.

full hard The temper of nonferrous alloys and some ferrous alloys corresponding approximately to a cold-worked state beyond which the material can no longer be formed by bending.

□ A full-hard temper is commonly defined in terms of minimum hardness or minimum tensile strength.

galena Lead ore which is a sulfide of lead (PbS) containing 86.5 percent lead.

galling A condition in which excessive friction between mating parts results in localized damage to the surfaces and subsequent spalling of the mating parts.

□ This generally results in the seizure or "welding" of the parts.

galvanize To coat a metal surface with zinc, generally by immersing the metal in molten zinc.

gamma iron The face-centered cubic form of pure iron, stable from 910 to 1400°C. *(See figure.)*

gas entrapment Gas bubbles trapped in molten metal as it freezes into solid ingots from the outside surface toward the center.

grain An individual crystal in a polycrystalline metal or alloy.

grain flow Fiberlike lines appearing on polished and etched cross sections of forgings.

□ These are caused by orientation of the constituents of the metal in the direction of workings during pressure forging where the heated metal is forced into the die under extremely high pressure.

□ Grain flow produced by proper die design can improve the required mechanical properties of forgings.

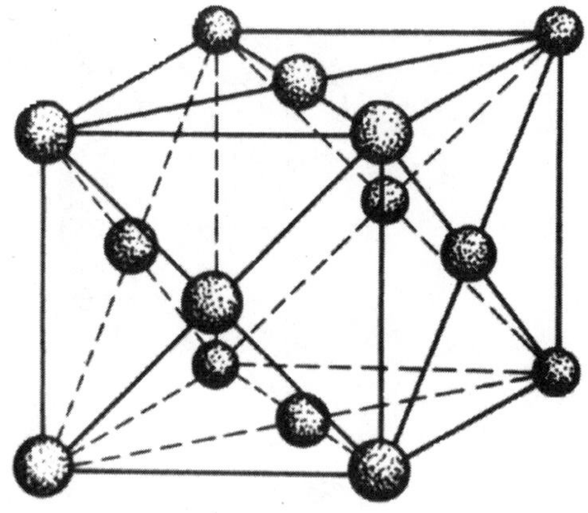

gamma iron. *(Courtesy Kelmar Associates.)*

grain growth An increase in the average grain size causes some crystals to absorb adjacent ones when the metal is heated to above the annealing temperature.

□ This results in an increase in grain size and a decrease in the number of grains, and occurs more rapidly at high temperatures.

granular fracture A type of irregular surface produced when metal is broken, characterized by a rough, grainlike appearance.

□ It is often called a *crystalline fracture,* inferring it cannot be assumed that the metal broke because it crystallized, since all metals are crystalline when in the solid state.

graphitic corrosion Deterioration of gray cast iron where the metallic constituents become corroded, leaving the graphite intact.

graphitization A metallurgical term describing the formation of graphite in iron or steel, usually from decomposition of iron carbide at elevated temperatures.

gray cast iron A cast iron that gives a gray fracture due to the presence of flake graphite; often called *gray iron.*

growth In cast iron, a permanent increase in dimensions resulting from repeated or prolonged heating at temperatures above 480°C that may be caused by the graphitization or by oxidation of carbides.

Guinier-Preston (G-P) zone A small precipitation domain in a supersaturated metallic solid solution. A G-P zone has no well-defined crystalline structure of its own and contains an abnormally high concentration of solute atoms.

hard chromium Chromium that has been applied to a metal surface by electroplating to increase the wear resistance between two sliding metal surfaces.

□ The chromium deposit in this case is thicker than decorative chrome plating, but is not necessarily harder.

harden To accurately heat a metal to above the lower critical temperature and then cool it rapidly, usually by quenching in oil or water. *(See figure.)*

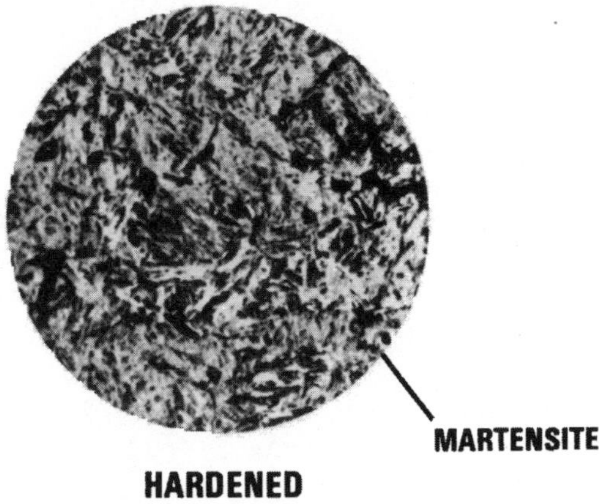

hardened tool steel. *(Courtesy Kelmar Associates.)*

□ Maximum hardness and the finest grain structure in steel are achieved when it is quenched from above the upper critical temperature.

hardenability The relative ability of a ferrous alloy to form martensite when quenched at a temperature above the upper critical temperature.

□ Hardenability is commonly measured as the distance below the quenched surface at which the metal indicates a specific hardness.

hardness A mechanical property of metal indicating relative resistance to penetration, cutting, or abrasion.

hardness test A destructive test that measures the resistance of the metal to a penetrator under a given load.

hardness tester An instrument used to record the amount of pressure required to form an indentation in a material.

□ A variety of scales are used to measure hardness; the Rockwell C, the Brinell, the Vickers, and the Scleroscope hardness scales are the most commonly used.

H-band steel Alloy steel produced to specified limits of hardenability; the chemical composition range may vary slightly from that of the corresponding grade of ordinary alloy steel.

hearth Fireplace shapes, made of refractory brick, used in the smelting process.

□ In the blast furnace, it is the lower basin into which the molten pig iron trickles. In other furnaces, it is the basin which holds the charge of molten metal being refined.

heat The term applied to a single batch of molten steel that is being refined.

heat-affected zone That portion of the base metal that was not melted during brazing, cutting, or welding, but whose microstructure and mechanical properties were altered by the heat.

heat check A pattern of parallel surface cracks that are formed by alternate rapid heating and cooling of the surface metal, often found on forging dies and piercing punches.

heat of fusion (latent heat) The additional heat required to permit the fusion of a metal, even though the metal has reached its melting point.

□ At this point, there is no change in the temperature of the solid metal, even though additional heat (latent heat of fusion) is being applied.

heat treatment The carefully controlled heating and cooling of a metal to obtain the desired structural changes required in the metal.

□ The primary purpose of heat treatment is to change the properties of the material.

hematite ore Iron oxide (69 percent iron), Fe_2O_3.

high-speed steel (HSS) High-carbon steel alloyed with tungsten, molybdenum, and certain other elements such as manganese, silicon, chromium, or vanadium. This alloy steel permits higher cutting speeds and feeds than do carbon steel cutting tools.

□ The addition of one or more of these elements adds certain qualities to the steel:

1. Chromium increases hardness and wear resistance.
2. Cobalt imparts red hardness.
3. Manganese increases hardness and wear resistance.
4. Molybdenum increases tensile strength, toughness, and red hardness.
5. Tungsten increases wear resistance and promotes red hardness.
6. Vanadium increases toughness, shock, and wear resistance.

high-strength, low-alloy (HSLA) steels Steels containing no more than 0.28 percent carbon and small amounts of vanadium, columbium, copper, and other alloying elements.

□ They provide higher hardness, toughness (impact strength), and fatigue failure limits, than do carbon steel bars.

homogenizing An annealing treatment at a temperature below the melting point designed to create uniformity in the structure of the metal.

hot crack A crack formed in a cast metal because of internal stress developed on cooling, following solidification.

hot-rolled steel Steel brought to white heat and passed through a series of rolls to reduce the cross section and increase the length. It is then cooled, cut to length, or coiled.

hot shortness A condition found in low-carbon steel, usually caused by the presence of sulfur, which causes the steel to crack at high temperatures during working, rolling, or forging operations.

hot tear A fracture formed in a metal during solidification because the shape of the object or the mold does not allow enough thermal contraction.

hot working Plastic deformation of a metal at temperatures above the recrystallizing temperatures.

- □ During hot working, grains are constantly being deformed and broken up while new ones are constantly formed, providing the metal remains at, or above, the lower critical (recrystallization) temperature.
- □ This prevents any work hardening or permanent distortion.

HSS *See* HIGH-SPEED STEEL.

hydrogen damage A general term for the embrittlement, cracking, blistering, and hydroxide formation that can occur when hydrogen is present in some metals.

hydrogen embrittlement The brittleness in steel caused by the absorption of hydrogen, usually from a pickling or plating operation.

hypereutectic alloy Any alloy whose composition has an excess of alloying element compared with the eutectic composition.

hypereutectoid steel Steel containing more carbon than will completely dissolve in the iron when the steel is heated to the critical range—similar to a supersaturated solution.

hypoeutectoid steel Steel containing less carbon than can be dissolved when the steel is heated to the critical range—similar to an unsaturated solution.

impact energy The amount of energy required to fracture a material, usually measured by means of an Izod or Charpy impact test. The type of specimen and test conditions affect the values and should be specified. *See* CHARPY IMPACT TEST.

impact test A destructive test used to measure the resistance of a metal to a sudden impact force.

inclusions Nonmetallic particles of impurities (usually oxides, sulfides, silicates, etc.), which are held in the base metal during solidification.

- □ In some grades of steel, inclusions such as sulfur are intentionally added to aid machinability.

induction hardening A form of hardening in which the heating is done by an induced electric current. *(See figure.)*

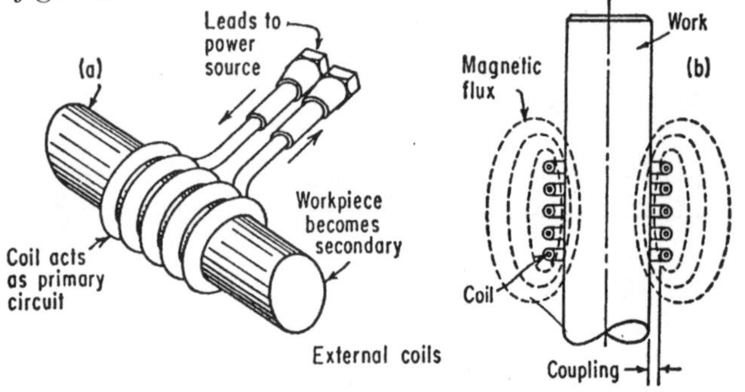

induction hardening.
(Courtesy Society of Manufacturing Engineers.)

ingot mold The thick-walled cast-iron mold into which molten steel is cast to form an ingot.

ingot steel The first solid form of steel from which other shapes or products are made by forging, rolling, or extrusion.

ingot stripping Process of removing ingots from ingot molds.

inherent defect An imperfection in a metal that sometimes occurs during production and before fabrication takes place.

integrated steel mill Generally a large steel manufacturer (big steel) that produces all types of products, from the basic raw materials to finished products.

intergranular corrosion Corrosion occurring usually at grain boundaries, with slight or negligible effect on the adjacent grains, and resulting in little or no loss to the mechanical properties of the metal.

internal oxidation The formation of isolated particles of corrosion products beneath the metal surface. This occurs as the result of oxidation of certain alloy constituents by the presence of oxygen, nitrogen, sulfur, and other elements.

interrupted quench The quenching in water of carbon tool steel at a temperature above the critical range to a predetermined temperature, followed by an oil quench to room temperature.

- □ This process will achieve maximum hardness with minimal cracking.

interstitial Minute changes in a metal's structure caused by the forcing of small, foreign atoms in between atoms of a crystal.

- □ May be produced by distortion during severe plastic deformation.

investment casting A precision-casting process which uses a wax pattern to produce a mold for casting the final product (usually metallic).

- □ A wax pattern is placed in a die and then filled with a slurry of finely ground refractory, a binder, and water.
- □ The die is vibrated in a high vacuum to harden the investment.
- □ It is heated to melt and burn out the wax, leaving a cavity that is used as an expendable mold to cast the final product.

iron It is produced in an impure form called *pig iron* in the blast furnace by the reduction of iron ore with coke and limestone.

- □ The pig iron—containing about 4 percent carbon and varying amounts of silicon, manganese, phosphorus, and sulfur—may be remelted in a cupola furnace for the production of iron castings or refined into wrought iron or steel.

iron carbide A compound containing iron and carbon, a source of metal as well as fuel, that is the equivalent of iron pellets and coking coal smelted in a blast furnace.

□ The chemical reaction—the heat-producing conversion of carbon monoxide into carbon dioxide—can be used to superheat liquid iron, refining it into steel.

isothermal transformation The structure changes that take place in eutectoid steel when quenched from above the upper critical temperature, where it exists as stable austenite, to where it is transformed into martensite below the lower critical temperature. *(See figure.)*

□ Aluminum-killed steels have a fine-grained structure that remains more stable at higher temperatures than do aluminum-free steels.

kinetic energy Energy possessed by a body in motion.

Knoop hardness The microhardness determined from the resistance of metal to indentation by a very narrow pyramidal diamond penetrator under a very light load.

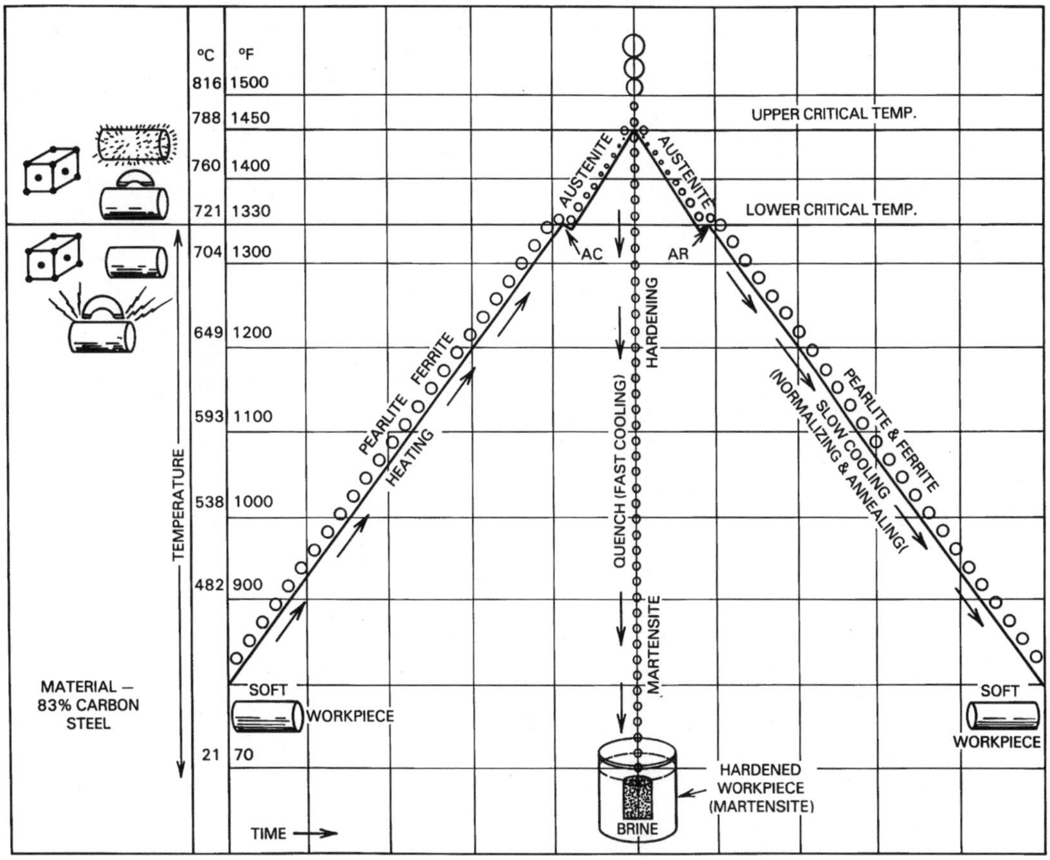

isothermal transformation. *(Courtesy Kelmar Associates.)*

isothermal treatment A type of heat treatment in which a part is quenched rapidly down to a given temperature and then held at that temperature until all transformation is complete.

isotropic Referring to crystalline structure exhibiting identical physical properties in all directions. The opposite of anisotropic.

Izod test A pendulum-type single-blow impact test in which the specimen, usually notched, is fixed at one end and broken by a swinging pendulum.

□ The energy required to fracture the material is measured by the height to which the pendulum rises on the scale after fracture, somewhat similar to the Charpy impact test.

killed steel Steel treated with a strong deoxidizing agent such as silicon, manganese, or aluminum, to reduce the oxygen content and prevent the formation of blow holes when the steel solidifies into ingots.

ladle metallurgy A family of in-ladle injection, stirring, bubbling, and degassing systems that constitute ladle metallurgy.

□ A development of new fluxes and refining technologies for the removal of phosphorous and nitrogen to ensure proper chemistry, temperature, and consistency of the molten steel.

ledeburite A cast-iron alloy, formed at 2065°F, that contains a eutectic mixture of austenite and cementite containing 4.3 percent carbon.

levitation melting An induction melting process in which the metal being melted is suspended by an electromagnetic field and is not in contact with a container.

liquid-penetrant inspection A nondestructive inspection method that uses a penetrating dye (usually red) or a fluorescent liquid penetrant to locate very fine surface cracks in nonporous materials.

liquidus The point at which liquid starts to solidify.

lost-wax process An investment casting process in which an expendable wax pattern is used.

machinability The ease or difficulty with which a metal can be machined.

- ☐ This is influenced by the cutting-tool life, surface finish produced, and power required.
- ☐ It is measured by the length of cutting-tool life in minutes, and the rate of stock removal (depth of cut) in relation to the cutting speed.

macroetching Subjecting a metal sample to a deep acid etch to show the structural details such as grain flow, segregation, porosity, or cracks, for visual inspection by the unaided eye or at a magnification of 10 times or less.

magnesite Magnesium ore, $MgCO_3$.

magnetic-particle testing Nondestructive inspection to detect surface cracks using an iron powder and an electromagnetic field.

magnetite Black iron oxide (72% iron), Fe_3O_4. Strongly magnetic—also called *lodestone.*

malleable cast iron A fibrous-grained cast iron produced by extended annealing of white cast iron. During the process, white cast iron (iron carbide) is converted into ferrite and nodules of temper carbon.

martempering or marquenching Martempering is a form of interrupted quenching in which the steel is quenched rapidly from its upper critical temperature to about 450°F, held at 450°F until the temperature is uniform, and then cooled slowly in air to room temperature.

- ☐ Actual hardening occurs when cooling starts and with a minimum temperature differential.
- ☐ Martempering is used to minimize distortion in low- and medium-alloy steels.

martensite The structure of fully hardened steel when austenite is quenched from or above its upper critical temperature. (See Figure.)

martensite. *(Courtesy Kelmar Associates.)*

- ☐ Martensite appears as an acicular or needle-like structure and can be considered to be a supersaturated solution of carbon in tetragonal (distorted cubic) iron.
- ☐ Hardness of martensite will vary from Rc 30 to 68 depending on the carbon content.

matte A compound of copper sulfide and iron sulfide when copper-bearing ore is smelted at 2600 to 2800°F.

mechanical properties The properties of a material that reveal its behavior when force is applied, thereby indicating its suitability for mechanical applications; examples are modulus of elasticity, tensile strength, elongation, hardness, and fatigue limit.

metallic glass A noncrystalline metal or amorphous alloy, commonly produced by drastic supercooling of a molten alloy by electrodeposition, or by vapor deposition.

metallizing **1.** Forming a metallic coating by atomized spraying with molten metal or by vacuum deposition to build up worn or undersized parts before remachining to size. **2.** Applying an electrically conductive metallic layer to the surface of a nonconductor.

metallography The process of preparing, polishing, and examining a metal sample by microscope.

metalloid A chemical element that resembles a metal in some properties and nonmetals in others. Examples are boron, carbon, silicon, phosphorus, sulfur, and manganese, which are commonly present in small amounts in steel.

metallurgy The scientific study of metals and their ores, with respect to their mechanical properties and processing, and their physical, chemical, and microscopic structures.

microscope A laboratory instrument or tool that uses an optical lens system to enlarge structures that are too small to be seen by the unaided human eye.

microstructure The structure of a metal as revealed by microscopic examination of the polished and etched surface of a specimen, using a magnification of 10 diameters or more.

mild steel Carbon steel with a maximum of 0.20 percent carbon. It cannot be hardened and tempered, but may be case-hardened.

mill scale The deposit of iron oxide that forms on the surface of steel during the heating and rolling operation.

mineral A metal ore as found in nature. The ore is commonly a chemically combined oxide or sulfide of the metal.

minimill A small, more efficient steel mill that produces steel and thin-slab casting faster, and at less cost than a large integrated steel mill.

mischmetal A natural mixture of rare-earth elements of magnesium–rare earth zirconium and magnesium–rare earth zinc-casting alloys.

- ☐ It contains about 50 percent cerium and is used for low-cost commercial alloys.

Modulus of elasticity (Young's modulus) A measure of stiffness or the ability of material to resist deflection. It is indicated on a stress-strain diagram by the slope of the initial straight-line portion, below the point *P* that indicates the proportional limit, or the point at which the metal will no longer return to its original length when the strain is removed from it.

Monel An alloy smelted from a naturally occurring mixture of nickel-copper iron ore from the Sudbury mines, Ontario, Canada.

***M* temperatures** In high-carbon alloy steels, the temperature at which the white needlelike (acicular) grains of martensite begin to form; on cooling it is known as M_s.

□ The temperature at the end of the acicular formation is known as M_f.

national standards Specifications commonly agreed on by professional associations after many years of cooperative effort.

neutron The electrically neutral (uncharged) particle of an atom.

nitriding The process of adding nitrogen to the surface of a steel, usually in a protected ammonia environment.

□ Nitriding develops a very hard case after a long time at comparatively low temperature (935 to 1000°F) without quenching.

nitrocarburizing Any processes in which both nitrogen and carbon are absorbed into the surface layers of a ferrous material at temperatures below the lower critical temperature.

□ Nitrocarburizing provides a tough surface layer and improves fatigue resistance.

noble metal A metal, such as platinum or gold, with marked resistance to oxidation when immersed in acidic solutions.

nondestructive testing Scientific inspection of materials for defects without serious damage to the material.

nonferrous metals All metals and alloys having some metal other than iron as a major component.

normalize To accurately heat a metal to above the lower critical temperature, and then cool it slowly in still air.

□ Normalizing may be used to refine grain structure before hardening the steel, to harden the steel slightly, or to reduce the separation of the constituents in castings or forgings.

nucleation Freezing of first crystals from liquid metal into solid.

nucleus The central part of an atom, containing protons and neutrons.

open-hearth process A little-used steelmaking process in which excess carbon is removed from pig iron by the addition of iron ore and limestone in a large dishlike hearth. *(See figure.)*

orange peel **1.** A surface roughening, in the form of a pebble-grained pattern, that occurs when a coarse-grained metal is stressed beyond its elastic limit. **2.** The pebble-like appearance on a coat of paint which has been sprayed on with too much air pressure.

orbit Path of electrons rotating around a nucleus.

ore dressing A process of mechanically separating the concentrated mineral from its impurities, such as rock and earth.

oxidation **1.** Chemical process of combining an element with oxygen. In some steelmaking processes, oxygen is used to "burn out" or remove impurities from the iron. **2.** The process of rusting when some metals such as steel are exposed to moisture and air.

oxygen converter or **furnace** *See* BASIC OXYGEN PROCESS.

passivity The decrease in chemical reaction of certain metals, after prolonged immersion in strong nitric acid or other oxidizing agents, due to the formation of an oxide on the surface of the metal.

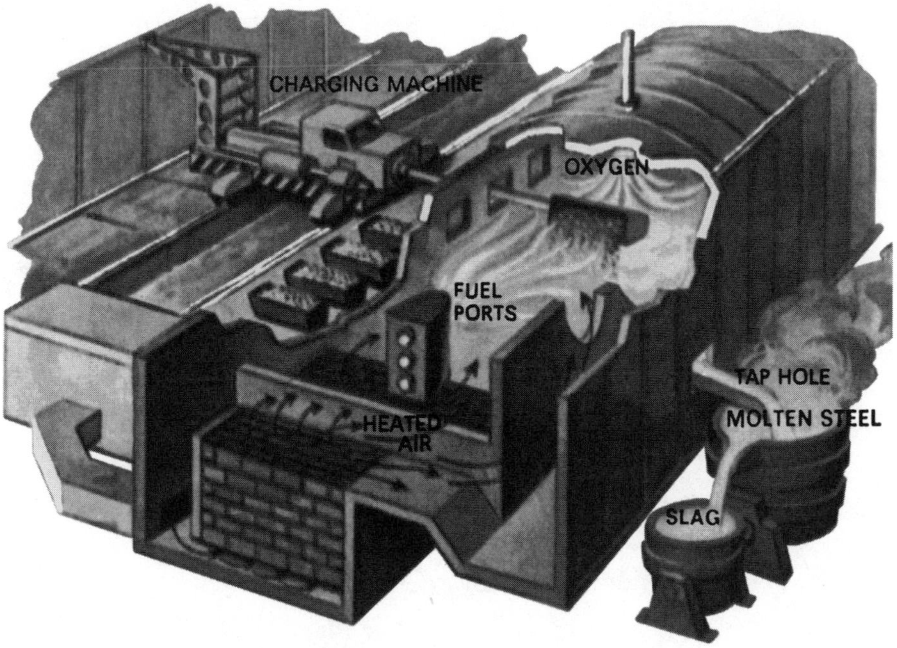

open-hearth process. *(Courtesy U.S. Steel Corp.)*

pearlite A eutectoid mixture of iron containing 0.80 percent carbon that is formed at 1333°F during slow cooling to form a laminated structure of ferrite and cementite. *(See figure.)*

pearlite. *(Courtesy Kelmar Associates.)*

- □ When viewed through a microscope, the pearlitic grain structure resembles human fingerprints.

peening The working of a metal by hammer blows or shot impingement.

periodic table A scientifically organized table of all chemical elements.

photomicrograph A photograph of a tiny area enlarged optically by a microscope to 100 to 500× magnification.

physical properties Properties of a metal or alloy that can be measured without the application of force; examples are density, electrical conductivity, coefficient of thermal expansion, and magnetic permeability.

pickling Removing surface oxides from metals by chemical or electrochemical reaction.

pig iron Iron produced in a blast furnace and poured into sand molds arranged around the main pouring channel. This pattern was said to resemble a litter of pigs around the mother—thus the name "pig" iron.

pitting Localized corrosion of a metal surface, confined to a point or small area, that takes the form of cavities.

plasma spraying A thermal spraying process in which the coating material is melted with heat from a plasma torch that generates heat up to 30,000°F.

- □ A molten coating material is propelled against the base metal by the hot, ionized gas issuing from the torch.
- □ This process is used for spraying high-temperature materials such as tungsten, cobalt, chromium, and refractory ceramics.

plastic deformation The permanent distortion of metals under applied stresses that stretch the material beyond its elastic limit.

plate A flat-rolled steel product $^1/_2$ in or thicker and 6 to 48 in wide.

potential energy Stored energy that is related to the distance between atoms.

powder metallurgy (PM) Processes in which metallic particles are fused under various combinations of heat and pressure to create solid metals. *(See figure.)*

- □ Specially prepared powdered metal particles are blended, then compacted under pressure (20 to 100 tons) in a precision die.
- □ The part is then sintered at 800 to 1500°C in a

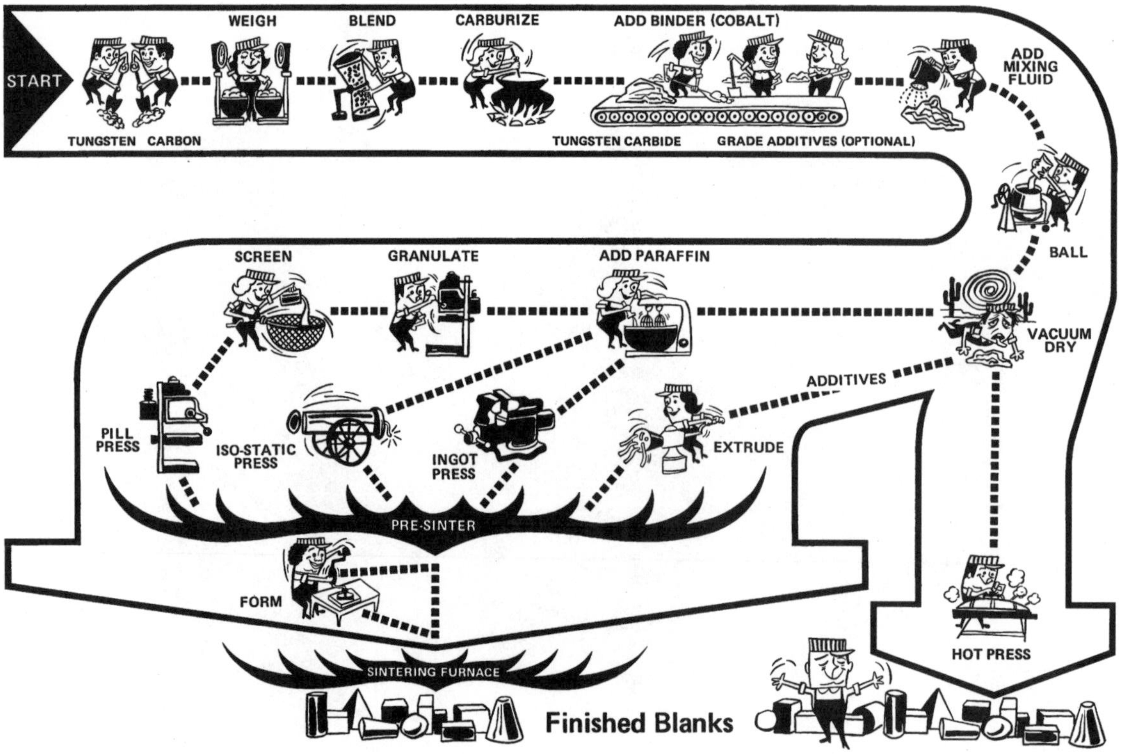

powder metallurgy. *(Courtesy Carboloy, Inc.)*

controlled atmosphere which bonds the metal particles together.

precipitation The settling or dropping out of particles.

primary processing defect An imperfection caused in the casting, forging, rolling, or welding of a metal.

proportional limit The range in a tensile test of a specimen in which the applied stress and the resulting strain are proportional.

□ It is within this range that the specimen will return to its original length when the force (stress) is removed.

□ Beyond this point (*P*), deformation occurs in the specimen. (*See* TENSILE TEST *and* MODULUS OF ELASTICITY.)

proton A positively charged particle in the nucleus of an atom around which the electrons revolve.

radiography A nondestructive inspection of metal by passing x rays or gamma rays through the part and then projecting them onto film, or a fluoroscope.

recrystallization The formation of a new strain-free grain structure from the highly deformed grains caused by extended cold working of the metal.

□ This is brought about by annealing the metal at the critical temperature, which changes the crystal structure.

recrystallization temperature The approximate minimum temperature at which complete recrystallization of a cold-worked metal occurs within a specified time.

red hardness The ability of a cutting tool to perform well even when the edge has become red through the friction of cutting. This quality is obtained by the addition of cobalt, molybdenum, and/or tungsten, to the tool steel.

red shortness The condition of wrought iron or some steels which makes it impossible to roll or forge at red heat, due to the presence of sulfur in the metal.

reduction of area The difference, expressed as a percentage, between the original cross-sectional area of a tensile test specimen and the minimal cross-sectional area after separation of the specimen has occurred.

refractory brick Specially processed brick, made from materials such as magnesite or alumina and having greater strength and durability than clay at high temperatures.

refractory metal A metal having an extremely high melting temperature such as tungsten, molybdenum, tantalum, niobium, chromium, vanadium, and rhenium.

□ This generally refers to metals having melting points above the range of iron.

residual elements Elements present in an alloy in small quantities, but not added intentionally, such as sulfur in steel.

residual stress Stresses that remain in the part after the force has been removed. It often occurs in parts having a nonuniform shape.

□ This may create a serious problem when the part is heat-treated, since distortion and cracking, or even failure of the part, may occur.

reversing mill A type of rolling mill in which the stock being rolled passes forward and backward between the same pair of rolls that are reversed between the passes.

Rockwell hardness test A hardness test that uses a diamond penetrator (brale) or steel ball under a specific load to measure the depth of penetration into a metal. *(See figure.)*

Rockwell hardness tester. *(Courtesy American Chain & Cable Co., Wilson Instrument Division.)*

□ A 120° conical diamond penetrator, under a 150-kg major load, is used for hardened metals, and the reading is taken on the C scale.

□ A 1.5-mm steel-ball penetrator, under a 100-kg load, is used for soft metals, and the reading is taken on the B scale.

rod (wire rod) Smaller round cross-sectional steel, produced by hot rolling and formed into coils or rods, which are then drawn through one or more dies to produce wire.

Scleroscope hardness test A test used to determine the hardness of a metal by means of a small diamond-tipped hammer that is dropped from a given height. The height of the rebound is measured and converted to a hardness reading.

scrap Metal left from the steelmaking process, manufacturing plants, and worn-out capital goods such as automobiles, ships, machinery, and other products and "tin" cans. This scrap metal is melted and used again in the production of steel (recycled).

scrubber Tanklike vessel containing water sprays that clean the gases blown through it.

secondary hardness The higher hardness developed by certain alloy steels when they are cooled from a tempering temperature. This is always be followed by a second tempering operation.

secondary processing defect An imperfection that develops in the heat treatment, grinding, or forming operations of the final processing of a product.

semifinished steel A rolled form such as blooms, billets, and slabs.

shear **1.** The resistance of a body to being cut or sheared by the action of two parallel forces acting in opposite directions, such as the action exerted on a rivet holding two metal plates together, each of which is being forced in parallel but opposite directions. **2.** A cutting tool that uses two parallel blades, in contact, to cut sheet metal, plates, rods, or wire. **3.** The cutting action of a tool on which rake has been ground, to produce a better shearing action.

shear strength The stress required to produce failure, when the forces acting in opposite directions on two parallel contiguous component parts produce separation.

sheet A thin, flat-rolled piece of steel usually over 12 in wide and limited in length.

shot blasting Blasting with metal shot to remove deposits or mill scale more rapidly or more effectively than possible by sand blasting.

shot peening Cold working a metal by metal-shot impingement on the surface.

sintering The bonding together of small particles or powdered metal at high temperatures (just below the melting point of the metal) to form a single mass. Sintering may be used

- □ In the powder metallurgy production of difficult-to-produce machine parts, and carbide cutting tools.
- □ To burn the sulfur out of the iron ore and turn it into an oxide that is used in the production of pig iron.

skelp Steel strip or plate from which pipes or tubes are made by welding.

slab A semifinished form of steel rolled from an ingot; it is rectangular in shape and always more than twice as wide as it is thick.

slag Rocklike waste, a by-product of the blast furnace and similar operations.

smelting The process used to remove the metal from the ore by means of heat and fluxing agents.

solid solution An alloy in which two (or more) metals remain dissolved in each other when solid.

- □ *Example 1.* Gold and silver, when melted together, will form a solid homogeneous mass.
- □ *Example 2.* During the heat treatment of 0.83 percent carbon steel, the carbon in the steel, which is in the form of iron carbide, will completely dissolve in the iron to form austenite at temperatures in the critical range.

solidus The point at which freezing is complete.

solution heat treatment Heating an alloy to a suitable temperature and holding at that temperature long enough to cause one or more constituents to enter into a solid solution.

- □ Rapid cooling from this temperature will hold these constituents in a solid solution.

spectrograph Any type of analyzing equipment that uses light or energy rays as a basis of identifying each chemical element.

sphalerite Zinc sulfide ore, ZnS.

spheroidizing A process of heating steel for an extended period of time to just below the lower critical temperature (Ac_1) followed by slow cooling.

- □ This process converts the needlelike particles of cementite into spherical or globular carbide, making the metal easier to machine.

stabilizing treatment A process used for the rapid aging of a semifinished precision part or tool such as a plug or ring gage.

- □ The part is subjected to several cycles of heating (to about 212°F) and refrigerated cooling for a similar time.
- □ After each cooling cycle, the part is allowed to return to room temperature.
- □ This process is continued for 5 to 6 times, depending on the shape and size of the part.

stainless steel A wide variety of corrosion-resistant steels commonly containing a high percentage of chromium and nickel.

steel An alloy of iron and up to 1.7 percent carbon plus other elements such as manganese, phosphorus, silicon, and sulfur. There are two types of steel:

- □ Carbon steel, which is a combination of iron and carbon
- □ Alloy steel, which is carbon steel plus manganese, molybdenum, chromium, nickel, or other alloying elements

steel specification number Numbering systems, developed by the American Iron and Steel Institute (AISI) and Society of Automotive Engineers (SAE), that use four or five digits to identify the compositions of a steel. *(See figure on next page.)*

- □ The first digit indicates the class of the steel.
- □ The second digit indicates the percentage of the major alloying element.
- □ The third and fourth digits indicate the carbon content.
- □ The fifth digit is used for steels with over 1 percent carbon.

strain The change in the size and shape of a body

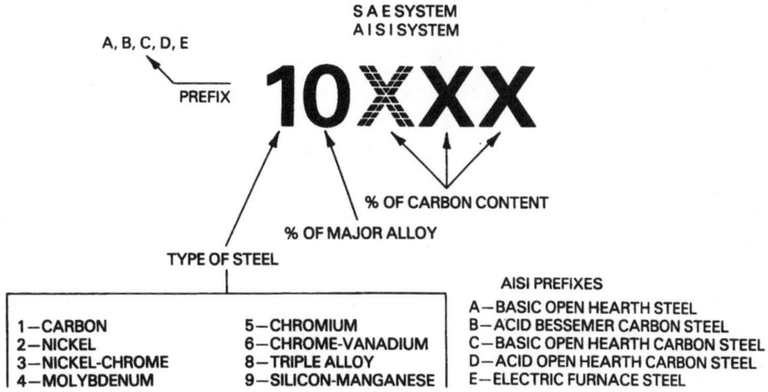

steel numbering system. *(Courtesy Kelmar Associates.)*

when an external force is applied.

- □ Linear strain is measured on a tensile test specimen by the increase in length of the specimen compared to its original length, and is stated in percent elongation. *See* TENSILE TEST.

strain hardening An increase in hardness and strength caused by distortion of the part due to cold working.

stress The internal resistance of a body to any applied force attempting to change the shape or size of the body. *See* TENSILE TEST.

stress-corrosion cracking Failure by cracking under the combined action of corrosion and applied (external) or internal (residual) stresses.

stress relieving Heating the part to a suitable temperature and holding it long enough at this temperature to reduce internal stresses. This is followed by slow cooling to minimize the development of new residual stresses.

stretcher strains Elongated markings that appear on the surfaces of some sheet metals when bent just past the yield point. These markings lie approximately parallel to the direction of maximum stress (bend) and are the result of localized yielding.

- □ Surface defects which may appear on temper-rolled steel, usually occurring at about a 45° angle to the stress axis.

strip Thin, flat-rolled steel in long continuous-strip form produced from slab, blooms, or billets by hot rolling, followed by pickling and cold rolling.

structural shapes Steel for use in construction, made by passing blooms through grooved rolls. The most important types are angle iron, I-beams, channels, T and Z structures, and piling.

superalloys Tough, hard-to-machine alloys; includes Hastelloys, Inconels, and Monels. These alloys are very corrosion-resistant, and are used a great deal in the chemical industry.

superplasticity Tough, hard-to-machine stainless steel and titanium alloys that show an unusually high degree of ductility, especially at elevated temperatures.

tapping **1.** The removal of molten iron or steel from the various iron- or steelmaking furnaces. **2.** To cut an internal thread in a workpiece, usually to accept a threaded bolt.

teeming The pouring of molten steel into ingot molds. *(See figure.)*

temper carbon Carbon in nodular form as found in malleable iron.

tempering Reheating quenched steel to a temperature below the critical range, followed by any desired rate of cooling (in water, oil, or air) to relieve quenching stresses, or to develop desired strength characteristics. *(See figure on next page.)*

temper rolling Light cold rolling of steel sheet to improve flatness, minimize the tendency toward formation of stretcher strains, and obtain the desired texture and mechanical properties.

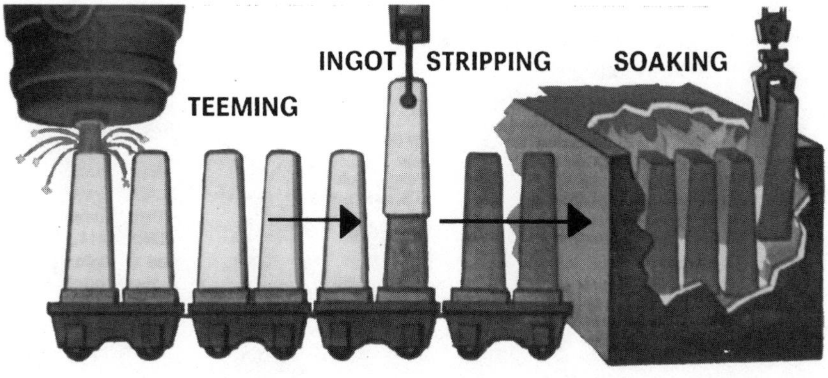

teeming. *(Courtesy U.S. Steel Corp.)*

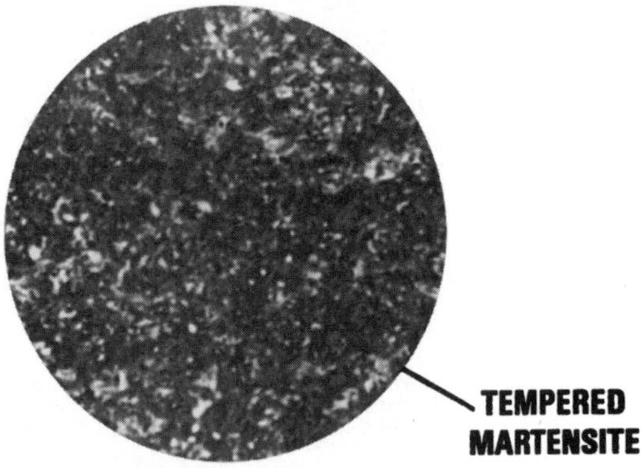

tempered tool steel. *(Courtesy Kelmar Associates.)*

tensile strength A mechanical property of a metal's strength relative to stress.

tensile test Destructive testing in which a sample of material is pulled on a tensile testing machine until it breaks. *(See figure.)*

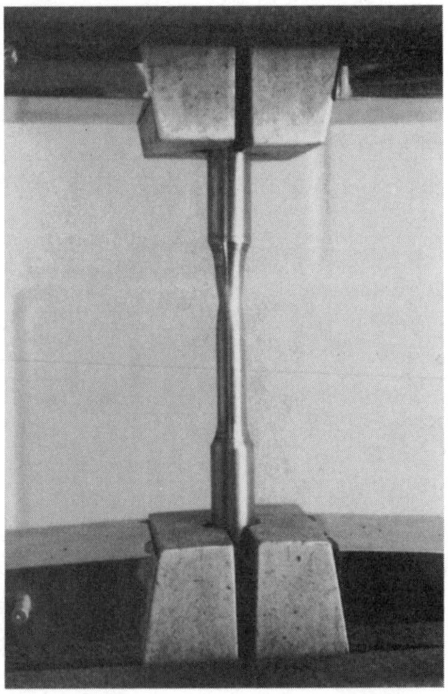

tensile test. *(Courtesy Kelmar Associates.)*

□ This will show the specimen's properties such as proportional limit, elastic limit, yield point, yield strength, ultimate strength, breaking strength, percent of elongation, and percent reduction in area.

thermometry The measurement of temperatures below 950°F by means of a thermometer.

tinplate Thin sheet steel covered with an adherent layer of tin by two methods:

1. The steel sheets are dipped in a bath of molten tin at about 600°F. In most cases, the tin coating is about .0001 in thick.
2. Electrotinning, in which the thin sheets are immersed in an electrolyte and a current is passed from a tin electrode to the plate. Here the deposited tin is only about .00003 in thick.

titanium carbide (TiC) An extremely hard material added to tungsten carbide during the manufacture of cemented carbide tools to reduce cratering and built-up edge.

□ Also used as a tool coating in which a very thin layer of TiC is fused to the cutting edge of the carbide insert to increase the wear resistance of the cutting edge by up to 500 percent.

titanium nitride (TiN) Added to titanium carbide tooling to permit machining of hard metals at high speeds.

□ Also used as a tool coating and offers greater wear resistance at speeds of up to 500 sf/min.

tool steel This term covers a wide range of high-carbon and alloy steels used to make various types of tools. Tool steels offer many advantages over plain carbon steels, such as

□ High hardness, toughness, and red hardness

□ Resistance to abrasion, wear, shock, deformation, and high temperatures

transformation temperature Also called *critical temperature*; the point at which a change in a metal's crystalline structure takes place.

ultimate strength The maximum conventional stress (tensile, compressive, or shear) that a material can withstand before breaking.

ultrasonic test Nondestructive inspection of a metal using high-frequency sound to detect cracks and internal imperfections.

use defect An imperfection in a metal caused by its actual use.

vacuum arc remelting A refining technique that achieves a degree of uniformity unattainable from a single melting operation.

□ Metal is remelted in a vacuum arc furnace, and combined gases and nonmetallic impurities are broken down and removed from the melt.

vacuum deposition The condensation of ultrathin metal coatings on the cool surface of work in a vacuum.

□ Articles to be plated may include metals, paper, wax, fabric, and glass.

vacuum induction melting A melting technique that occurs in a vacuum to produce cleaner, sounder alloys with lower inclusions and more uniform mechanical and physical properties.

□ The vacuum greatly reduces the levels of trapped or dissolved oxygen, nitrogen, and hydrogen gases.

□ This technique is used to melt special alloys containing highly reactive metals such as titanium and zirconium that would be lost in conventional furnaces.

vacuum processing A process used to remove gasses such as oxygen, nitrogen, and hydrogen from molten steel in order to make high purity steel. *(See figure on next page.)*

Vacuum degassing. Where conventionally melted steel is fed into a vacuum chamber and is broken into droplets and the unwanted gases are drawn off.

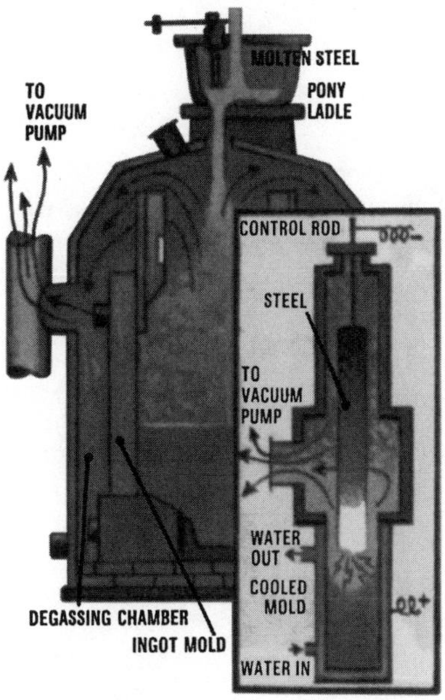

vacuum processing. *(Courtesy U.S. Steel Corp.)*

Vacuum melting. Where an electric arc remelts the end of a solid steel billet in a vacuum chamber to remove harmful gases and eliminate most of the nonmetallic inclusions in the billet.

vapor plating Deposition of a metal or compound on a heated surface by reduction or decomposition of a volatile compound at a temperature below the melting points of the deposit and the base material.

□ The reduction is usually accomplished by a gaseous reducing agent such as hydrogen.

white cast iron Cast iron that shows a white fracture because the carbon is in combined form as cementite.

white rust Zinc oxide; the powdery product of corrosion of zinc or zinc-coated surfaces.

work hardness Hardness developed in metal resulting from cold working.

wrought iron A high-purity iron and smelting slag mixture (iron silicate) that forms a fibrous structure. It is easily worked and shaped, resists rusting, and is often used in products such as hand railings and porch posts.

wustite An iron oxide, containing 23.5 percent oxygen, that is an intermediate product during the reduction of hematite and magnetite ores.

yield point The point just beyond the proportional limit (during a tensile test) at which the metal starts to stretch without a corresponding increase in stress.

yield strength The maximum useful strength of the material; the stress at which a metal sample begins to permanently deform.

zircon Zirconium silicate ore, $ZrSiO_4$.

Metalworking Fluids and Coolants

Steve F. Krar
Consultant
Kelmar Associates
Welland, Ontario

Cutting fluids affect the performance of the cutting tool by increasing the material-removal rate, prolonging tool life, improving part quality and accuracy, and reducing manufacturing costs. Excessive heat generated during a grinding or machining operation will shorten tool life, and can cause thermal damage to the microstructure of the workpiece. The selection of the proper cutting fluid to suit the workpiece material, and the cutting operation and its application, are essential for a successful manufacturing operation.

active cutting oils The oils that will darken a copper strip immersed in them for 3 h at a temperature of 212°F (100°C) or 24 h at 60°C.

- □ The oils may be dark or transparent; dark oils may contain more sulfur and are better for heavy-duty jobs.
- □ The most common are sulfurized mineral oils, sulfochlorinated mineral oils, and sulfochlorinated fatty mineral oil blends.

additive Any substance such as sulfur, chlorine, or other agents that are added to cutting fluids to reduce friction, improve lubricity, and prevent a built-up edge from forming during a machining operation.

admixture A mixture of chemical compounds that is used to refill a cutting fluid reservoir or bring the fluid back to its original strength.

amines and nitrites Chemical agents added to a cutting fluid for rust inhibition.

aqueous fluids Cutting fluids that are water-miscible and are either true solutions, surface-active agents (surfactants), emulsions, or mixtures of these.

bactericides Chemical agents that are added to cutting fluid to prevent or retard the growth of bacteria.

- □ Once bacteria start to grow, the cutting fluid may become rancid and have a disagreeable odor.

boundary additives These agents are considered to be fats or fatty acids or esters that promote wetting and go to the source of heat.

- □ The boundary film helps prevent a built-up edge from forming at low cutting speeds and produces a good surface finish.

built-up edge A layer of compressed metal from the material being cut that sticks to and piles up on the edge of the cutting tool during a machining operation. *(See figure.)*

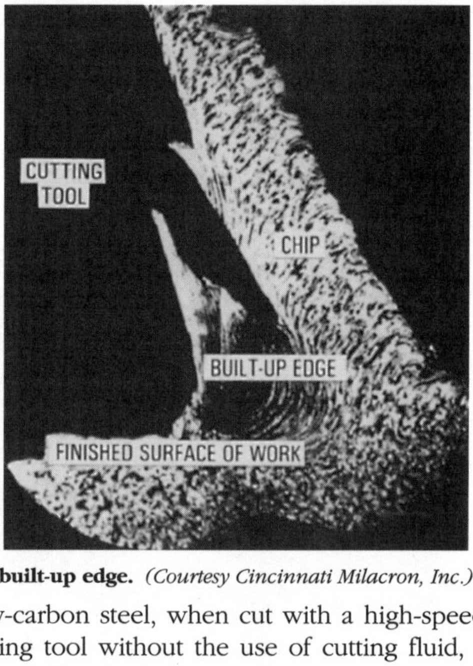

built-up edge. *(Courtesy Cincinnati Milacron, Inc.)*

- □ Low-carbon steel, when cut with a high-speed steel cutting tool without the use of cutting fluid, generally produces a built-up edge.
- □ A built-up edge produces a rough surface finish on the work, shortens tool life, and results in severe work hardening of the chip and work surface.

chemical cutting fluids Sometimes called *synthetic fluids*; fluids that contain very little oil, mix easily with water, and depend on chemical additives for lubrication and friction reduction.

□ Some types contain extreme-pressure (EP) lubricants to reduce the heat caused by the plastic deformation of the metal and the heat of friction between the chip and the tool face.

□ Most common are true solution fluids, wetting-agent types, and wetting-agent types with EP lubricants.

chip The material removed from a workpiece by a cutting tool through either a grinding or machining operation.

chip-tool interface The portion of the cutting-tool face that the chip slides on as it is separated from the metal.

concentrate The base of a cutting fluid, containing the chemical agents or additives, which, when added to the correct percentage of diluent (diluting liquid) such as oil or water, produces a cutting fluid.

coolant Another term for a cutting fluid; coolants operate according to the heat-transfer principle. *(See figure.)*

□ Cutting fluid reduces the friction and temperature buildup at the chip-tool interface during metal-removal operations.

□ Water is the best cooling medium but cannot be used alone because it causes workpieces and machine parts to rust.

□ Most cutting fluids are water- or oil-based, and chemical compounds are added to provide the desired qualities in a fluid.

□ In some cases pressurized or refrigerated air can be used as a coolant for specific operations.

coolant management The monitoring and control of all coolant variables to optimize tool life, increase productivity, reduce costs, and improve the shop environment.

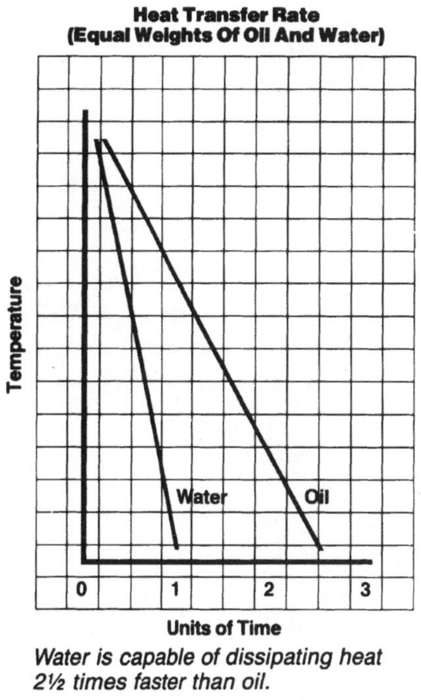

heat transfer. *(Courtesy Master Chemical Corp.)*

corrosion The gradual breakdown of a metal by surface disintegration caused mainly by a chemical action.

□ This action will weaken and damage the metal.

crystal elongation The distortion of the crystal structure of the work material that occurs as a result of the pressure of the machining operation. *(See figure.)*

curtain application A method of using multiple nozzles to apply a cutting fluid to a large cutting area, as in milling and sawing operations.

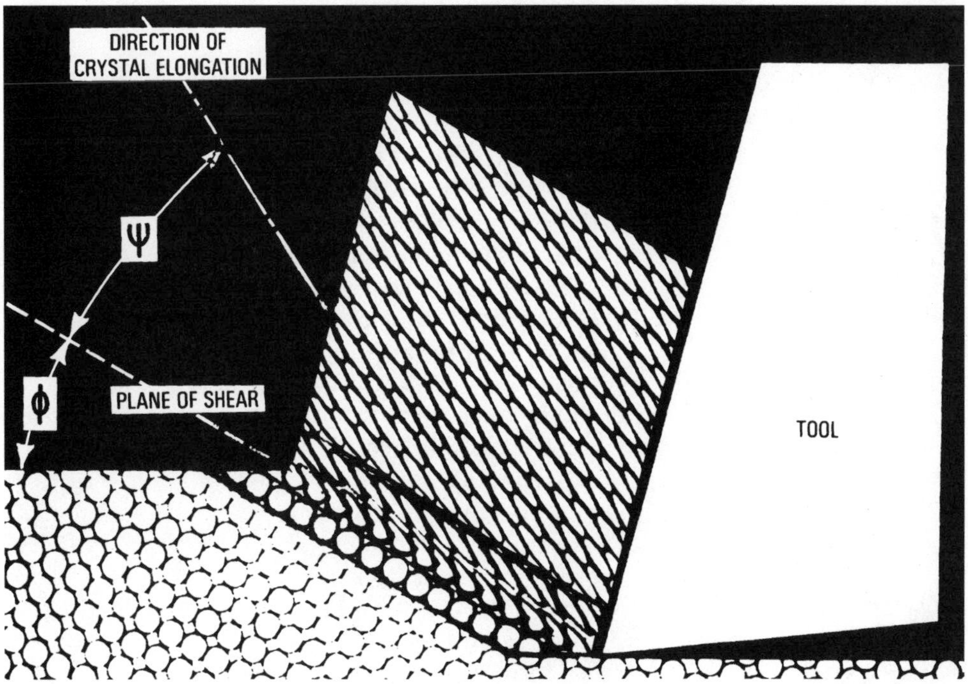

crystal elongation. *(Courtesy Cincinnati Milacron, Inc.)*

cutting fluid Also called *coolant*; a liquid mixture used to increase metal-removal rates, extend cutting-tool life, wash away the chips, and cool the workpiece and the cutting tool. *(See figure.)*

□ Four types of cutting fluids are straight oils, emulsifiable oils that mix in water, chemical true solution fluid, and surface-active fluids.

cutting fluid. *(Courtesy Modern Machine Shop.)*

□ Each type of cutting fluid has certain qualities that make it suitable for certain machining or grinding operations.

deionization The process involving treatment of water with minerals to remove polyvalent minerals and salts that can cause difficulties in preparing and using good cutting-fluid solutions.

emulsifiable mineral oil Mineral oil to which various compounds have been added to make the oil miscible in water.

emulsion A liquid mixture in which an oil is suspended as globules in another mixture, such as oil in water. *(See figure.)*

extreme-pressure (EP) additives Cutting-fluid additives (chlorine, sulfur, or phosphorous compounds) that chemically react with the workpiece material to minimize chip welding.

emulsifiable oils. *(Courtesy Modern Machine Shop.)*

□ The soluble-oil mixture that turns water into a milky cutting-fluid solution.

extreme-pressure (EP) emulsifiable oils Emulsifiable oils that contain sulfur, chlorine, phosphorus, and fatty oils to provide extra lubrication qualities for tough machining operations.

extreme-pressure (EP) lubricants A term referring to any cutting fluid or medium that provides high lubrication qualities for difficult metal-removal operations.

fan-shaped nozzle Wide nozzle with a relatively thin opening to spread the cutting fluid over a wide area of the workpiece. *(See figure.)*

fan-shaped nozzle. *(Courtesy Cincinnati Milacron, Inc.)*

fatty–mineral oil blends Combinations of fatty and mineral oils that provide better wetting and penetrating qualities than does straight mineral oil.

fatty oils Substances such as animal, vegetable, or marine oils find limited applications as cutting fluids.

□ They are usually blended with various chemical additives.

film strength The quality that a cutting fluid has to form a film between the workpiece and the cutting tool to reduce friction and metal-to-metal contact.

flood application A large flow of cutting fluid that is applied through a wide nozzle to flood the cutting tool and the area of the material being cut.

□ This application generally results in the highest metal-removal rates and the longest cutting-tool life.

fluid recycling The process of removing tramp oil and metal fines from a dirty cutting fluid by a filtration process.

□ Fresh, clean coolant is then added to the recycled coolant and put back into the machine.

foam depressants Special wetting agents used to control the amount of air whipped into water-miscible fluids because the sump does not have sufficient capacity to control foaming.

foaming The condition in which air has leaked into a cooling system and is beat into the coolant mixture to cause foam. *(See figure on next page.)*

foaming. *(Courtesy Modern Machine Shop.)*

friction The resistance that results when the chip being removed moves across the face (note long shear plane in A *(see figure)* and short shear plane in B of the cutting tool.

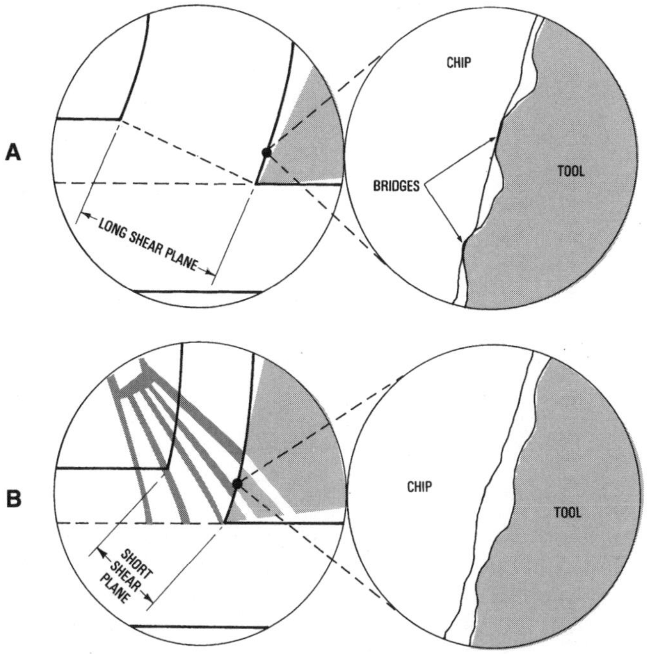

shear cutting angle. *(Courtesy Cincinnati Milacron, Inc.)*

□ This action will change the microstructure of a metal and shorten the life of a cutting tool.

□ Ninety-seven percent of the power used in this process results in heat.

fungicide Material added to water-miscible cutting fluids to inhibit the growth of fungi.

gaseous liquids Air or other gasses such as argon, carbon dioxide, helium, and nitrogen that are used for both cooling and lubricating functions.

□ Compressed or refrigerated air directed to the cutting zone can cool and blow away the chips.

□ Tool life is increased, but the cost of gaseous coolants is extremely high.

germicides Agents added to cutting fluids to destroy germs and microorganisms.

□ They are generally added to prevent foul odors and corrosion.

glycol A sweet, oily, colorless alcohol added to cutting fluids to act as a blending agent.

hard water Water that contains varying amounts of calcium, magnesium, and other polyvalent minerals that make it unfit for coolant usage.

□ It should be deionized or distilled to make it suitable for coolant use.

heat dissipation The ability of a cutting fluid to remove heat from the chip-tool interface during a machining operation *(see figure)* (note dry cutting in A and wet cutting in B).

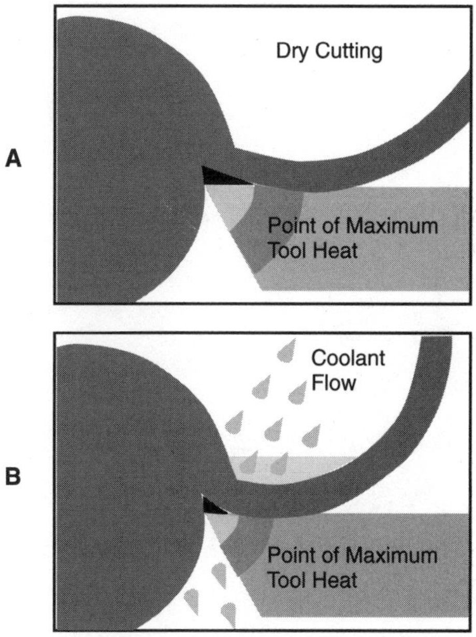

heat dissipation. *(Courtesy Master Chemical Corp.)*

□ Water is the best cooling medium; however, it cannot be used by itself because it causes work and machine parts to rust.

inactive cutting oils The oils that will not darken a copper strip immersed in them for 3 h at a temperature of 212°F (100°C) or 4 h at 60°C.

□ The sulfur in these oils is so firmly attached to the oil that very little is released during the cutting operation.

□ Most common types are straight mineral oils, fatty oils, fatty–mineral oil blends, and sufurized fatty–mineral oil blends.

lubricant Any substance that reduces friction by metal-to-metal contact between moving surfaces such as the cutting tool during the cutting process.

lubricity The effectiveness of any material to reduce friction between any surfaces coming into contact with each other.

miscibility The ability of any liquid to mix and blend with another liquid.

mist application A cooling system in which a spray of atomized fluid is applied to the chip-tool interface.

□ Used in contour bandsawing, where it is necessary to have a clear view of the layout lines and the point of the cut.

mixture ratio The exact ratio of the concentrate to the amount of diluent (diluting liquid) that provides the best cutting-fluid qualities of cooling, lubrication, and corrosion prevention.

nonflammable An important quality of a cutting fluid that will not ignite as a result of the high heat created at the chip-tool interface during machining.

oil-feed drill A type of drill, used for deep-hole drilling, where the cutting fluid is directed to the cutting edges of a drill through holes in the drill body. *(See figure.)*

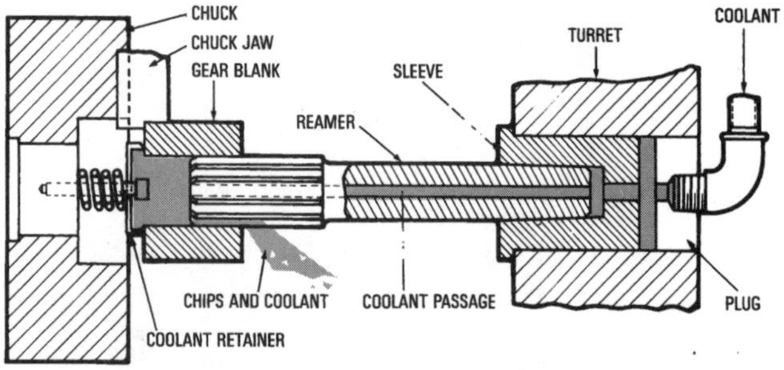

oil-feed drill. *(Courtesy Cincinnati Milacron, Inc.)*

passivating film A quality of certain cutting fluids in which an insulating blanket is formed on the metal surface to prevent corrosion.

phosphates and borates The chemical agents added to cutting fluids to reduce corrosion.

plastic deformation The compression of a metal's crystal structure that occurs in the shear zone during a metal-removal operation.

polar additives Animal, vegetable, or marine oils or esters that, when added to a mineral oil, improve its ability to penetrate the work-tool interface.

polar film A quality of certain types of cutting fluids in which a protective film is formed on a metal surface to prevent corrosion. *(See figure.)*

□ The polar film consists of negatively charged, long, thin molecules that are attracted to and firmly bond themselves to the metal.

rancidity Refers to the condition of a cutting fluid that has begun to go rancid as a result of the growth of bacteria and fungus.

□ Other signs of a cutting fluid starting to become rancid are unpleasant odors and stained workpieces.

□ To prevent this condition, cutting-fluid reservoirs should be cleaned and replaced with new fluid at regular intervals.

refrigerated air An effective cooling method that uses a refrigerated air system in which dry machining is necessary or preferred. *(See figure.)*

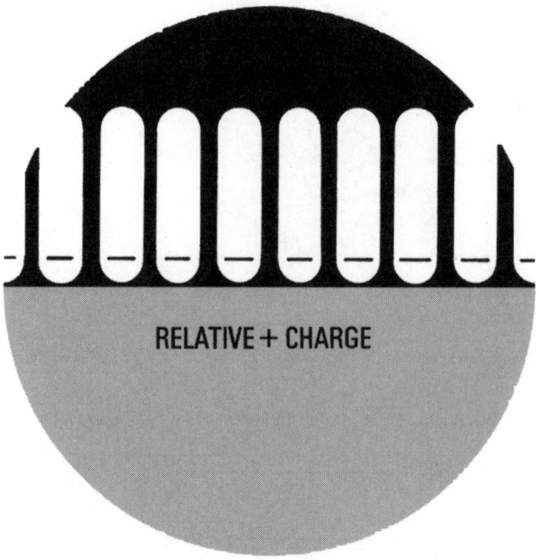

polar film. *(Courtesy Cincinnati Milacron, Inc.)*

□ Compressed air is cooled in a vortex chamber and is directed to the chip-tool interface to blow away the chips.

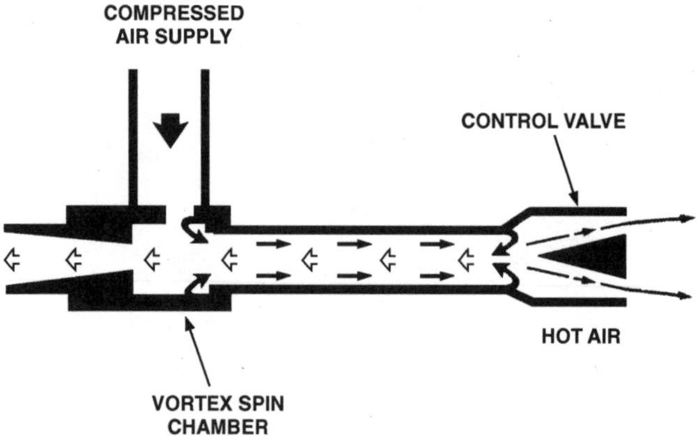

refrigerated air. *(Courtesy Exxair Corp.)*

residual film The material left behind on the workpiece and machine tool after the water evaporates from the coolant system.

ring-type distributor A flood-cooling device that surrounds face-milling cutters to keep each cutter tooth immersed in cutting fluid at all times.

□ It increases metal-removal rates and prolongs the milling cutter life by almost 100 percent.

rust Oxidized iron, or iron that has chemically reacted with oxygen and water. *(See figure on next page.)*

□ There are a number of types of iron rust.

rust inhibitors Chemical agents that are added to cutting fluids to prevent rust from forming.

semisynthetic cutting fluid A water-based chemical solution that contains some oil.

shear angle or plane The angle of the area of material in which plastic deformation occurs. *(See figure on next page.)*

□ A positive rake angle on a cutting tool, and use of cutting fluid, allow the chip to easily slide over the

rust. *(Courtesy Modern Machine Shop.)*

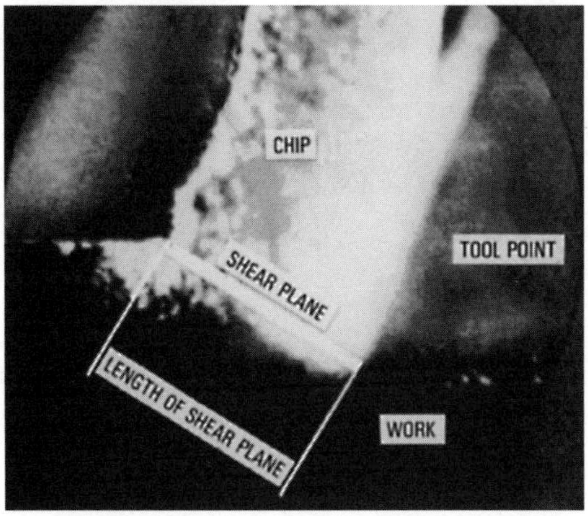

shear angle or plane. *(Courtesy Cincinnati Milacron, Inc.)*

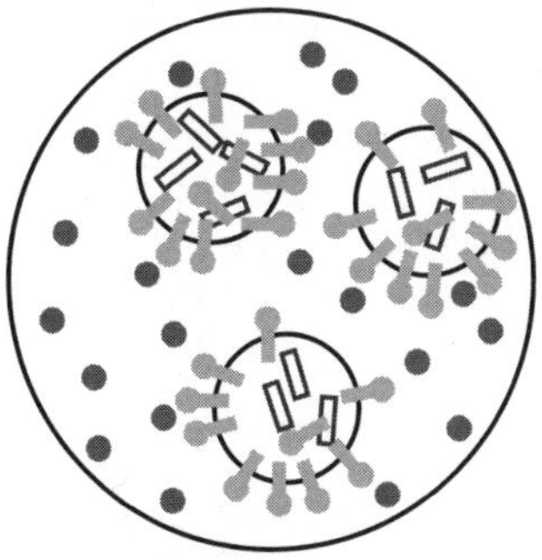

soluble oils (emulsions). *(Courtesy Master Chemical Corp.)*

tool face and results in a short shear plane.

□ A negative rake angle on a cutting tool makes it more difficult for the chip to slide over the tool face and results in a long shear plane.

shear zone The area in a metal where plastic deformation, due to the pressure of the machining operation, occurs.

soft water Water from rain is soft because it contains almost no minerals; well water usually contains minerals.

□ Water used for coolant mixtures should be as free as possible of minerals for the most economical and trouble-free coolant.

soluble-oil cutting fluid Also called *emulsifiable oils,* these are mineral oils containing emulsifying agents *(see figure)* that make them miscible in water.

□ Other additives could include chemical lubricating agents that have wetting qualities, provide lubricity and rust protection, and prevent the formation of a built-up edge.

□ Soluble-oil mixtures of from 1 to 5 parts of concentrate to 100 parts of water are used for light machining operations.

□ A higher percentage of concentrate is used when more lubrication is essential.

spray-mist cooling An effective cooling system that uses the atomizer principle to spray a small amount of coolant into the chip-tool interface.

□ Cooling results from the action of the compressed air and the evaporation of the vapor mist.

□ The compressed air blows away the chips from the machining area to allow the operator to see clearly.

stability The ability of a cutting fluid to maintain its cooling and lubricating qualities, resistance to bacterial and fungal attack, and corrosion effects.

straight mineral oil Mineral oil (paraffinic, naphthenic, or aromatic) of low or higher viscosity.

□ It may be used alone or blended with fats, esters, or chemical additives to make more complete cutting and grinding fluids.

sulfurized fatty–mineral oil blends These mixtures consist of sulfur and fatty oils combined with mineral oils.

□ Used for tough, abrasive machining operations where tool pressures are high.

superfatted emulsifiable oil An emulsifiable mineral oil to which some fatty oil has been added to provide extra lubrication qualities.

synthetic cutting fluid Generally refers to chemical cutting fluids that contain very little oil and depend on chemical agents to provide lubrication, reduce friction, and prevent corrosion and rancidity. *(See figure on next page.)*

thermal damage The damage or changes to the microstructure of a material due to excessive heat of a grinding or machining operation.

□ Often caused by a dull tool, improper tool rakes and clearances, and poor use or application of cutting fluid.

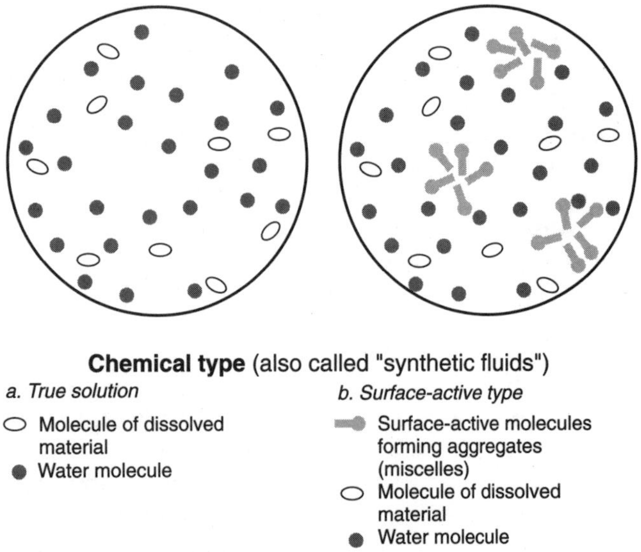

synthetic fluids. *(Courtesy Master Chemical Corp.)*

through-the-wheel cooling A system in which cutting fluid is fed to a special grinding wheel flange and is forced through the wheel voids (pores) to the periphery of the wheel by centrifugal force. *(See figure.)*

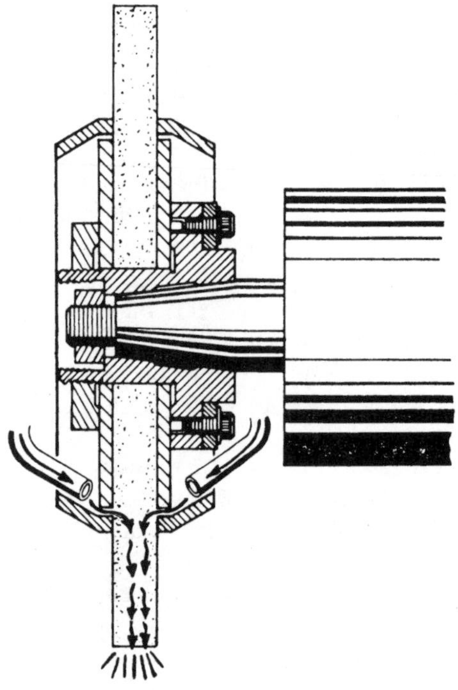

through-the-wheel cooling. *(Courtesy DoAll Co.)*

tramp oil Hydraulic and gear oils, and machine way lubricants that have leaked into the coolant system.

true solution fluids These contain mostly rust inhibitors and are used primarily to prevent rust and provide for rapid heat removal during grinding operations.

□ These types have a tendency to form hard crystalline deposits when the water evaporates, which may interfere with machine functioning.

viscosity Measure of a fluid's tendency to flow; varies with temperature.

water-miscible fluids These contain concentrates that thoroughly mix with the water base.

wetting A condition in which the coolant is made up of surface-active agents that can penetrate to the chip-tool interface during a machining-and-grinding operation: (*A*) Poor wetting or spreading shown by the large contact angle of the solution drop on the surface; (*B*) superior wetting or spreading shown by the low contact angle of the solution drop on the surface. *(See figure.)*

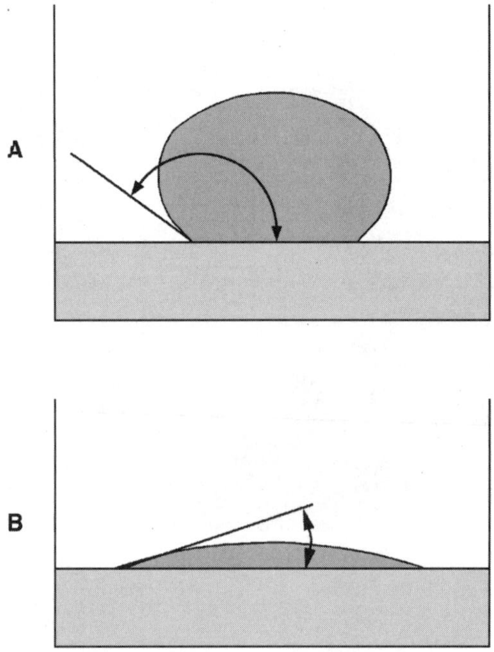

wetting. *(Courtesy Master Chemical Corp.)*

wetting-agent types Agents that improve the wetting action of water, providing more uniform heat dissipation and antirust action.

□ Used for general-purpose machining with high-speed steel and carbide tools.

wetting-agent types with EP lubricants Similar to wetting-agent types but also contain chlorine, sulfur, or phosphorus additives to provide EP or boundary lubrication effects.

□ Used for tough machining jobs with high-speed steel and carbide tools.

SECTION 12

Nontraditional Machining Processes

Paul J. Wanner
Manufacturing Technology Department
Advanced Technology Center
Clackamas Community College
Wilsonville, Oregon

Nontraditional machining processes are methods that are relatively new to the manufacturing of parts or do not fall into the category of conventional machining because they do not always involve machining forces and perishable tooling. Over the past 40 years these developments have included processes such as electrodischarge, electrochemical procedures, laser techniques, and stereolithography. These processes have made it possible to cut materials that were difficult, or almost impossible, to machine by conventional methods. These material-removal processes use a combination of computer technology with electrical energy, chemical reaction, hydrodynamic energy, laser light energy, and ultrasound to remove material to accurate dimensions. In some cases, the material-removal rate is slower; however, the time to get the part manufactured is greatly reduced overall compared to conventional methods.

abrasive flow machining (AFM) A controlled removal of workpiece material using two vertically opposed cylinders to force a semisolid abrasive medium back and forth through or around the workpiece to precisely remove material.

□ The machining action is similar to a filing, grinding, or lapping operation; it does not correct out-of-round or out-of-flat conditions.

□ AFM is used mainly on materials such as soft aluminum, nickel alloys, and ceramics.

abrasive jet machining (AJM) A material-removal process that uses fine abrasive particles that are blasted at the workpiece through a high-velocity gas stream (compressed air or CO_2) to cut, deburr, and clean parts. *(See figure.)*

□ Types of abrasives used are silicon carbide,

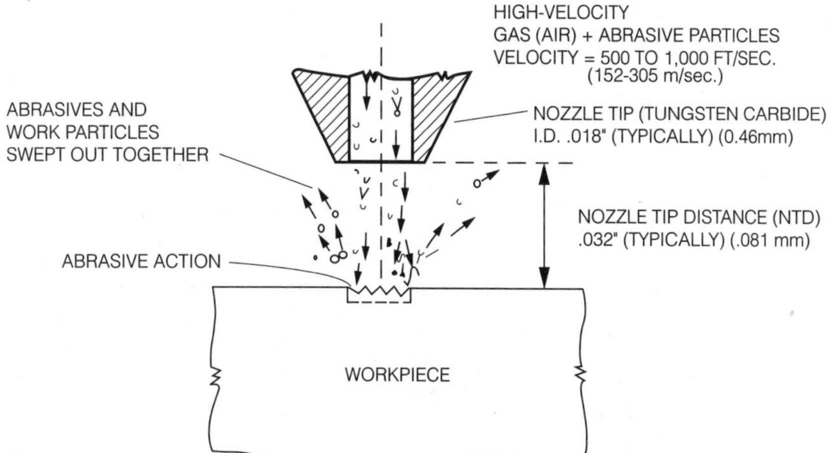

abrasive jet machining.
(Courtesy Society of Manufacturing Engineers.)

aluminum oxide, boron carbide, or glass beads, depending on the material being machined.

□ The carrier gas helps keep the workpiece cool.

altered metal zone The mechanical and structural properties of a metal that are changed on the surface of the metal by the electrical discharge machining (EDM) process.

alternating current (AC) Refers to the polarity of the electrical current switching from negative to positive and back again.

ammeter A meter located at the power supply to monitor the amperage draw in the machining gap of the EDM process.

amperage The amount of average current that the ammeter measures during the EDM cutting process.

ball burnishing The forcing of oversized tungsten carbide balls through an undersized hole to bring it to size, improve the surface finish, and work-harden the hole.

□ High speed is used to force a lubricated ball through the hole to prevent galling, seizing, and distorting.

ballistic particle manufacturing A process that ejects streams of material (e.g., plastic, ceramic, metals, or wax) through a small orifice at a surface using an inkjet mechanism. It uses a piezoelectric pump at 50 μm/10,000 Hz, and a three-axis robot.

barrel finishing/tumbling A versatile method of using loose abrasive materials for edge and surface conditioning that is a low-cost but slow operation. *(See figure.)*

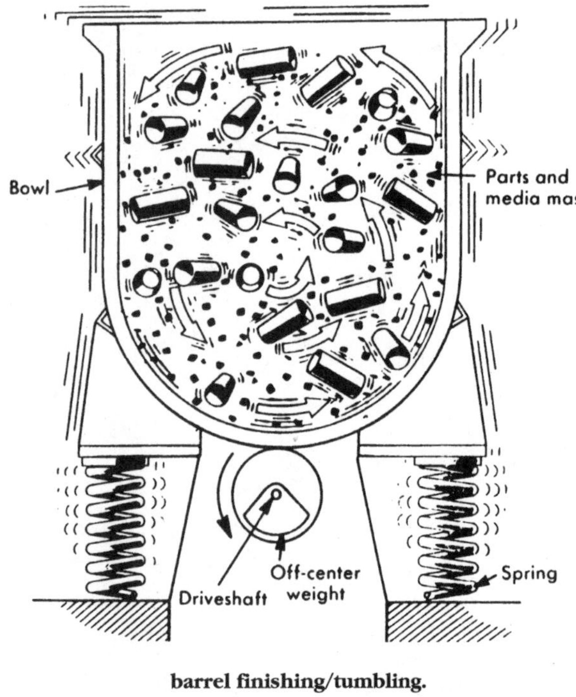

barrel finishing/tumbling.
(Courtesy Society of Manufacturing Engineers.)

□ The parts and abrasive are put into a barrel-finishing machine, with wear-resistant, replaceable linings, and slowly rotated or vibrated until the parts are finished.

burn, burning The common name given to electrical discharge machining (EDM).

CAD casting *See* COMPUTER-ASSISTED DESIGN CASTING.

chemical milling (CHM) The process used to shape metals to an exact size through controlled chemical action or etching rather than by conventional milling machine operations.

chemical processes These include the various chemical material-removal processes that use an acid or alkaline solution to remove (dissolve) unwanted material.

CNC sinker *See* COMPUTER NUMERICAL CONTROL SINKER.

coated-abrasive grinding The use of coated abrasives in the form of sheets, disks, rolls, belts, and other components to perform operations of deburring, deflashing, blending, surface finishing, and polishing.

□ Coarse-coated abrasives are used for fast stock removal.

□ Fine-coated abrasives are used for finishing operations.

computer-assisted design (CAD) casting A term used to define a casting process in which the mold is produced directly from a computer model with no intervening steps.

computer numerical control (CNC) sinker A ram-type electrical discharge machining (EDM) that has computerized numerical control of the *X-Y* table and the *Z* (ram) axis from which the electrode is held.

□ This is a more versatile machine than a die sinker, which is used to produce complex cavities by moving the electrode in the *X* or *Y* directions.

□ It can also be equipped with tool changers so that complex cavities can be burned in a single setup.

dielectric fluid A nonconductive fluid (oil) in which the part and electrode are submersed to control the arc discharge during the EDM burning process. It has two purposes:

□ The oil insulates the electrode/workpiece gap to prevent a spark from forming until the gap and voltage are correct.

□ It also flushes away particles during the burning process and helps cool and solidify melted particles.

die sinker The term given to ram-type (plunge) EDM machines that either lower the electrode or raise the workpiece to produce a cavity.

dressing The sharpening or remachining of an electrode's face after it has been used to EDM a mold cavity, primarily to reproduce a sharp corner.

EDM *See* ELECTRICAL DISCHARGE MACHINING.

electrical discharge grinding (EDG) A surface-grinding-type machine whose wheel (cathode) is made of an electrode material and in which electrical current is used to erode away the surface metal (anode) by an arcing process similar to EDM.

- □ No abrasive action occurs, but there is a slight wheel wear due to the arcing between the wheel and the workpiece.
- □ Used when low machining forces are required to grind material with thin sections.

electrical discharge machining (EDM) A process that vaporizes conductive materials by controlled application of pulsed electrical current that flows between a workpiece and an electrode (tool) in a dielectric fluid. *(See figure.)*

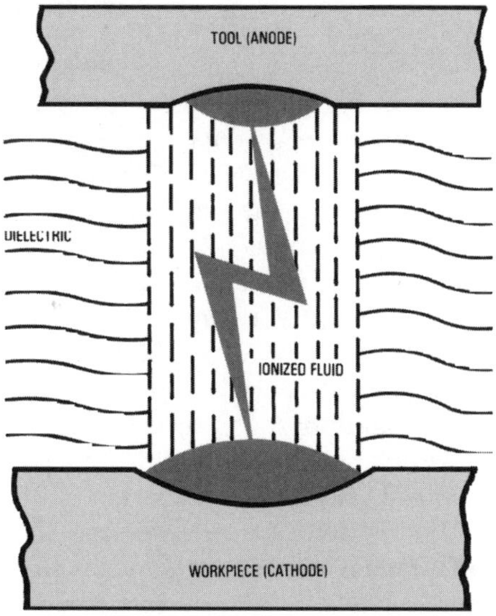

electrical discharge machining. *(Cincinnati Milacron, Inc.)*

- □ The two most common EDM machines are the ram type, used for producing cavities in moldmaking, and the wirecut, used for producing intricate shapes in diemaking.
- □ This process is used to machine shapes to tight accuracies without the internal stresses that conventional machining often produces.

electrical discharge wire cutting (EDWC) A through-cutting process of a continuously spooling conductive wire that produces an arc with the material and cuts the shape that is programmed for the part. *(See figure.)*

electrical energy In nontraditional machining, this process is based on a reverse electroplating principle involving material removal by a high-speed electrolyte flowing between the tool and the workpiece.

- □ Electrochemical machining and grinding are examples.

electrochemical deburring (ECD) A process, a specialized form of electrochemical machining, that is used to dissolve burrs from metallic workpieces electrochemically and flush them away by pressurized electrolyte.

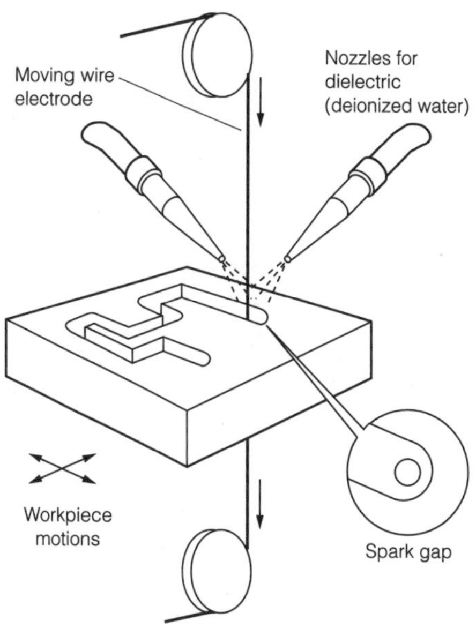

electrical discharge wire cutting. *(Courtesy Society of Manufacturing Engineers.)*

- □ Proper insulation of the tools and protective shielding (masking) are used to limit the electrochemical action to the desired areas of the part.

electrochemical discharge grinding (ECDG) This process combines the material-removal actions of both electrical discharge and electrochemical grinding to remove material through electrolysis by low-level, direct-current voltage.

- □ There is no direct contact between the graphite wheel and the workpiece; therefore, no abrasive action occurs.
- □ This process is effective on most electrically conductive materials, and the hardness of the workpiece does not affect removal rates.

electrochemical grinding (ECG) A process that uses a combination of abrasion and electrochemical action to remove material from an electrically conductive workpiece. *(See figure on next page.)*

- □ Direct-current power is used between the part and the conductive bond of the grinding wheel.
- □ It uses a negatively charged grinding wheel (cathode) with an insulating abrasive that is set in a conductive bonding material.
- □ Most of the material is removed by electrolysis; however, some material is removed by the abrasive which is in contact with the positively charged workpiece (anode).

electrochemical honing (ECH) A process that uses the same principles of ECM regarding current, voltage, and electrolyte and materials processed to size a part and improve its surface finish.

- □ The abrasive honing stones are rotated and recip-

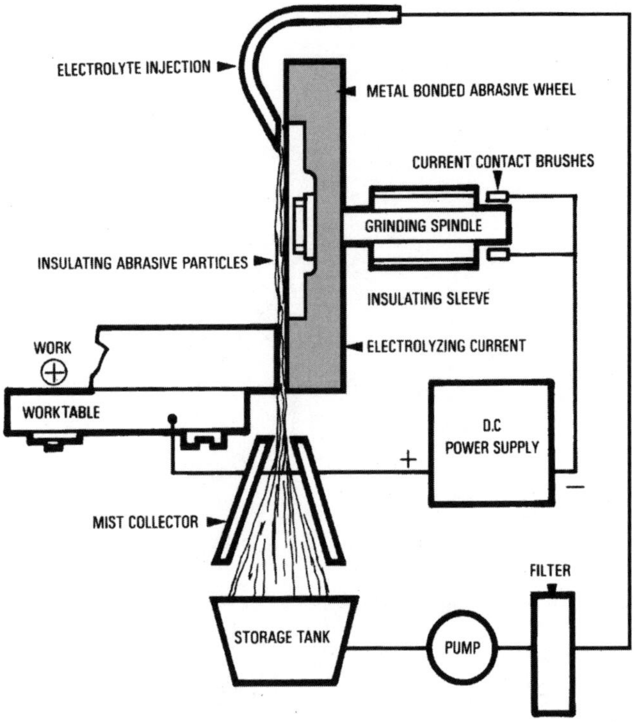

electrochemical grinding. *(Courtesy Kelmar Associates.)*

rocated through the part or cylinder, and the electrolyte flows equally in all directions.

□ The nonconductive stones assist in the electrochemical action to produce a round, straight cylinder even if the cylinder is tapered, out of round, or wavy.

electrochemical machining (ECM) A process used to remove electrically conductive workpiece material through the use of electrical energy and an electrolyte bath to form a reaction of reverse plating. *(See figure.)*

□ Direct current at high amperage and low voltage is continuously passed between the workpiece (anode) and the tool (cathode) through a conductive electrolyte.

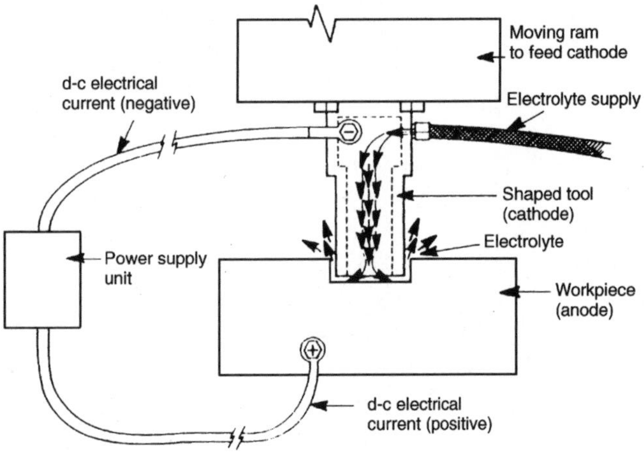

electrochemical machining.
(Courtesy Society of Manufacturing Engineers.)

□ Material leaves the anode (positively charged workpiece) and is attracted to the cathode (negatively charged tool).

□ The ECM procedure is used for machining forms beyond the capacity of conventional machining processes.

electrode In an EDM process, the cutting tool that the current flows through to arc with the part and erode away material with an electrode that is a mirror image of the cavity.

□ Common electrode materials are graphite, copper, and tungsten.

□ A brass wire, sometimes coated to improve performance, is the electrode of the wire EDM process.

electron-beam machining (EBM) A pulsating stream of high-speed electrons bombards the surface of the metal in a vacuum chamber with appropriate shielding from the radiation being generated.

□ As the electrons strike the work surface, their kinetic energy changes into thermal energy that melts or evaporates the material.

electropolishing A process (similar to electroplating) in which the metal is removed and not deposited through an electrochemical polishing action.

□ When the current is applied, a polarized film forms to brighten and polish the metal.

□ Metal particles pass through the polarized anodic film as metal salts and enter the electrolyte.

etching The removal of material by the use of acids.

fine nylon *Fine Nylon* and *Fine Nylon, Medical Grade* (MG) are fine-grade nylon powders recommended for use in selective laser sintering (SLS) applications, where durable parts with good thermal stability and chemical resistance are required.

□ Parts made from nylon benefit from enhanced feature details, strength of small features, and finer-textured surface finishes.

□ Fine Nylon MG has been certified U.S. Pharmacopeia (USP) level VI for limited use in an artificial environment; it can be steam-sterilized for medical applications.

fused deposition modeling (FDM) The FDM process forms three-dimensional objects from CAD-generated solid or surface models. FDM patterns are generally used when an acrylonitrile-butadiene-styrene (ABS) thermal plastic part is required for a working prototype. *(See figure on next page.)*

□ Thermoplastic modeling material is fed into the temperature-controlled FDM extrusion head, where it is heated to a semiliquid state, then extruded and deposited onto a base in place in ultrathin layers.

□ The temperature-controlled head extrudes thermoplastic material layer by layer as it solidifies, laminating to the preceding layer.

□ The designed object is produced as a solid three-dimensional part without the need for tooling.

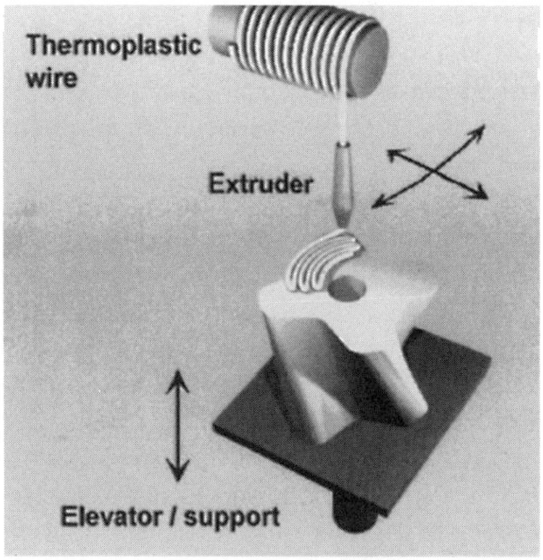

fused deposition modeling.
(Courtesy www.designinsite.dk.)

glass-filled nylon A reinforced fine-grade nylon powder recommended for use in SLS applications where improved stiffness, heat resistance, finish ability, and ease of processing are desired. It also offers small-feature definition and strength.

graphite A form of carbon that has high heat resistance and efficiently transfers electric current; used in the production of electrodes for the EDM process.

hardfacing The deposition of filler material on a metal surface to improve wear resistance or increase its dimensions.

- ☐ Hardfacing is used to increase part life or to replace metal that has worn or corroded away.
- ☐ Filler materials include metal alloys, ceramics, or a combination of these.
- ☐ Hardfacing is generally applied by oxyacetylene, shielded-metal arc, submerged arc, gas tungsten, plasma arc welding, and lasers.

hexapod A revolutionary CNC machine tool that can move and position the spindle in almost any direction, giving the machine six-axis contouring capabilities for milling applications. *(See figure.)*

- ☐ It has a parallel kinematic link mechanism, commonly known as the *Stewart platform,* which is used in flight simulators and fairground rides.
- ☐ It consists of a lower platform to hold the part, and an upper platform containing what appears to be a free-floating spindle.
- ☐ Six legs containing ball screws connect the bed to the head; they can be extended or retracted to change the spindle position.
- ☐ The hexapod is more accurate, faster, more rigid, and has higher acceleration rates than do standard CNC machining centers.

laminated-object manufacturing (LOM) This process creates models from inexpensive, solid-sheet materials. LOM is similar to stereolithography (SL) in that it slices a three-dimensional electronic file from the computer to the LOM machine to produce parts for visualization models, casting patterns, and designs. *(See figure.)*

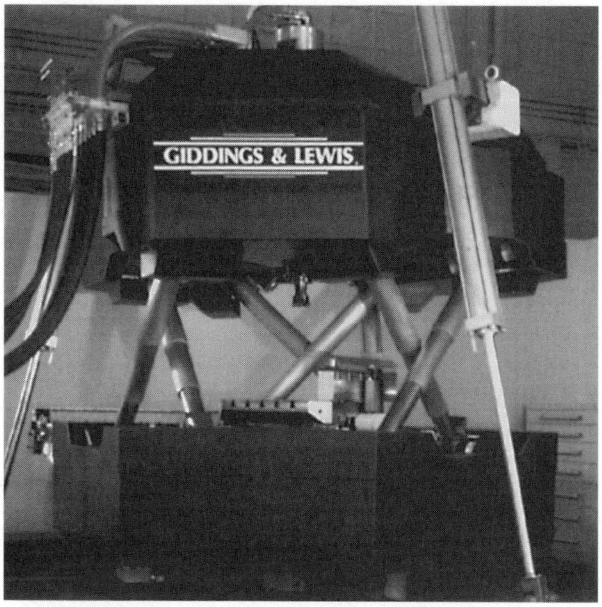

hexapod. *(Courtesy Giddings and Lewis Inc.)*

- ☐ Sheets of plastic-coated paper are thermally laminated and laser-cut to create a geometrically stable model.
- ☐ The three-dimensional file is electronically fed to the LOM machine one layer at a time.
- ☐ The laser cuts the cross-sectional outline in the top layer of the paper, and crosshatches the excess material for later removal.
- ☐ Each layer is bonded to the previous layer with a thermal roller and cut until the design resembles a blocklike shape.

laser Standard short term for *light amplification by*

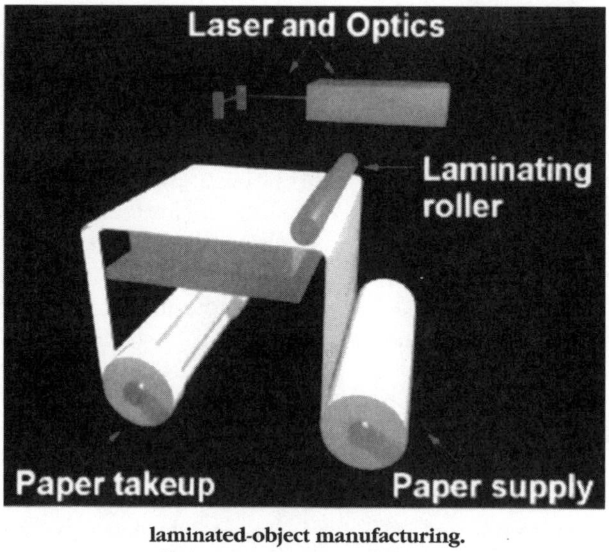

laminated-object manufacturing.
(Courtesy 3-Dimensional Services.)

stimulated emission of radiation; lasers are used in applications such as material processing communications, precision measurement, and identification and tracking. Under computer control, lasers can drill, weld, engrave, mark, and case-harden; lasers are also used for in-process quality-control monitoring systems.

□ For machining, intense, pulsed beams of light generated by carbon dioxide (CO_2) or neodymium-doped yttrium aluminum garnet (Nd:YAG) lasers are used.

□ Assist gasses such as compressed air, oxygen, and nitrogen are most common. Argon is used to assist in the laser welding process.

lasercaving A CO_2 laser with a .008-in light beam that can control the laser's cut depth within .004 in when machining cavities common to the tool-and-die and moldmaking processes. *(See figure.)*

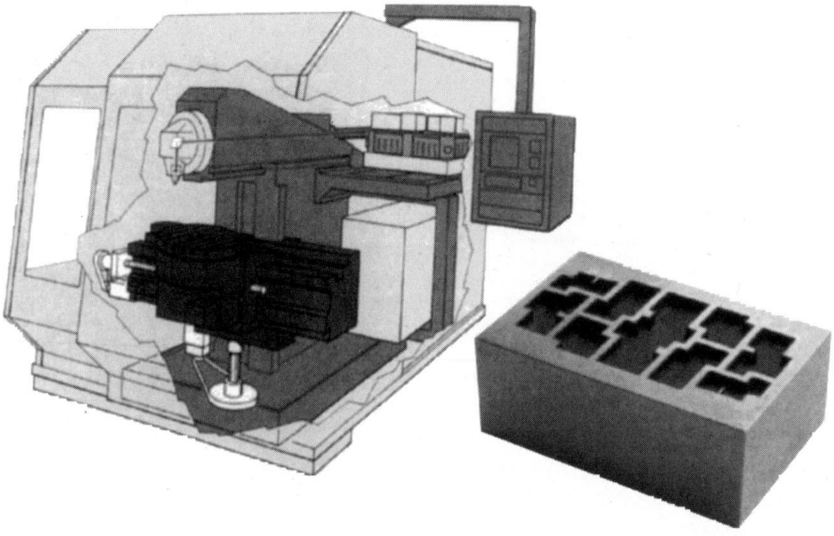

lasercaving. *(Courtesy Modern Machine Shop.)*

□ A newly developed process for machining hard or nonconductive materials that are difficult to EDM such as ceramics, composites, carbide, quartz glass, and titanium-based alloys without force or perishable tooling.

laser digitizing (scanning) The use of lasers to scan a surface of a part to produce a three-dimensional image; also referred to as *reverse engineering*.

□ An accurate method of measuring parts after they are produced, for comparison to a master CAD file.

laser etching The use of a laser to etch designs, lettering, numbers, or symbols onto the surface of a material.

laser machining The cutting of a material with a CO_2 or YAG laser to produce intricate part shapes and holes; it is usually a through-cut process.

laser marking The use of a YAG laser to produce marks and lettering or numbering on the surface of a part by burning the plating, such as an anodized surface.

mechanical energy Used in nontraditional machining where erosion is the principal mechanism for material removal.

□ Ultrasonic and abrasive jet machining are examples of mechanical energy being used to produce parts.

no-wear When the EDM settings are adjusted so that the electrode wear is reduced to 2 percent or less by volume, the machine is burning in a no-wear process.

optical fabrication This system is similar to stereolithography except that it uses a visible-light argon-ion laser, and the part is built on a stationary platform.

orbital grinding A process in which an abrasive master produces its full three-dimensional shape into a workpiece, usually graphite electrodes for EDM.

photochemical machining The process of producing metallic and nonmetallic parts by placing a chemical-resistant image of a part on a metal sheet and exposing the sheet to chemical action which dissolves all the metal except the desired part. *(See figure.)*

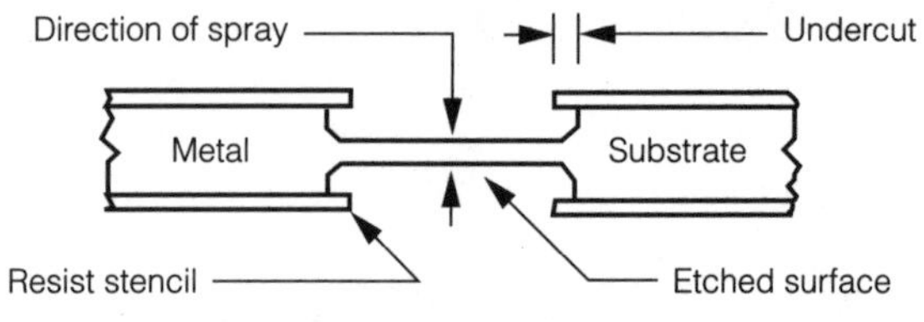

photochemical machining. *(Courtesy Vaga Industries.)*

□ This process is generally used to produce complex shapes on thin materials that must be burr-free, and hardened or brittle materials.

plasma arc machining (PAM) This process is used

for cutoff or rough shaping of metal plates or bars with an arc, then blowing the molten metal out of the kerf with a high-velocity jet of ionized gas.

- ☐ *Plasma* is defined as a gas that has been heated to a high temperature to become ionized and penetrate through the metal thickness.

postcuring apparatus (PCA) A device used to cure a stereolithography part with ultraviolet (UV) light.

- ☐ STL parts must be cleaned in an ether-based solvent, washed, and then placed in the PCA.
- ☐ The time needed for curing parts may range from a few minutes to several hours, according to size, because the parts are soft and easily distorted before they are cured.

powder metallurgy The compressing (compacting) of powdered metals into a die representing the part shape and sintering (baking) in a furnace to produce a hardened metal part.

- ☐ Powder metallurgy consists of four operations: powder production, blending, compacting, and sintering and finishing.
- ☐ The process is rapidly becoming competitive with casting, forging, and machining processes for complex parts made of high-strength and hard alloys.

QQC process A diamond-coating process developed by Pravin Mistry of Turchan Industries that uses four lasers to produce the energy and carbon dioxide from the air as the source of carbon for the diamond reaction. *(See figure.)*

QQC diamond process.
(Courtesy QQC Inc. Division of Turchan Industries.)

- ☐ Laser energy is directed at the substrate (base material to be coated) to mobilize, vaporize, and react with the primary element, such as carbon, to change the crystalline structure of the substrate.
- ☐ Beneath the substrate surface, a conversion zone is created which changes to a composition of the diamond coating formed on its surface, producing a superior metallurgical bond.
- ☐ Deposition rates exceed 1 μm/s compared to 1 to 5 μm/h with chemical vapor deposition (CVD).
- ☐ The diamond coating can be applied to stainless steels, high-speed steel, iron, plastic, glass, copper, aluminum, titanium, and silicon.

rapid part prototyping Rapid part prototyping (making a full scale model) is the process of taking a part from a design concept, such as a three-dimensional computer model, through to a finished part that is automatically made directly from CAD data in a matter of days, rather than weeks.

- ☐ Models can be made quickly and cost-effectively so that the part design can be evaluated before investing in expensive tooling.
- ☐ Parts made with rapid prototyping technology can be used for three-dimensional database verification before making tooling commitments, and for use as models for presentations or tool-and-die development.

rapid tooling A prototyping process that produces parts and tooling in relatively short lead times with a CNC program produced by computer systems such as CAD/CAM.

- ☐ The CNC machine will produce one part or one tool for production of the part.
- ☐ No drawings are ever required; all the information is transmitted through computer files.

reverse engineering The use of lasers, CMM, or other measuring devices to create a CAD file from an existing part.

- ☐ The reverse-engineered image can be altered to produce a new or modified product.
- ☐ A copy or duplicate of the original part may be manufactured from the data gathered.

roller finishing A cold-forming process that uses rollers or balls under high force to plastically deform a work surface to increase the material's strength and improve its accuracy.

rotary ultrasonic-assisted machining (RUM) A process in which the tool rotates at high speeds (around 5000 r/min) and vibrates axially at high frequency to process hard, difficult materials such as ceramics, ferrites, glass, quartz, and zirconium. *(See figure on next page.)*

- ☐ The tool (diamond)–workpiece contact is often assisted with coolant.
- ☐ Small-diameter holes are possible without tool backout.

selective laser sintering (SLS) A flexible technology that uses a CO_2 laser beam to fuse (sinter) layers of nylon, metal, or trueform powdered materials into a three-dimensional model.

- ☐ It is a leading rapid prototyping technology, providing more choices of materials for flexibility, and more applications than other technologies.

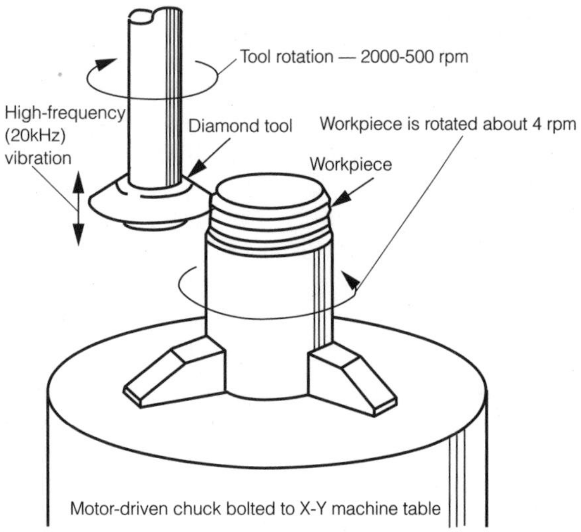

rotary ultrasonic machining.
(Courtesy Society of Manufacturing Engineers.)

□ The SLS process quickly creates parts strong enough to be used as functional prototypes and investment casting patterns.
□ A three-dimensional program is sent to the SLS machine, where a thin layer of heat-fusible powder is deposited into the part-building cylinder.
□ A cross section of the object is drawn on the powder with a CO_2 laser, and the powder is fused into a strong plastic.
□ Another layer of powder is deposited on top of the previous layer, and the process is repeated until the part is completed.

shaped-tube electrolytic machining (STEM) A special application of ECM incorporating the same principles but used where it is desirable to accurately produce deep small-diameter holes. *(See figure.)*

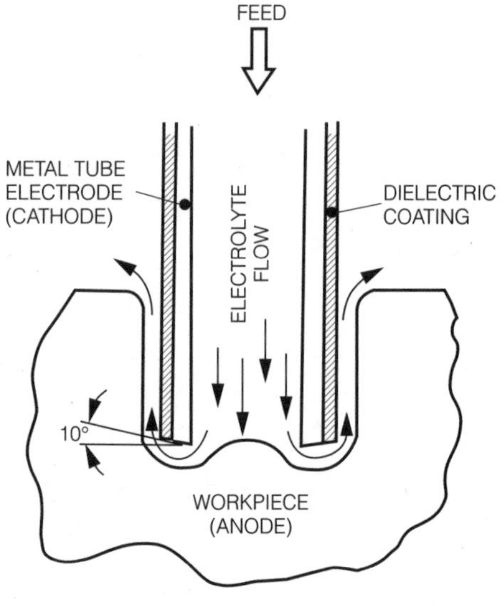

shaped-tube electrolytic machining.
(Courtesy Society of Manufacturing Engineers.)

□ An electrolytic fluid flows through the metal tube electrode (cathode) and washes out the particles produced by the electrolysis between the tool and workpiece.
□ This process is capable of producing shaped or round holes up to a 200:1 depth-to-diameter ratio.

shot peening The operation of cold-working a metal surface with a stream of high-velocity spherical shot particles aimed at the work surface.
□ This action produces a uniform compressive stress pattern and eliminates microscopic defects in the thin surface of a part.

solid-ground curing (SGC) A physical-imaging technology used to produce accurate, durable prototypes. Models are built in a solid environment, eliminating curling, warping, support structures, and any need for final curing.
□ A computer produces a CAD file as a stack of slices that are printed on a glass photomask using an electrostatic process with the part of the slice representing solid material remaining transparent.
□ A thin layer of photoreactive polymer is spread evenly on the work surface; then an ultraviolet floodlight is projected through the photomask to harden and solidify the exposed resin.
□ Liquid wax is spread across the work area, filling the cavities previously occupied by the unexposed liquid polymer resin.
□ A chilling plate hardens the wax. The entire layer, wax and polymer, is now solid.

stereolithography (SL) A process that uses a combination of laser, photochemistry, optical scanning, and computer technology to make a three-dimensional prototype (model) from a CAD file one layer at a time. *(See figure.)*
□ The 3D file is electronically broken (sliced) into layers, and each time the ultraviolet laser passes over

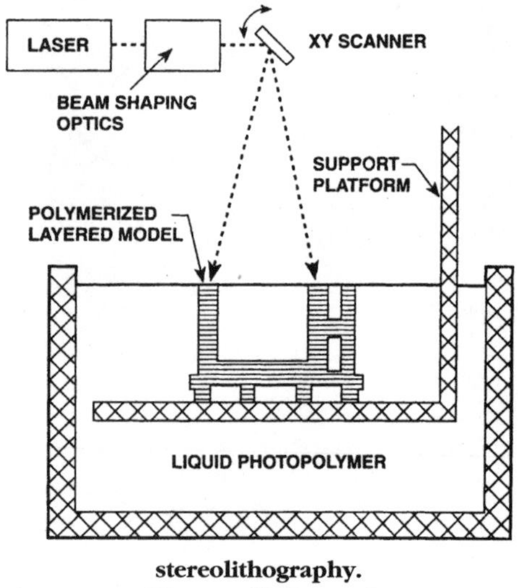

stereolithography.
(Courtesy 3D Systems.)

the bath of uncured polymer (plastic), the liquid resin hardens forming a layer.

- □ The workpiece is lowered a small amount, and another layer of resin is applied.
- □ The SLA creates layer upon layer until the design becomes a 3D prototype within hours rather than weeks.
- □ Completed prototypes are useful for engineering design verification of fit and function, sales and marketing evaluations, as well as master patterns for urethane molding, sand, and investment casting.

stereolithography apparatus (SLA) SLA is the actual machine, computer, and tool that produces models which are formed layer by layer from a liquid resin. An ultraviolet laser passes over the resin, curing it into solid geometry that is defined directly from the CAD file.

thermal energy In nontraditional machining, this is the removal of metal through vaporization and part building through fusion.

- □ Electrical discharge machining is a vaporization process.
- □ Selective laser sintering is a fusion process.

thermal energy finishing (TEM) A very fast (20-ms) method of burning away burrs and flash from internal and external surfaces using thermal energy.

- □ It is ideal for metal parts through which fluids or gases must flow.
- □ Advantages of this process are consistent quality, reduced need for inspection, less rejection and rework, and parts are free of contaminants.

thermal spraying The process of depositing molten or semimolten materials such as metals, alloys, and ceramic coatings onto a part so that they solidify and bond to the surface. *(See figure.)*

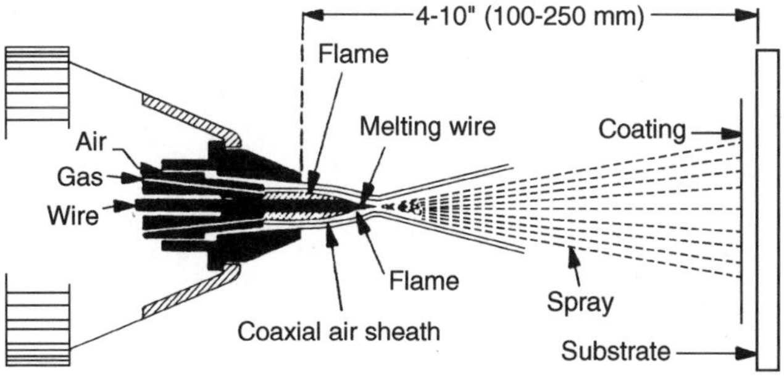

thermal spraying.
(Courtesy Society of Manufacturing Engineers.)

- □ The material, wire, rod, cord, or powder, is passed through a spray unit, where it is heated to a molten state, atomized, and projected onto the work surface.
- □ Electric arc, gas flame, or detonation of a combustible gas mixture are methods of heating the materials.

three-dimensional printers Three-dimensional printers produce three-dimensional prototype parts from CAD-generated STL files.

three-dimensional printing Three-dimensional printing is a manufacturing process for rapid and flexible production of prototype parts and tooling, layer by layer, directly from a CAD model.

- □ It can create parts of any geometry, including undercuts, overhangs, and internal volumes.
- □ A thin distribution of powder is spread over the surface of a powder bed and the computer calculates the information for the layers.
- □ Using a technology similar to inkjet printing, a binder material joins particles where the object is to be formed.
- □ A piston then lowers so that the next powder layer can be spread and selectively joined; this process is repeated until the part is completed.
- □ Three-dimensional printing greatly reduces the time to develop new products, reduces the tooling cost, and improves product quality.

trueform PM polymer An acrylic-based polymer used for SLS parts when excellent features and edge definitions with extremely smooth surface finishes are required.

- □ Common applications for the trueform PM polymer include the creation of concept models, pattern masters for secondary tooling processes [especially room-temperature vulcanizing (RTV) tooling], and pattern masters for investment castings.

tungsten A high-melting metal (3380°C) used in making electrodes for the EDM process.

ultrasonic-assisted machining (UAM) A mechanical process in which an outside source of vibrating energy is added to conventional drills or toolholders and insert assemblies used for drilling and machining operations.

- □ This assisted machining reduces machining forces, and has increased cutting quality and speed in the processing of hard, tough materials.
- □ It has almost eliminated subsurface tearing and plastic flow of the work material.

ultrasonic deburring A process that uses ultrasound to produce high-energy shock waves to force a slurry of weak etching solution and small quantities of abrasive particles against the part surface.

- □ Ultrasonic deburring is used to deburr precision gear teeth, ends of needles, and small precision stampings.

ultrasonic machining (USM) A mechanical process performed by a cutting tool that oscillates at high frequency (20,000 Hz) in an abrasive slurry. It is used to machine hard, brittle materials such as carbide, stainless steel, ceramics, and glass. *(See figure on next page.)*

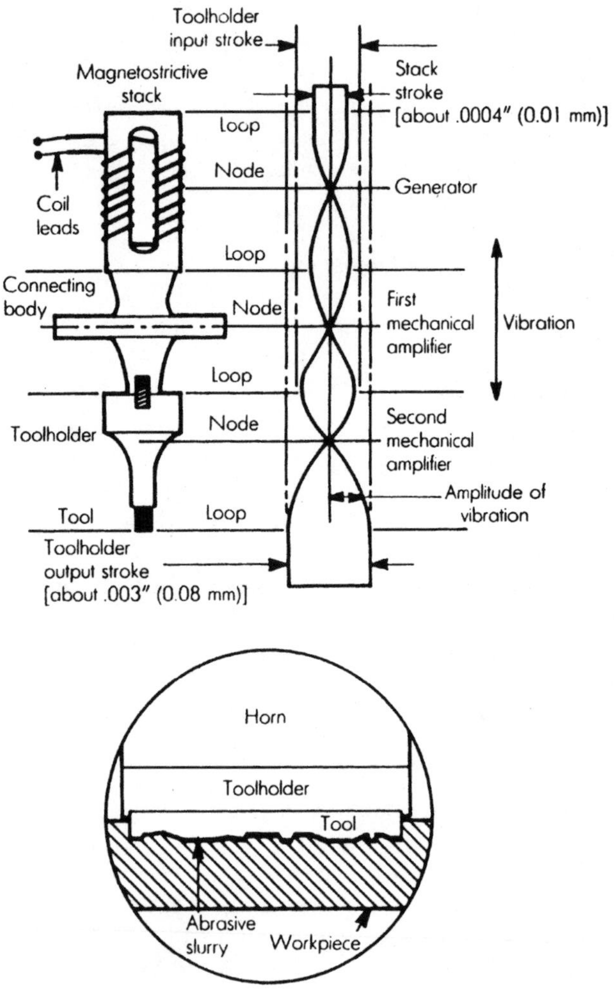

ultrasonic machining.
(Courtesy Society of Manufacturing Engineers.)

□ The high-speed cycle of the tool drives the abrasive grains across a small gap (a few thousandths of an inch) against the workpiece.

□ The impact of the abrasive against the part removes material, duplicating the shape of the tool in the part.

urethane molding Urethane cast parts for show samples, manufacturing reviews, qualification testing, and design verification can be generated from a silicone mold created from the master pattern.

□ A pattern is produced by stereolithography or from machinable wax, and RTV rubber is poured around the pattern to produce a mold.

□ Urethane and epoxy are mixed and poured into the rubber mold to produce the final parts.

waterjet cutting (WJC) A fine, high-pressure, high-velocity jet of water directed by a small nozzle is used to cut hard or soft materials. Abrasives such as micro-grain diamond may be added to the process to assist in cutting materials such as glass or jade.

□ The small nozzle opening [between .004 and .016 in (0.10 to 0.41 mm)] produces a vary narrow kerf.

□ An intensifier pump pressurizes water to 55,000 psi and forces it through the nozzle, producing a high-velocity, coherent stream of water traveling at speeds of up to 3000 fps (ft/s).

□ A stream of water will easily cut soft materials such as rubber, foam, or plastic.

□ For hard materials such as stainless steel, titanium, or some aluminum types, an abrasive (80- to 120-grit garnet) is mixed into the water stream.

SECTION 13

Plastics

Steve F. Krar
Consultant
Kelmar Associates
Welland, Ontario

Plastic, or synthetic polymer, is an artificial material of high-volume molecular weight composed of long chain-like repeating organic chemical units. Plastics originated in the mid-1800s with the discovery of hard rubber by Charles Goodyear. Today there are over three dozen families of plastics with many grades in each family to suit a large variety of product requirements.

Plastics generally fall into two main classes:

1. Thermoset plastics, such as epoxies, melamine, phenolics, polyester, and ureas, change chemically during the curing process and cannot be resoftened by additional heating.
2. Thermoplastics, such as acrylics, cellulose, nylon, polyethelene, Teflon, and vinyl, can be softened and reshaped repeatedly by the application of heat.

Over the past 100 years, plastics have become one of the common and useful engineering materials because of their many properties and versatility.

abherent A release agent such as a coating or film that is applied to a surface to prevent or reduce its adhesion to another surface with which it may come into contact.
☐ Abherents applied to plastic films are usually antiblocking agents; the ones applied to molds and calender rolls are usually called *release agents.*

abhesive **1.** A material that resists adhesion. **2.** A film or coating applied to surfaces to prevent sticking and heat sealing by the presence of a parting agent, mold release, or antistick material.

ablative plastics Plastic composites that are used as heat shields on aerospace vehicles to protect against the penetration of intense heat.
☐ The intense heat encountered erodes and chars the top layers, which in turn insulates the inner areas against further penetration of the intense heat.

accumulator A cylindrical reservoir, mounted on machines (injection, blow molding, etc.), that is fed melted resin from the extruder to provide extremely fast molding cycles or output rates. *(See figure.)*
☐ The accumulator cylinder accepts a predetermined amount of resin and then delivers this stored material to the die opening, pushed forward by a hydraulic ram to deliver the parison as required.

acetal A polymer with the molecular structure of a linear acetal (polyformaldehyde), which consists of unbranched polyoxymethylene chains.

acrylamide plastic A specialty plastic that is used mainly in water treatment, mining, and paper manufacture.

acrylate plastic A type of plastic for sparkling crystal clarity and surface hardness, together with superior weatherability (with optical retention) and good chemical resistance.
☐ Acrylic is supplied as monomer for coating plastics and casting sheets, rods, and tubes.
☐ It is also used for beads or pellets for extrusion (sheets, profiles, etc.), injection molding (automotive tail lights, lighting lenses, etc.), and other processes.

acrylics A family of commercial plastics produced from acrylic acid or its derivative.

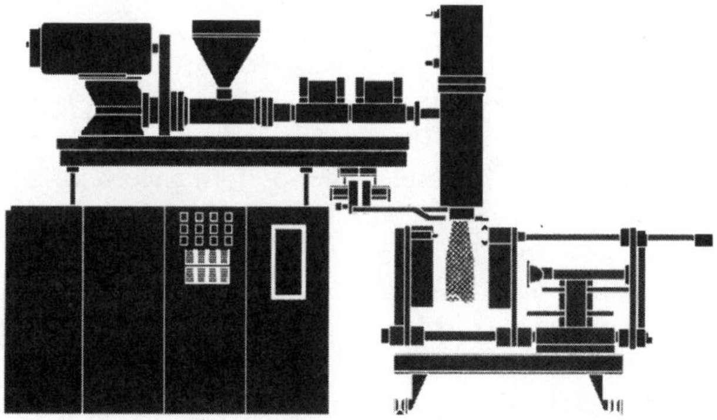

accumulator head. *(Courtesy Hartig Plastic Machinery.)*

acrylonitrile A monomer that is very useful in copolymers. Its copolymer with butadiene is nitrile rubber, and several copolymers with styrene exist that are tougher than polystyrene.

acrylonitrile-butadiene-styrene (ABS) Acrylonitrile and styrene liquids and butadiene gas are polymerized together in various ratios to produce the family of ABS resins. *(See figure.)*

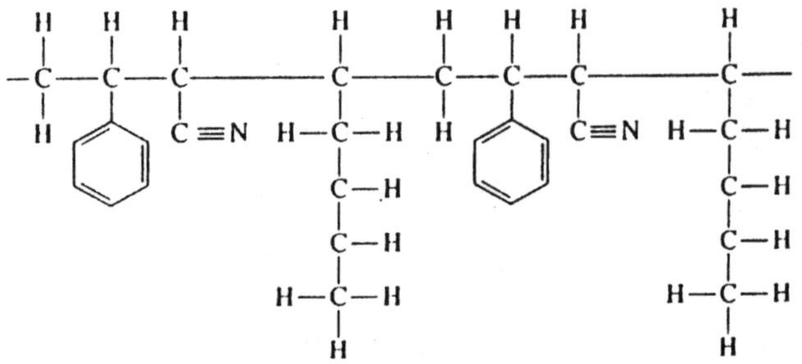

acrylonitrile-butadiene-styrene.
(Courtesy Equistar Chemicals, LP.)

acrylonitrile-styrene plastic A high nitrile plastic combining styrene-acrylonitrile or other monomer acrylonitrile copolymer with a high acrylonitrile content of about 70 to 80 percent.

□ These copolymers have a very low permeability to gases, carbon dioxide, and liquids; thus they are called *barrier plastics.*

□ They are used in packaging, especially for bottles containing carbonated drinks.

acute toxicity The adverse effect on a human or animal body such as dizziness, inflammation, nausea, skin rashes, tearing of the eyes, unconsciousness, and even death caused by exposure to some chemical.

additives Additives, such as cross-linking agents and catalysts, are generally added to a plastic matrix without significantly affecting the molecular structure of the plastic (polymer).

□ Additives can act as processing aids; mechanical, surface, physical, electrical, and optical property modifiers; antiaging additives; and to reduce formulation and manufacturing costs.

adhesion The state in which two surfaces are held together by interfacial forces that may consist of interlocking action.

adhesive A substance capable of holding materials together by surface attachment. It is one of the most important and least thought-of applications of plastic.

advanced plastics The general term used in the plastics industry to indicate higher-performance reinforced plastics.

□ A more specific definition is advanced reinforced plastics or advanced plastic composites pertaining to mechanical properties such as modulus of elasticity.

aging **1.** The effect of exposure of plastics to an environment for a period of time. **2.** Exposing plastics to an environment for a period of time, resulting in improvement or deterioration of properties on the basis of the type of plastic with or without additives.

air-assist forming A thermoforming method that uses airflow or air pressure to partially preform the sheet just before the final pulldown into the mold by vacuum. *(See figure.)*

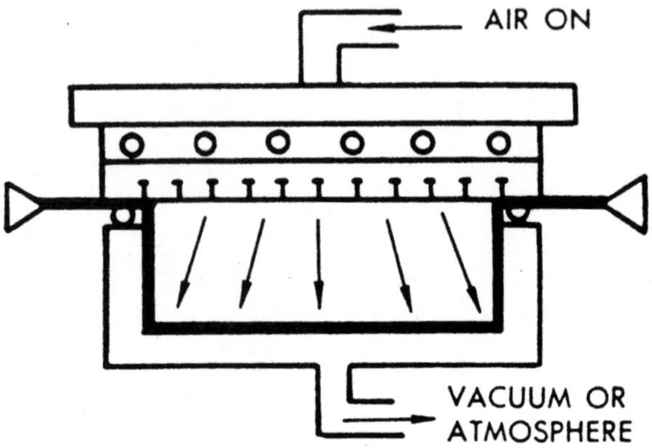

air-assist forming. *(Courtesy Atlas Vac Co.)*

air atomization A high-velocity airstream coming out of an airgun to break up liquid plastic into very fine particles, producing a fine spray in which all droplets are less than 5 μm in diameter.

□ Fine-particle size allows good control of the coating thickness.

air pollution Releasing into the atmosphere undesirable particles, some of which may be toxic.

air-slip forming A thermoforming process in which air pressure forms a bubble and then a vacuum forms the hot plastics against the mold.

alignment pins Devices used in molds and dies to maintain proper cavity alignment as the mold or die closes.

alkyd A polyester resin made with some fatty acid as a modifier.

alkyd plastic A formaldehyde-based plastic (phenolic, etc.) that has dimensional stability, possesses superior electrical qualities, and does not condense water during cure. *(See figure.)*

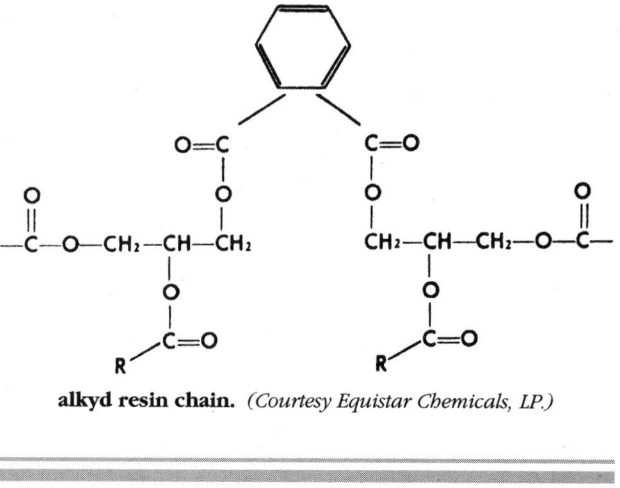

alkyd resin chain. *(Courtesy Equistar Chemicals, LP.)*

□ Alkyds, formulated with a variety of fillers and reinforcements, are cured with pressures from contact to 10 MPa (1450 psi); however, some types can be cured without pressure.

□ Plastics, which permit molding of relatively complicated shapes, have high heat resistance, rapid cure cycle, and good mold flow.

allowances The intentional difference in dimensions that allows for a variety of quality fits between mating parts.

allyl plastic One of the more versatile thermoset plastics that can be homopolymerized or copolymerized with alkyds.

□ By the proper selection of peroxide, or at lower temperatures when cross-linked with alkyd, they can be cured at any temperature above 93°C (200°F) as homopolymers.

□ Using UV-sensitive accelerators, they may be sunlight-cured as alkyd copolymers.

□ Molding processes include contact, vacuum and pressure bag, compression, and transfer molding.

alpha particle A subatomic particle composed of two neutrons and two protons, making it the same as the nucleus of a helium atom.

alphatic molecules These are organic compounds whose molecules do not have their carbon atoms arranged in a ring structure.

American National Standards Institute (ANSI) A privately funded, voluntary membership organization that coordinates and develops industrial and national standards.

American Society for Testing and Materials (ASTM) A voluntary organization consisting of individuals, agencies, and industries concerned with materials.

□ ASTM is a resource for sampling and testing methods, health and safety aspects of materials, safe performance guidelines, and effects of physical and biological agents and chemicals.

amides Compounds produced from organic acids.

amines Organic derivatives of ammonia obtained by substituting hydrocarbon radicals for one or more hydrogen atoms.

amino plastic Plastic that condenses with amino compounds to form methylol derivatives, which, when heated, condense further to form hard, colorless, transparent plastics.

□ Amino monomer is characteristically present as amide in urea formaldehyde and melamine formaldehyde plastics.

amorphous plastic Thermoplastic in which the molecular chains exist in random coil conformation similar to water-boiled spaghetti. *(See figure.)*

□ The structure of amorphous plastic is characterized by the absence of a regular three-dimensional arrangement of molecules or subunits of molecules extending over distances that are large compared to atomic dimensions.

amortization The gradual repayment of the cost of equipment such as presses and molds that may be done by contribution to a sinking fund at the time of each interest payment.

anisotropic Showing different properties when tested along different axis directions.

annealing Removing internal stresses and strains set up during the fabrication cycle without shape distortion.

□ The material is held at a temperature near, but below, its melting point, for a period of time.

ANSI *See* AMERICAN NATIONAL STANDARDS INSTITUTE.

antioxidant A stabilizer that slows the breakdown of the plastic by oxidation.

antistatic An additive that reduces static charges on a plastics surface.

apparent density The mass per unit volume of a material including the voids present in the material.

aqueous polymerization Vinyl polymerization using water as the medium with the monomer present within its inherent solubility limit.

□ This polymerization process is very important in preparing emulsifier-free latexes where the size distribution among the dispersed particles is fairly sharp, and in preparing special plastics.

aromatic hydrocarbons Hydrocarbons derived from or characterized by their unsaturated ring structures.

ashing Using wet abrasives on wheels to sand and polish plastics.

A-stage An early stage in the reaction of thermosetting resin where the material is fusible and still soluble in certain liquids.

ASTM *See* AMERICAN SOCIETY FOR TESTING AND MATERIALS.

atactic plastic Plastics with molecules where substitute groups or atoms are arranged in a random pattern above and below the backbone chain of atoms, when the latter are arranged in the same plane.

atactic stereoisomerism A random arrangement of molecular chains in a polymer.

atom The smallest particle of an element that can combine with particles of other elements to produce the molecules of compounds.

□ Atoms consist of a complex arrangement of electrons revolving around a positively charged nucleus containing particles called *neutrons* and *protons*.

amorphous chain.
(Courtesy The Society of the Plastics Industries, Inc.)

atomic number (atomic mass number) A number equal to the number of protons in the nucleus of an atom of the element.

autoclave A closed strong-steel pressure vessel that can maintain the temperature and pressure of a desired air or gas for the curing of organic-matrix composite materials. *(See figure.)*

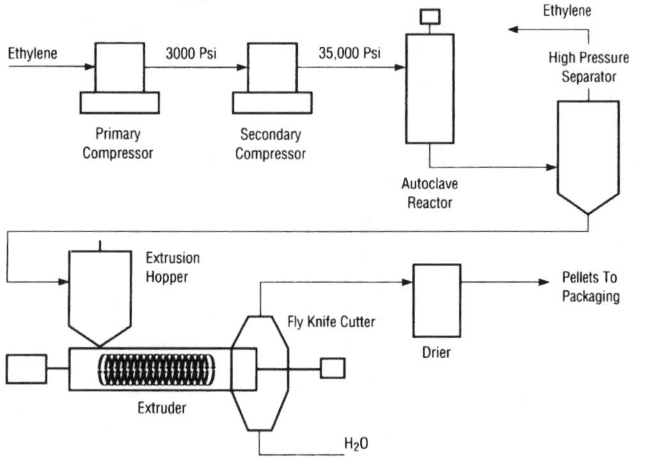

autoclave process. *(Courtesy Equistar Chemicals, LP.)*

autoclave molding A molding process in which an entire assembly is put into a steam or electrically heated autoclave at elevated pressure.

automatic cycle A machine operation that will continue to repeat its cycle without human intervention.

auxiliary equipment Equipment such as filters, vents, ovens, and take-up reels that are needed to help control or form the product.

average molecular weight Most synthetic polymers are a mixture of individual chains of many different sizes; therefore, a molecular weight for such a mixture is an average molecular weight.

Avogadro's number The number of atoms in a gram-atom mass of an element: $6.024\ 86 \times 10^{23}$.

backing plate A plate used as a support for the mold cavity block, guide pins, bushings, and other structures.

back pressure The resistance of a material to continued flow when a mold is closed. In extrusion, it is the resistance to the forward flow of molten material.

baffle A device used to restrict or divert the passage of fluid or gases through a channel or pipeline.

bag manufacturing A process in which extruded blown film is drawn from the extruder or collection roll and formed into a flat tube as it passes over a forming mandrel or former. *(See figure.)*

- □ At the same time, heat sealing or a suitable adhesive is applied to the edge of the web, and the seam is completed by pressure when the joint passes between a set of rollers.
- □ As the web continues, gussets are added and a cutoff converts the web into bag lengths.
- □ Bottom heat sealing or adhesive is automatically applied to the trailing edge of each bag-length tube.

Bakelite A proprietary name for phenolic and other plastic materials that are made by the condensation of cresol or phenol with formaldehyde.

- □ The name is derived from Dr. Leo Baekeland, who developed phenolic plastic in 1909.

Banbury A machine containing a pair of contrarotating rotors that masticate and blend compounding materials.

barrel The cylindrical housing of the plasticating chamber of extrusion, blow-molding, and injection-molding machines in which the extruder screw rotates or the plunger moves backward and forward.

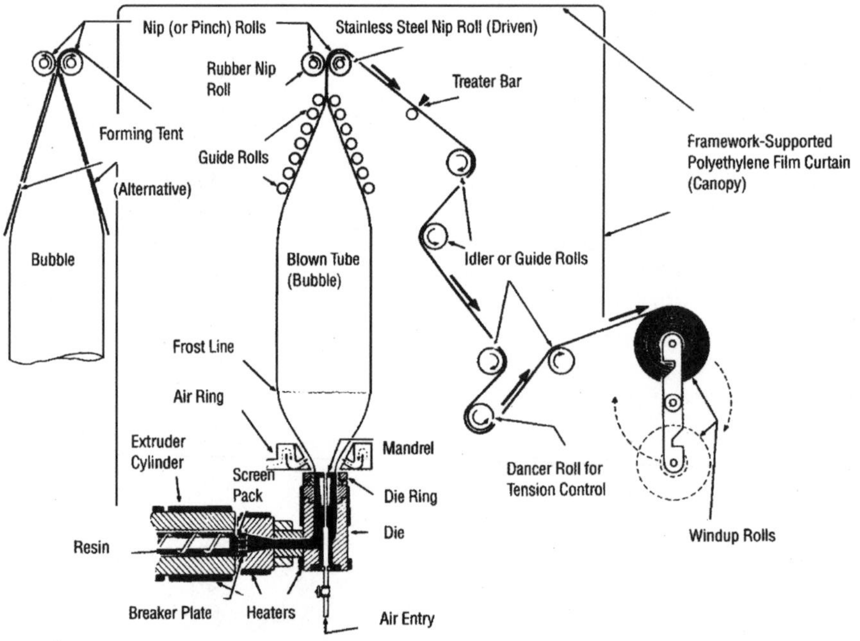

bag manufacturing. *(Courtesy Equistar Chemicals, LP.)*

barrel vent An opening in a barrel wall that allows air and volatile matter to escape from the material being processed.

barrier plastic A term that applies to a group of plastics which have barrier properties against gas, aroma, and flavor.

□ Common barrier plastics are used in various plastics, paper, and foils that protect against vapors and leakage.

batch A general term referring to a quantity of materials, having identical characteristics, formed during the same process or in one continuous process.

biaxial winding A type of filament winding in which the helical band is laid in sequence, side by side, with no crossover fibers.

bifunctional A molecule with two active functional groups.

binder A bonding plastic that holds together fibers and strands in a mat or preform before the manufacture of a molded part.

biocide Natural and synthetic plastics that can attack biological agents.

□ Active chemical compounds such as plasticizers, antistats, and stabilizers are used for preservative control to change the biological resistance of the compounded plastics.

□ Plastics such as polyolefins, polyesters, or vinyls are considered to be resistant to biological attack.

biodegradability The ability to break down or decompose rapidly under natural conditions or processes.

□ Some plastics contain additives or organic compounds that can be digested by microorganisms in the environment.

□ These biodegradable plastics are used for sutures, surgical implants, controlled release of drugs and agricultural chemicals, agricultural mulch, and so on.

bismaleimide plastic A type of polyimide plastic that cures by an addition, rather than a condensation polymerization reaction.

□ This avoids the problem of volatile formations that are produced by a vinyl-type polymerization of a prepolymer terminated with two maleimide groups.

blanket Plies or fibers that have been laid up in a complete assembly and placed on or in the mold all at once as in bag molding.

blanking The cutting of flat-sheet stock to shape in a punch press by striking it sharply with a punch-and-mating die.

bleeder cloth A nonstructural layer of material used in the manufacture of composite parts that allows the escape of excess gas and resin during the curing process.

blend A mixture of two or more types of plastics to produce products that contain the best properties of each component.

block copolymer A copolymer whose polymeric chain consists of shorter homopolymeric chains which are linked together.

□ A linear copolymer consists of a smaller number of repeated sequences of polymeric segments having different chemical structures.

block polymer A polymer molecule that is made up of alternating sections of one chemical composition that is separated by sections of a different chemical character or by a coupling group of low molecular weight.

blow holes Holes or blisters caused by trapped water or air, decomposition gases, contamination, or unplasticized resin, that show up in the wall surfaces of blown containers.

blowing agents Chemical additives that produce inert gases when heated and cause the polymer to assume a cellular structure.

blow molding The process of fabricating hollow products by forcing a hot plastic (melt) into the shape of the mold cavity by internal air pressure. *(See figure.)* There are three main types of blow molding:

1. *Extrusion blow molding* (EBM), which uses an unsupported parison
2. *Injection blow molding* (IBM), which uses a preform supported by a metal core pin
3. *Stretched blow molding* (SBM), which is used for EBM or IBM to obtain bioriented products

boss A projection on a part designed to add strength, provide support for other components, or aid in assembly.

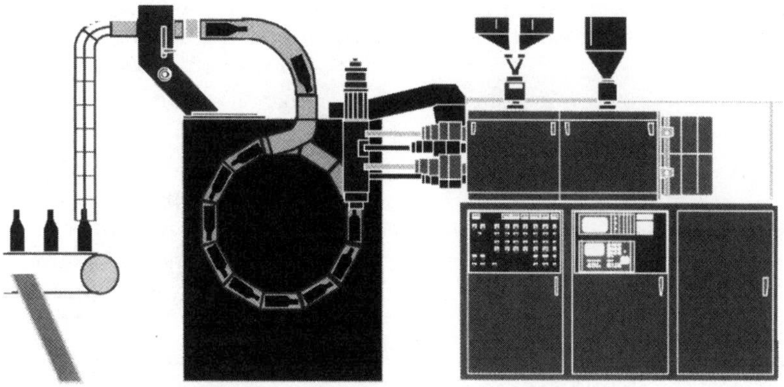

blow molding. *(Courtesy Equistar Chemicals, LP.)*

□ Bosses, sometimes called *studs,* may be hollow or solid to provide rigidity and help reduce or eliminate warpage.

braiding The process of weaving fibers into a tubular shape instead of a flat fabric. It is widely used in reinforced plastic products such as antenna poles, golf-club shafts, fishing rods, and cherry-picker high-rise booms.

branching The growth of long or short side chains that attach to the main chain of the polymer.

breaker plate A perforated metal plate located between the end of the screw and the die head.

breather Porous material, such as a fabric or cilitate, for removal of air, moisture, and volatiles during curing.

breathing The opening and closing of a mold to allow gases to escape early in the molding cycle. Also called *degassing.*

brittleness temperature The temperature at which plastics and elastomers rupture by impact under certain conditions.

B stage An intermediate stage in the reaction of a thermosetting molding compound.

□ In this stage, the material softens when heated and swells in contact with certain liquids, but it does not entirely fuse or dissolve.

bubble pack A laminated thermoplastic film that incorporates air-bubble pockets to cushion and protect products from breakage in the packaging and shipping industries.

buckling An elastic instability-type failure characterized by a sharp discontinuity in the material.

□ Crimping of the fibers in a reinforced plastic, often glass-reinforced thermosets, due to plastic shrinkage during cure.

□ In reinforced plastic (RP), a failure characterized by fiber deflection rather than breaking because of compressive action.

□ A failure characterized by an unstable lateral material deflection due to compressive action on the structural element.

buffing An operation, usually following polishing, that produces a high luster on a surface but does not remove much material.

bulk factor The volume ratio of any given weight of loose plastics to the volume of the same weight of the material after forming or molding.

bulk polymerization The polymerization in bulk of a monomer from undiluted low-molecular-weight starting materials is a method used for the synthesis of macromolecules.

□ The advantages of this method are

- ▶ The equipment or apparatus required is fairly simple.
- ▶ The reaction is fast, and the yield is good.
- ▶ Plastics of high purity are formed from the process.
- ▶ The plastics formed are immediately processable.

butadiene A gas soluble in alcohol and ether that is obtained from the cracking of petroleum, from coal-tar benzene, or from acetylene produced from coke or lime.

□ It is used in the formation of copolymers with styrene, acrylonitrile, vinyl chloride, and other monomeric substances, and imparts flexibility to the products molded.

butylene plastic Plastics produced from materials made by the polymerization of butene or the copolymerization of butene with one or more unsaturated compounds.

calendering The process of forming continuous sheets of plastic film by squeezing the material between two or more parallel rolls to produce the required finish or to ensure uniform thickness. *(See figure.)*

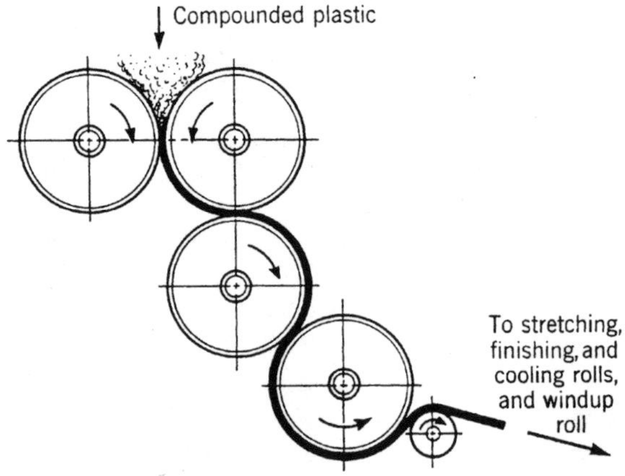

calendering. *(Courtesy The Society of the Plastics Industries, Inc.)*

□ The pastelike plastic melt passes through nips of a series of precision-heated and rotating speed-controlled rolls.

□ Although the basic plastic forming operation occurs in the calender itself, additional equipment must be used for the production of thermoplastic film or sheet.

CAMM *See* COMPUTER-ASSISTED MOLDMAKING.

caprolactam A cyclic amide–type compound that contains six carbon atoms. When the ring is opened, caprolactam is polymerizable into a nylon plastic known as type 6 nylon or polycaprolactam plastic.

carbon Carbon is the element that provides the backbone for all organic plastics; amorphous carbon is used in large quantities in the field of plastics technology.

□ *Amorphous carbon* commonly refers to a wide range of natural and artificial carbon such as coal, active carbon, and carbon black.

carbon fiber A fiber used for reinforced plastics because of its light weight, high strength, high stiffness, and other properties.

□ Carbon fibers are produced by pyrolysis of the organic precursor fiber in an inert atmosphere at temperatures above 980°C (1800°F).

□ The material may also be graphitized by heat treatment above 1650°C (3000°F).

□ Fibers in the form of filament and tape are the most widely used for reinforced plastics.

casein A protein material produced from skimmed milk by the action of either rennin or dilute acid.

casting The process of pouring heated plastic or other liquid resin into a mold to solidify and take the shape of the mold by cooling, loss of solvent, or completing polymerization. *(See figure.)*

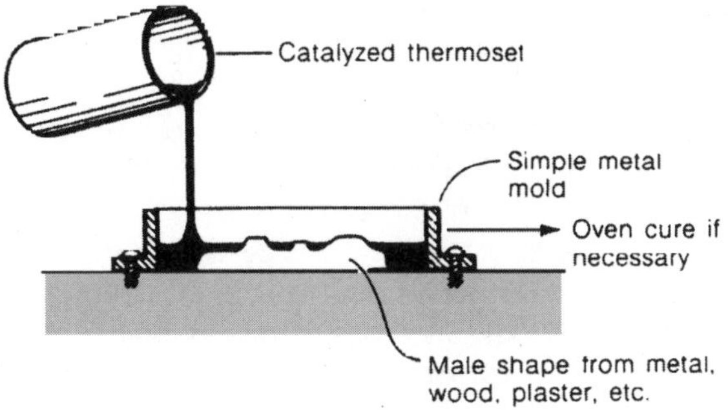

casting. *(Courtesy The Society of the Plastics Industries, Inc.)*

catalyst A chemical material that changes the rate of a chemical reaction without changing itself permanently in composition or becoming a part of the molecular structure of a product.

□ It speeds up the cure or polymerization of a compound when added in relatively small quantity, compared to the amounts of primary reactants.

caul plate A smooth-metal plate that is used in contact with the layup during curing to apply normal pressure and provide a smooth surface to the finished part.

cavity A depression in a mold made by casting, machining, or hobbing.

cellophane The name of a thin transparent film that consists of a base sheet of cellulose, regenerated from viscous materials, containing varying amounts of softener and water and coated to make it moistureproof and capable of being sealed with heat or solvent.

□ Cellophane is manufactured from wood pulp, sodium hydroxide solution, and carbon disulfide.

□ By varying sheet thickness, type, softener, color, plastic treatment, and coating, over 100 different varieties of cellophane have been produced.

□ Cellophane is widely used in the packaging and food industries.

cellular or foamed A sponge form that may be flexible or rigid, with the cells closed or interconnected.

□ The density of various cellular plastics may range from the solid parent resin to 32 kg/m^3 (2 lb/ft^3).

□ *Cellular, expanded,* and *foamed plastics* are the terms used; however, *cellular* is most descriptive of this product.

Celluloid A trade name for some strong, elastic plastics made from nitrocellulose, camphor, and alcohol.

cellulose acetate plastic This plastic is widely used in coat linings, and in the molding of appliance and tool handles, telephones, and containers.

□ The raw cellulose is dried and then reacted with glacial acetic acid, acetic anhydride, and a sulfuric acid catalyst.

□ After hydrolysis to reach a controlled level of chain degradation, the plastic is separated from the chemical solution, washed, and dried.

cellulose nitrate plastic A relatively brittle thermoplastic that is usually blended with plasticizers to give it properties so that products such as ping-pong balls and fountain-pen barrels can be made.

□ When properly formulated with stabilizers, its characteristics are toughness, dimensional stability, and low water absorption.

cellulosics A family of plastics with the polymeric carbohydrate cellulose as the main element.

cement The process of bonding together with a liquid adhesive using a solvent base of the synthetic elastomer or resin.

centipoise One one-hundredth of a poise equals a unit of viscosity. Water at room temperature has a viscosity of about one centipoise.

centrifugal casting Also called *rotational casting*; a method of forming plastic in which dry or liquid plastic is placed in a rotatable container and heated to a molten condition by heat entering through the walls of the container. (*See Figure.*) It is rotated so that the centrifugal force causes the molten plastic to conform to the shape of the container's interior surface.

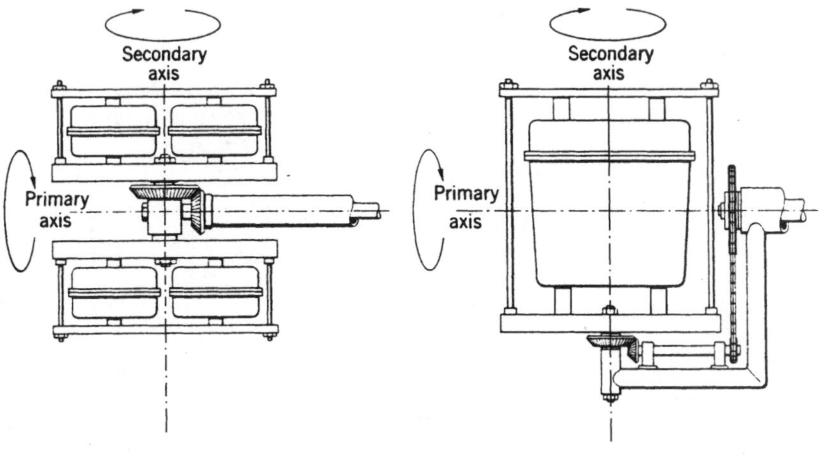

centrifugal casting.
(Courtesy U.S. Industrial Chemicals.)

☐ Used for producing fabricating tanks, large-diameter pipe with or without fiber reinforcement, and other applications.

ceramic fiber A type of reinforcement that is used for various plastics because of its unique wear and corrosion resistance, and its high temperature stability.

☐ It consists of approximately 50 percent alumina and 50 percent silica with traces of other inorganic materials.

☐ The fibers are made by atomizing a molten ceramic stream using high-pressure air or spinning wheels.

chain polymerization The most common type of polymerization process is the chain reaction.

☐ Chain reactions are as follows:

- Plastic of high molecular volume is formed at all stages even during the first fraction of a second of reaction.
- The polymer molecules formed do not react with one another to produce material of higher molecular weight.
- The "active centers" responsible for the reaction are free radicals or ions.

charge The weight or measurement of material necessary to form a usable parison to produce a complete blow-molded item.

chlorinated polyether A corrosion- and chemical-resistant thermoplastic made from pentaerythritol by preparing a chlorinated oxethane and polymerizing it to a polyether by opening the ring structure.

☐ Its heat-insulating characteristics, dimensional stability, and outdoor exposure resistance are excellent.

☐ Its primary use is in the manufacture of products and equipment for chemical and processing industries; also used for pumps and water meters, pump gears, bearing surfaces, and other applications.

chlorinated polyethylene elastomer A rubberlike material, made by moderate random chlorination of polyethylene, that can be cross-linked with organic peroxides.

☐ This rubber-type material has good heat, oil, and ozone resistance.

☐ It is also used as a plasticizer for polyvinyl chloride.

chlorosulfonated polyethylene plastic A thermoset elastomer that can be made into a wide range of colors because carbon black is not required.

☐ It has properties such as

- Total resistance to ozone.
- Excellent resistance to abrasion, weather, heat, flame, oxidizing chemicals, crack growth, and dielectric properties.
- Low moisture absorption.
- Resistance to oil similar to neoprene.
- Low temperature flexibility.

cladding Applying a plastic coating on different materials such as plastics, metals, and aluminum, to improve specific properties of the base material. *(See figure.)*

☐ The properties obtained can be abrasion and corrosion resistance, color, and so on, depending on the plastic used and methods of processing.

closed-cell condition The condition of individual cells that make up cellular or foamed plastics when cells are not interconnected.

coating The placement of a permanent layer of material on a substrate.

cohesion A substance sticking to itself or the internal attraction of molecular particles toward each other.

co-injection molding A special multimaterial injection process in which a mold cavity is first partially filled with one plastic and then a second shot is injected to enclose the first shot.

cold forming A process used to change the shape of a room-temperature thermoplastic sheet or billet in the solid phase through plastic deformation with the use of pressure dies.

cold molding A procedure in which a composition is shaped at room temperature and then cured by baking.

colorants Dyes or pigments that give color to plastics.

comonomer A monomer that is mixed with a different monomer for a polymerization reaction to produce a copolymer.

composite A combination of two or more materials, generally a polymer matrix with reinforcements, with properties that the component materials do not have by themselves.

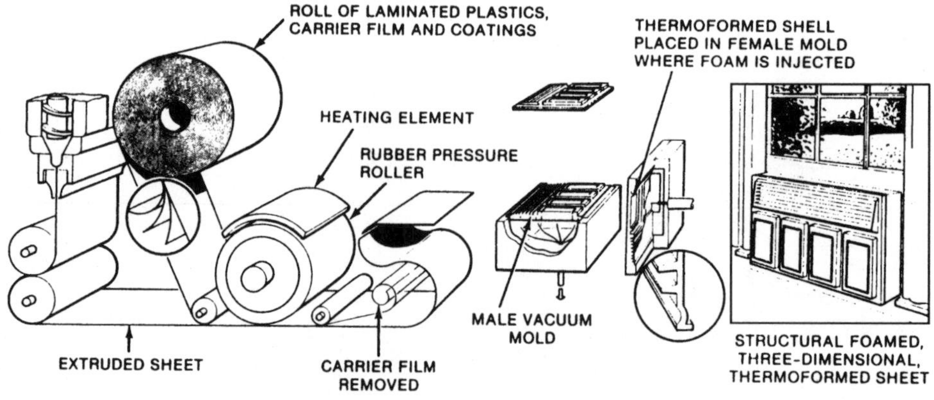

cladding. *(Courtesy Dri-Print Foils.)*

□ The structural components of the composites are sometimes subdivided into fibrous, flake, laminar, particulate, and skeletal.

□ Composites provide almost unlimited potential for corrosion resistance, high strength, stiffness, and other properties; they may be the steels of tomorrow.

compound A composition of two or more elements joined together in definite proportions.

compression The ratio in the extruder screw; the ratio of the channel volume in the first flight at the hopper to that of the last flight at the end of the screw.

compression mold A mold that is open when the mix is loaded and then shapes the material by the heat and pressure of closing.

compression molding A thermoset plastic molding technique in which the preheated molding compound is placed in a heated open-mold cavity. The mold is closed under pressure, causing the material to flow and completely fill the cavity where the pressure is held until the material has cured. *(See figure.)*

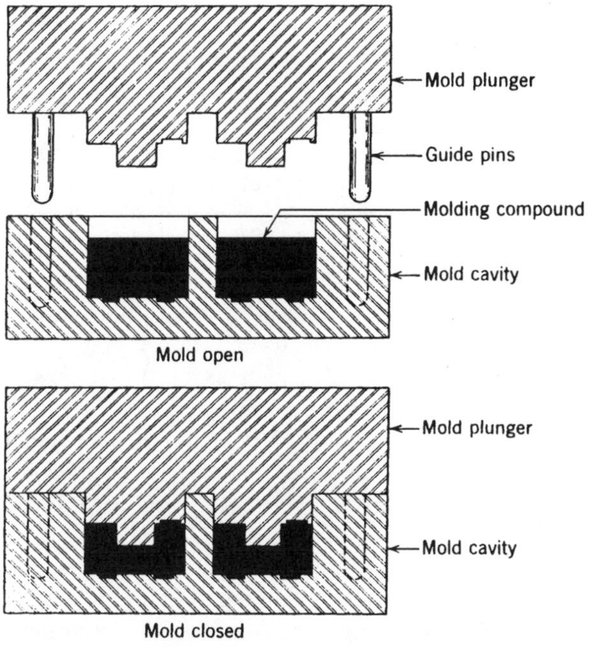

compression molding. *(Courtesy HPM, Division Koehring Co.)*

compressive strength The maximum stress in compression that a material will withstand before rupturing or breaking.

□ The amount of force and the cross-sectional area are used to calculate compressive strength.

computer-assisted moldmaking (CAMM) A computer system that analyzes mold temperatures and flows through finite element or boundary element analysis.

concentrate A specific amount of additive (dye, fiber reinforcement, flame retardant, foaming agent, etc.) that is added to a predetermined amount of plastic.

□ This concentrate can then be mixed into larger amounts of plastic to produce the desired mix and required property for a processed part.

condensation A chemical reaction in which two or more molecules combine with the separation of water or some other simple substance.

□ The condensation process is called *polycondensation* if a polymer is formed.

condensation polymerization Polymerization by chemical reaction that produces a by-product such as water, ammonia, carbon dioxide, hydrogen chloride, methanol, nitrogen, or sodium bromide.

Consumer Products Safety Commission (CPSC) A federal agency responsible for regulating hazardous materials when they appear in consumer goods.

cooling channels The channels located within the mold body through which the cooling medium is circulated to control the mold surface temperature. *(See figure.)*

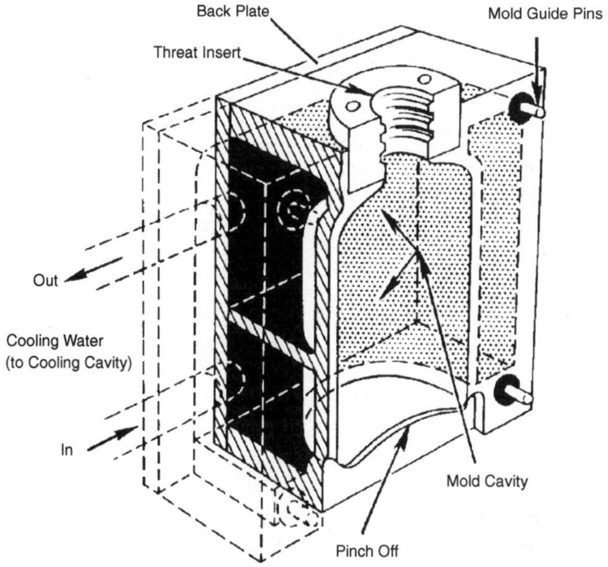

cooling channels. *(Courtesy Equistar Chemicals, LP.)*

copolymer A polymeric system that contains two or more monomeric units and has a functional chemical group added for improved properties.

core The male element in a die or mold that produces a hole, recess, or undercut in a finished product.

□ In blow molding, it is the center portion of the blowing head that controls the inside diameter of the parison.

corona discharge A method of oxidizing a plastic film to make it printable by passing it between electrodes and subjecting it to a high-voltage discharge.

coumarone A compound found in coal tar, and polymerized with indene to form thermoplastic resins that are used in coatings and printing inks.

CPSC *See* CONSUMER PRODUCTS SAFETY COMMISSION.

cracking The thermal or catalytic decomposition of organic compounds to break down high-boiling compounds into lower-boiling fractions.

creep The permanent deformation of a material resulting from prolonged application of a stress below the elastic limit.

□ A plastic subjected to a load for a period of time deforms more than it would have from the same load removed immediately after application.

□ The amount of the deformation depends on the load duration.

cross-linking The formation of primary valance bonds between polymer molecules by tying together adjacent polymer chains, making an infusible super molecule that results in a marked increase in melt viscosity.

crystallization The process or state of molecular structure in some plastics that denotes uniformity and compactness of the molecular chains forming the polymer. Polyolefin regions are shown as follows: (*A*) crystalline; (*B*) amorphous. *(See figure.)*

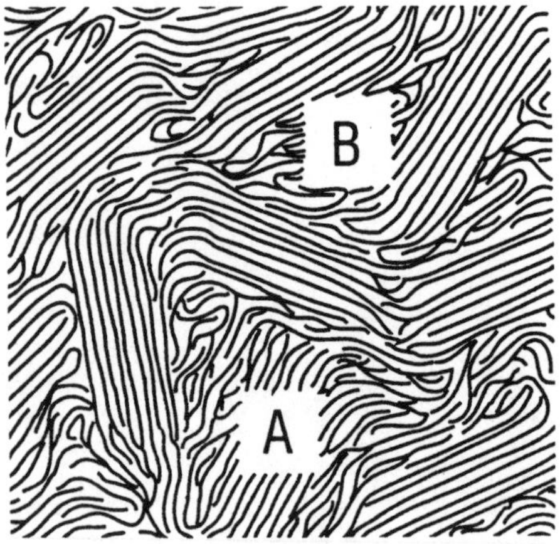

crystallization. *(Courtesy Equistar Chemicals, LP.)*

□ It usually causes the formation of solid crystals with a definite geometric form.

C stage The last stage in the reaction of a thermosetting resin where the material is relatively insoluble, infusible, and fully cured.

cure The process of changing physical properties of a material by a chemical reaction such as condensation, polymerization, or vulcanization.

□ It usually occurs through the action of heat and catalysts, alone or in combination with or without pressure.

□ This reaction is usually associated with thermoset plastics where a permanent change occurs.

curling A condition in which the parison curls upward and sticks to the outer face of the die; sometimes called "doughnuting."

□ The adjustment of the mandrel or die-face levels or adjusting the temperature can usually correct this problem.

cyanate plastic A thermoset plastic produced from bisphenols or polyphenols, which are available as blends, monomers, oligomers, and solutions.

cycle The total or elapsed time between a certain point in the molding cycle of one part to the same point in the next part.

□ The complete, repeating sequence of operations in a process or part of a process.

Dacron Du Pont's trade name for thermoplastic (TP) polyester fiber made from polyethylene terephthalate. It is available as filament, yarn, tow, and fiberfill.

damping The variations in properties that result from dynamic loading conditions (vibrations).

□ Damping dissipates energy without excessive temperature rise, prevents premature brittle fracture, and improves fatigue performance.

daylight opening The clearance between platens when the press is fully opened. This must be large enough to allow a part to be ejected when the mold is fully opened.

deaerate To obtain the maximum performance (strength, aesthetics, permeability resistance, etc.) of a plastic by removing air that would cause the formation of bubbles or blisters in a finished part.

□ This important step in the production of vinyl plastisols is accomplished by subjecting the fluid to a high vacuum with or without agitation.

debond The area of separation with or between plies in a laminate, within a bonded joint, caused by contamination, improper adhesion during processing, or damaging interlaminar stresses.

deformation The change in the dimensions of a product caused by flow and elasticity.

□ Flow is the permanent deformation since the material cannot return to its original shape when the stress is removed.

□ Elasticity is a reversible deformation because the deformed body returns to its original shape.

□ Viscoelastic materials, such as plastics, show both flow and elasticity.

dehydration The removal of water from a material through drying or heating, absorption, chemical reaction, condensation of water vapor, centrifugal force, or hydraulic pressure.

delamination A debonding process resulting primarily from unfavorable interlaminar stresses.

□ Edge delamination can be prevented by a wraparound reinforcement.

density Mass per unit volume of a substance, expressed in grams per cubic centimeter, or kilograms per cubic meter.

diallyl phthalate plastic Diallyl phthalate (DAP) and diallyl isophthalate (DAIP), the principal thermosets in the allyl family, are used for preimpregnated glass cloth and paper.

□ Molding compounds are reinforced with fibers to improve mechanical and physical properties.

□ DAP's major use is in electrical connectors for communications, computer, and aerospace systems.

diamines Compounds containing two amino groups.

dibasic acid An acid that has two replaceable hydrogen atoms.

die A die is a molding tool, usually made of steel, that has a cavity or opening of a shape or design of the part it is to produce, which it imparts to plastics using different methods of processing, such as extrusion, impact stamping, casting, and cutting. *(See figure.)*

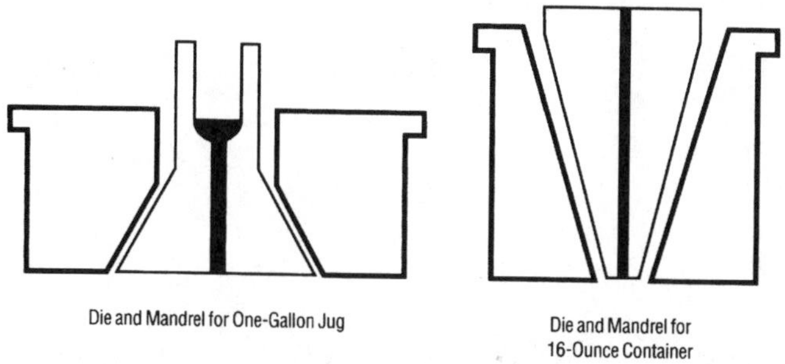

die. *(Courtesy Equistar Chemicals, LP.)*

□ In plastics technology, the terms *die, mold,* and *tool* are sometimes considered the same in that they have a female or negative cavity through, or into, which a molten plastic moves under heat and pressure.

die gap In blow molding, it is the distance between the mandrel and die in the blowing head that determines the extruded parison thickness and the thickness of the finished-part walls.

dimensional stability The ability of a plastic to keep the exact shape in which it was molded, fabricated, or cast.

dip casting The process of submerging a hot mold into a resin and then allowing it to cool and be removed from the mold.

dip coating The process of applying a coating by dipping a part into a tank of melted resin or plastisol, followed by cooling. The part may be heated and powders used for the coating melt as they hit the hot part.

dip forming A process similar to dip coating that is used for making vinyl plastisol products but in which the fused, cured, or dried deposit is stripped from the dipping mandrel. *(See figure.)*

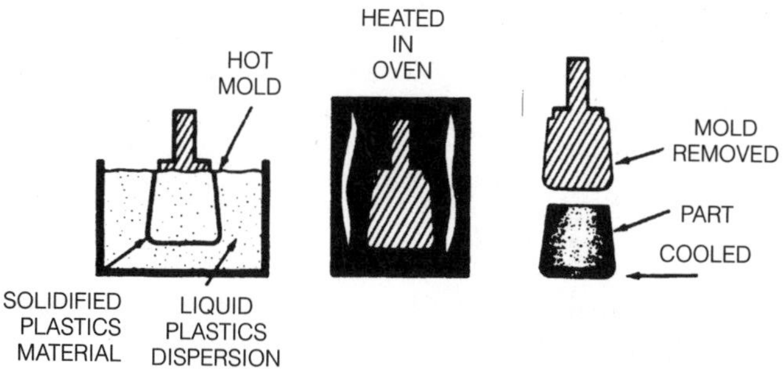

dip forming. *(Courtesy The Society of the Plastics Industries, Inc.)*

□ A preheated form shaped to the desired inside dimensions of the finished product is dipped into a gel which forms a layer on the surface of the form.

□ The coated plastic is withdrawn, heated to fuse the layer, cooled, and the deposit stripped off.

double-shot molding Molding parts in two colors or two materials in a single mold, or set of molds, by injecting material into a closed mold, transferring half of the mold to mate with another mold half of different cavity shape, and injecting the second material around the first material.

draft The degree of taper of a mold cavity side wall or the angle of clearance provided to remove parts from a mold.

drape forming The process of forming a thermoplastic sheet in a movable frame by heating and draping it over high points of a male mold and using a vacuum to complete the forming.

drawing Stretching a thermoplastic sheet, rod, film, or filament to reduce its cross-sectional area and change the physical properties.

ductility of plastic The ability of a material to be stretched, pulled, or rolled into shapes without deforming or fracturing.

□ The ductility of plastics ranges from very little to the extreme amounts as with thermoplastic elastomers.

□ Bendability, crushability, elongation, flattening, kinking, and twisting are some indications of ductility.

dwell A pause in applying pressure to a mold before it is completely closed to allow gas to escape from the molding material.

dyes Soluble colorants that generally form a chemical bond with the substrate or become closely associated with it by a physical process.

□ They usually have good transparency, high tinctorial strength, and low specific gravity.

□ Important dye families are the anthraquinone, azo, sulfur, reactive, and vat dyes.

E-glass fiber A family of glasses composed of calcium aluminoborosilicate and a maximum 2 percent alkali content most often used in reinforced plastics.

□ These fiber types are drawn during their manufacture, making their properties very different from bulk glass.

□ Because of its high electrical resistivity, E-glass fiber is suitable for electrical laminates.

ejector pins Pins that are pushed into a mold cavity from the rear as the mold opens to force out the finished part. Also called "knockout" pins. *(See figure on next page.)*

elasticity The property of a material that allows it to return to its original size and shape after it has been deformed.

elastic limit The amount that a material can be stretched or deformed before taking on a permanent set.

□ A permanent set occurs when a stretched material does not return to its original dimensions after the stress is removed.

elastomer A natural or synthetic material that exhibits

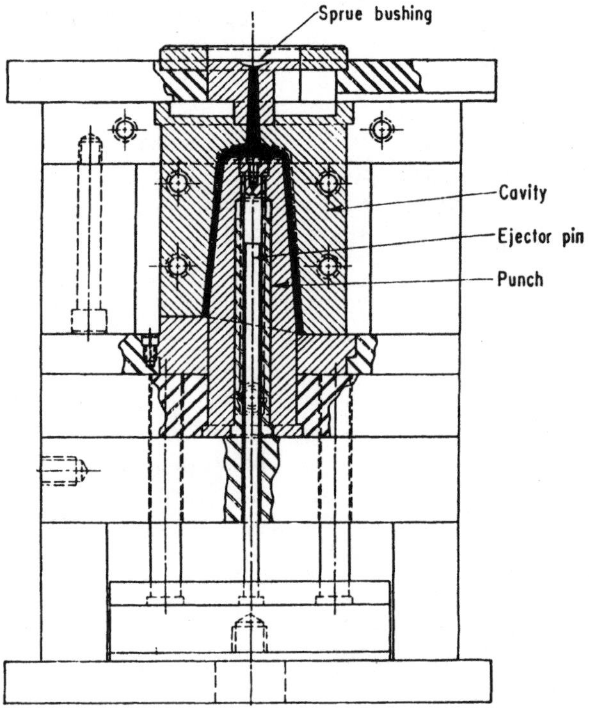

ejector pin. *(Courtesy Michigan Panelyte Molded Plastics.)*

rubberlike properties of high extensibility and flexibility.

- □ Elastomer materials include neoprene, nitrile, styrene, butadiene, and natural rubber.
- □ Elastomers are used for shock absorption, noise and vibration control, sealing, corrosion protection, abrasion and friction resistance, electrical and thermal insulation, waterproofing, and all types of load-bearing products.

elastoplastic A material that has a greater or lesser degree of resilience and will return to, or close to, its original size and shape if deformed a little below its elastic limit.

electrically conductive plastic A class of plastics that can be oxidized or reduced more easily and more reversibly than conventional plastics.

- □ Dopants (charge-transfers agents) affect oxidation or reduction and convert an insulating plastic to a conducting plastic with near-metallic conductivity.
- □ They have chemical structures (conjugated π-electron backbones) that display unusual electronic conductive properties such as low-energy optical transitions, low ionization potentials, and high electron affinities.

electromagnetic bonding *See* INDUCTION BONDING.

electromagnetic interference (EMI) The use of conductive materials in composites to make them conductive and capable of protecting (shielding) electronic devices from interferences such as lightning, radiation, and static electricity.

electron A negatively charged particle that is present in every atom.

electroplating A method of applying a thin coating of one metal to another by electrodeposition for either protection against corrosion or appearance.

electrostatic printing The process of depositing ink on a plastic surface where electrostatic potential is used to attract the dry ink through an open area defined by opaquing. *(See figure.)*

electrostatic printing. *(Courtesy Equistar Chemicals, LP.)*

embedding Embedding, casting, potting, molding, impregnation, and encapsulation, often used interchangeably, involve some form of complete covering of a uniform external shape.

- □ Embedment is used for protecting or decorating a component or assembly.
- □ Applications range from packaging for electrical and electronic devices to providing protection from oxygen, moisture, temperature, electrical flashover, current leakage, mechanical shock, vibration, and other factors.
- □ Plastic embedding materials include allylic, epoxy, polyester, polysulfide, polyurethane, and silicone.

embedment Enclosing a component in a transparent plastic envelope by immersing it in a casting resin and allowing the resin to polymerize.

embossing A technique that uses heat and pressure to produce permanent depressions of a pattern or texture on thermoplastic films or sheeting.

- □ The plastic is preheated and drawn by vacuum into a mold of engraved design. It is then cooled, and the shape is retained.

EMI *See* ELECTROMAGNETIC INTERFERENCE.

emulsion polymerization A technique in which additional polymerizations are carried out in a water medium containing an emulsifier (soap) and a water-soluble initiator.

- □ The reaction in the emulsion type takes place inside a small hollow sphere composed of a film of soap molecules, called *micelle.*

encapsulating Enclosing an item, usually an electronic component, in a plastic envelope by immersing it in a casting resin and allowing the plastic to polymerize or, if hot, to cool.

engraving The process of cutting figures, letters, or symbols into a surface by interposing a resilient offset roll between an engraved roll and the web.

EPA *See* U.S. ENVIRONMENTAL PROTECTION AGENCY.

epichlorohydrin rubber The basic epoxidizing plastic intermediate in the production of epoxy plastics for seals, gaskets, and wire covering.

□ Because of specifications regarding the resistance to fuels, ozone, weathering, and permeation by liquids, the use of this rubber is greatly increased, especially for fuel hoses.

epoxy plastic The family of epoxy thermoset resins (EPS) that includes epichlorohydrin with bisphenol-A. These most widely used epoxies range from low-viscosity liquids to high-molecular-weight solids. Their general properties include toughness, less shrinkage, weatherability, good wetting and adhesion, good mechanical and thermal properties, fatigue resistance, outstanding electrical properties, and water and corrosion resistance.

□ The *novolacs* are an important class that offer high thermal properties and improved chemical resistance.

□ The *cycloaliphatics* are used for applications that require high resistance to arc tracking and weathering.

ester A compound of hot-formed organic acid and alcohol, with the elimination of water, by replacement of the acidic hydrogen of an organic acid by a hydrocarbon radical.

estimating The act of calculating, from statistical records, experience, or other facts, the final cost of producing a product or service.

ether oxide plastic The common name of some important plastics that include the words *ether* or *oxide,* which reflects the presence of an ether or oxygen linkage (–O–) in the backbone of the repeat unit.

□ Plastics include acetal, polyphenylene ether, polyphenylene oxide, polyethylene oxide, and polypropylene oxide.

ethyl cellulose plastic A plastic based on polymers of ethylene or copolymers of ethylene with other monomers; the ethylene is in greatest amount by mass.

□ It is used because of its toughness over a wide temperature range, dimensional stability, and lack of odor.

□ Product applications include flashlights, gears, helmets, slides, and tool handles.

ethylene acrylic plastic This family of ethylene/acrylics are moderately priced, heat- and fluid-resistant, and surpassed only by the expensive specialty types such as the fluorocarbons and fluorosilicones.

□ A special feature is the nearly constant damping characteristic over a broad range of temperatures and frequencies.

□ They have good resistance to hot oils and hydrocarbon, glycol-based proprietary lubricants, and automotive transmission and power-steering fluids.

ethylene plastics A plastic, commonly called *polyethelyene,* based on polymers of ethylene or copolymers of ethylene with other monomers; the ethylene is in the greatest amount by mass.

ethylene-propylene rubber These copolymers are unsaturated, have excellent ozone and weathering resistance, and have good heat aging.

□ By incorporating the monomer diene D, these copolymers allow conventional vulcanization and correspond to EPDM rubbers.

□ They have low resistance to oil and fuel, poor adhesion qualities, and low compatibility to other rubbers.

ethylene vinyl acetate A copolymer of ethylene and vinyl acetate that has many properties of polyethylene, but shows increased flexibility, elongation, and impact resistance.

expandable polystyrene EPS molding shows the use of blowing agents where resin beads, about 0.1 to 0.3 mm in diameter and containing a small amount of hydrocarbon liquid (usually pentane), are supplied to the molder in solid form.

□ The first step is a preexpansion of the virgin beads by heat (steam, hot air, radiant heat, or hot water).

□ The next step takes the beads, usually by an air-transport tube, to the mold cavity(ies).

□ The final step occurs in the mold, where the beads melt together, adhering to each other and forming a relatively smooth skin, and filling the cavity or cavities.

□ These plastics provide improved sound insulation, resistance to additional heat deformation, and better recovery of shapes in moldings.

expanded (foamed) plastics Plastics that are cellular or spongelike.

extruder The part of the blow-molding machine in which the polylefin resin is melted and pushed forward toward the die. *(See figure.)*

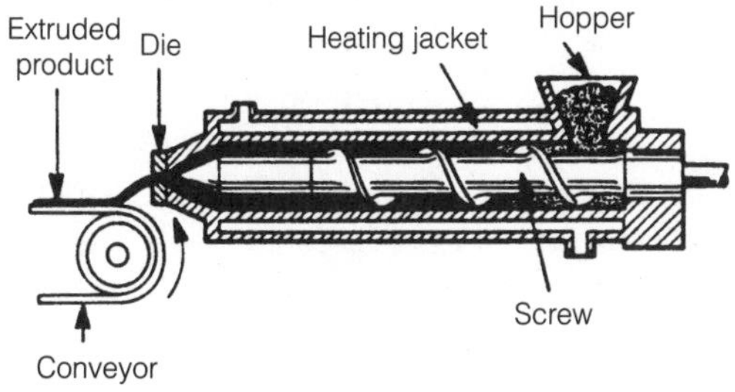

extruder. *(Courtesy Society of Manufacturing Engineers.)*

□ It consists of a barrel containing a turning screw, heaters, thermocouples to measure melt temperatures, and a screen pack.

□ This low-cost, continuously operating process is the main one used in the manufacture of films, sheets, tapes, filaments, pipes, rods, and other products.

□ The screw compresses, melts, and homogenizes the material.

□ As the melt reaches the cylinder end, it is forced through a screen pack before entering a die, which gives it the desired shape.

extrusion The compacting and forcing of heated plastic material through a shaping orifice (a die) in one continuous flow.

extrusion coating The resin coating created by extruding a thin film of molten resin and pressing it into or onto the substrate (or both), without the use of adhesives. *(See figure.)*

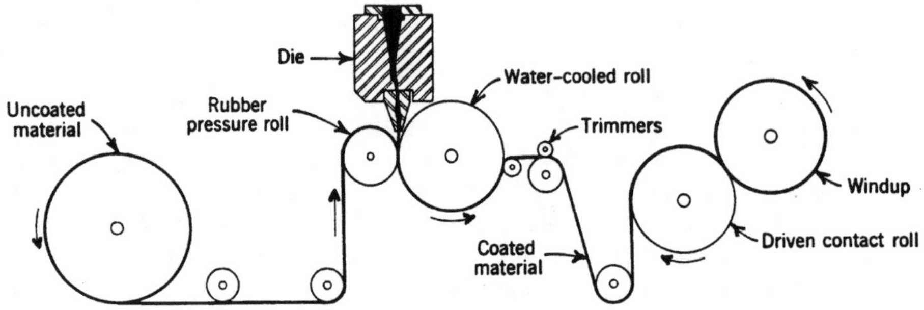

extrusion coating. *(Courtesy Eastman Chemical Products, Inc.)*

fatigue strength The highest stress that a material can withstand, for a given number of cycles, before failure occurs.

FDA *See* U.S. FOOD AND DRUG ADMINISTRATION.

feeder A piece of equipment that provides controlled flow of materials (from powders to pellets) to or from processing operations. Common types of feeders are gravimetric, vibrator, and volumetric.

feedstocks Raw materials, such as crude oil or natural gas, from which polymers are made.

□ Feedstocks used to make plastics are methane, ethylene, benzene, acetylene, naphthalene, toluene, and xylene. Other feedstocks include coal and vegetation.

female mold The indented half of a mold or the cavity that receives the male half.

fiber A term basically used to refer to filamentary materials. Fibers are often called *filament, monofilament,* "whisker," and "yarn."

□ Natural and manufactured (organic and inorganic synthetics) fibers are used as additives and reinforcements to lower material cost and produce strong materials.

fiber orientation Fiber alignment in a nonwoven or a mat laminate where most of the fibers are in the same direction, which results in higher strength in that direction.

fiber-reinforced plastic (FRP) A term for plastic that is reinforced with cloth, mat, strands, or other fiber forms.

filament The smallest unit of a fibrous material characterized by extreme length, with little or no twist.

filament winding A process that consists of winding a continuous reinforcing fiber (impregnated with resin) around a rotating and removable form (mandrel). *(See figure.)*

filler An inert substance added to plastic to make it less costly and improve the physical properties such as hardness, stiffness, and impact strength.

□ Fillers are usually small and do not improve the tensile strength as do reinforcements.

□ Common fillers are clays, silicates, talcs, carbonates, and wood flour.

film Films are produced by calendering, chemical conversion, melt extrusion using flat or circular dies, skiving, or solvent casting. They may be rolled to change their properties.

film coating A process that involves applying plastic (solution, emulsion, or extrusion coating) onto the base film to provide it with new and unique properties such as heat sealability, impermeability, energy barrier, and optical or electrical properties.

fire-resistant A term characterizing a product that does not easily burn by reducing the flame spread on its surface.

fixing agent **1.** A mechanical substance (albumin) that holds pigments permanently on textile fibers. **2.** Gums and starches that can mechanically hold dyes and other materials on textile fibers long enough to be processed. **3.** A material that aids the fixation of mordants on textiles by chemically uniting with them and holding them on the fiber until the dyes can react with them.

flame-retarded plastic Plastic that is mixed with certain chemicals to reduce or eliminate its tendency to burn.

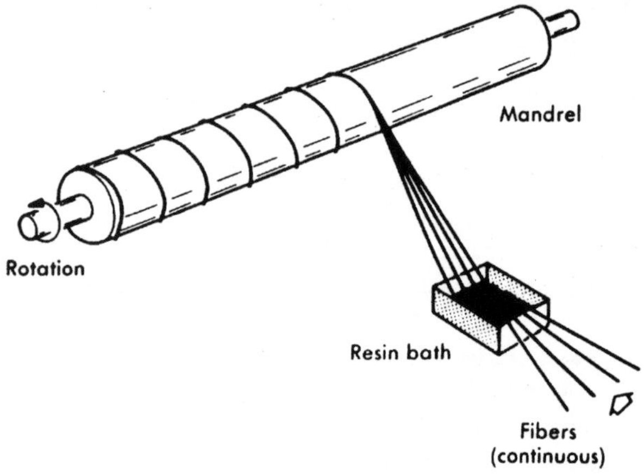

filament winding. *(Courtesy Polytech Ltd.)*

flame spraying A method of applying a plastic coating where finely powdered plastics and fluxes are projected, through the cone of a flame, onto a surface.

flash Extra plastic that is attached to a molding along the parting line. It must be removed to finish a part.

flexural modulus The ratio, within the elastic limit, of the applied stress on a test sample in flexure to the matching strain in the outermost fibers of the sample.

flow The motion of a plastic material that flows during processing in molds or dies, or a measure of its moldability.

fluidized bed A method of coating heated items by immersion in a dense-phase fluidized bed of powdered resin.

fluorescence A property of a material that causes it to produce light while it is being acted on by ultraviolet light or x rays.

fluoroplastics A group of plastic materials containing the monomer fluorine.

□ They have outstanding chemical inertness, resistance to temperatures from −425 to 500°F (−220 to 260°C), low coefficient of friction, good electric properties, low permeability, practically zero moisture absorption, and good resistance to weathering and ozone.

foamed plastic Thermoplastics or thermoset plastics with a density that is decreased by the presence of numerous cells throughout its mass caused by an expanding gas foam blowing agent. *(See figure.)*

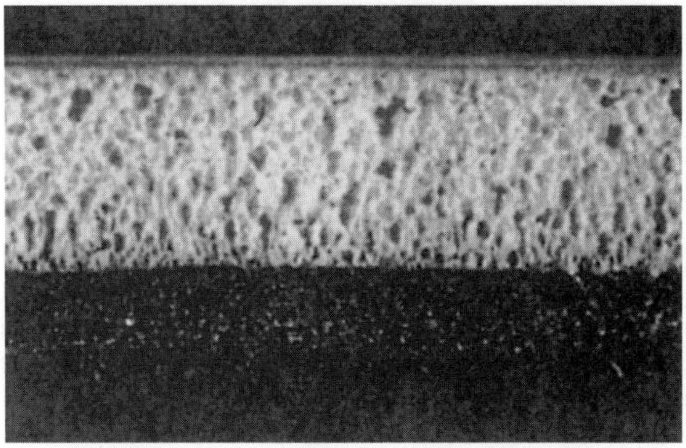

foamed plastic. *(Courtesy Firestone Plastics Co.)*

foaming agents Chemicals that produce inert gases on heating or chemical reaction, causing the resin to take on a cellular structure.

□ The gases, when injected into a plastic melt under pressure (higher than the melt pressure), form a cellular structure when the melt is released to atmospheric or low pressure.

free-forming Characterizing a process in which air pressure is used to blow a heated sheet of plastic, with its edges held in a frame, until the desired shape or height is obtained.

furan plastic (FUN) A generic term for a thermoset-resinous product that contains a heterocyclic unsaturated furan ring in its molecular structures.

□ Used in the manufacture of chemical-resistant cements, grinding wheels, and foundry molds.

□ Furan-based composites possess good heat and chemical resistance and excellent surface hardness, and are nonflammable.

fusion The process of melting two or more materials to produce a homogeneous mixture.

gas-phase polymerization The process of having a gaseous monomer and catalyst mixed in a continuous reactor, followed by the removal of dry, free-flowing plastic powder ready to be pelletized and sold.

gate The orifice or passageway through which the melt enters the cavity in injection and transfer molding. *(See figure.)*

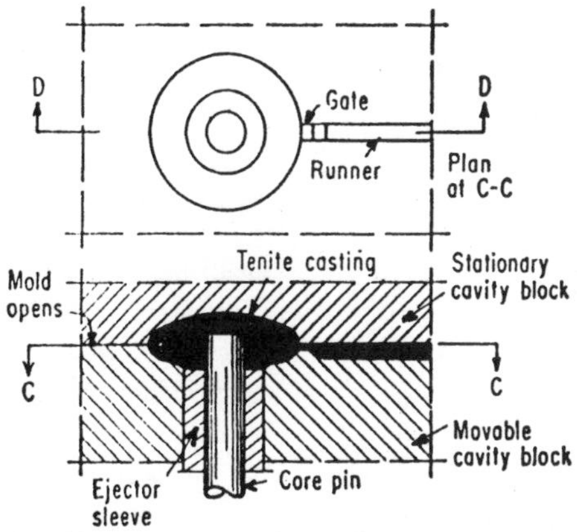

gate. *(Courtesy Society of Manufacturing Engineers.)*

gel A cross-linked plastic network swollen in a liquid medium whose properties depend on the interaction of the two components.

□ The liquid prevents the plastic network from collapsing into a compact mass, while the network retains the liquid.

□ Gels are important in the manufacture of adhesives, films, plastics, membranes, and rubbers.

□ They are used as absorbents in disposable diapers, for water purification, in implants in plastic surgery, and for artificial and soft contact lenses.

geotextiles The primary plastics in geosynthetics are filament-extruded polyester, nylon, polypropylene, and high-density polyethylene. Geotextiles (also geonets, geogrids, and geomembranes) are widely used in all types of civil works, from roads to canals and from landfills to landscaping.

□ Geogrids are fairly rigid latticelike fabrics that are used in reinforcement work; geomembranes provide impermeable barriers.

□ Reinforcement and repair of roadways are major uses of geotextiles.

glass fiber A fiber melt, spun from various types of glass, that has cooled to a rigid condition without crystallizing. Glass fibers are usually hard and relatively brittle, and have a shell-like fracture.

glass transition The change in an amorphous polymer or in amorphous regions of a partially crystalline polymer from a viscous or rubbery condition to a hard and relatively brittle one.

graft copolymer A combination of two or more chains of different features, one of which serves as a backbone main chain, and at least one of which is bonded at some point(s) along the backbone to make a side chain.

granulating A process used to reclaim reprocessable thermoplastic scrap, flash, and rejected parts to eliminate scrap and conserve materials.

gravimetric feeders Devices that measure the weight of material fed to the extrusion from a special weigh hopper and regulate the rate at which it is used.

gutta percha A rubberlike product obtained from certain tropical trees.

halocarbon plastic Plastics made by polymerizing monomers consisting only of carbon and a halogen or halogens.

halogens The elements astatine, bromine, chlorine, fluorine, and iodine.

hand layup A method of positioning successive layers of reinforcement mat or web on a mold by hand.

□ Resin is used to coat the reinforcement, followed by curing of the resin to permanently fix the formed shape.

head In blow molding, it is the end section of the molding machine that consists of the core, die, mandrel, and other parts required to form resin into a hollow tube (parison) with the correct dimensions and thickness. *(See figure.)*

□ The head delivers this parison to the area where it can be picked up and transferred to the blowing mold.

heated-tool bonding Joining plastics by applying heat and pressure to areas in contact.

heat-resistant plastic Materials that retain their mechanical properties for thousands of hours at 446°F (230°C), hundreds of hours at 572°F (300°C), minutes at 1004°F (540°C), or seconds up to 1400°F (760°C).

heat-transfer decorating The process of transferring an image from the carrier film to the product by stamping with rigid or flexible shapes, using heat and pressure.

hob A hardened master model of the required form that is pressed into a soft metal block to produce the required cavity shape.

homopolymer The result of polmerizing a single monomer.

hopper dryer A feeding and drying device that is used for extrusion- and injection-molding thermoplastics.

hopper loader A piece of equipment used for automatically loading resin pellets into a machine hopper.

hot-gas welding A process of joining thermoplastic materials where the materials are softened by a jet of hot air from a welding torch and joined together at the softened points.

hot-leaf stamping A decorating operation for marking plastics by bringing a heated die into contact with the plastic while a stamping leaf foil is between the die and the plastic.

hot melt A term referring to thermoplastic synthetic resins composed of 100 percent solids and used as adhesives at temperatures between 248 and 392°F (120 and 200°C).

hydrocarbon An organic compound, found in petroleum, natural gas, coal, and bitumens, that contains only carbon and hydrogen.

hydrocarbon plastics Plastics that are based on resins made by polymerizing monomers composed only of carbon and hydrogen.

hydrogel Hydrogels or water-containing gels are plas-

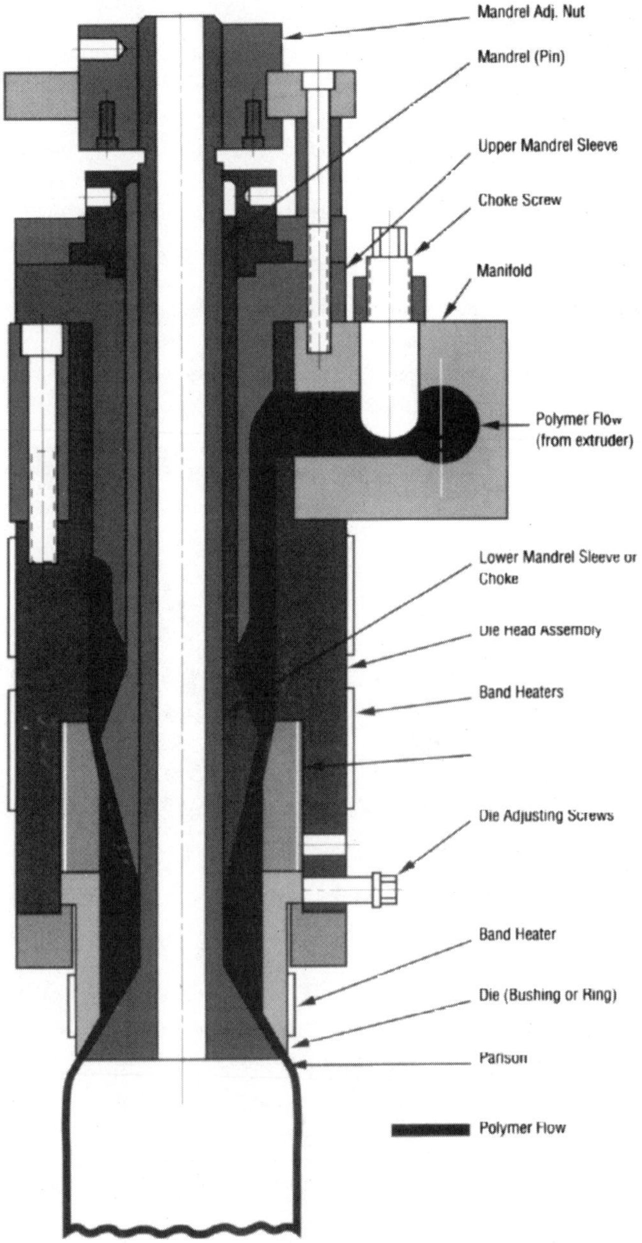

head (die). *(Courtesy Equistar Chemicals, LP.)*

tics characterized by hydrophilicity and insolubility in water.

- □ Natural hydrogels are used in pulp and paper production, artificial silk, cellulose membranes, and biomedical applications.
- □ Synthetic hydrogels are used in prosthetic materials, soft contact lenses, and membranes for controlled drug release because of their compatibility with living tissues.

hygroscopic plastic A plastic group that can attract, absorb, and retain atmospheric moisture. These include thermoplastics such as acrylics, nylons, polycarbonates, and polyurethanes.

impact strength The ability of a material to withstand applied shock.

impingement A method of mixing where two or more materials collide.

impregnation The process of soaking and filling the voids of porous material such as wood, paper, or fabric with synthetic resin. The porous material serves as a reinforcement for the plastic binder after curing.

induction bonding The use of high-frequency electromagnetic fields to excite the molecules of metallic inserts placed in the plastic to fuse them in place.

inert (rare) gases Gases such as argon, helium, neon, krypton, radon, and xenon that do not combine with other elements.

inhibitor A substance that slows down a chemical reaction where certain monomers and resins tend to prolong storage life.

initiation phase The first of three steps in addition polymerization that produces a reactive state of the molecules, usually by some high-energy source catalysts, or radiation.

initiator An agent that is required to start polymerization, especially in emulsion-polymerization processes.

injection molding A molding procedure in which heat-softened plastic is forced from a cylinder into a relatively cool cavity that gives the item the desired shape. *(See figure.)*

- □ Plastic moves from the hopper into the feeding portion of the reciprocating extruder screw, which rotates, causing the material to move through a heated extruder barrel, where it is softened.
- □ When the shot size is reached, the screw stops rotating, and at a preset time the screw acts as a ram to push the melt into the mold.

in-mold decorating The process of making patterns or decorations on molded products by placing the pattern or image in the mold cavity before the actual molding cycle. The pattern becomes a part of the plastic item as it is fused by heat and pressure.

insert A part of a plastic product which could be metal, plastic, or other material that may be molded into position or pressed into the part after molding.

inspection A term indicating that during the manufacture of a part, someone will check, measure, examine, test, gage, and perform other functions on a procedure or a part to detect errors or check the quality.

interlaminar shear The shear strength at rupture where the plane of fracture is between the layers of reinforcement of a laminate.

interpenetrating polymer network (IPN) An entangled combination of two cross-linked polymers that are not bonded to each other.

ion An atom or group of atoms that carries a positive or negative electrical charge.

ionic bonding Atomic bonding by electrical attraction of unlike ions.

ionomer A polymer that has ethylene as its major component, but contains both covalent and ionic bonds. These resins have high transparency, resilience, tenacity, and many of the characteristics of polyethylene.

ionomer plastic A member of the polyolefin family whose ionic cross-links occur randomly between long-chain molecules to produce properties usually associated with high-molecular-weight materials.

- □ Ionomers are abrasion-resistant and extremely tough, with tensile impact strengths as high as 600 ft · lb/in) and tensile strengths as high as 35,000 kPa (5000 psi).

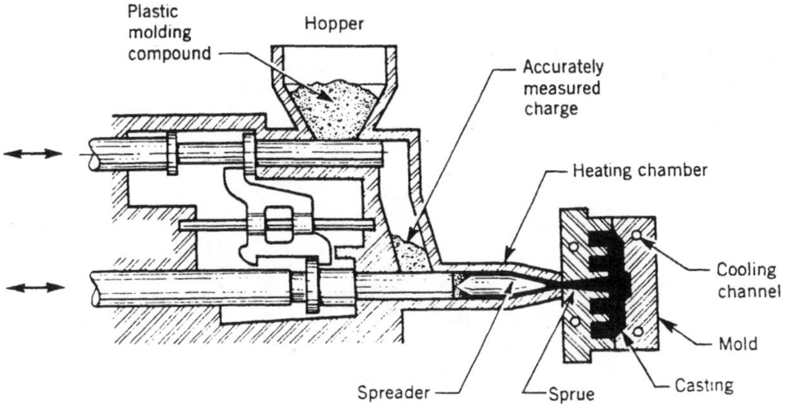

injection molding.
(Courtesy The Society of the Plastics Industries, Inc.)

□ The clarity, strength, and good adhesion of ionomer films to metal surfaces led to their wide use in food packaging.

□ Ionomers are used in bowling-pin and golf-ball covers and bumper guards because of their high impact strength and cut resistance.

irradiation Bombardment with various forms of ionizing and nonionizing radiation to begin polymerization and copolymerization of plastics; in some cases irradiation changes the physical properties of plastics.

isocyanate resins Resins synthesized from isocyanates and alcohols; mostly used in combination with polyols.

isomer A compound that has the same number of atoms of the same elements but with different structures and properties.

isoprene rubber This elastomer (plastic), made from a colorless, volatile liquid derived from propylene or from coal gases or tars, is the closest synthetic rubber comes to the natural rubber.

isotactic A plastic molecular structure that has a sequence of regularly spaced asymmetrical atoms arranged in like configuration in a polymer chain.

isotropic A term denoting that the properties of a material are equal in all directions.

joint The point at which two parts or surfaces are held together with a layer of adhesive. Joints used for the same or different materials, such as plastic to plastic or plastic to aluminum, are classified as having high- to low-strength test values.

kirksite An alloy of aluminum and zinc used for molds because of its high thermal conductivity.

knife coating A method of coating a continuously moving web that has been coated with a plastic and whose thickness is controlled by an adjustable knife or bar set at a suitable angle to the substrate.

lamellar A term referring to the aligned, looped molecular structure of crystalline polymers that are sheet- or platelike in shape.

laminar composite A term that refers to a composite composed of layers of materials (laminates and sandwiches) that are held together by the polymer matrix.

laminate A product made by bonding together two or more layers of material or materials, such as plastic film, sheet, and tape; foils of aluminum, steel, paper, and other materials; and different types of woven and nonwoven fabrics using synthetic and natural fibers. *(See figure.)*

□ In reinforced plastics, the term *laminates* refers mainly to superimposed layers of plastic-impregnated or plastic-coated fabrics, or fibrous reinforcements which have been bonded together.

laminated plastics A dense, tough solid produced by bonding together layers of sheet materials impregnated with a resin and curing them by application of heat or heat and pressure.

laminating The process of producing a composite laminate.

laser cutting A method of cutting materials by using laser energy.

latex An emulsion of natural or synthetic resin particles dispersed in a watery medium.

□ Synthetic latexes are made by emulsion polymerization from styrene-butadiene comolymer, acrylate, polyvinyl acetate, and other compounds.

□ Foamed latex rubber is produced by foaming a compounded rubber latex by mechanical or chemical means, and then converting the latex phase to a rubbery continuum.

leaching The process of removing a soluble component from a polymer mix with solvents.

linear This refers to a long straight-chain molecule, as compared with one having many side chains or branches.

liquid-crystal polymer plastics (LCPs) Sometimes called *superpolymers,* these compounds are known to be self-reinforcing plastics because of their densely packed fibrous polymer chains.

□ They are noted for their superior strength at extreme temperatures, excellent mechanical-property retention, low coefficient of thermal expansion, flame resistance, and easy processibility.

lubrication bloom An irregular, cloudy, greasy film on a plastic surface caused by excess lubricants.

luminescence Light emission by the radiation of photons after luminescent pigments are activated by ultraviolet radiation.

macrocyclic plastic Macromolecules are commonly shown as long, flexible, randomly coiled chains that may be branched. Ring-chain equilibria occur in some polycondensation systems, suggesting the presence of functional links in the polymer chain.

macromolecules The large (giant) molecules that make up the high polymers.

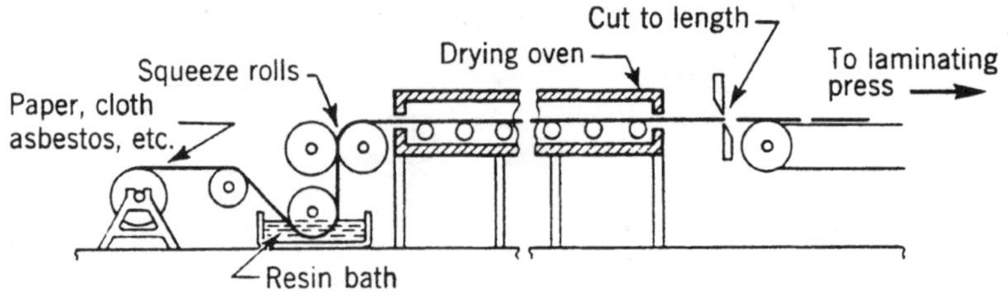

laminating process.

(Courtesy The Society of the Plastics Industries, Inc.)

mandrel A form around which filament-wound and pultruded composite structures are shaped.

matched-mold forming The forming of hot sheets between matched male and female molds.

matrix The polymer material used to bind reinforcements together in a composite.

mechanical fastening Mechanical means of joining plastics with machine screws, self-tapping screws, drive screws, rivets, spring clips, clips, dowels, catches, and so on.

mechanical forming A process in which heated plastic sheets are shaped or formed with the aid of jigs and fixtures.

melamine plastic Melamine formaldehyde (MF), one of two major thermoset resins in the amino family, uses various fillers to make MF compounds to meet different requirements. It is rigid, possesses a hard surface, and is not affected by most organic solvents, greases, oils, weak acids, and alkalis.

□ MF is widely used in food containers and dishes because it does not transfer odor or taste to foods, is not affected by heat, and is highly flame-resistant. MF with alpha cellulose is used in heavy-duty dishware.

□ Melamines are excellent for most electrical applications, especially where arc resistance is desired.

melt flow rate A measure of the molten viscosity, in grams per 10 minutes, of a polymer by determining the weight of polymer forced through an orifice under specified conditions of pressure and temperature.

mer The smallest repetitive unit along the molecular chain in a polymer.

metallizing plastic Plastic can be coated with metals, 0.01 to 3 mils (0.03 to 0.8 mm), for decorative and/or functional purposes. *(See figure.)*

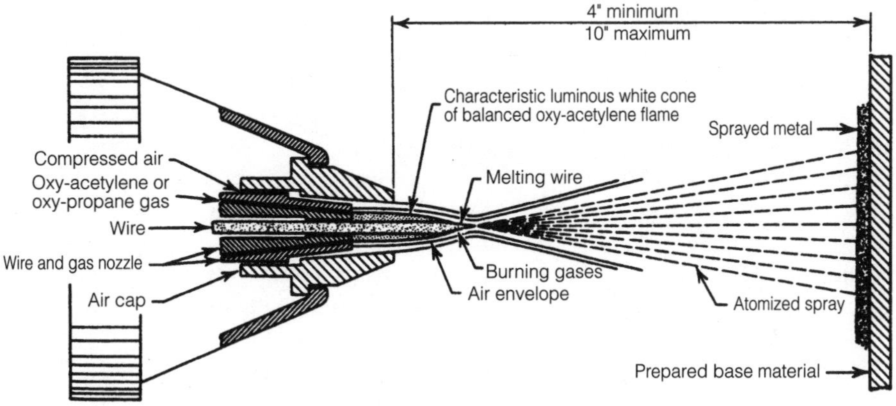

metallizing plastic. *(Courtesy Metco, Inc.)*

□ The main commercial process is electroless plating, which is used for automobiles, printed circuits, and other products.

□ Flame and arc (metal) spraying are used for electromagnetic interference shielding.

□ Sputtering and vacuum metallizing are used for continuous coating of plastic film.

methyl methacrylate A colorless, volatile liquid used in the production of acrylic resins and produced from acetone cyanohydrin, methanol, and dilute sulfuric acid.

microballoons Hollow glass spheres, also called *microspheres.*

miscibility A term used to describe a mixture of two or more plastics combined to form a single-phase (solid or liquid) solution.

□ In the research for new plastics, compounding or mixing two or more plastics is a way to develop new property combinations without synthesizing novel structures.

mixer Mixers are used to mix materials so that the lumps, pools, or aggregates of each material are formed with little or no particle cohesion.

□ Various mixers are designed for processing medium- to high-viscosity formulations, wetout and disperse solids into a liquid vehicle, solids dispersion, and other applications.

□ Common mixers include the ball mill, Banbury, centrifugal impact, conical dry blender, drum tumbler, extruder compounder, hopper blender, kneader, paddle agitator, propeller, roll mill, screw, static, and tumbling agitator.

mold The cavity in which plastics or resins are formed into finished products by heat, or heat and pressure. *(See figure on next page.)*

□ It is a controllable complex device that must be an efficient heat exchanger.

molding compounds Plastic or resin materials in various stages of formulation (powder, granular, or preform) consisting of resin, filler, pigments, plasticizers, and other compounds ready for molding.

molecular mass The sum of the atomic mass of all atoms in a molecule. Measurement methods include light scattering, osmotic pressure, sedimentation equilibrium, solution pressure, and viscosity.

□ In high polymers, the molecular masses of individual molecules vary widely, so they are expressed

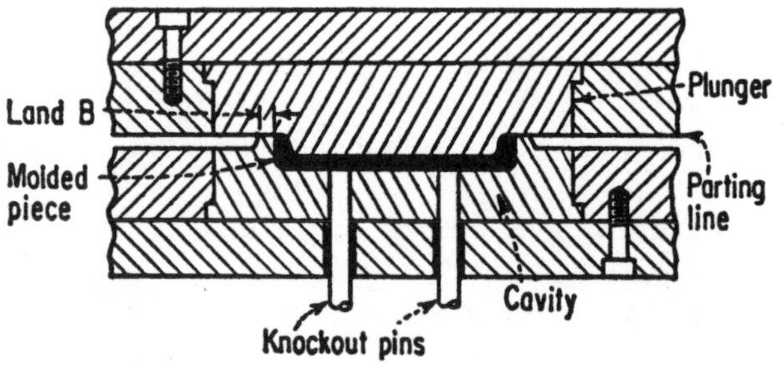

mold. *(Courtesy Society of Manufacturing Engineers.)*

as averages.

□ Average molecular mass of polymers is expressed as number-average molecular mass (M_n) or mass-average molecular weight (M_w).

molecule The smallest particle of a substance that can exist independently while retaining its chemical identity.

monomer A simple molecule capable of reacting with like or unlike molecules to form a polymer or the smallest repeating structure of a polymer, also called a *mer.* It is the basic material from which plastics are made. *(See figure.)*

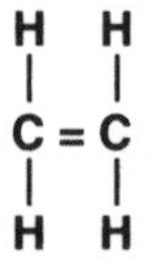

monomer. *(Courtesy Equistar Chemicals, LP.)*

multicavity mold A mold having two or more impressions forming finished items in one cycle of the molding press.

National Institute for Occupational Safety and Health (NIOSH) The federal agency that recommends occupational exposure limits for various substances.

needle blow A sharp, small-diameter blow pin that is forced through the parison as the mold closes around it or after the die closes.

neoprene rubber (CR) Neoprene, one of the family of elastomers, has good resistance to oils, ozone, oxidation, and flame. It is the most rubberlike of all materials, especially with regard to its dynamic response.

NIOSH *See* NATIONAL INSTITUTE FOR OCCUPATIONAL SAFETY AND HEALTH.

nitrile rubber (NBR) Nitriles are copolymers of butadiene (B) and acrylonitrile (AN) that are used mainly for parts requiring resistance to petroleum oils and gasoline, such as in automobiles.

nitrocellulose (cellulose nitrate) Material formed by the action of a mixture of sulfuric acid and nitric acid on cellulose.

novolac A phenolic-aldehyde-type resin that remains permanently thermoplastic unless a source of methylene is added.

□ It is used as a molding compound, for bonding materials, as an abrasive for grinding wheels, and in other applications.

nylon plastic Nylon, or polyamide (PA), the first of the so-called thermoplastic engineering plastics, was originally developed as high-strength textile fiber for stockings. Its main characteristics are strength, stiffness, and toughness.

□ Other types of nylons include the castable types, liquid monomers that polymerize and become solid, and castable liquid monomer, a moldable transparent material.

Occupational Safety and Health Administration (OSHA) A Department of Labor agency with safety and health regulatory and enforcement authorities for most U.S. industry and business.

offset printing A printing technique in which ink is transferred from the printing plate to a roller that in turn transfers the ink to the object to be printed.

oil-soluble plastic A type of plastic that, at moderate temperature, will dissolve or disperse in, or react with, drying oils to give a homogeneous film of modified characteristics.

open-celled This refers to cells interconnecting in cellular or foamed plastics.

organometallic plastic These plastics contain metals either in the backbone chains or pendent to them. The metals may be connected to the plastic by bonds to carbon.

□ They can be classified as addition, condensation, coordination, or other plastics.

organosol A breakup usually of vinyl or polyamide in a liquid phase containing one or more organic solvents.

orthophthalic plastic A thermoset polyester plastic using phthalic anhydride as one of the chemical components.

OSHA *See* OCCUPATIONAL SAFETY AND HEALTH ADMINISTRATION.

oxygen index A test for the minimum oxygen concentration in a mixture of oxygen and nitrogen that will support a burning polymer flame.

parameters A term usually applied to a specific range of variables, characteristics, or properties related to the subject.

parison A hollow plastic tube that is extruded from the die head and expanded within the cavity by air pressure to produce blown objects. *(See figure on next page.)*

particulate Small particles of various shapes and sizes that are used to reinforce a polymer matrix.

parting lines The marks on a molded or cast part where the two halves of the mold meet in closing.

pellets Formulations of molding compounds that are in tablets or granules of uniform size.

perfluorcarbon plastic (PFA) A fluorocarbon plastic that does not scatter and reflect light, to allow for the production of water-clear film.

□ A major use is for wafer baskets used in the automated production of microcomputer chips.

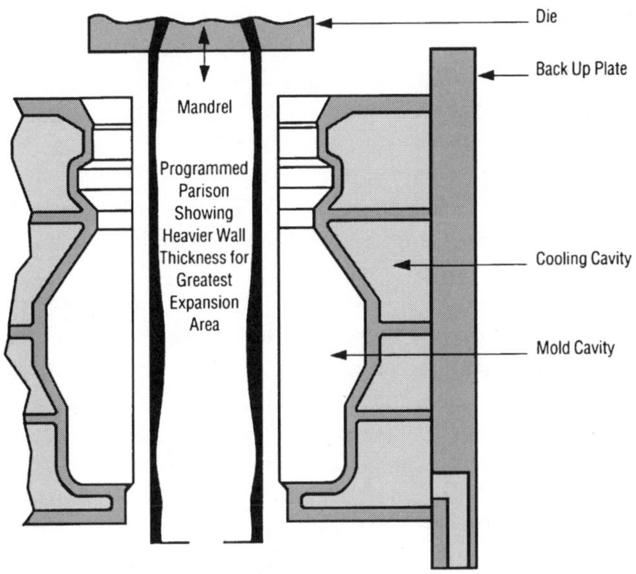

parison. *(Courtesy Equistar Chemicals, LP.)*

periodic table A fundamental framework for the systematic organization of chemistry. An arrangement of the elements in order of increasing atomic number, forming groups of members that show similar physical and chemical properties.

phenolic A synthetic resin produced by the condensation of aromatic alcohol with an aldehyde, particularly of phenol with formaldehyde.

phenoxy plastic A high-molecular-weight thermoplastic polyester based on bisphenol-A and epichlorohydrin; an outgrowth of the epoxy plastic technology.

- □ Their combination of good impact resistance, strength, clarity, and impermeability makes them useful for molding, including blow molding of cosmetic, foodstuff, and household-chemical bottles.
- □ Phenoxies are used in automotive and marine primers as well as in heavy-duty maintenance primers.

photon The least amount of electromagnetic energy that can exist at a given wavelength.

photopolymer A polymer (plastic) that undergoes a change on exposure to light, which causes further polymerization or cross-linking.

- □ They are used for printing and lithography plates for photographic prints and microfilm copying.

photosynthesis The synthesis of chemicals with the aid of radiant energy from the light of the sun.

piezoelectric plastic (PVDF) A polyvinylidene fluoride plastic that automatically gives off an electric charge when mechanically stressed or develop a mechanical response when an electric field is introduced.

- □ Their structures are irregular so that their centers of positive and negative charges are sensitive to pressures which change the dipole distance and the polarization.
- □ These plastics are used as transducers or acoustic sensors.

pinch-off A raised edge around the mold cavity that seals off the part and separates the excess material as the mold closes around the parison.

plastic A synthetic or natural product (excluding rubber) that contains as an essential ingredient an organic substance of large molecular weight. At some stage of working under heat and pressure, it is capable of flowing, being formed, and being held in a desired shape when cooled.

- □ The terms *plastic, resin, elastomer,* and *polymer* are often used interchangeably; *resins, elastomers,* and *polymers* refer to basic material as polymerized.
- □ Plastic properties range from rigid to rubbery, poor to high heat resistance, poor to high chemical resistance, and so on.
- □ The term also refers to a flow behavior of all materials, including plastics, metals, and aluminum.

plasticizer A chemical agent that is added to plastics to make them softer and more flexible.

- □ Plasticizers allow the long molecule chains to move more easily in relation to each other when under a strain, giving the material lower strength but making it easier to process.
- □ Low-cost plasticizers, often called *extenders,* are used to reduce the cost of compounds when the stiffness level is not important.

plastic strain The strain permanently given to a material by stresses that exceed the elastic limit.

plastic, synthetic A plastic produced by the polymerization of its monomer or monomers by a controlled chemical action as opposed to a plastic produced in nature by biosynthesis.

- □ The most important commercial synthetic materials are called *plastics, elastomers* or *rubbers, fibers, coatings,* and *adhesives.*

platens The mounting plates of a press to which the mold assembly is bolted.

pneumatics A branch of science dealing with the mechanical properties of gases.

polyacrylate A thermoplastic resin made by the polymerization of an acrylic compound.

polyacrylonitrile plastic (PAN) Polyvinylcyanide homopolymer is a polar crystallizing polymer that undergoes chemical decomposition before crystalline melting at temperatures above 572°F (300°C).

- □ It is not a true thermoplastic material but can be processed into fiber (acrylic fibers) through its spinning from solutions.
- □ PAN is used as a base material in the manufacture of certain carbon and graphite fibers.

polyallomers Crystalline polymers that are produced from two or more olefin monomers.

polyamide (PA) Commonly known as *nylon*; a tough, strong plastic that can be processed into fibers,

polyamide parts. *(Courtesy DuPont Co.)*

filaments, and molded parts. It is widely used for parts that require impact, abrasion, and solvent resistance such as automobile radiators and plumbing fittings. *(See figure.)*

polyarylate plastic (PAR) These are often defined as copolyesters involving bisphenol-A (BA) and a mixture of terephthalic acid (TA) and isopthalic acid (IA).
- □ Performance includes good resistance to heat, steam, radiation, and good weatherability and fire resistance without additives.

polyblends Plastics that have been modified by the mechanical mixture of two or more polymers, such as polypropylene and rubber.

polycarbonate Polymers obtained from direct reaction between aromatic and aliphatic dihydroxy compounds with phosgene, or by ester-exchange reaction with appropriate phosgene-derived materials.

polychloroprene rubber (CR) CR or neoprene was one of the first synthetic rubbers that had desirable properties in many applications. It was used for exterior applications in vehicles and buildings, and for cables.

polyester A resin formed by the reaction between an organic dibasic acid and an organic dihydroxy alcohol.
- □ Thermosetting resins have been modified with multifunctional acids, or acids and bases and some unsaturated reactants that permit cross-linking.
- □ Polyesters that have been modified with fatty acids are called *alkyds.*

polyethylene A thermoplastic material composed of ethylene polymers.

polyimide A group of resins that are made by reacting pyromellitic dianhydride with aromatic diamines. This polymer is characterized by tightly bound-together rings of four carbon atoms.

polyimide plastics (PIS) Polyimides were the first so-called high-heat-resistant plastics that keep much of their room-temperature mechanical properties from −400 to +600°F (−240 to 315°C) in air.
- □ These materials have good wear resistance, have low coefficient of friction, and are unaffected by exposure to dilute acids, aromatic and aliphatic hydrocarbons, esters, alcohols, hydraulic fluids, and kerosene.

polymer A compound with high molecular mass (weight), either natural or synthetic, formed by a chain of chemically linked units called *monomers,* whose structure can be represented by repeating small units.
- □ Some polymers are elastic, while others are plastic.

polyolefin A term used for a family of polymers made from hydrocarbons with double carbon-to-carbon bonds. These include polyethylene, polymethylpentene, and polypropylene.

polypropylene A tough, lightweight, rigid plastic material made by polymerization of high-purity propylene gas in the presence of an organometallic catalyst at relatively low pressures and temperatures.

polypropylene plastic (PP) These plastics, in the polyolefin family of plastics, are produced by a stereoselective catalyst that puts order in its molecular configuration so that the basic resin has a predominantly regular, uniform structure. *(See figure.)*

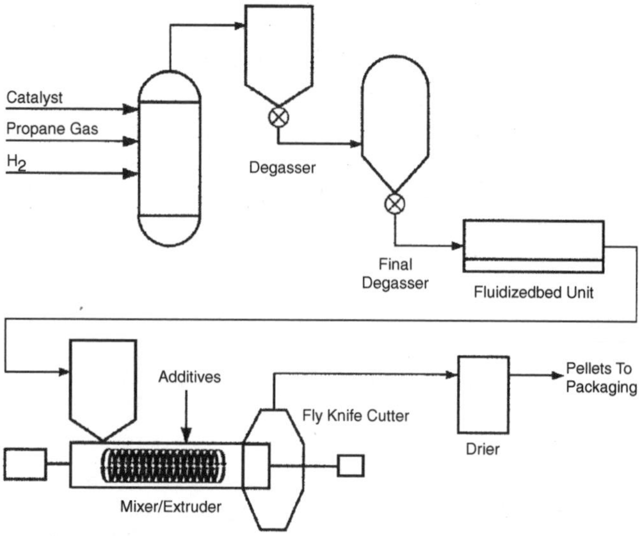

polypropylene plastic. *(Courtesy Equistar Chemicals, LP.)*

- □ This means that the molecules crystallize into compact bundles, making them stronger, more heat-resistant, and more rigid than other members of the polyolefin family.
- □ They are semitranslucent and milky white in color, with excellent colorability, and are widely used for indoor and outdoor applications.

polystyrene A thermoplastic material produced by polymerization of styrene (vinyl benzene).

polystyrene plastic (PS) An amphorous thermoplastic that is noted for its sparkling clarity, hardness, ease of processing, excellent colorability, dimensional stability, and low cost.

polysulfone A thermoplastic consisting of benzene rings that are connected by a sulfone group, an isopropylidene group, and an ether linkage.

polyurethane A family of resins produced by reacting diisocyanate with organic compounds that contain two or more active hydrogens to form polymers having free isocyanate groups. These groups, under influence of heat or certain catalysts, react with each other or with water, glycols, or other materials to form a thermoset.

polyurethane plastic (PUR) This plastic, produced by the reaction of polyisocyanates with polyester or polyester-based resins, can vary in form and physical or mechanical properties of toughness, flexibility, and abrasion resistance.

□ The *rigid foam type* is widely used for insulation material in buildings, appliances, coolers, and other structures.

□ The *flexible foam* is an excellent cushioning material for furniture.

□ The *elastomeric type* is used in solid tires, shock absorbers, and other products.

polyvinyl Vinyl plastics that have a basic molecule with two carbon atoms along with three pendant hydrogens, and a fourth pendant group that is a unique group of atoms, depending on the particular vinyl. The fourth group of atoms provides the special qualities of each vinyl plastic.

polyvinyl plastic (PVB) These plastics are soluble in esters, lectones, alcohols, and chlorinated hydrocarbons but insoluble in the aliphatic hydrocarbons. They are stable in dilute alkalis but tend to decompose in dilute acids.

□ They are used as safety-glass interlayers and between sheets of acrylic to protect pressurized cabin enclosures in aircraft against shattering.

postcure An additional elevated-temperature treatment to improve final properties and/or complete the final cure. This process can decrease volatiles in the part, relieve stresses, and/or improve dimensional stability.

potting An embedding process for parts, similar to encapsulating except that the object is covered and not surrounded by an envelope of plastics.

preexpansion A process of partially expanding polymer beads or granules before they are molded into cellular parts.

preplasticator A device that softens material before forcing it into the mold, another molding machine, or an accumulator.

prepuffs The preexpanded pieces of polymers used to make cellular polymer parts.

pressure-bag molding A process for molding reinforced plastics, where a tailored flexible bag is placed over the contact layup on the mold, sealed, and clamped in place. Compressed air forces the bag against the part to apply pressure while the part cures.

primary bonds A strong association (interatomic attraction) between atoms.

primary recycling The processing of scrap plastics into the same or similar types of product from which it was originally made, using standard plastics processing methods.

promoter A chemical, itself a feeble catalyst, that speeds up the activity of a given catalyst.

propagation phase The second step in addition polymerization that refers to rapid growth or addition of monomer units to the molecular chain.

proportional limit The greatest stress a material can withstand without deviation from proportionality of stress and strain (Hooke's law), or the point at which elastic strain becomes plastic strain.

pultrusion A continuous process for manufacturing composites in which a combination of liquid plastic and continuous fibers is pulled through a heated shaping die. *(See figure.)*

□ Reinforced plastic shapes include structural I-beams, L-channels, tubes, angles, rods, and sheets.

□ Longitudinal fibers are generally continuous glass-fiber rovings, and glass-fiber material in mat or woven form is added for cross-ply properties.

purging The process of forcing out one molding material or color by another from the machine cylinder before molding a new material.

pyrolysis Chemical decomposition of a substance by heat and pressure to change waste into usable compounds.

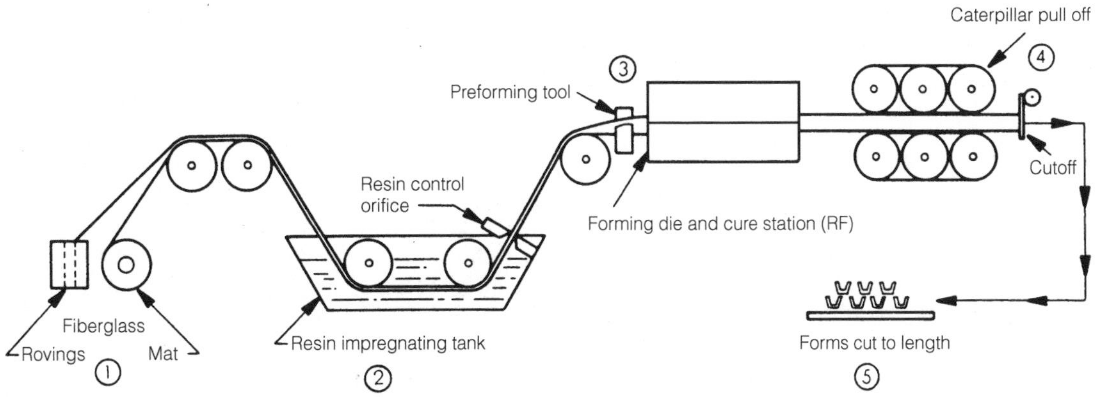

pultrusion process.
(Courtesy Society of Manufacturing Engineers.)

quality control A technique that includes the operation of inspection to ensure that a product is being manufactured to specifications; a management tool for achieving quality.

quartz fiber Fibers produced from high-purity, natural-quartz crystals that are formed into rods from which very small diameter (1/5 diameter of human hair) fibers are drawn.

□ Up to 240 filaments are combined to form a flexible, high-strength fiber that can be made (for reinforced plastics) into yarn and then woven into fabric.

radiation-resistant plastic Ionizing radiation can greatly change the molecular structure and macroscopic properties of plastics such as elasticity, light weight, and formability.

□ These plastics are as radiation-resistant as possible; they are used in space vehicles and nuclear power plants.

radical A group of atoms of different elements that behave as a single atom in chemical reactions.

ram The rod or plunger of an extrusion system that pushes or forces molten resin from an accumulator or feed area through a barrel and into the mold of an injection-molding machine or an extruder. *(See figure.)*

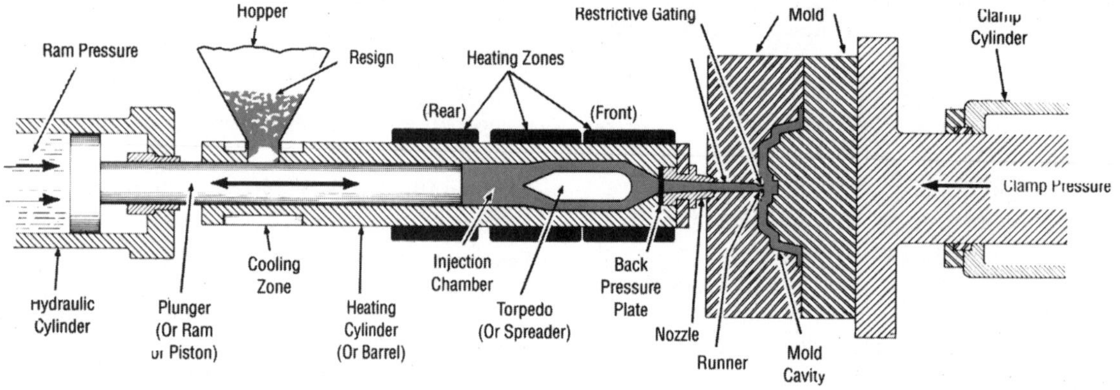

ram. *(Courtesy Equistar Chemicals, LP.)*

reaction injection molding (RIM) A molding process in which two or more liquid polymers are mixed by impingement atomizing in a mixing chamber and then injected into a closed mold.

reactive polymer A reactive polymer is used to alloy different materials by changing their molecular structure inside a compounding machine.

□ They use a reactive agent or compatibilizer to cause a molecular change in one or more of the blend's components and as a result assist bonding.

reciprocating screw The screw inside the extruder barrel that moves backward as the melt is produced in front of the tip.

□ When the desired melt volume is reached, the screw is pushed forward, ejecting the melt into a parison die head or into an injection blow mold.

recycling Collection and reprocessing of waste plastic materials.

reinforced plastics (RP) Plastics with strength increased by adding reinforcing materials such as filler and reinforcing fibers, fabrics, or mats to the base resin. *(See figure.)*

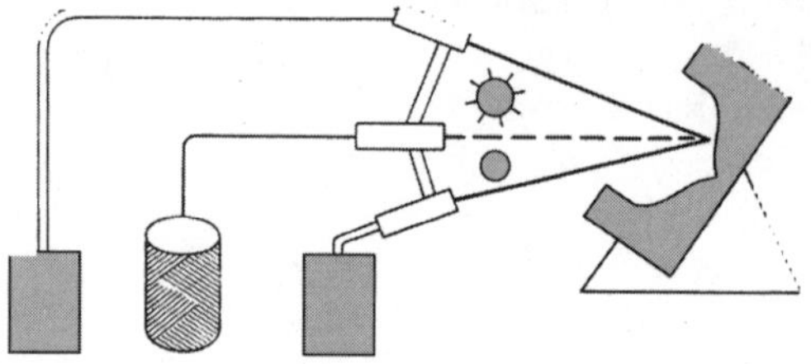

reinforced plastic. *(Courtesy Owens-Corning Fiberglass Corp.)*

□ It is possible to produce RP products whose mechanical properties in any direction will be predictable and controllable by selecting the proper plastic and reinforcement.

relative density The density of any material divided by water density usually at 68 or 73°F (20 or 23°C). Since water density is almost 1.00 g/cm^3, the density in grams per cubic centimeter and the relative density are numerically equal.

release agent Also called parting agent, adherent, dusting agent, lubricant dusting agent, and mold release agent depending on material type and/or process involved.

□ They are lubricating liquids or solids (including dusting powder) that are applied to a mold to prevent plastic from sticking during molding or casting.

□ Common types include calcium stearate, clay, flour, long-chain alkyl quaternary ammonium compounds, mica, silicones, sodium dioctylsulfosuccinate, talc, waxes, and zinc stearate.

resin A gumlike solid or semisolid substance that may be obtained from certain plants and trees or made from synthetic materials; an alternative term for plastic.

resin transfer molding (RTM) The transfer of catalyzed resin into an enclosed mold where fiber reinforcing material has been placed. Also called *resin injection molding* and *liquid resin molding* (LRM).

resole plastic A plastic made by adding formaldehyde to phenol in the presence of a basic acid to give a linear phenolic plastic produced by alkaline condensate of phenol and formaldehyde. It is also called a *one-step* or *one-stage plastic.*

retractable cores These devices are used when molding parts that have cavities not perpendicular to the direction in which the part is ejected from the mold.

□ The cores are automatically pulled from the mold prior to the mold opening and reinserted when the mold closes again and prior to injection.

rib A reinforcing member that adds strength and stiffness to a fabricated or molded part.

□ Ribs (ridges or raised sections) provide lateral, longitudinal, or horizontal support and reduce the bulk weight of parts that require high strength and rigidity.

rotary molding A term sometimes used for a type of injection, transfer, compression, blow molding, or other operation that uses a mold cavity mounted on a rotating table or dial. *(See figure.)*

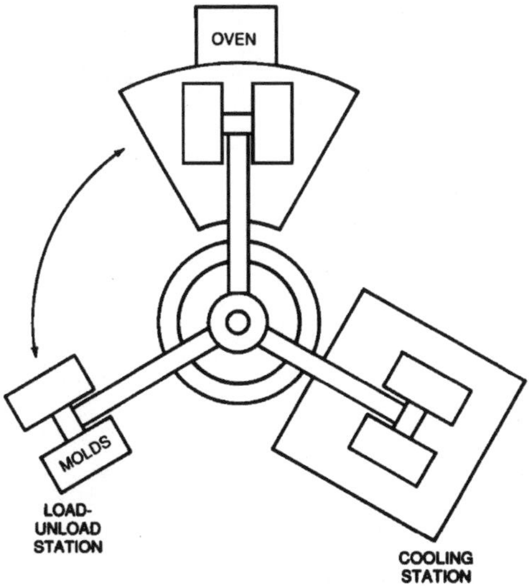

rotary molding. *(Courtesy Equistar Chemicals, LP.)*

rotational casting A method used to make hollow objects from plastisols or powders by charging the mold and rotating it in one or more planes.

□ The hot mold fuses the material into a gel during rotation and covers all surfaces. The mold is then chilled and the product removed.

rubber, natural This latex, a milklike serum, is produced by the tropical tree *Hevea brasiliensis* and contains about 35 percent natural-rubber hydrocarbon, and about 5 percent nonrubber components consisting of protein, lipids, salt, and sugar.

□ Most latexes are thickened by the addition of acetic or formic acid to produce solid natural rubber.

runner The channel that connects the sprue with the gate for transferring melt to the mold cavities. *(See figure.)*

safety factor The ultimate strength ratio of the material to the allowable working stress.

sandwich A class of laminar composites of a lightweight core material (honeycomb, foamed plastics, etc.) to which two thin, dense, high-strength faces or skins are adhered.

sandwich panel A panel consisting of two thin face sheets bonded to a thick, lightweight honeycomb or foam core.

saturated compounds Organic compounds that do not contain double or triple bonds and therefore compounds or elements cannot be added.

scrap plastics Waste plastics that can be reprocessed into acceptable plastics products.

screw A hardened-steel shaft having a rotating helical form inside a plasticizing barrel to advance plastic into an extrusion die or injection mold. *(See figure on next page.)*

screw travel The distance the screw travels forward in filling the mold cavity.

secondary bonds Forces of attraction, other than primary bonds, that cause many molecules to join.

secondary color A color produced by mixing two or more primary colors.

self-extinguishing A term used to describe the ability of a material to stop burning when the flame has been removed.

shell A region about the nucleus of an atom in which electrons move; each electron shell corresponds to a definite energy level.

shrinkage The dimensional differences or the reduction in size between a molded part and the actual mold cavity due to the cooling of plastic from elevated temperatures.

□ The factors that can affect shrinkage are the type of resin, melt temperature, mold design, part thickness, and temperature of the part when removed from the mold.

shrink fit An assembly method in which an insert is put into a heated plastic part and when the plastic cools, it shrinks around the insert.

□ Shrink fitting takes advantage of plastics expanding when heated and shrinking on cooling.

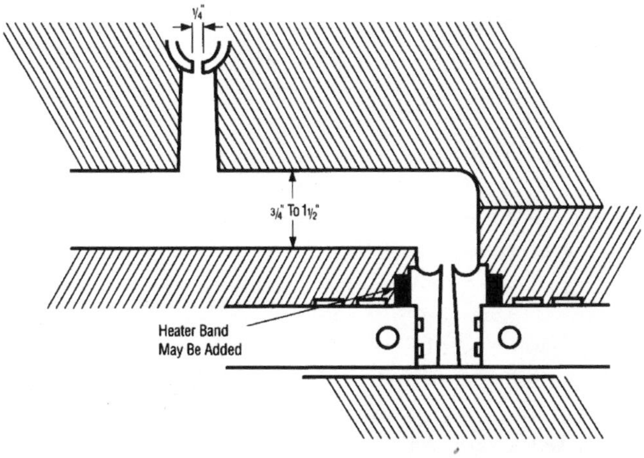

runner system. *(Courtesy Equistar Chemicals, LP.)*

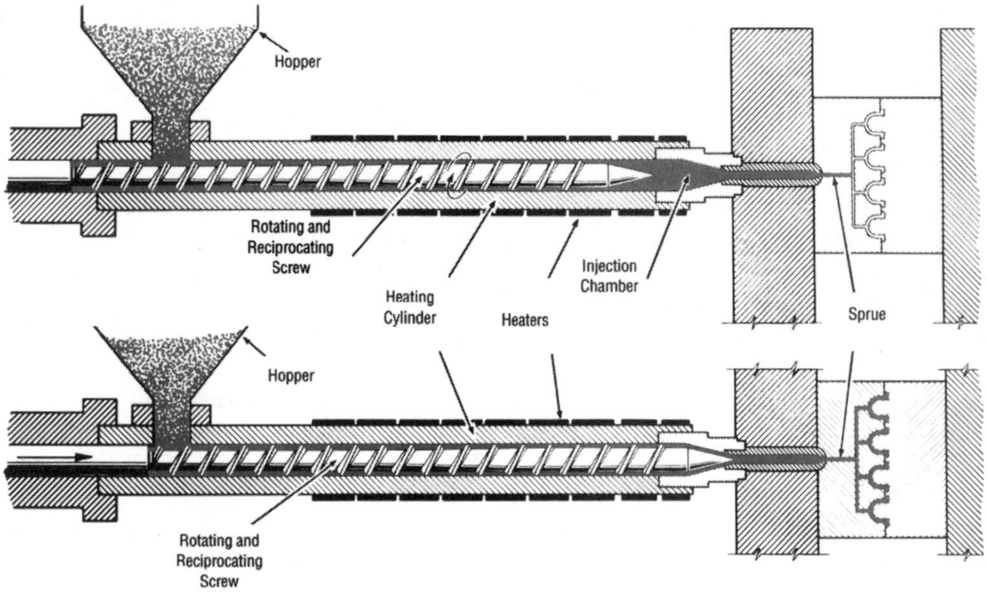

screw injection molding. *(Courtesy Equistar Chemicals, LP.)*

shrink wrapping A process of packaging where the strains in a plastics film are released by raising the temperature of the film, causing it to shrink over the package.

□ Shrink characteristics are built into the film during manufacture by stretching it under controlled temperatures to orient its molecules.

□ Shrink film has excellent clarity and provides protection to the products packaged.

silicone elastomer A high-molecular-weight polyorganosiloxane rubber, also called *silicone rubber,* used where it is important to retain the properties at high and low temperatures.

□ Temperature ranges from −70 to +600°F (−51 to 316°C) have very little effect on its properties.

□ Silicone elastomers have excellent electrical properties under extreme temperature and moisture, but have poor abrasion resistance.

silicone plastic This plastic is based on main polymer chains that consist of alternating silicone and oxygen atoms. It is similar to silicone elastomers but has better mechanical properties.

sink mark A shallow depression or dimple on the surface of a finished part caused by shrinkage or low fill of the cavity.

skiving A process of shaving off a thin layer from a large block of solid plastic to produce film that cannot be made by extrusion, calendering, or casting.

□ Continuous film (or sheeting) is obtained by skiving in a lathe-type cutting operation.

slush casting Pouring liquid or powdered resin into a hot mold which forms a viscous skin. The excess material is drained off, the mold is cooled, and the casting is removed.

snap-back forming A process in which a plastic sheet is stretched to a bubble shape by vacuum or air pressure, a male mold is inserted into a bubble, and the vacuum or air pressure is released, allowing the plastics to snap back over the mold.

soluble-core molding The soluble-core technology (SCT), also called *lost-wax* or *lost-core molding,* is similar to the lost-wax molding used since ancient times. *(See figure.)*

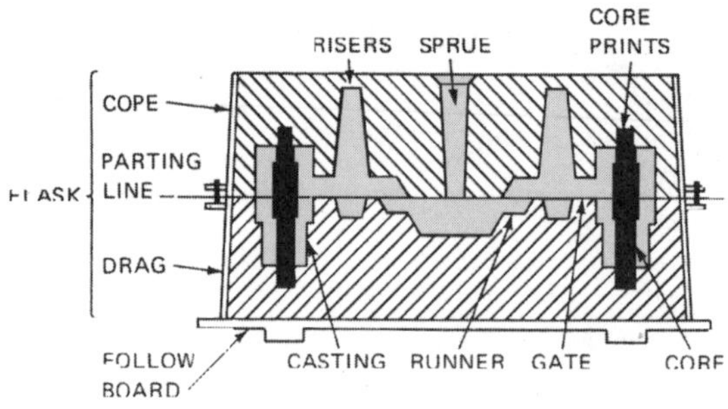

soluble-core molding.
(Courtesy The Society of the Plastics Industries, Inc.)

□ A core is usually molded out of water-soluble thermoplastic, wax formation, low-melting-point alloy (including zinc alloy and tin-bismuth), and similar.

□ The core is inserted into a mold, like an injection mold, and the plastic is injected around the core.

□ When the part has solidified, the core is removed by heating at a temperature below the melting point of the plastic.

solution polymerization A technique in which a chemically inert medium is used to cause monomer solutions to polymerize.

solvent A substance, usually water, in which other substances can be dissolved.

solvent resistance The ability of plastic material to withstand exposure to a solvent.

spherulite A rounded aggregate of radiating crystals with fibrous appearance that is present in most crystalline plastics ranging in diameter from a few tenths of a micrometer to several millimeters.

spin bonding or welding A process in which two objects are fused together while one or both are spinning, until frictional heat melts the interface. The operation is then stopped and pressure held until the parts are frozen together.

spinneret A type of extrusion die containing many tiny holes where plastic melt is forced through to make fine fibers and filaments.

sprayup A term that covers processes in which a spray gun is used to apply some plastic material to the surface of a part.

□ In reinforced plastics, it is the simultaneous spraying of resin and chopped reinforcing fibers onto the mold or mandrel.

sprue The feed opening or channel through which the plastic melt can flow into the mold cavity.

stabilizer A material or ingredient that is used in forming some plastics to slow or retard degradation, generally caused by heat or ultraviolet radiation.

□ They assist in holding the physical and chemical properties of the materials at their initial values throughout the processing and service life of the material.

stereoisomer An isomer in which atoms are joined in the same order but are in different arrangements; side atoms or side charges are arranged on the same side of a double bond present in a chain of atoms.

stiffness The ability of a material to resist a bending force.

strand An untwisted bundle of filaments or an assembly of continuous filaments used as a unit, including ends, slivers, tows, yarn, and similar.

stress The force that produces deformation of a substance. It is expressed as the ratio of applied load to the original cross-sectional area.

stripper plate A plate that operates when a mold is opened to strip a molded piece from core pins or force plugs.

styrene-acrylonitrile (SAN) copolymer An amorphous, transparent plastic that is prepared by emulsion, suspension, or bulk-polymerization processes. Its properties are determined by the amount of acrylonitrile (AN), its molecular weight, and its molecular distribution.

□ It differs from standard polystyrene in properties such as toughness, heat resistance under load, chemical resistance, and resistance to stress cracking (crazing).

styrenic plastics The family of plastics that include acrylic styrene-acrylonitrile, acrylonitrile-butadiene- and acrylonitrile-chlorinated polyethylenestyrene, expandable and impact-resistant polystyrene, polyparamethylstyrene, polyalphamethylstyrene, styrene-acrylonitrile, styrene-butadiene, and styrene-maleic anhydride.

substrate A base to which an adhesive or similar substance is applied.

suspension polymerization A process in which liquid monomers are polymerized as liquid droplets suspended in water.

syndiotactic stereoisomer A polymer molecule in which atoms that are not part of the primary structure alternate regularly on opposite sides of the chain.

synergism The use of slip additives or stabilizers in a polymer to improve its stability more than would be expected from the additive effect of each stabilizer alone.

syntactic foam Cellular resins or plastics where preformed bubbles or microspheres of glass, ceramics, plastics, and the like are embedded in the thermoset matrix mass. *(See figure.)*

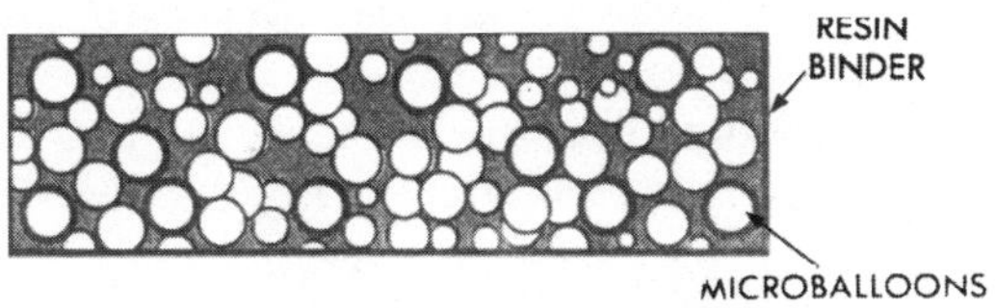

syntactic foam. *(Courtesy Firestone Plastics Co.)*

□ The lightweight materials are used as buoyants, energy absorbents, electrical/electronic waveguides, cores for high-performance sandwich structures, and so on.

synthetic Materials that are produced by chemical, rather than natural, means.

synthetic rubber Synthetic elastomers that come close to one or more properties of natural rubber.

tampoprint A process of transferring ink from an engraved ink-filled surface to a product surface using a flexible printing pad.

telomer A polymer that consists of molecules with terminal groups not capable of reacting with other monomers, under the conditions of the synthesis, to form larger polymer molecules of the same chemical type.

termination phase The last of three steps in addition polymerization that refers to ending molecular growth of polymers by adding chemicals.

terpolymer A copolymer, such as ABS, made from three monomers that can be polymerized simultaneously, or there can be one monomer grafted to the copolymer of two different monomers.

□ ABS is made from acrylonitrile, butadiene, and styrene.

testing Methods or procedures to determine physical, mechanical, chemical, optical, electrical, or other properties of a part.

tex An ISO standard unit used for expressing linear density equal to the mass of weight in grams per kilometer or micrograms per millimeter.

□ One tex is the linear density of a fabric that has a mass of 1 g and a length of 1 km and is equal to 10^{-6} kg/m.

thermal expansion resin transfer molding A variation of the RTM process in which, after the resin is injected, heat causes the cellular core material to expand, forcing reinforcements and matrix against the mold walls.

thermoforming Any process that forms thermoplastic sheet by heating the sheet and forcing it onto the mold surface by vacuum or air pressure, mechanical means, or a combination of both.

□ The ease of forming depends on material thickness, pinholes, its ability to retain heat across the surface and the thickness, and the applied stress.

thermoplastic (TP) Plastic that repeatedly softens when heated and hardens when cooled. Softening temperatures vary with the polymer type and grade.

□ Its qualities include higher impact strength, easier processing, and better adaptability to complex designs.

thermoset or thermosetting A network polymer that undergoes or has undergone a chemical reaction due to heat, catalysts, ultraviolet light, or other factors, leading to a permanent infusible state.

thermoset molding Compression and transfer molding (CM and TM) are the two main methods used to produce low-cost, heat-resistant, and dimensionally accurate molded parts from thermoset (TS) resins. *(See figure.)*

□ In compression molding, material is compressed into the desired shape using a press containing a two-part closed mold, and then it is cured with heat and pressure.

□ CM and TM are classified as high-pressure processes, requiring 2000- to 10,000-psi molding pressures.

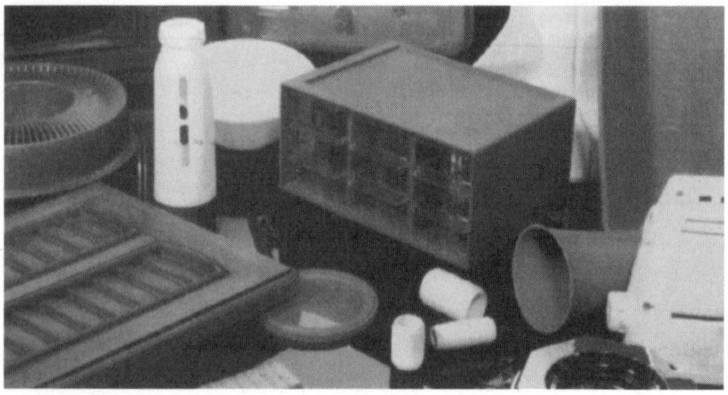

thermoset molding. *(Courtesy Cincinnati Milacron, Inc.)*

thixotropy The state of materials, such as liquids containing suspended solids, that are gel-like at rest but fluid when agitated.

tools Fixtures, molds, dies, or other devices that are used to manufacture parts.

toughness The property of material to withstand shock or impact. It generally denotes a condition between brittleness and softness.

trade name A registered name for a company or product to make it distinctive and easy to recognize, spell, and pronounce. A trade name is used in the plastics industry to identify a particular resin or product.

transfer molding (TM) A method of molding plastics in which the material is softened by heat and pressure in a transfer chamber, then forced by high pressure through the sprues, runners, and gates in a closed mold for final curing. *(See figure.)*

□ This process allows the molding of intricate parts with small holes, numerous metal inserts, and other complex features.

transparency The quality of a material or substance that is capable of a high degree of light transmission and allows objects to be seen clearly through it.

tumbling A finishing operation for small plastic parts where the gates, flash, and fins are removed and the surfaces are polished by rotating the parts in a barrel containing some type of polishing or deburring material.

ultrasonic bonding A joining method that uses vibratory mechanical pressure at ultrasonic frequencies where electrical energy is changed to ultrasonic vibrations using either a magnetostrictive or a piezoelectric transducer.

□ The ultrasonic vibrations produce frictional heat to melt the plastics, allowing them to join.

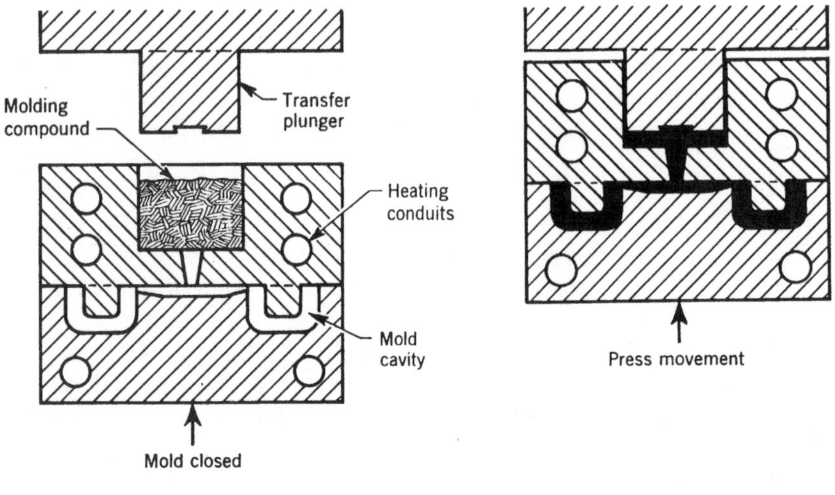

transfer molding. *(Courtesy HPM Division of Koehring Co.)*

undercut An indention that can prevent the removal of a part from a two-piece rigid mold. Flexible materials can be ejected intact with slight undercuts.

urea formaldehyde (UF) plastic A thermoset plastic compound in the amino family of plastics, which is available in a range of colors, from translucent colorless to a lustrous black.

□ They are translucent, giving them a brightness and depth of color that equals or betters opal glass.

□ These noninflammable (self-extinguishing), odorless, and tasteless materials char at about 395°F (200°C); they have good electrical properties, high dielectric strength, and high arc resistance.

U.S. Environmental Protection Agency (EPA) A federal agency with environmental protection regulatory and enforcement authority that administers the Clean Air Act, the Clean Water Act, and other federal environmental laws.

U.S. Food and Drug Administration (FDA) The federal agency that establishes requirements for labeling foods and drugs to protect consumers from misbranded, unwholesome, ineffective, and hazardous products.

vacuum casting This process is used primarily to enclose high-voltage components in a void-free, dielectric compound based on a plastic casting material such as epoxy, polyurethane, silicone, or polyester.

□ This process provides

1. A homogeneous structure of the plastic, resulting in a high-mechanical-strength casting.
2. A blister-free surface that requires only little post-casting treatment.
3. Consistent quality of castings.
4. Reliable protection of the enclosed parts from environmental influences.

vacuum forming A method of sheet forming in which the edges of the plastic sheet are clamped in a stationary frame, and the plastic is heated and drawn into a mold by a vacuum. *(See figure.)*

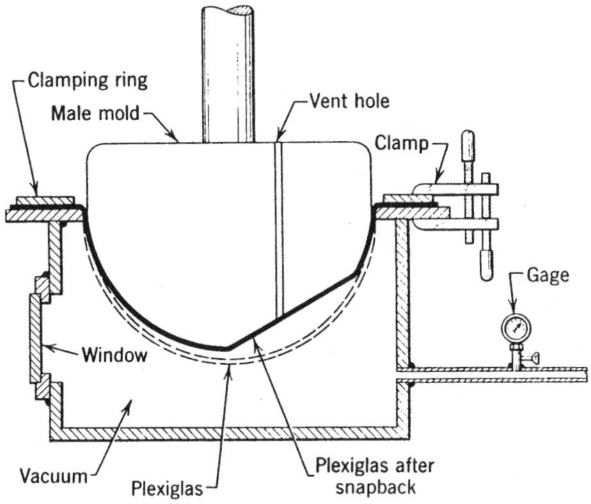

vacuum forming. *(Courtesy Rohm & Haas Co.)*

vacuum metallizing A process in which surfaces are thinly coated by exposing them to a metal vapor under vacuum.

valve gating A gate in which a pin is held in the gate or channel by spring tension, and as the injection stroke moves forward, it compresses the plastic in the runner.

□ When the pressure buildup is sufficient to overcome the spring tension, the pin is then pushed back (pulled) and the fast decompression of the melt fills the cavity at extremely high speed.

vibrational microlamination A casting process in which heated molds are vibrated in a bed of polymer pellets or powder.

Vicat softening point The temperature at which a flat-ended needle of 1 mm^2 circular or square cross section will penetrate a thermoplastic specimen to a depth of 1 mm under a specified load using a uniform rate of temperature rise.

vinyl ester plastic A thermoset plastic, used mainly as the matrix in glass-fiber-reinforced plastics, that is stiff, brittle, and tough.

□ It has exceptional high-strength properties in highly corrosive or chemical environments when compared to other glass-fiber polyesters.

viscosity A measure of internal friction or resistance to the flow of a liquid.

□ The coefficient of viscosity is the value of the tangential force per unit area that is necessary to maintain unit relative velocity between two parallel planes a unit distance apart.

vulcanization A process in which heat is applied to a rubber or elastomer compound that changes its chemical structure (cross-linking) to become less plastic and more resistant to swelling by organic liquids. Elastic properties are improved or increased over a wider range of temperature, but this process cannot be reversed.

warp The lengthwise direction of the weave in cloth or roving; also the dimensional distortion of a plastic object.

waste plastics A plastic resin or product that must be reprocessed or disposed.

water-soluble plastic Any high-molecular-weight plastic that swells or dissolves in water at normal room temperature. The main types are certain ethylene oxides, polyethylene imides, polyvinyl alcohols, and polyvinyl pyrrolidones.

weight In technical practice, the word *weight* should be replaced by gravitational force exerted by a mass on an object, measured in newtons.

wet layup molding A method of making a reinforced plastic product by applying the plastic liquid when the dry reinforcement is put in place before curing by techniques such as bag molding and compression molding. *(See figure on next page.)*

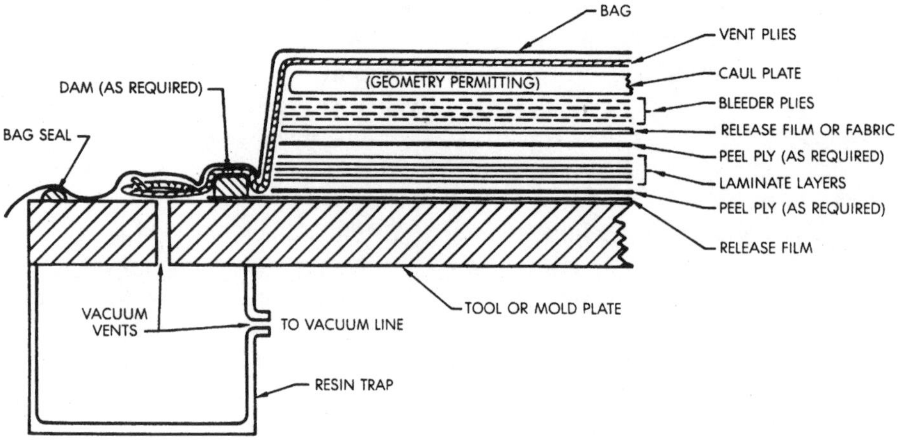

wet layup molding.
(Courtesy The Society of the Plastics Industries, Inc.)

whisker reinforcement Whiskers are metallic and nonmetallic single crystals of ultrahigh strength and modulus and micrometer-size diameters.

□ The extremely high strength is due to their near-perfect crystal structure, chemically pure nature, and fine diameters that minimize defects.

□ They have a higher resistance to fracture (toughness) than do all other types of reinforcing fibers.

□ Because of the high cost of high-performance types, these crystals are used in reinforced plastic for the aerospace and medical (dental) markets.

XT plastic This terpolymer is a member of the family called *impact acrylics* or *multipolymers*. It is naturally pale transparent yellow; however, its production materials have a glass tint, permitting a wide range of transparent or opaque colors.

□ The products are tough and rigid, having several times the notched Izod impact strength of many of the familiar transparent and glassy plastics.

xylylene plastic Used for making pinhole-free coatings of outstanding conformality and thickness uniformity due to the chemistry of the xylylene monomer.

□ The coating process is a vapor-deposition polymerization. On condensation, the monomer spontaneously polymerizes to produce a high-molecular-weight coating.

yarn A bundle of twisted strands.

SECTION 14

Lasers and Robotics

Ken (Kun) Li
Manufacturing Engineer
GEC Precision Corporation
Wellington, Kansas

Steve F. Krar
Consultant
Kelmar Associates
Welland, Ontario

The use of lasers in manufacturing has continually expanded since its development in the early 1960s. Lasers are used for measurements and alignment where accuracies of micrometers (millionth of a meter) are necessary. Other applications include cutting, producing holes, welding, heat treating, and engraving almost any type of material.

Lasers in Manufacturing

carbon dioxide laser A type of gas laser that uses a mixture of CO_2, nitrogen, and helium to produce a continuous output of laser light at a wavelength of 10.6 mm. *(See figure.)*

□ CO_2 lasers are used to produce holes in plastic pipes, baby-bottle nipples, and aerosol can spray tips.

continuous-wave laser A laser that produces light beams continuously rather than as a series of pulses.

□ The continuous mode is preferred for straight and mildly contoured cuts.

excimer laser A type of pulsed-gas laser that emits light in the ultraviolet (UV) region of the spectrum that produces high-quality edges on parts with little or no microcracking or thermal damage. *(See figure on next page.)*

□ Lasing gases are a mixture of a noble gas (argon, krypton, or xenon) and a halogen gas.

□ The entire process takes only a few hundred nanoseconds (one-billionth of a second), and most of the laser energy and the heat created is carried away by the ejected fragments.

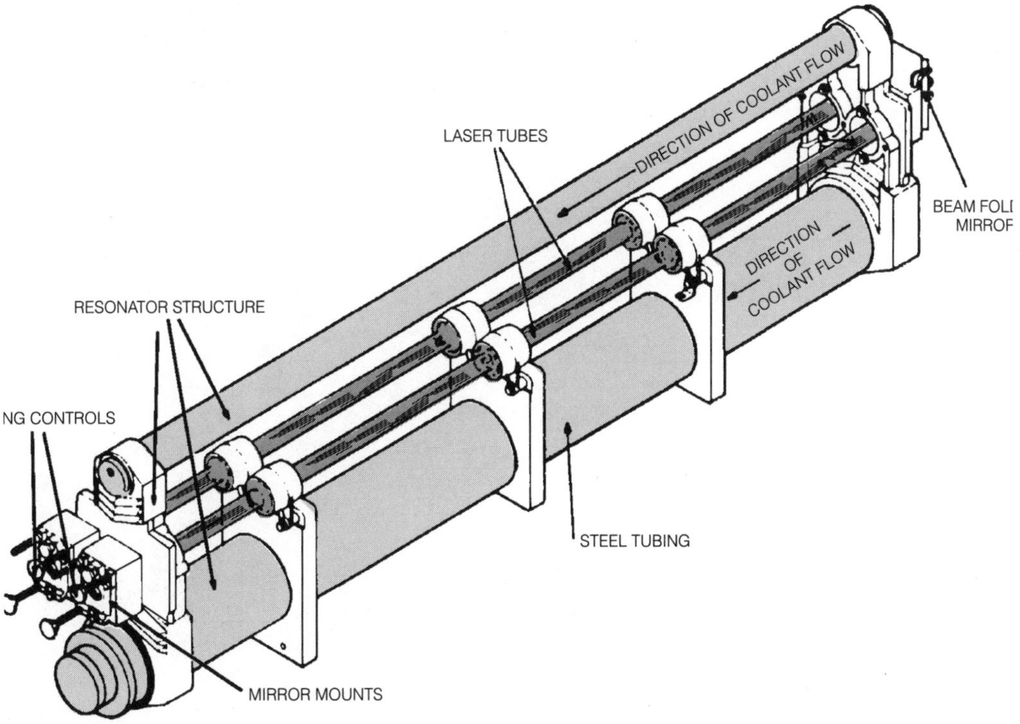

carbon dioxide laser. *(Courtesy Society of Manufacturing Engineers.)*

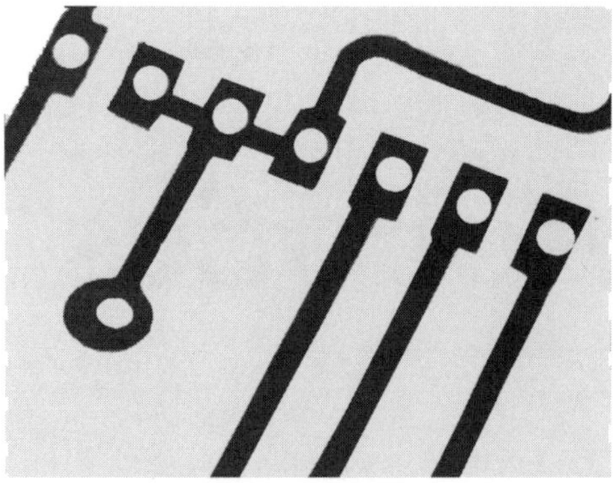

excimer laser cut forms.
(Courtesy Society of Manufacturing Engineers.)

gas laser A laser that uses a gas or gas mixture such as carbon dioxide (CO_2), argon, or other gases as the lasing medium.

- □ Because of the high reflectivity of metals at the CO_2 wavelength, most of the laser energy is reflected and lost; therefore the cutting efficiency is low.
- □ Gas lasers can be used for hole production, perforating, and welding.

helium-neon (HeNe) laser A gas laser that uses a mixture of helium and neon as the lasing medium, generating a bright-red beam.

- □ It is most commonly used in bar-code scanners.

kerf width The width of cut made by a laser or other cutting tool in a workpiece material.

laser *L*ight *a*mplification by *s*timulated *e*mission of *r*adiation, an extremely narrow and coherent monochromatic (one wavelength) beam of electromagnetic energy in the visible light spectrum or a device producing such a light. *(See figure.)*

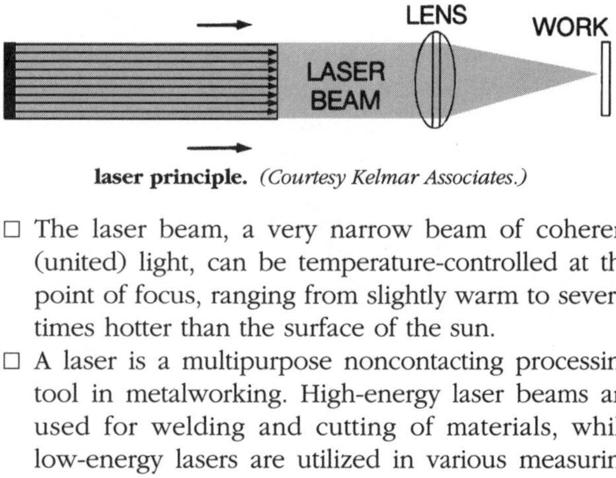

laser principle. *(Courtesy Kelmar Associates.)*

- □ The laser beam, a very narrow beam of coherent (united) light, can be temperature-controlled at the point of focus, ranging from slightly warm to several times hotter than the surface of the sun.
- □ A laser is a multipurpose noncontacting processing tool in metalworking. High-energy laser beams are used for welding and cutting of materials, while low-energy lasers are utilized in various measuring and gaging situations.

laser-assisted machining The use of a laser beam to heat and soften a metal workpiece just ahead of a cutting tool to make the workpiece easier to machine.

laser-beam machining (LBM) A laser is focused on the surface of the workpiece, and the high-density energy melts and evaporates portions of the workpiece in a controlled manner. *(See figure.)*

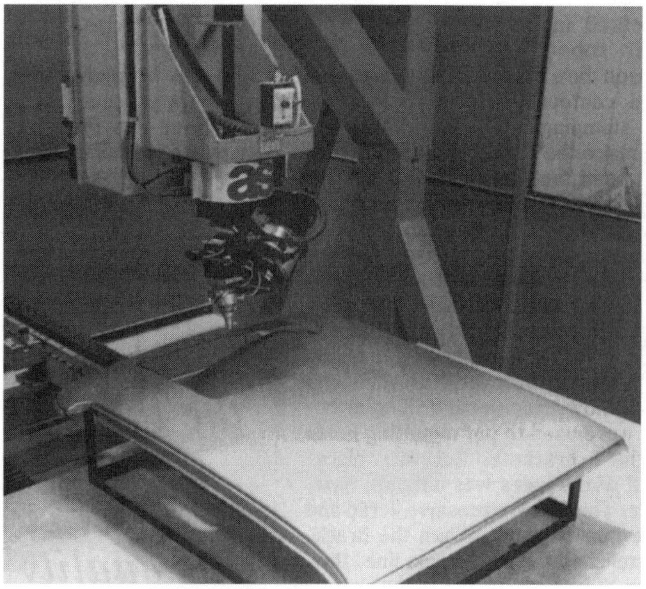

laser-beam machining.
(Courtesy Society of Manufacturing Engineers.)

- □ Both CO_2 and Nd:YAG (*see* NEODYMIUM-DOPED YTTRIUM ALUMINUM GARNET) lasers are used in LBM.
- □ LBM is used to machine a variety of metallic and nonmetallic materials for small-scale cutting operations, such as slitting and drilling holes as small as 0.005 mm with hole depth-to-diameter ratios of 50:1.

laser-beam torch (LBT) A torch that combines a laser beam and a gas stream such oxygen, nitrogen, or argon for cutting thin-sheet materials. Gas streams blow away molten and vaporized material from the workpiece surface.

laser-beam welding (LBW) A welding process with deep penetration power that uses a laser (Nd:YAG laser or CO_2 laser) beam as the source of heat. *(See figure.)*

laser-beam welding. *(Courtesy Robomatix International, Inc.)*

□ It can be directed, shaped, and focused precisely on the workpiece and a very thin column of vaporized metal within a surrounding liquid pool is produced.

□ It is particularly suitable for welding narrow and deep joints on a variety of materials up to 25 mm thick.

□ Welding of transmission components with a high-powered laser is the most widespread application in the automotive industry.

□ LBW produces welds of good quality, with minimum shrinkage and distortion, and with depth-to-width ratios as high as 30:1.

lasercaving A laser application that makes it possible to machine cavities in work that is too hard to machine, or cannot be machined by the EDM process because the material is nonconductive. *(See figure.)*

laser hardening A laser high-speed transformation hardening technique for the hardening of carbon steels with over 0.2 percent carbon content. It allows selected areas of a part to be hardened quickly and reduces or eliminates workpiece distortion.

□ The process induces residual compressive stresses into the part, allows for the high-speed hardening of various part thicknesses, and can be automated.

laser interferometer A laser-light measuring device that measures changes in position (alignment) by means of lightwave interference. It is used for very precise linear measurement and alignment in the production of large machines, to calibrate precision machines and measuring devices, and to check machine setups. *(See figure on next page.)*

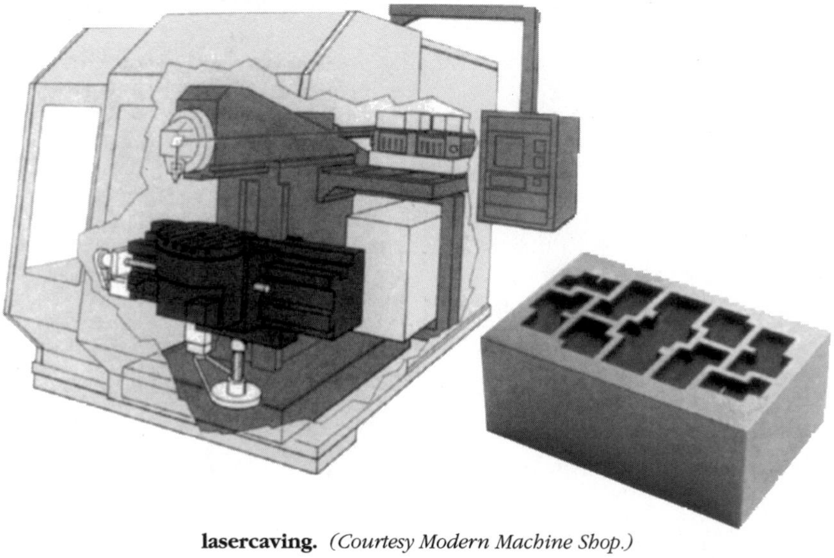

lasercaving. *(Courtesy Modern Machine Shop.)*

□ The control of the beam is similar to the milling process where the laser tool is run back and forth through the area to be machined.

□ It is very effective in materials such as ceramics, composites, carbides, quartz, glass, and titanium-based alloys.

laser coding A marking system that is used in industry to identify parts, control inventory, avoid errors, and identify defects in the manufacturing process. *(See figure.)*

laser energy modes Depending on the laser medium, only pulsating and continuous energy modes are possible.

□ The *pulsating mode* is possible with ruby and neodymium gas lasers. Giant pulses at the rate of 1000 to 10,000 pulses per second produce holes or lines on parts.

□ The *continuous mode*, produced by YAG or CO_2 lasers, emits photons in a continuous beam that is ideal for cutting and welding operations.

laser machining center/machine tool A numerically controlled machining center/machine tool with the laser as a processing tool to perform a variety of processing tasks.

laser coding. *(Courtesy Society of Manufacturing Engineers.)*

laser interferometer. *(Courtesy Cincinnati Milacron, Inc.)*

LaserMike An optical micrometer that uses a helium-neon laser beam to provide instant readouts, even on moving parts, to a high degree of accuracy. *(See figure.)*

□ *See* SCANNING LASER DEVICE for operating principle.

laser probes Laser-scanning devices used on machine tools and coordinate measuring machines to improve the machine's accuracy and to be able to scan contoured surfaces. *(See figure.)*

laser spot welding A spot welding process that uses a fine-focused laser beam to scan the area of the weld and produce a small hole through the molten puddle. As the beam is removed, molten metal flows into the hole and solidifies, forming a fusion-type nugget.

□ It is a noncontact process and produces no indentations.

lasing The process of generating a laser beam through stimulated emission of radiation.

latent heat The thermal energy absorbed or released when a substance undergoes a phase change.

liquid laser A laser that uses an organic dye in a solvent as the lasing medium. *(See figure.)*

□ Liquid lasers are not used for metalworking applications.

neodymium-doped yttrium aluminum garnet (Nd:YAG) laser A solid-state laser that uses a crystal of yttrium aluminum garnet (YAG) doped with neodymium (Nd) as the lasing medium.

nondestructive testing A process, commonly called *holographic interferometry,* that uses a beam of laser light to produce three-dimensional images to detect strains and cracks that are invisible to the unaided eye.

□ Widely used to inspect machine tools, tires, computer components, and automobile and aircraft parts for defects.

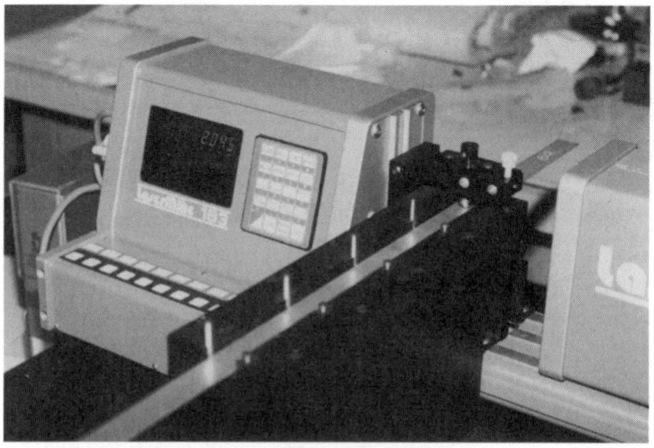

LaserMike. *(Courtesy LaserMike, Inc.)*

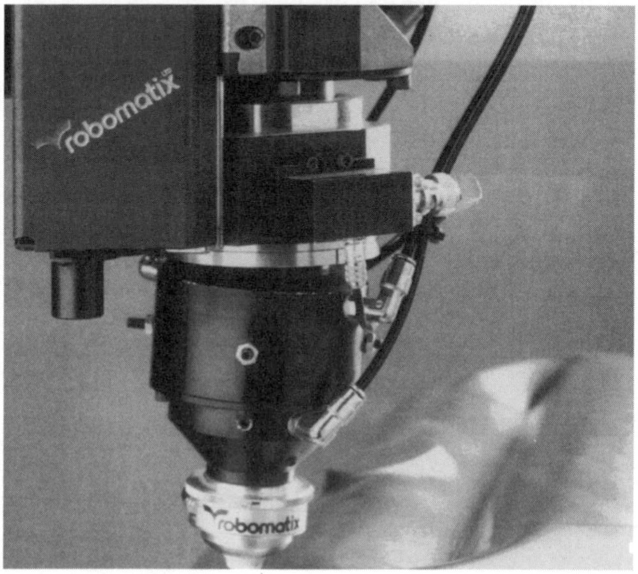

laser probe. *(Courtesy Robomatix International, Inc.)*

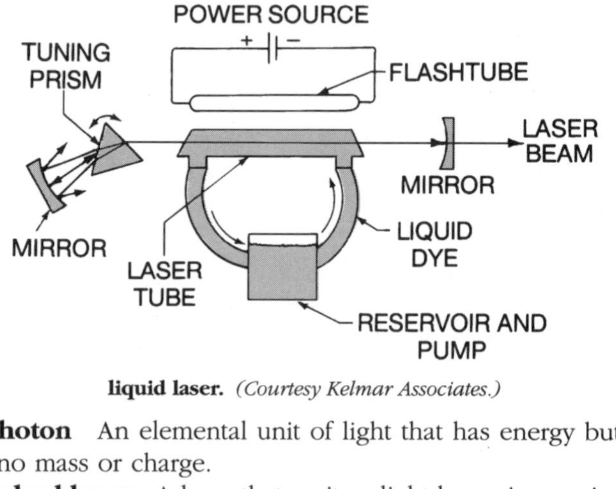

liquid laser. *(Courtesy Kelmar Associates.)*

photon An elemental unit of light that has energy but no mass or charge.

pulsed laser A laser that emits a light beam in a series of pulses rather than continuously.

□ The pulsed mode is preferred for thin materials because it allows tight corners and intricate details to be cut without excessive burning.

reflectivity The capability of a material to reflect light.

□ Reflectivity of the workpiece surface is an important consideration in laser-beam machining.

ruby laser A solid-state laser that uses a synthetic ruby crystal doped with a chromium impurity as the lasing medium. *(See figure.)*

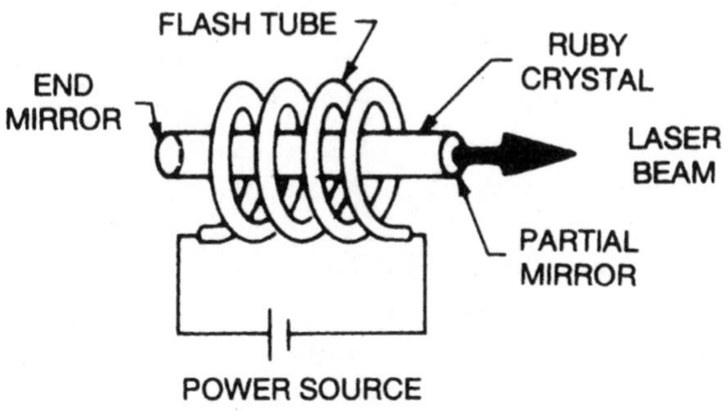

ruby laser. *(Courtesy Kelmar Associates.)*

□ Ruby lasers are used to produce precise holes in diamonds for wire-drawing dies.

□ Ruby lasers (crystalline aluminum oxide or sapphire) are restricted in their applications because of their relatively short wavelength (0.69 μm).

scanning laser device A measuring device that uses a coherent beam of light which can be projected with minimum diffusion in high-production online/post-process inspection and gaging. *(See figure.)* Applications of the scanning laser technique include rolling mill operations, wire extrusion, and some machining and grinding processes.

□ A laser beam is deflected by a rotating mirror to produce a beam of light that can be focused to sweep past an object.

□ A photodetector on the far side of the object senses the light beam, except for the period during the sweep, when it is interrupted by the object. This period can be timed with great accuracy and related to the size of the object in the path of the laser beam.

□ A microprocessor-based system counts the time interruption of the scanning-laser beam as it sweeps past the object, and makes the conversion from time to a linear dimension.

solid-state laser A laser that uses a crystal or glass as a host to an impurity, such as neodymium or chromium, which produces the lasing action.

□ Solid-state lasers have proved very useful in the fields of surgery, atomic fusion, drilling diamond dies, measurement, and spot welding.

specific heat The energy required to raise the temperature of a unit mass of material by one degree.

thermal conductivity The ease with which heat flows within and through the material.

yttrium aluminum garnet (YAG) lasers These lasers are used in manufacturing, and their power ranges from a few milliwatts to more than 400 W (100 W is most common).

□ Low-power models are used for the manufacture of integrated circuit-mounted resistors.

□ High-power YAG lasers are used for cutting, drilling, heat-treating, and welding operations.

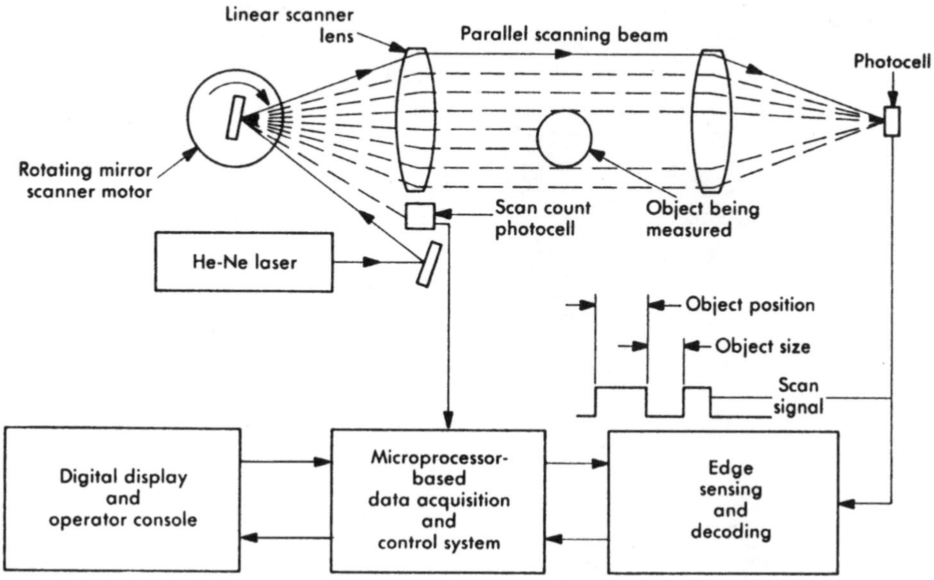

scanning laser device. *(Courtesy LaserMike, Inc.)*

Industrial Robots

Industrial robots can manipulate parts or tools automatically through a sequence of operations, or motions, as directed by the computer program. Robots are generally used for repetitive operations and to perform tasks that are monotonous, physically difficult, or environmentally unpleasant.

accuracy Proper positioning, as measured by the difference between the position a robot is commanded to reach and the position it actually reaches.

actuator A motor or other device for changing electrical, hydraulic, pneumatic, or other energy, into motion for a robot.

alpha level intelligence *See* BUILT-IN INTELLIGENCE.

American Standard Code for Information Interchange (ASCII) A 7-bit code used to represent 128 characters for communications control of computers.

AML A high-level programming language developed by IBM for robot applications.

arm *See* MANIPULATOR.

arm language (AL) A research language developed by Stanford University that coordinates a robot's vision and hand movements.

articulate robot *See* JOINTED-ARM ROBOT.

artificial intelligence The operational basis for a machine, computer, or other sensory mechanism to perform some function that is normally associated with human intelligence. *(See figure.)*

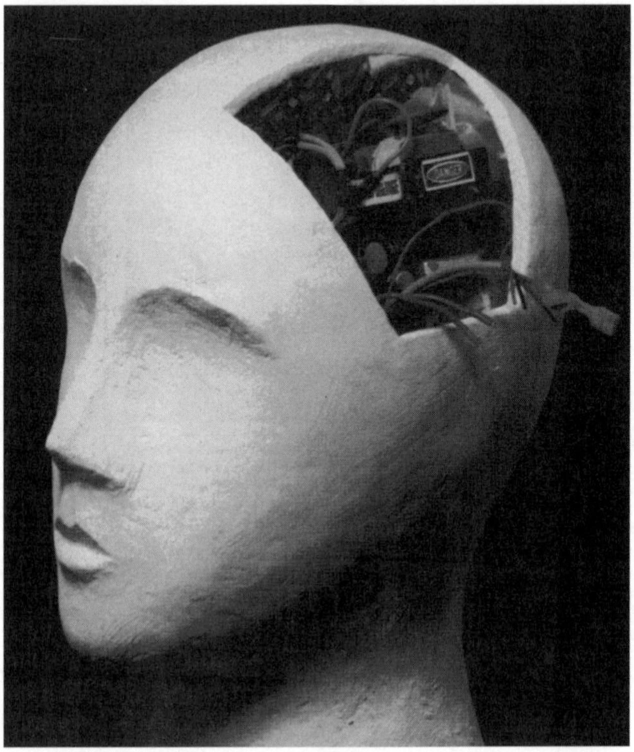

artificial intelligence. *(Courtesy Society of Manufacturing Engineers.)*

ASCII *See* AMERICAN STANDARD CODE FOR INFORMATION INTERCHANGE.

assembly language A medium-level computer language that acts as an intermediary between a human language (e.g., English) and a machine language.

automated factory A factory that is capable of producing goods with a minimum of human labor.

automated guided vehicle A computer-controlled vehicle that is guided to perform operations without human assistance. *(See figure on next page.)*

automated storage-and-retrieval system A computer-controlled system used for storing and retrieving parts from a warehouse as they are required in manufacturing, through the use of automated guided vehicles. *(See figure on next page.)*

automatic mode The automatic operation of a robot through the programmed or taught cycle.

automation The use of machines to replace human labor in performing repetitive, environmentally unpleasant, or dangerous operations.

axis A degree of freedom or basic motion around a point, along a straight line, or through an angle of rotation.

bang-bang robot A pneumatic-powered pick-and-place robot which makes noise when it contacts any mechanical stop that determines its extreme limits.

base The rigid foundation for a stationary robot which supports all other parts.

basic robot design The five basic types of industrial robots are *(See figure on next page.)*

- ☐ *Overhead, or gantry robot.* This robot moves on a crane or bridge-type support; the arm can have a number of axial movements.
- ☐ *Rectilinear, or cartesian robot.* This robot moves in straight lines, the *X, Y,* and *Z* axes.
- ☐ *Polar, or spherical envelope robot.* This robot rotates about a perpendicular axis; its arm is capable of moving in and out and through an up-and-down arc.
- ☐ *Anthropomorphic, or articulated robot.* This robot has movements such as the human arm, with rotating shoulder, bending elbow, and rotating wrist.
- ☐ *Cylindrical robot.* This is similar to the cartesian model except that it rotates about a stationary base instead of moving from side to side.

beta-level intelligence Intelligence that can learn from its experiences, or build on previously learned knowledge.

automated guided vehicle. *(Courtesy Giddings & Lewis Inc.)*

automated storage-and-retrieval system. *(Courtesy Giddings & Lewis Inc.)*

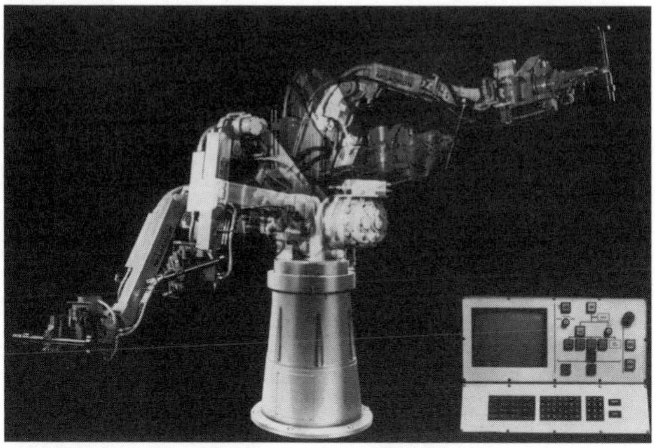

basic robot. *(Courtesy Cincinnati Milacron, Inc.)*

built-in intelligence Nonlearning and probably nonlearned intelligence, of the type placed in the read-only memory of a computer.

CAM *See* COMPUTER-AIDED MANUFACTURING.

CAR *See* COMPUTER-AIDED ROBOTICS.

cartesian-coordinates robot A robot whose arm moves along all three axes only in straight lines. This type of robot has a rectangular work cell, accounting for its other name, *rectangular-coordinates robot.*

cell (robotic) The work area, work cell, or work envelope of a robot, which includes all the space that the robot can reach. *(See figure on next page.)*

center of gravity The point at which a robot's, or any object's, mass is centered.

closed-loop servo-controlled system A control system which compares the system output to the system input, and can make corrections for any variations that exist.

coefficient of friction A measure of how well or firmly a gripper holds a part.

compiler A program that is capable of taking statements written in a high-level computer language and changing them into the machine language used by the computer or robot.

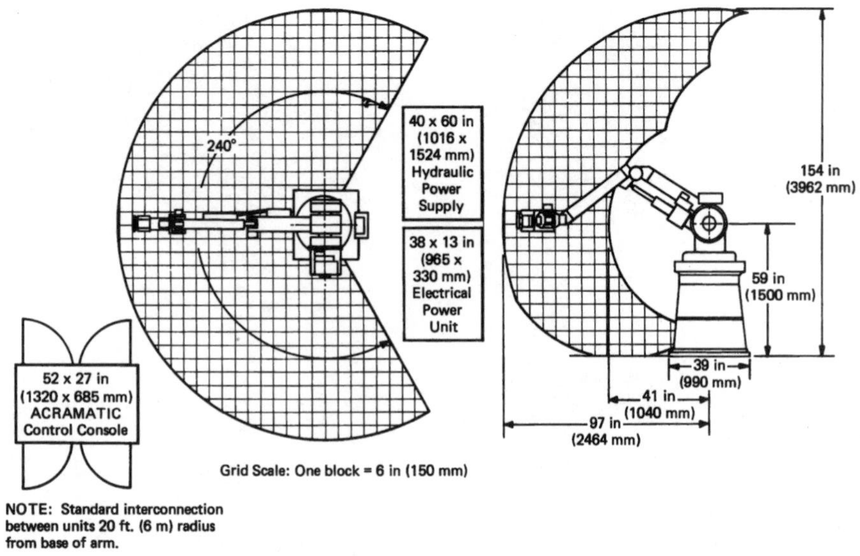

cell (robotic). *(Courtesy Cincinnati Milacron, Inc.)*

compliance surface A padded or soft surface that allows for minor mispositioning of a robot's gripper.

computer-aided manufacturing (CAM) A method that uses computers at various stages of manufacturing to improve quality and lower costs.

computer-aided robotics (CAR) A method using simulations of robots and robot systems to design robots, train operators, and program robots off line.

controlled-path motion The ability of the system to provide coordinated control of all axes during the teaching and automatic operations.

controller The brain of the robot.

cylindrical-coordinates robot A robot whose cell is cylindrical and allows the robot's arm to rotate about its base and move in horizontal and vertical planes.

data highways A local area network (LAN) for exchanging electronic information, or data, with computers or other devices.

degrees of freedom The number of joints or axes of motion that an industrial robot possesses. Most robots move along the three basic cartesian axes (*X, Y, Z*), although some models may move in six or more axes. *(See figure.)*

educational robot A robot which can be used to teach the principles of robotics.

effectuating device An electronic or mechanical device, such as a hand, gripper, toolholder, magnet, spray gun, drill, welding head, or other item, that is used for holding parts and/or performing some function.

elbow The first joint out from the shoulder of a robot or human arm.

ELIZA A program prepared by Joseph Weizenbaun to meet Turing's definition of artificial intelligence; it uses alpha-level intelligence.

end effector A device, tool, gripper, or hand, located at the end of a manipulator or arm, that is used to perform work or movement on a part. *(See figure on next page.)*

end-of-arm tooling Any tool or device mounted to a robot manipulator. *See* EFFECTUATING DEVICE.

external sensor A device that supplies a robot with information about its surroundings.

feedback Sensory information that is taken back to a comparison source to see how well a task is being done.

finger The jaws or fingers of a gripper that is used to handle a tool or part.

first-generation industrial robot A robot produced during the design period when robots were of a fixed-sequence type.

flexible automation The use of reprogramnable, general-purpose machines such as the robot to introduce changes in the manufacturing process.

four Ds of robotic applications Jobs which consist mainly of work that is too dirty, dangerous, dull, and difficult for humans and is best suited for robots.

free-standing robot A robot that is suitable for working with any machine or workstation.

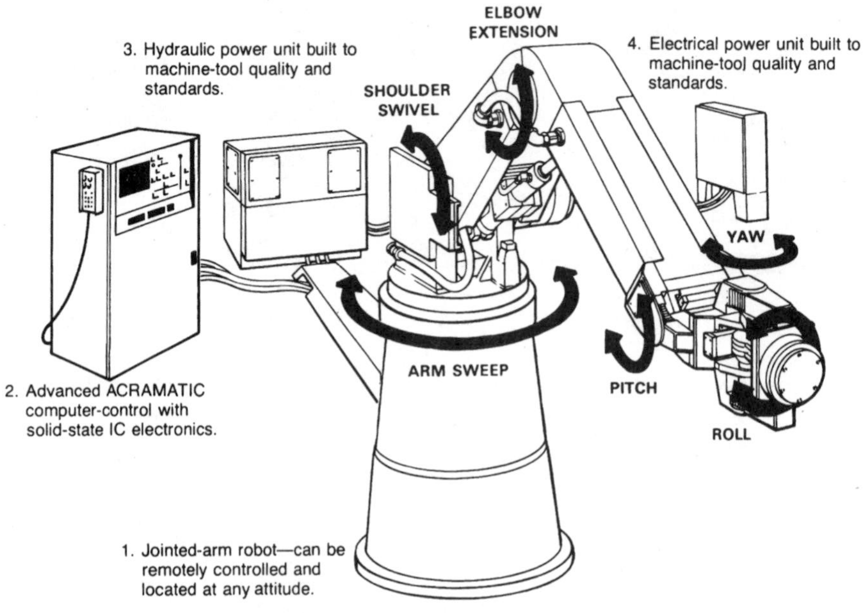

degrees of freedom. *(Courtesy Cincinnati Milacron, Inc.)*

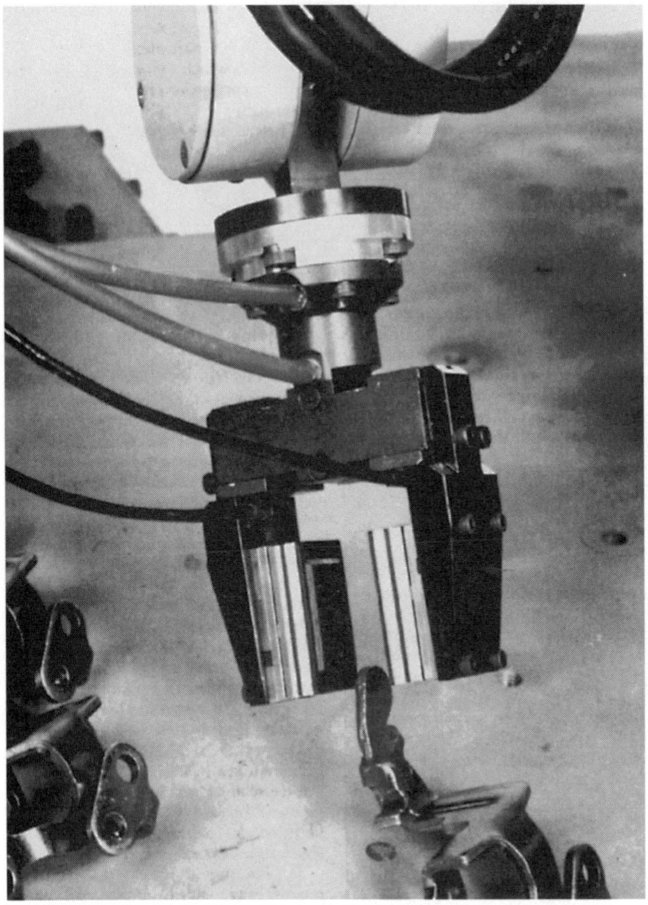

end effector. *(Courtesy Society of Manufacturing Engineers.)*

gantry robot A robot that is positioned by an overhead supporting structure such as a crane or bridge-type support.

generation of robots A group of robots classified together because they were produced during a particular design period.

gripper A specialized attachment fitted to the end of a robot arm for handling parts, tools, materials, or special instruments.

hard automation The use of specialized tools and machinery (not robots) to perform specific operations efficiently.

HELP A high-level programming language developed for use with General Electric's Allegro assembly robots.

high-technology robot A continuous-path control unit robot.

HOME A reference point from which all the robot's movements are measured.

hydraulic power Power produced through the use of a noncompressible fluid such as hydraulic oil.

industrial robot A robot designed especially for industrial use such as materials handling, tool changing, assembly, welding *(See figure)*, and measuring. □ An industrial robot that can be programmed or taught by using a digitizing system which translates movements into commands for the robot to understand.

input device A teaching pendant, control panel, or sensor that can gather or enter information the robot requires to perform its duties.

input signal Information or programmed instructions that are received by a robot to perform some task.

intelligence The ability to learn from experience and to adapt to a changing environment.

internal sensor A device that furnishes a robot with information about itself.

jointed-arm robot A robot whose manipulator (arm) has a bend or joint in it so that it resembles the human arm. This robot has the largest work cell of any robot type for a given amount of floor space occupied.

laws of robotics Principles created by Isaac Asimov (author of the classic work *I, Robot*) to govern relations between humans and robots.

LERT *See* LINEAR, EXTENSIONAL, ROTATIONAL, AND TWIST.

limit switch A device that tells the robot it is at the end of travel on an axis.

linear, extensional, rotational, and twist (LERT) A robot classification system based on the type of motion produced by each robot axis.

industrial welding robot. *(Courtesy ABB Robotics, Inc.)*

line tracking The ability of a robot to perform operations on parts mounted on a continuously moving conveyor. *(See figure.)*

- □ *Moving-base line tracking* is where the robot is mounted on some form of transport system that moves parallel to the conveyor line and at line speed.
- □ *Stationary-base line tracking* is where the robot is mounted in a fixed position in relation to the conveyor line.

medium-technology robot A point-to-point control robot.

modular robot A robot with interchangeable end effectors or gripping hands that can be changed quickly to suit the size and shape of the part to be held or moved.

1.5-generation industrial robot A robot produced during the design period when robots were in the initial development of sensory-controlled actions.

line tracking control. *(Courtesy Cincinnati Milacron, Inc.)*

low-technology robot A pick-and-place robot that operates on an open-loop system and receives no positional feedback information.

machine language The language of a computer that a robot's control unit can understand to follow programmed instructions. An electronic computer operates on the binary system, which consists of a series of ones and zeros.

manipulator An arm, having hands or gripping devices attached to its end, that produces motion and is used as an object-transferring device.

manual mode A robotic mode that enables an operator to move each axis of the arm independent of the others, through the use of the buttons on the teach pendant.

- □ Its prime purpose is to allow alignment of the six coordinates to the HOME position.

Manufacturing Automation Protocol (MAP) A communications standard developed for General Motors.

Manufacturing Control Language (MCL) A high-level robot application programming language developed by McDonnell-Douglas.

mechanical stop A device that physically stops the motion of an axis of a robot.

operator override A feature of some robot controls that uses the operator as the feedback mechanism for adjusting deviations in the programmed path. At the end of the move, the robot returns to the programmed endpoint, canceling out the operator's deviations.

output device A device for executing, reproducing, or storing input or program information and instructions. The main output device for a robot is the actuator, which opens or closes the gripper.

parallel processing A processing approach in which each of the 16 sensor pads in an array has its own signal amplifier, analog-to-digital convertor, and microprocessor. These signals are sent to another microprocessor which sorts the data and sends it to the host computer for processing.

payback period The length of time it takes a robot to pay for itself through the savings it provides. This period is generally 2 years or less.

payload, workload The maximum load that the robot has been designed to handle safely.

pick-and-place robot A robot that is capable of the simple action of picking up an object from one position and placing it at another position.

pin-lifting device A pneumatic gripper that grips a part by inflating around the outside of it.

pitch One of the wrist's axes of motion that permits the up-and-down movement of the robot gripper. *(See figure.)*

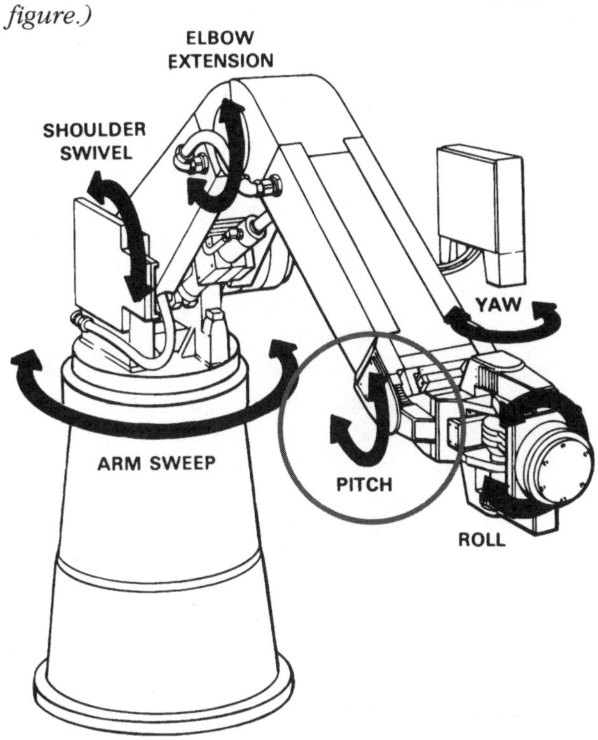

pitch. *(Courtesy Cincinnati Milacron, Inc.)*

pneumatic finger A holding device that grips a part by bending its joint when compressed air is forced into it.

pneumatic power Power provided by a compressible fluid such as air.

polar-coordinates robot A robot whose manipulator or arm can rotate about its head and base, as well as moving in and out. The work cell for the robot takes the shape of the space between two hemispheres.

proximity detector A device which senses the presence of objects without actually touching them.

RAIL A high-level programming language developed by Automatix for use with robots and vision systems.

random-access memory (RAM) A computer-memory storage device in which data can be entered, and from which data can be retrieved in a nonsequential manner.

range finder A device used to measure the distance from it to some other object.

remote-center compliance device A multiaxis float mechanism, built into the hand or gripper of a robot, that can adjust sideways, or up and down, to allow it to adjust for small misalignments.

remote control Any mechanical or electrical device controlled from a distance by mechanical, wire, or radio means.

repeatability The ability of a robot to perform the same operation any number of times to a specified degree of accuracy.

robot A computerized or electromechanical automatic, general-purpose device, whose primary function is to produce motion in order to accomplish some task. *(See figure.)*

industrial robot. *(Courtesy ABB Robotics, Inc.)*

☐ Computerized robots usually consist of three main components: the machinery or mechanical parts, the controller/computer system, and the software.

☐ The six main types of robots are educational, hobbyist, industrial, medical, military, and show.

robotics **1.** The techniques used in designing, building, and using robots. **2.** Computer-aided manufacturing, where robots are used for repetitive manufacturing operations such as assembly, painting, and welding.

Robot Off-line Programming System (ROPS) A coordinated package of software modules that provides simpler, faster, more efficient, and more productive programming.

Robot Programming Language (RPL) A high-level robot programming language developed by SRI.

roll One of a robot wrist's axes of motion that involves the twisting or rotational movement of a hand or gripper. *(See figure.)*

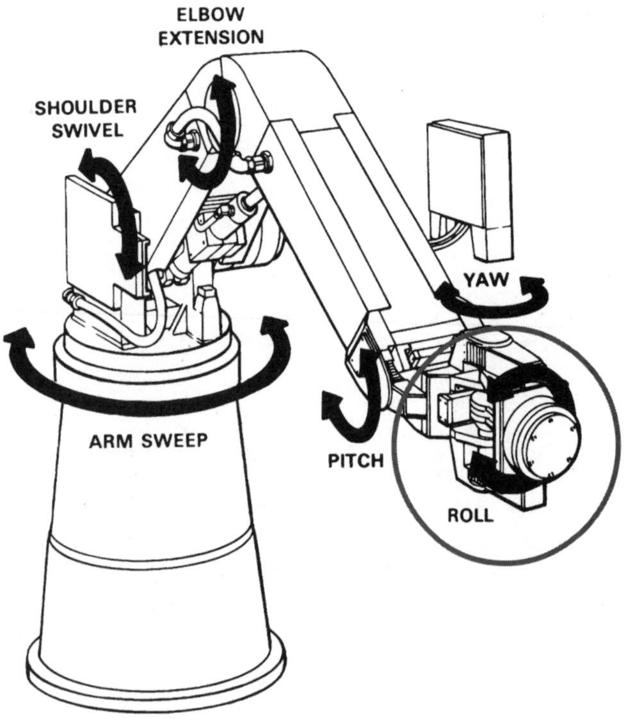

roll. *(Courtesy Cincinnati Milacron, Inc.)*

ROPS *See* ROBOT OFF-LINE PROGRAMMING SYSTEM.

RPL *See* ROBOT PROGRAMMING LANGUAGE.

SCARA *See* SELECTIVE COMPLIANCE ARTICULATED ROBOT ARM.

searching A feature that can be used when objects must be stacked or unstacked. Two points, one below and one above the stack, are programmed, and a sensor on the robot's hand sends a signal to indicate misalignment.

second-generation industrial robot A robot from the design period when robots have hand-to-eye coordination control.

selective compliance articulated robot arm (SCARA) A robot arm shape that looks like a cylindrical robot, but has a linear-reach axis instead of a rotational one.

sensor A device that provides a robot with information about itself or its surroundings.

sensory-input interpretation The process of understanding sensory information.

servo control An industrial robot control system in which sensing devices monitor movement, and report any deviation between commands as issued and movement as monitored. Deviations will automatically cause corrective action to be taken.

shaft encoder A unit that changes rotational movement into pulses or direct degree readings.

shoulder The first joint of the arm, or the point at which the arm is attached to the base of a robot or human. *(See figure.)*

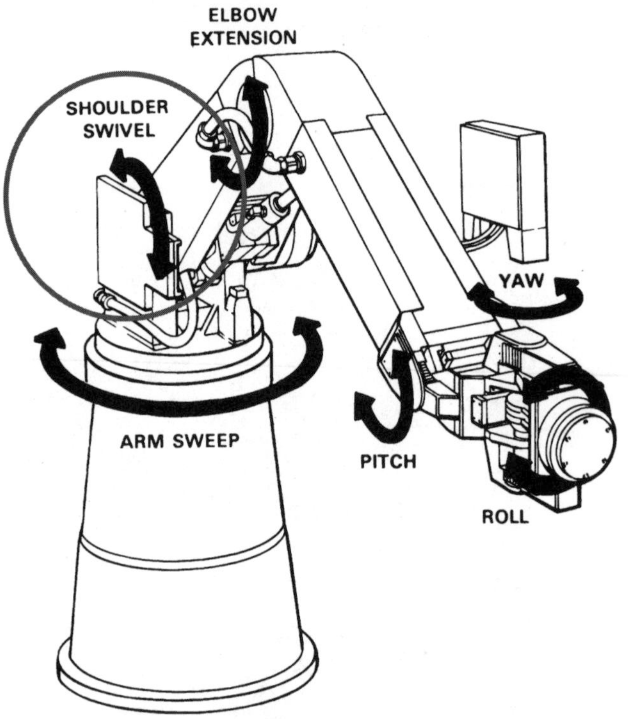

shoulder. *(Courtesy Cincinnati Milacron, Inc.)*

signal-error detection The process of detecting bad information resulting from electrical noise or other environmental causes.

speech recognition The ability of a robot to understand and act on meaningful speech input.

speech synthesis The ability of a robot to produce speech that is understandable to humans.

tactile sensing A method of continuous and variable ability to grope and identify shape, surface features, texture, force, and slippage of objects.

TCP *See* TOOL CENTER POINT.

teach mode The operational mode that allows the operator to teach the robot arm to perform designated tasks.

teaching The process of giving a robot a series of instructions for doing a task.

teaching pendant A device used to position a robot manually at each point in its task, and cause the robot to record these points for future use. *(See figure.)*

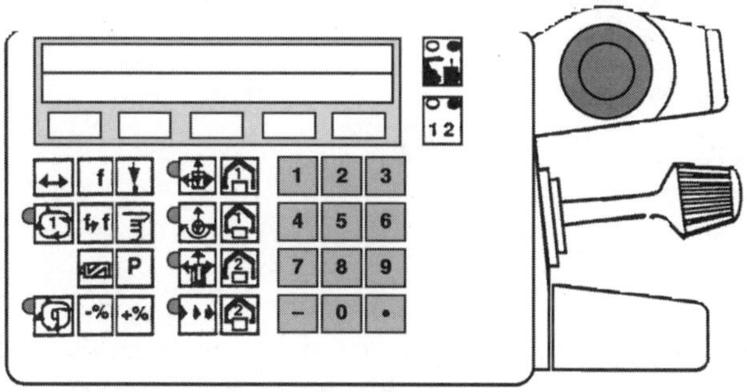

teaching pendant. *(Courtesy ABB Robotics, Inc.)*

third-generation industrial robot A robot manufactured during the design period when robots will be able to make selective or best choices during parts assembly.

tool center point (TCP) The point at the center of the flange at the end of the manipulator, where the tool attaches to the manipulator.

2.5-generation industrial robot A robot manufactured from the design period when robots have perceptual motor functions.

VAL A high-level robot application programming language developed for Unimation's Unimate and Puma lines of robots.

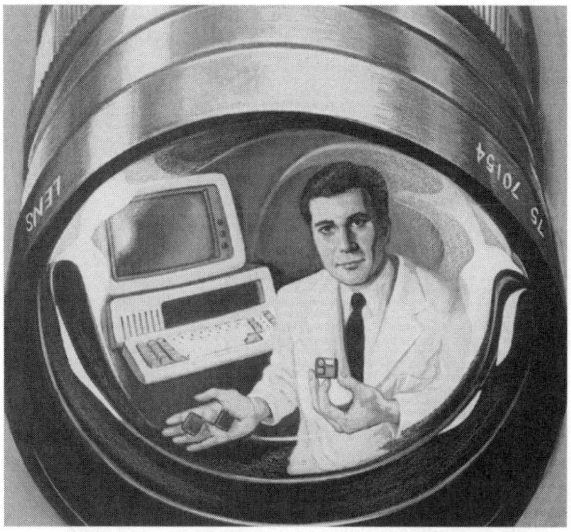

vision system. *(Courtesy Society of Manufacturing Engineers.)*

vehicle A mobile base that enables a robot to move to a location where it will use its manipulator.

vision interpretation The process of understanding the information from one, or a series, of vision sensors.

vision system A system in which information from visual sensors is processed to allow machines to react to changes in the manufacturing process. *(See figure.)*

voice recognition The ability of a machine to hear and understand spoken words.

voice synthesis The process in which a machine can produce and speak understandable words.

WAVE A robot research high-level programming language developed by Stanford Artificial Laboratory in 1973.

work envelope The cube, sphere, cylinder, or other physical space in which a robot operates or is capable of reaching.

wrist The multiaxis joint between a robot's hand or end effector.

yaw One of the robot wrist's axes of motion; the sideways or side-to-side movement of a hand. *(See figure.)*

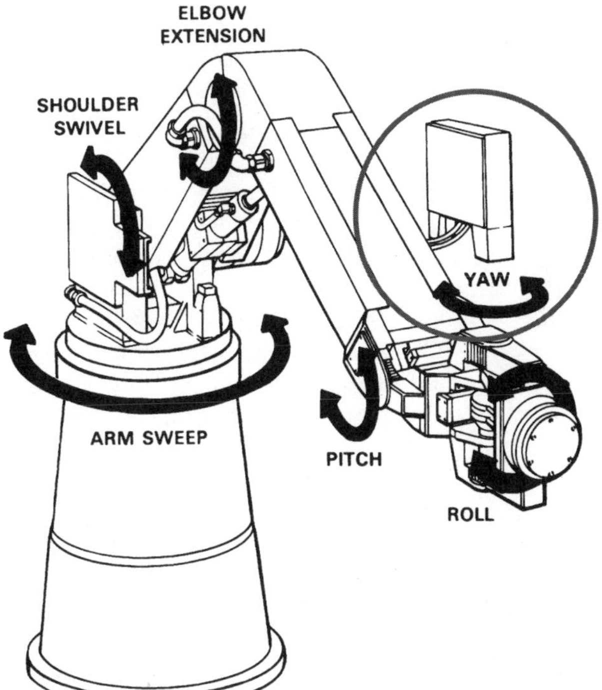

yaw. *(Courtesy Cincinnati Milacron, Inc.)*

SECTION 15

Welding Processes

Steve F. Krar
Consultant
Kelmar Associates
Welland, Ontario

Welding is a joining process used throughout the ages to fasten tools and implements together. Early metal-joining methods used molten metal poured between two pieces of metal that was allowed to harden and bond. Forge or hammer welding was used from the mid-1750s to 1886, when the resistance welding process was developed.

Welding may be defined as a process that produces coalescence (the fusion or growing together of the grain structures of metals) by heating the metals to the proper welding temperature to cause them to soften, melt, and flow together. The most common welding processes are the oxyfuel, shielded-metal arc, gas-tungsten arc, gas-metal arc, flux-cored arc, electron-beam, and laser-beam welding.

abrasion soldering A soldering process during which the faying surface of the base metal is mechanically abraded.

accelerating potential (electron-beam welding and cutting) The potential that imparts velocity to the electrons.

acetylene A fuel gas used for cutting and welding processes.
- □ Acetylene is produced as a result of the chemical action between calcium carbide and water.
- □ When burned in the presence of oxygen, it produces a high-heat, high-temperature flame.

acetylene feather The intense, white, feathery-edged portion next to the cone of a carburizing oxyacetylene flame.

acid-core solder A solder wire or bar containing acid flux as a core.

activated-rosin flux A rosin-base flux containing an additive that increases wetting by the solder.

active flux (submerged arc welding) A flux from which the amount of elements deposited in the weld metal is dependent on the welding conditions, primarily the arc voltage.
- □ *See also* NEUTRAL FLUX.

actual throat The shortest distance between the weld root and the face of a fillet weld.

adaptive control welding A process control system that automatically determines changes in welding conditions and directs the equipment to take appropriate action. *(See figure.)*
- □ Variations of this term are adaptive control brazing, adaptive control soldering, adaptive control thermal cutting, and adaptive control thermal spraying.
- □ *See also* AUTOMATIC WELDING, MANUAL WELDING, MECHANIZED WELDING, ROBOTIC WELDING, *and* SEMIAUTOMATIC WELDING.

air-acetylene welding (AAW) An oxyfuel gas welding process that uses an air-acetylene flame to coalesce metals.
- □ The process is used without the application of pressure, and with or without the use of filler material.
- □ This is an obsolete or seldom-used process.

air carbon arc cutting (CAC-A) A carbon arc cutting process where molten metal is removed by a blast of air. *(See figure on next page.)*
- □ The torch combines the carbon electrode holder and the air stream in the same unit.
- □ A wide variety of metals can be cut.

air feed A thermal spraying process where an airstream carries the powdered surfacing material through the gun and into the heat source.

aligned discontinuities Three or more discontinuities aligned approximately parallel to the weld axis, spaced

adaptive control welding. *(Courtesy Lincoln Electric Co.)*

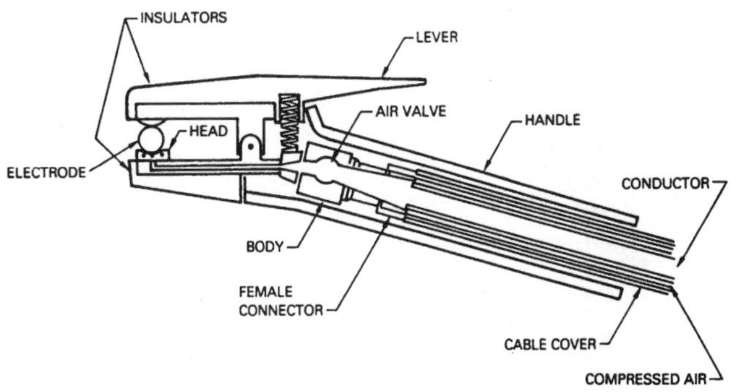

air carbon arc cutting. *(Courtesy American Welding Society.)*

sufficiently close together to be considered a single intermittent discontinuity.

alloy A substance with metallic properties composed of two or more chemical elements, at least one of which is a metal.

alloy powder Powder prepared from a homogeneous molten alloy or from the solidification product of such an alloy.

arc blow The deflection of an arc from its normal path during a weld due to the residual magnetic field in the part or the flowing current.

□ Arc blow is noticeable in corners, at the end of a plate, and when the lead is connected to one side of a plate.

arc braze welding (ABW) A braze welding process that uses an electric arc as the heat source.

arc cutting (AC) A group of thermal cutting processes that cut or remove metal by melting with the heat of an arc between an electrode and the workpiece.

□ *See also* AIR CARBON ARC CUTTING, CARBON ARC CUTTING, GAS-METAL ARC CUTTING, GAS-TUNGSTEN ARC CUTTING, *and* PLASMA ARC CUTTING.

arc-cutting gun A device used to transfer current to a continuously fed cutting electrode, guide the electrode, and direct the shielding gas.

arc gouging Thermal gouging that uses an arc-cutting process to form a bevel or groove. *(See figure.)*

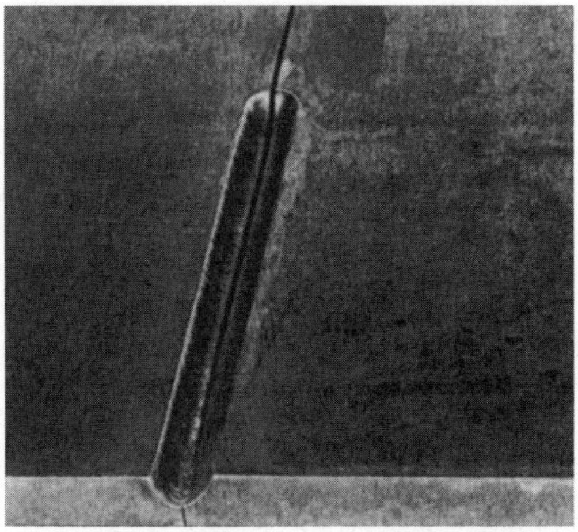

arc gouging. *(Courtesy Lincoln Electric Co.)*

arc length The distance the arc must jump from the tip of the welding electrode to the surface of the plate or weld pool.

□ A short arc length may short-circuit the electrode and cause it to stick to the plate.

□ A long arc length produces splatter, making the weld wider, but with little buildup.

arc plasma A gas that has been heated by an arc to at least a partially ionized condition, enabling it to conduct an electric current.

arc seam weld A seam weld made by an arc-welding process.

arc spraying (ASP) A thermal spraying process using an arc between two consumable electrodes of surfacing materials as a heat source and a compressed gas to atomize and propel the surfacing material to the substrate.

arc strike A gap or break resulting from an arc, consisting of any localized remelted metal, heat-affected metal, or change in the surface profile of any metal object.

arc stud welding (SW) An arc-welding process that uses an arc between a metal stud, or similar part, and the other workpiece. *(See figure on next page.)*

□ This process is used without filler metal, with or without shielding gas or flux, with or without partial shielding from a ceramic or graphite ferrule surrounding the stud, and with the application of pressure after the faying surfaces are sufficiently heated.

arc time The time during which an arc is maintained in making an arc weld.

arc voltage The voltage across the welding arc.

arc welding (AW) A group of welding processes that produce coalescence of workpieces by the heat produced with an electric arc. *(See figure.)*

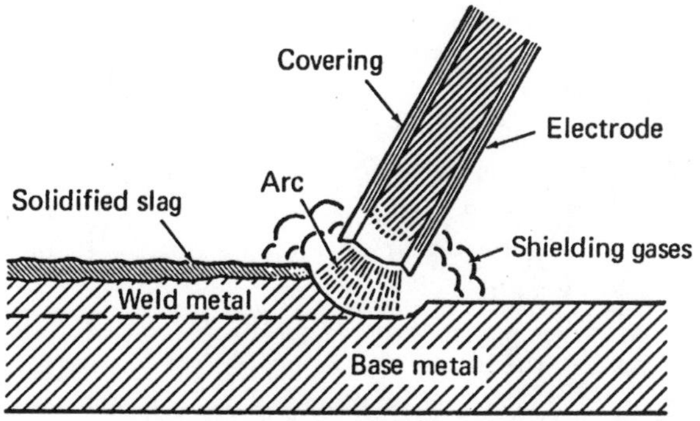

arc welding. *(Courtesy Lincoln Electric Co.)*

□ The electrical energy from the arc creates intensive heat (approximately 5500°C) to join the metal parts.

□ This process is used with or without the application of pressure and with or without filler metal.

arc-welding deposition efficiency The ratio of the weight of filler metal deposited in the weld metal to

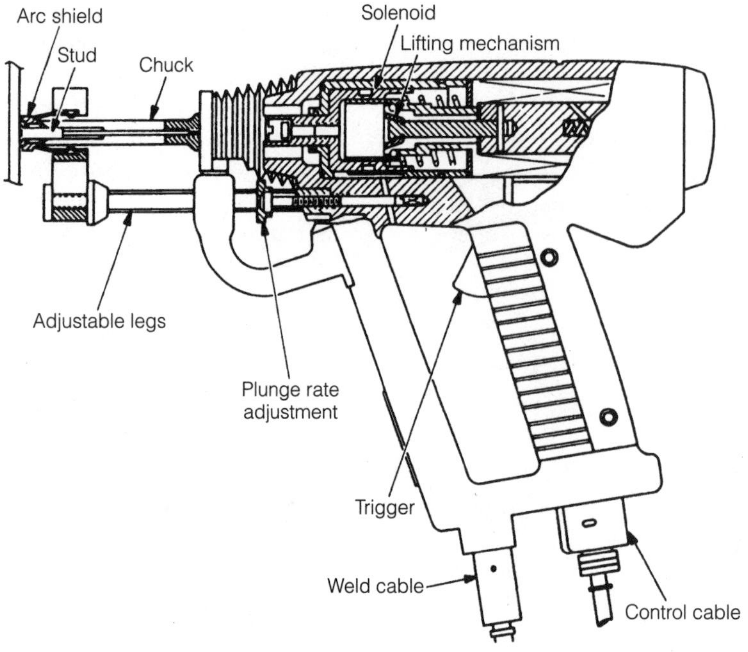

arc stud welding. *(Courtesy KSM Fastening Systems.)*

the weight of filler metal melted, expressed in percent.

arc-welding electrode A component of the welding circuit through which current is conducted and that ends at the arc.

arc-welding gun A device used to transfer current to a continuously fed consumable electrode, guide the electrode, and direct the shielding gas.

arc-welding torch A device used to transfer current to a fixed welding electrode, position the electrode, and direct the flow of shielding gas.

arm (resistance welding) A projecting beam extending from the frame of a resistance-welding machine that transmits the electrode force and may conduct the welding current.

as-brazed A term referring to the condition of brazements after brazing, before any further thermal, mechanical, or chemical treatments.

assist gas A gas used to blow molten metal away to form the kerf in laser-beam inert-gas cutting, or to blow vaporized metal away from the beam path in laser-beam evaporative cutting.

☐ Nonreactive gases do not add heat to the cutting process; they blow the molten metal away.

☐ Exothermic (heat-producing) gases produce additional heat and help blow the molten metal out of the kerf.

as-welded Term referring to the condition of weld metal, welded joints, and weldments after welding, but before any further thermal, mechanical, or chemical treatments.

atomic hydrogen welding (AHW) An arc-welding process that uses an arc between two metal electrodes in a shielding atmosphere of hydrogen and without the application of pressure.

☐ This is an obsolete or seldom-used process.

autogenous weld A fusion weld made without filler metal.

automatic The control of a process with equipment that requires only occasional or no observation of the welding, and no manual adjustment of the equipment controls.

automatic arc-welding current The current in the welding circuit during the making of a weld, but excluding upslope, downslope, and crater fill current.

automatic arc-welding weld time The time interval from the end of start time or end of upslope to beginning of crater fill time or beginning of downslope.

automatic welding Welding with equipment that requires only occasional or no observation of the welding, and no manual adjustment of the equipment controls.

☐ Variations of this term are *automatic brazing, automatic soldering, automatic thermal cutting,* and *automatic thermal spraying.*

☐ *See also* ADAPTIVE CONTROL WELDING, MANUAL WELDING, MECHANIZED WELDING, ROBOTIC WELDING, *and* SEMIAUTOMATIC WELDING.

back cap A device used to exert pressure on the collet in a gas-tungsten arc-welding torch and create a seal to prevent air from entering the back of the torch.

backfire The momentary recession of the flame into the welding tip, cutting tip, or flame-spraying gun followed by immediate reappearance or complete extinction of the flame, accompanied by a loud, sharp noise.

backgouging The removal of weld metal and base

metal from the weld-root side of a welded joint to facilitate complete fusion and complete joint penetration on subsequent welding from that side.

backhand welding A welding technique in which the arc force is directed into the molten weld pool of metal. *(See figure.)*

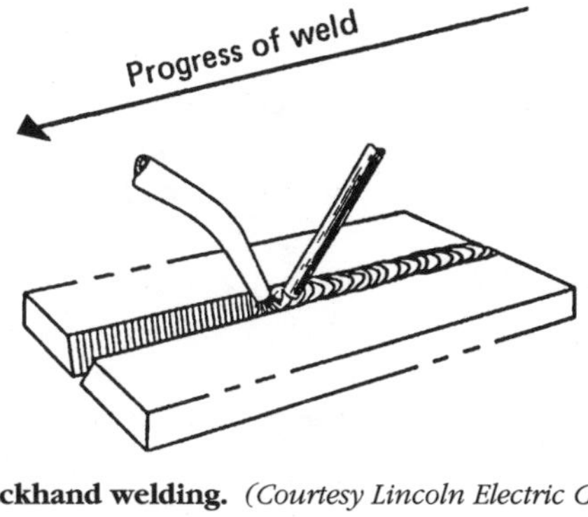

backhand welding. *(Courtesy Lincoln Electric Co.)*

□ It is opposite to the process of welding.
□ The digging action of the process pushes the penetration deeper into the base metal while building up the weldhead.
□ *See also* TRAVEL ANGLE *and* DRAG ANGLE.

backing A material or device placed against the backside of the joint, or at both sides of a weld in electroslag and electrogas welding, to support and retain molten weld metal.
□ The material may be partially fused or remain unfused during welding and may be either metal or nonmetal.

balling up The formation of globules of molten filler metal or flux due to lack of wetting of the base metal.

bare electrode A filler-metal electrode that has been produced as a wire, strip, or bar with no coating or covering other than that used for its manufacture or preservation.

bare-metal arc welding (BMAW) An arc-welding process that uses an arc between a bare or lightly coated electrode and the weld pool.
□ The process is used without shielding and without the application of pressure, and filler metal is obtained from the electrode.
□ This is an obsolete or seldom-used process.

base metal The metal or alloy that is welded, brazed, soldered, or cut.
□ *See also* SUBSTRATE.

bead A weld resulting from a single pass.

bead weld A nonstandard term for surfacing weld.

beam divergence The expansion of a beam's cross section as the beam comes from its source.

bend test A test in which a specimen is bent to a specified bend radius.

bevel angle The angle between the bevel of a joint member and a plane perpendicular to the surface of the member. *(See figure.)*

bevel-groove weld A type of groove weld.

bit The part of the soldering iron, usually made of copper, that directly transfers heat (and sometimes solder) to the joint.

blind joint A joint of which no portion is visible.

block sequence A combined longitudinal and cross-sectional sequence for a continuous multiple-pass weld where separated increments are completely or partially welded before intervening increments are welded.

blowpipe *See* BRAZING BLOWPIPE.

bond coat (thermal spraying) A preliminary (or prime) coat of material that improves adherence of the subsequent thermal spray deposit.

bonding A nonstandard term used for welding, brazing, or soldering.

bonding force The force that holds two atoms together; it results from a decrease in energy as two atoms are brought closer to one another.

bond line (thermal spraying) The cross section of the interface between a thermal spray deposit and the substrate.

bottle A nonstandard term used for a gas cylinder.

boxing The continuation of a fillet weld around a corner of a member as an extension of the principal weld.

brazability The capacity of a material to be brazed under the imposed fabrication conditions into a specific, suitably designed structure, and to perform satisfactorily in the intended service.

braze welding (BW) A welding process that uses a filler metal with a liquidus above 450°C (840°F) and below the solidus of the base metal which is not melted.
□ In braze welding the filler metal is not distributed in the joint by capillary action.
□ *See also* FLOW WELDING.

brazing (B) A group of welding processes that produce coalescence of materials by heating them to the brazing temperature in the presence of a filler metal having a liquidus above 450°C (840°F) and below the solidus of the base metal. *(See figure on next page.)*
□ The filler metal is distributed between the closely fitted faying surfaces of the joint by capillary action.

brazing blowpipe A device used to obtain a small, accurately directed flame for fine work.
□ A portion of any flame blown to the desired location by the blowpipe, which is usually mouth-operated.

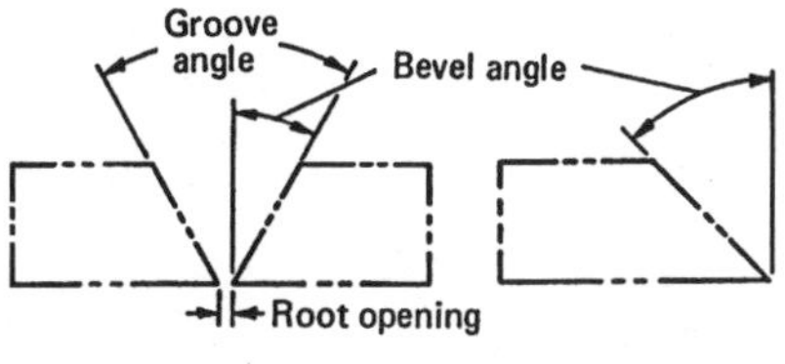

bevel angle. *(Courtesy Lincoln Electric Co.)*

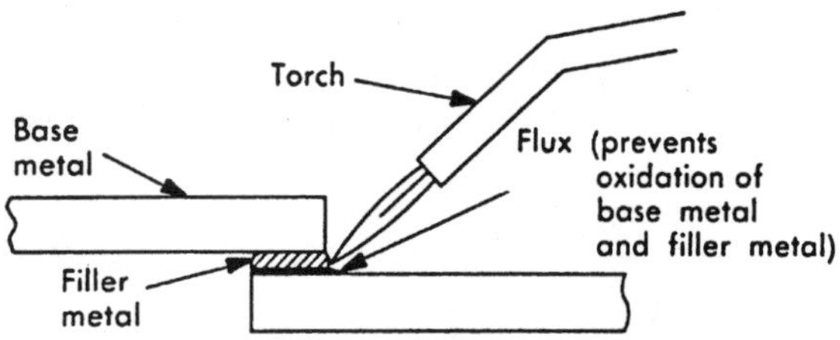

brazing. *(Courtesy Lincoln Electric Co.)*

brazing filler metal The metal or alloy used as a filler metal in brazing, which has a liquidus above 450°C (840°F) and below the solidus of the base metal.

brazing temperature The temperature to which the base metal is heated to allow the filler metal to wet the base metal and form a brazed joint.

brittle nugget A nonstandard term used to describe a faying plane failure in a resistance weld peel test.

buildup A surfacing variation in which surfacing material is deposited to achieve the required dimensions.

□ *See also* BUTTERING, CLADDING, *and* HARDFACING.

burn-through A nonstandard term used for excessive melt-through or a hole through a root bead.

buttering A surfacing process that deposits surfacing metal on one or more surfaces to provide metallurgically compatible weld metal for the completion of the weld.

□ *See also* BUILDUP, CLADDING, *and* HARDFACING.

butting member A joint member that is prevented, by the other member, from movement in one direction perpendicular to its thickness dimension.

□ Examples are both members of a butt joint, or one member of a T-joint or corner joint.

butt joint A joint between two members aligned approximately in the same plane. *(See figure.)*

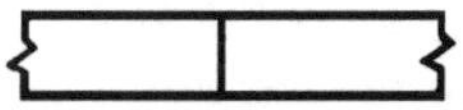

butt joint. *(Courtesy Lincoln Electric Co.)*

□ A "V" is generally ground on edges to be joined for better fusion between the weld and the base material.

button The part of a weld, including all or part of the nugget, that tears out in the destructive testing of spot, seam, or projection-welded specimens.

capillary action The force by which liquid, in contact with a solid, is distributed between closely fitted faying surfaces of the joint to be brazed or soldered.

carbon arc braze welding (CABW) A braze welding process that uses an arc between a carbon electrode and the base metal as the heat source.

carbon arc brazing A nonstandard term for twin carbon arc brazing.

carbon arc cutting (CAC) An arc-cutting process that uses a carbon electrode.

□ *See also* AIR CARBON ARC CUTTING.

carbon arc welding (CAW) An arc-welding process that uses an arc between a carbon electrode and the weld pool.

□ The process is used with or without shielding and without the application of pressure.

□ *See also* SHIELDED CARBON ARC WELDING.

carbon dioxide welding A nonstandard term for gas metal arc welding with carbon dioxide (CO_2) shielding gas.

carbon electrode A nonfiller metal electrode used in arc welding and cutting, consisting of a carbon or graphite rod that may be coated with copper or other materials.

carburizing flame A reducing oxyfuel gas flame containing an excess of fuel gas, resulting in a carbon-rich zone extending around and beyond the cone.

□ It shows as a long white feather inside the flame.

□ This is also called a *neutral, oxidizing,* or *reducing flame.*

carrier gas The gas used to move powdered material from the feeder or hopper to a thermal spraying gun or a thermal cutting torch.

cascade sequence A combined longitudinal and cross-sectional sequence in which weld passes are made in overlapping layers.

caulking Plastic deformation of weld and adjacent base-metal surfaces by mechanical means to seal or obscure discontinuities.

ceramic rod flame spraying A thermal spraying process in which the surfacing material is in rod form.

chain intermittent weld An intermittent weld on both sides of a joint where the weld increments on one side are approximately opposite those on the other side.

clad brazing sheet A metal sheet on which one or both sides are clad with brazing filler metal.

□ *See also* CLAD METAL.

cladding A process that deposits or applies surfacing material usually to improve corrosion or heat resistance. *(See figure.)*

□ *See also* BUILDUP, BUTTERING, *and* HARDFACING.

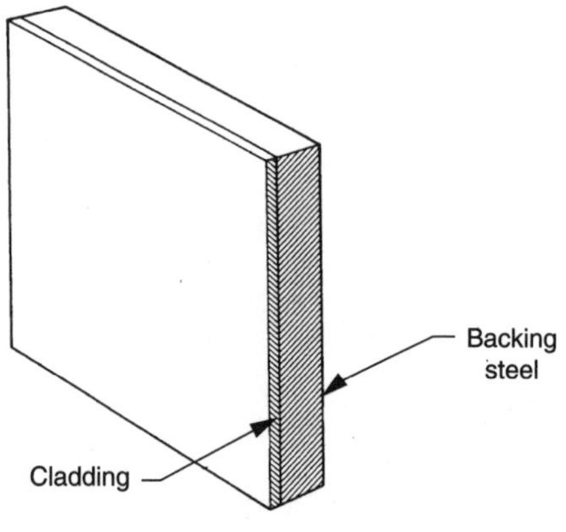

cladding. *(Courtesy Lincoln Electric Co.)*

clad metal A laminar composite consisting of a metal or alloy of varying chemical composition applied to one or more sides by casting, drawing, rolling, surfacing, thick chemical deposition, or thick electroplating.

coalescence The growing together or growth into one body of the materials being welded.

coated electrode A nonstandard term for a covered electrode or lightly coated electrode.

coextrusion welding (CEW) A solid-state welding process produced by heating to the welding temperature and forcing the workpieces through an extrusion die.

coil without support A filler-metal package consisting of a continuous length of welding wire in coil form without an internal support.

coil with support A filler-metal package consisting of a continuous length of welding wire in coil form wound on a simple cylinder without flanges.

cold crack A crack that forms after solidification of the joint is complete.

cold soldered joint A joint with incomplete coalescence caused by insufficient application of heat to the base metal during soldering.

cold welding (CW) A solid-state welding process in which only mechanical pressure is used to produce a weld at room temperature with substantial deformation at the weld.

□ *See also* DIFFUSION WELDING, FORGE WELDING, *and* HOT-PRESSURE WELDING.

collaring (thermal spraying) Adding a shoulder to a shaft or similar component as a protective confining wall for the thermal spray deposit.

commutator-controlled welding The making of multiple groups of resistance spot or projection welds sequentially with the same welding contractor using a commutating device.

complete fusion Fusion over the entire fusion face and between all adjoining weld beads.

□ *See also* INCOMPLETE FUSION.

complete joint penetration weld A groove weld in which weld metal extends through the joint thickness. *(See figure.)*

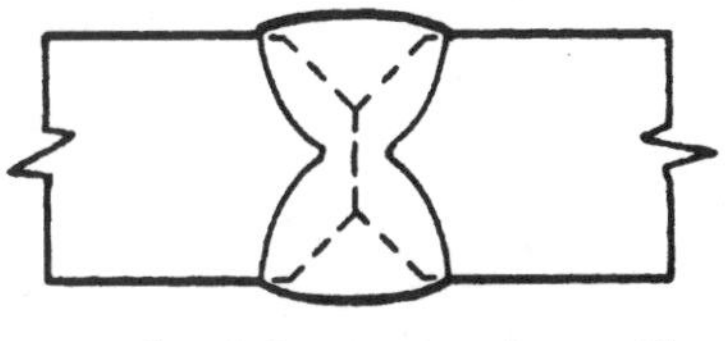

complete joint penetration weld.
(Courtesy Lincoln Electric Co.)

□ *See also* INCOMPLETE JOINT PENETRATION WELD, JOINT PENETRATION WELD, *and* PARTIAL JOINT PENETRATION WELD.

composite A material consisting of two or more different materials with each material retaining its physical identity.

□ *See also* CLAD METAL, COMPOSITE ELECTRODE, *and* COMPOSITE THERMAL SPRAY DEPOSIT.

composite electrode A generic term for multicomponent filler metal electrodes in various physical forms such as stranded wires, tubes, and covered wire.

□ *See also* COVERED ELECTRODE, FLUX-CORED ELECTRODE, METAL-CORED ELECTRODE, *and* STRANDED ELECTRODE.

composite thermal spray deposit A thermal spray deposit made with two or more dissimilar surfacing materials that may be formed in layers.

concave fillet weld A fillet weld having a concave face.

concurrent heating The application of supplemental heat to a structure before or during welding or cutting.

cone The conical part of an oxyfuel gas flame next to the carbon tip orifice.

connection A nonstandard term used for a welded, brazed, or soldered joint.

constant power source An arc-welding power source with a volt-ampere relationship yielding a large or small welding current change from a large or small arc voltage change.

constricted arc A plasma arc column that is shaped by the constricting orifice in the nozzle of the plasma arc torch or plasma spraying gun.

constricting nozzle A device at the exit end of a plasma arc torch or plasma spraying gun, containing the constricting orifice.

constricting orifice The hole in the constricting nozzle of the plasma arc torch or plasma spraying gun through which the arc plasma passes.

consumable electrode An electrode that provides filler metal.

consumable guide electroslag welding An electroslag welding process in which filler metal is supplied by an electrode and its guiding member.

consumable insert Filler metal that is placed at the joint root before welding, and is intended to be completely fused into the joint root to become part of the weld.

contact resistance, resistance welding Resistance to the flow of electric current between two workpieces or an electrode and a workpiece.

continuous-wave laser A laser having an output that operates in a continuous rather than a pulsed mode.

□ A laser operating with a continuous output for a period greater than 25 ms is regarded as a continuous-wave laser.

continuous weld A weld that extends continuously from one end of a joint to the other.

□ Where the joint is essentially circular, it extends completely around the joint.

convex fillet weld A fillet weld having a convex face.

convexity The maximum distance from the face of a

convex fillet weld perpendicular to a line joining the weld toes.

cool time (resistance welding) The time interval between successive heat times in multiple-impulse welding or in the making of seam welds.

copper brazing A nonstandard term used for brazing with a copper filler metal.

cord (thermal spraying) Surfacing material in the form of a plastic tube filled with powder extruded to a compact, flexible cord with characteristics similar to a wire.

cored solder A solder wire or bar containing flux as a core.

corner-flange weld A nonstandard term for an edge weld in a flanged corner joint.

corner joint A joint between two members located approximately at right angles to each other in the form of the letter L. *(See figure.)*

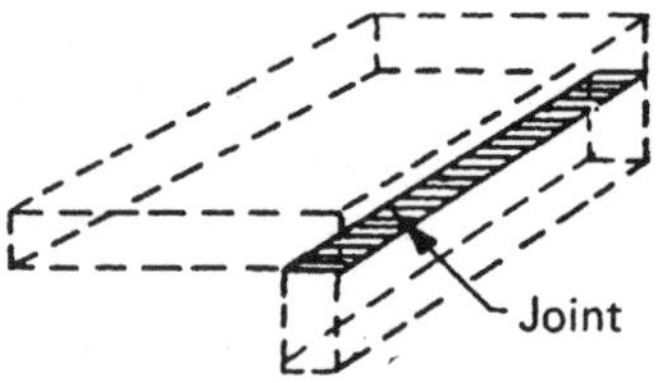

corner joint. *(Courtesy Lincoln Electric Co.)*

corona (resistance welding) The area sometimes surrounding the nugget of a spot weld at the faying surfaces that provides a degree of solid-state welding.

corrosive flux A flux with a residue that chemically attacks the base metal. It may be composed of inorganic salts and acids, organic salts and acids, or activated rosin.

CO_2 welding *See* CARBON DIOXIDE WELDING.

covalent bond A primary bond arising from the reduction in energy associated with overlapping half-filled orbitals of two atoms.

covered electrode A composite filler-metal electrode consisting of a core of a bare electrode or metal-cored electrode to which a covering sufficient to provide a slag layer on the weld metal has been applied.

□ The covering may provide functions such as shielding from the atmosphere, deoxidation, and arc stabilization, and can serve as a source of metallic additions to the weld.

crack A fracture-type discontinuity characterized by a sharp tip and high length:width ratio to the opening displacement. *(See figure.)*

crater A depression in the weld face at the end of a weld bead.

crater fill time The time interval following weld time but before meltback time during which arc voltage or current reaches a preset value greater or less than welding values.

□ Weld travel may or may not stop at this point.

cross-sectional sequence The order in which the weld passes of a multiple-pass weld are made in relation to the cross section of the weld.

□ *See also* BLOCK SEQUENCE *and* CASCADE SEQUENCE.

cross-wire welding A common type of projection welding in which the localization of the welding current is achieved by the intersection contact of wires; it is usually accompanied by considerable embedding of one wire into another.

cutting attachment A device for converting an oxyfuel gas welding torch into an oxygen cutting torch.

cutting blowpipe A nonstandard term for an oxyfuel gas cutting torch.

cutting electrode A nonfiller-metal electrode used in arc cutting.

□ *See also* CARBON ELECTRODE *and* METAL ELECTRODE.

cutting head The part of a cutting machine in which a cutting torch or tip is incorporated.

cutting tip The part of an oxygen cutting torch from which the gases issue.

cutting torch *See* AIR CARBON ARC CUTTING, GAS-TUNGSTEN ARC CUTTING, PLASMA ARC CUTTING, *and* OXYFUEL GAS CUTTING.

cylinder manifold A multiple header for interconnection of gas sources with distribution points.

defect A discontinuity or discontinuities that by nature or accumulated effect (e.g., total crack length) produce a part or product that is unable to meet minimum acceptance standards or specifications.

□ *See also* DISCONTINUITY *and* FLAW.

deposited metal (surfacing) Metal that has been added during a surfacing operation.

deposited metal (welding, brazing, and soldering) Filler metal that has been added during welding, brazing, or soldering.

deposition rate The weight of material deposited in a unit of time.

depth of fusion The distance that fusion extends into the base metal or previous bead from the surface melted during welding.

□ *See also* JOINT PENETRATION.

detonation flame spraying A thermal spraying process in which the controlled explosion of a mixture

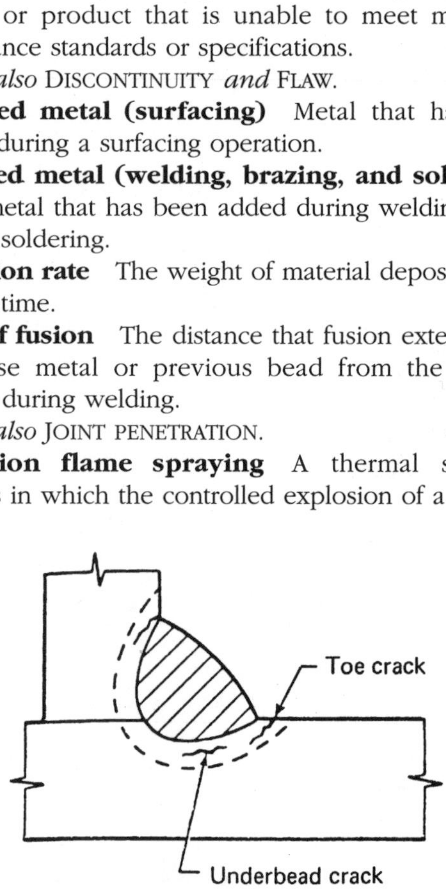

cracks. *(Courtesy Lincoln Electric Co.)*

of fuel gas, oxygen, and powdered surfacing material is used to melt and propel the surfacing material to the substrate.

diffusion brazing (DFB) A brazing process that produces coalescence of metals by heating them to brazing temperature and using a filler metal or a liquid stage.

□ The filler metal may be distributed by capillary attraction or may be placed or formed at the faying surfaces.

□ The filler metal is diffused with the base metal to the extent that the joint properties have been changed to approach those of the base metal.

diffusion welding (DFW) A solid-state welding process that produces a weld by applying pressure at elevated temperature with no macroscopic deformation or relative motion of the workpieces.

□ A solid filler metal may be inserted between the faying surfaces.

□ *See also* COLD WELDING, DIFFUSION BRAZING, FORGE WELDING, *and* HOT-PRESSURE WELDING.

dilution The change in chemical composition of a welding filler metal caused by the admixture of the base metal or previous weld metal in the weld bead.

□ It is measured by the percentage of base metal or previous weld metal in the weld bead.

dip brazing (DB) A brazing process that uses heat from a molten salt or metal bath.

□ When a molten salt is used, the bath may act as a flux.

□ When a molten metal is used, the bath provides the filler metal.

dip soldering (DS) A soldering process using the heat supplied by a molten-metal bath to provide the solder filler metal.

direct-drive friction welding A type of friction welding in which the energy required to make the weld is supplied to the welding machine through a direct motor connection for a preset period of the welding cycle.

direct welding (resistance welding) A resistance welding secondary circuit in which the welding current and electrode force are applied to the workpieces by directly opposed electrodes, wheels, or conductor bars for spot, seam, or projection welding.

discontinuity An interruption, but not necessarily a defect, of the typical structure of a material, such as a lack of homogeneity in its mechanical, metallurgical, or physical characteristics.

□ *See also* DEFECT *and* FLAW.

double arcing A condition in which the welding or cutting arc of a plasma arc torch does not pass through the constricting orifice but transfers to the inside surface of the nozzle.

□ A secondary arc is established at the same time between the nozzle outer surface and the workpiece.

double-flare V-groove weld A weld in grooves formed by two members with curved surfaces.

double-groove weld (fusion welding) A groove weld that is made from both sides. *(See figure.)*

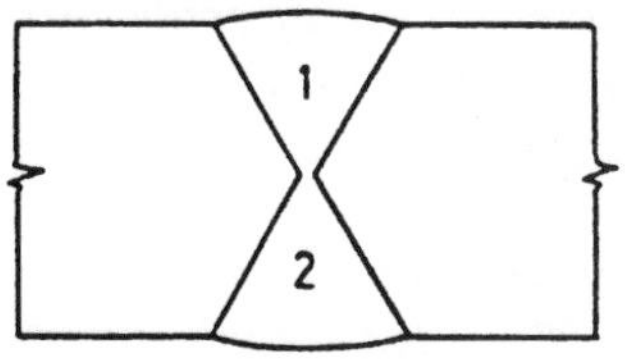

double-groove (fusion) weld. *(Courtesy Lincoln Electric Co.)*

double-welded joint (fusion welding) A joint that is welded from both sides.

dovetailing (thermal spraying) A method of surface roughening involving angular undercutting to interlock the thermal spray deposit.

drag (thermal cutting) The offset distance between the actual and straight-line exit points of the gas stream or cutting beam measured on the exit surface of the base metal.

drag angle The travel angle when the electrode is pointing in a direction opposite to the welding progression. *(See figure.)*

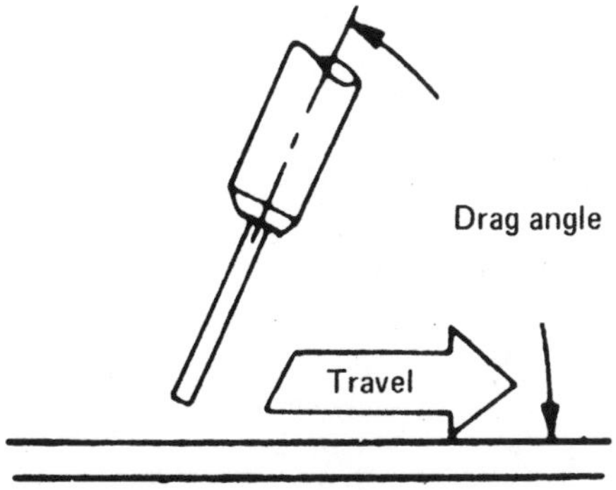

drag angle. *(Courtesy Lincoln Electric Co.)*

□ This angle can also be used to partially define the position of guns, torches, rods, and beams.

□ *See also* BACKHAND WELDING *and* TRAVEL ANGLE.

drop-through An undesirable sagging or surface irregularity, usually encountered when brazing or welding near the solidus of the base metal.

□ It is caused by overheating with rapid diffusion or alloying between the filler metal and the base metal.

drum A filler-metal package consisting of a continuous length of welding wire wound or coiled in a cylindrical container.

dwell time (thermal spraying) The length of time that the surfacing material is exposed to the heat zone of the thermal spraying gun.

dynamic electrode force The force created by electrodes on the workpieces during the actual welding cycle in making spot, seam, or projection welds by resistance welding.

edge effect (thermal spraying) Loosening of the bond between the thermal spray deposit and the substrate at the edge of the thermal spray deposit.

edge joint A joint between the edges of two or more parallel or nearly parallel parts.

edge preparation The preparation of the edges of the joint members by cutting, cleaning, plating, or other means.

edge weld A weld in an edge joint, a flanged butt joint, or a flanged corner joint where the full thicknesses of the members are fused.

effective throat The minimum distance minus any convexity between the weld root and the face of a fillet weld.

electric-arc spraying A nonstandard term for arc spraying.

electric bonding A nonstandard term for surfacing by thermal spraying.

electric brazing A nonstandard term for arc brazing and resistance brazing.

electrode A component of the electric circuit that ends at the arc, molten conductive slag, or base metal.
- ☐ *See also* WELDING ELECTRODE.

electrode force The force applied to the electrodes in making spot, seam, or projection welds by resistance welding.
- ☐ *See also* DYNAMIC ELECTRODE FORCE *and* STATIC ELECTRODE FORCE.

electrode holder A device used for holding and conducting current to an electrode during welding or cutting.

electrode mushrooming The enlargement of a resistance spot or projection welding electrode tip due to heat or pressure so that it resembles a mushroom in shape.

electrode pickup Contamination of the electrode tips or wheel faces by the base metal or its coating during resistance spot, seam, or projection welding.

electrode skid The sliding of a resistance welding electrode along the surface of the workpiece when making spot, seam, or projection welds.

electrogas welding (EGW) An arc-welding process that uses an arc between a continuous filler-metal electrode and the weld pool, employing approximately vertical welding progression with backing to confine the molten weld metal. *(See figure.)*
- ☐ The process is used with or without an externally supplied shielding gas and without the application of pressure.

electron-beam braze welding (EBBW) A braze welding process that uses an electron beam as the heat source.

electron-beam cutting (EBC) A thermal cutting process that cuts metal by melting it with the heat from a concentrated beam, consisting mainly of high-velocity electrons striking the workpiece.

electron-beam gun A device for producing and accelerating electrons.
- ☐ Typical components include the *emitter* (also called the *filament* or *cathode*) that is heated to produce electrons via thermionic emission, a *cup* (also called the *grid* or *grid cup*), and the *anode.*

electron-beam welding (EBW) A process that uses energy from a fast-moving beam of electrons to produce a strong, very clean and narrow weld.
- ☐ High temperatures [180,000°F (100,000°C)] vaporize metals and ceramics to join dissimilar materials.
- ☐ The process is used without shielding gas and without the application of pressure.
- ☐ This method is used in automotive, electronics, pipeline, aerospace, and high-speed welded tubing industries.

electroslag welding (ESW) A welding process that produces coalescence of metals with molten slag that melts the filler metal and the surfaces of the workpieces. *(See figure on next page.)*
- ☐ The weld pool is shielded by the slag as it moves along the full cross section of the joint during welding.
- ☐ The process is started by an arc that heats the slag.
- ☐ The arc is then extinguished by the conductive slag, which is kept molten by its resistance to electric current passing between the electrode and the workpieces.
- ☐ *See also* CONSUMABLE GUIDE ELECTROSLAG WELDING *and* ELECTROSLAG WELDING ELECTRODE.

electroslag welding electrode A filler-metal component of the welding circuit through which current is conducted from the electrode guiding member to the molten slag.

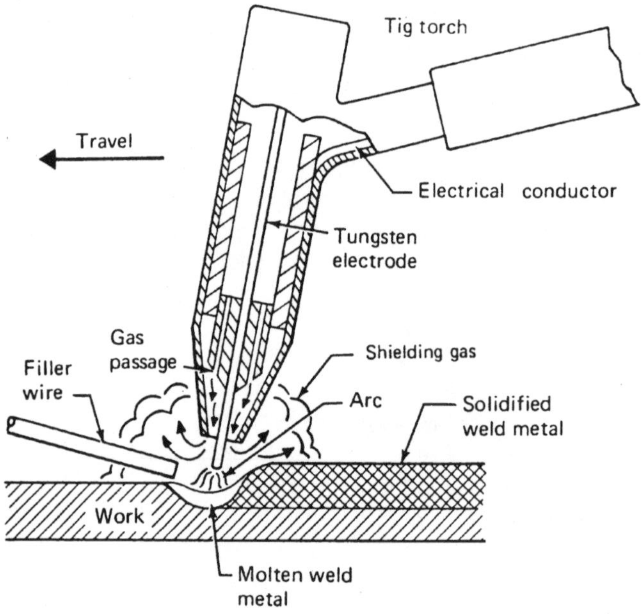

electrogas welding. *(Courtesy Lincoln Electric Co.)*

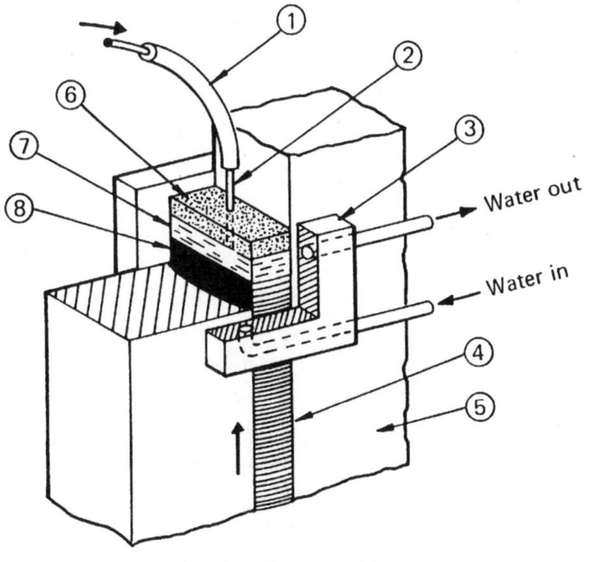

electroslag welding. (1) Electrode guide tube, (2) electrode, (3) water-cooled copper shoes, (4) finished weld, (5) base metal, (6) molten slag, (7) molten weld metal, (8) solidified weld metal. *(Courtesy Lincoln Electric Co.)*

erosion (brazing) A condition caused by dissolution of the base metal by molten filler metal, resulting in a reduction in the thickness of the base metal.

exothermic braze welding (EXBW) A braze welding process that uses an exothermic chemical reaction between a metal oxide and a metal or inorganic nonmetal as the heat source, with a reaction product as the filler metal.

exothermic brazing (EXB) A brazing process using an exothermic chemical reaction between a metal oxide and a metal or inorganic nonmetal as the heat source, with filler metal preplaced in the joint.

explosion welding (EXW) A solid-state welding process that produces a weld by high-velocity impact of the workpieces as the result of controlled detonation.

□ Often used to clad structural steel, plate, or tubing with a corrosion-resistant material.

expulsion The forceful ejection of molten metal from a resistance spot, seam, or projection weld usually at the faying surface.

face shield A device positioned in front of the eyes and over all or a portion of the face to protect the eyes and face.

□ Also called *hand shield* and *helmet.*

faying surface The mating surface of a member that is in contact with or near to another member to be joined.

ferrite number (FN) An arbitrary, standardized value for the ferrite content of an austenitic stainless-steel weld metal.

□ It should be used in place of percent ferrite or volume percent ferrite on a direct replacement basis.

ferrule (arc stud welding) A ceramic device that surrounds the stud base to contain the molten metal and shield the arc.

filler material The material to be added in making a welded, brazed, or soldered joint.

□ *See also* BRAZING FILLER METAL, CONSUMABLE INSERT, FILLER METAL, SOLDER, WELDING ELECTRODE, *and* WELDING WIRE.

filler metal The metal added in making a brazed, soldered, or welded joint.

fillet weld A weld of approximately triangular cross section joining two surfaces approximately at right angles to each other in a lap joint, T-joint, or corner joint. *(See figure.)*

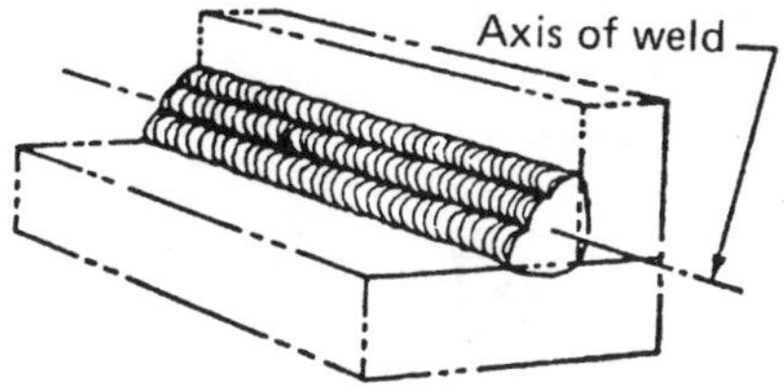

fillet weld. *(Courtesy Lincoln Electric Co.)*

□ This type of weld seldom uses a prepared groove.

fillet weld size For equal-leg fillet welds, the leg lengths of the largest isosceles right triangle that can be inscribed within the fillet weld cross section. For unequal-leg fillet welds, the leg lengths of the largest right triangle that can be inscribed within the fillet weld cross section.

filter plate An optical material that protects the eyes against excessive ultraviolet, infrared, and visible radiation.

fisheye A discontinuity found on the fracture surface of a weld in steel that consists of a small pore or inclusion surrounded by an approximately round, bright area.

5F A welding test position designation for a circumferential fillet weld applied to a joint in pipe, with its axis approximately horizontal, where the weld is made in the horizontal, vertical, and overhead welding positions.

□ The pipe remains fixed until the welding is complete.

5G A welding test position designation for a circumferential groove weld applied to a joint in a pipe with its axis horizontal, where the weld is made in the flat, vertical, and overhead welding positions. *(See figure.)*

□ The pipe remains fixed until the welding is complete.

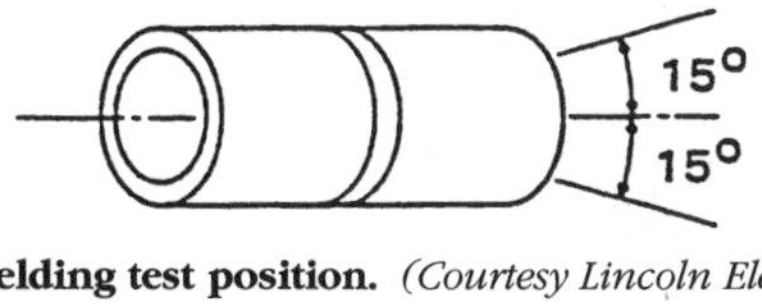

the 5G welding test position. *(Courtesy Lincoln Electric Co.)*

fixture A device designed to hold and maintain parts in proper relation to each other.

flame spraying (FLSP) A thermal spraying process in which an oxyfuel gas flame is the heat source for melting the surfacing material.

□ Compressed gas may or may not be used for atomizing and propelling the surfacing material to the substrate.

flanged butt joint A form of a butt joint in which at least one of the members has a flanged edge shape at the joint.

flanged corner joint A form of a corner joint in which the butting member has a flanged edge shape at the joint, and an edge weld is applicable.

flanged edge joint A form of an edge joint in which at least one of the members has a flanged edge shape at the joint. *(See figure.)*

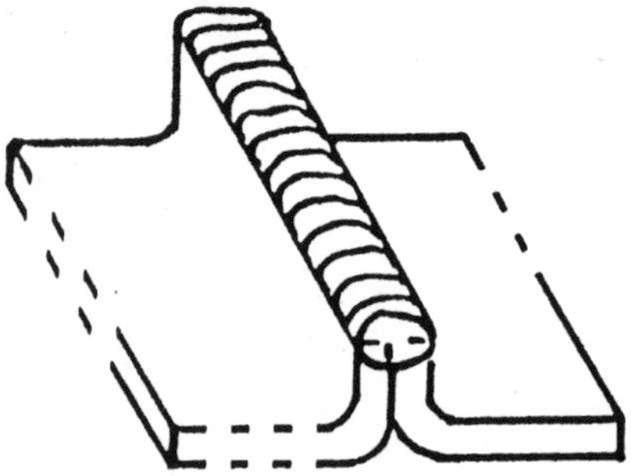

flanged edge joint. *(Courtesy Lincoln Electric Co.)*

flanged lap joint A form of a lap joint in which at least one of the members has a flanged edge shape at the joint, and an edge weld is not applicable.

flanged T-joint A form of a T-joint in which the butting member has a flanged edge shape at the joint, and an edge weld is not applicable.

flare-bevel-groove weld A weld in the groove formed between a joint member with a curved surface and another with a flat surface.

flare-groove weld A weld in the groove formed between a joint member with a curved surface and another with a flat surface, or between two joint members with curved surfaces.

□ *See also* FLARE-BEVEL-GROOVE WELD *and* FLARE-V-GROOVE WELD.

flare-V-groove weld A weld in a groove formed by two members with curved surfaces.

flash Material that is expelled from a flash weld prior to the upset portion of the welding cycle.

flashback A recession of the flame into or back of the mixing chamber of the oxyfuel gas torch or flame-spraying gun.

□ A flashback arrester is used to prevent spreading of the flame front.

flashing action The phenomenon in flash welding in which the points of contact, formed by light pressure across faying surfaces, are melted and explosively ejected because of the extremely high current density at contact points.

flash welding (FW) A resistance welding process in which fusion occurs over the ends of stock by the heat produced from the resistance to the electric current flow between two surfaces.

□ The flashing action, caused by the very high current densities at small contact points between the workpieces, forcibly expels the material from the joint as the workpieces are slowly moved together.

□ This process can be used to join dissimilar aluminum alloys and join aluminum to other metals.

flat welding position The welding position used to weld from the upper side of the joint at a point where the weld axis is approximately horizontal, and the weld face lies in an approximately horizontal plane.

flaw An undesirable discontinuity. *See also* DEFECT.

flood cooling (resistance seam welding) The application of liquid coolant directly on the work and the contacting electrodes.

flowability The ability of molten filler metal to flow or spread over a metal surface.

flow welding (flow) A braze welding process that uses molten filler metal poured over the fusion faces as the heat source.

□ This is an obsolete or seldom-used process.

flux A material used to prevent the formation of oxides and other undesirable substances in molten metal and on solid metal surfaces, and to dissolve or facilitate the removal of such substances.

□ *See also* ACTIVE FLUX *and* NEUTRAL FLUX.

flux-cored arc welding (FCAW) An arc-welding process that uses an arc between a continuous filler metal electrode and the weld pool. *(See figure.)*

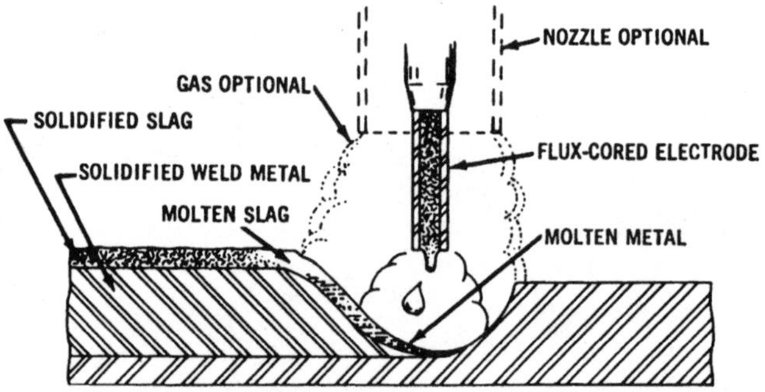

flux-cored arc welding. *(Courtesy Hobart Brothers Co.)*

□ This process is used with shielding gas from a flux contained within the tubular electrode, with or without additional shielding from an externally supplied gas, and without the application of pressure.

□ *See also* FLUX-CORED ELECTRODE *and* GAS-SHIELDED FLUX-CORED ARC WELDING.

flux-cored electrode A composite tubular filler-metal electrode consisting of a metal sheath and a core of various powdered materials, producing an extensive slag cover on the face of a weld bead.

□ External shielding may be required.

flux cutting (FOC) An oxygen cutting process that uses heat from an oxyfuel gas flame, with a flux in the flame to aid cutting.

flux oxygen cutting A nonstandard term for flux cutting.

focal spot (electron-beam welding and cutting, and laser-beam welding and cutting) A location where the beam has the most concentrated energy and the smallest cross-sectional area.

forehand welding A welding technique in which the welding torch, gun, or coated electrode is directed toward the progress of welding and keeps the weld puddle in front of the torch or electrode. *(See figure.)*

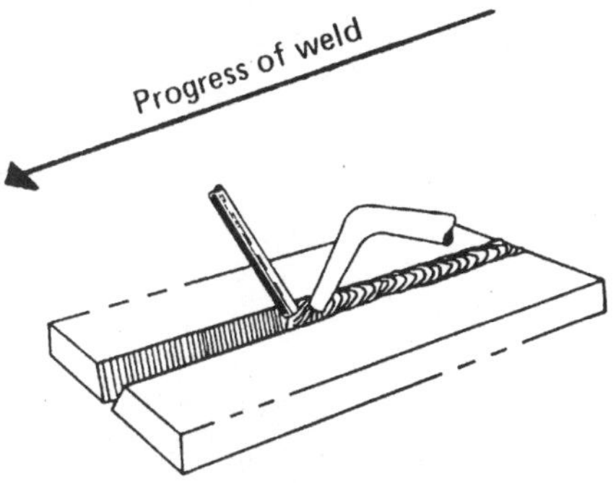

forehand welding. *(Courtesy Lincoln Electric Co.)*

□ *See also* PUSH ANGLE, TRAVEL ANGLE, *and* WORK ANGLE.

forge welding (FOW) A solid-state welding process that produces a weld by heating the workpieces to welding temperature and applying blows sufficient to cause permanent deformation at the faying surfaces.

□ *See also* COLD WELDING, DIFFUSION WELDING, *and* HOT-PRESSURE WELDING.

4F (plate) A welding test position designation for a linear fillet weld applied to a joint where the weld is made in the overhead welding position.

4F (pipe) A welding test position designation for a circumferential fillet weld applied to a joint in pipe, with its axis vertical, where the weld is made in the overhead welding position.

4G A welding test position designation for a linear groove weld applied to a joint where the weld is made in the overhead welding position.

friction welding (FRW) A solid-state welding process that produces a weld under compressive force contact of workpieces rotating or moving relative to one another to produce heat and plastically displace material from the faying surfaces.

□ *See also* DIRECT-DRIVE FRICTION WELDING.

fuel gas A gas such as acetylene, natural gas, hydrogen, propane, stabilized methylacetylene propadiene, and other fuels normally used with oxygen in one of the oxyfuel processes and for heating.

furnace brazing (FB) A brazing process in which the workpieces are placed in a furnace and heated to the brazing temperature.

fused thermal spray deposit A self-fluxing thermal spray deposit that is subsequently heated to coalescence within itself and with the substrate.

fusion welding Any welding process that uses fusion of the base metal to produce the weld.

gas-metal arc cutting (GMAC) An arc-cutting process that uses a continuous consumable electrode and a shielding gas.

gas-metal arc welding (GMAW) An arc-welding process that uses an arc between a continuous filler-metal electrode and the weld pool. *(See figure.)*

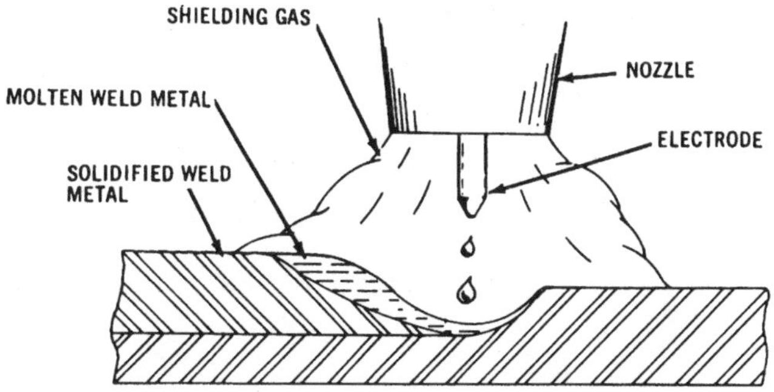

gas-metal arc welding. *(Courtesy Hobart Brothers Co.)*

□ This process is used with shielding from an externally supplied gas and without the application of pressure.

□ *See also* PULSED GAS-METAL ARC WELDING.

gas-shielded arc welding A group of processes including electrogas, flux-cored arc, gas-metal arc, gas-tungsten arc, and plasma arc welding.

gas-shielded flux-cored arc welding (FCAW-G) A flux-cored arc-welding process in which shielding gas is supplied through the gas nozzle, in addition to that obtained from the flux in the electrode.

gas-tungsten arc cutting (GTAC) An arc-cutting process that uses a single tungsten electrode with gas shielding.

gas-tungsten arc welding (GTAW) An arc-welding process that uses an arc between a tungsten electrode (nonconsumable) and the weld pool. *(See figure.)*

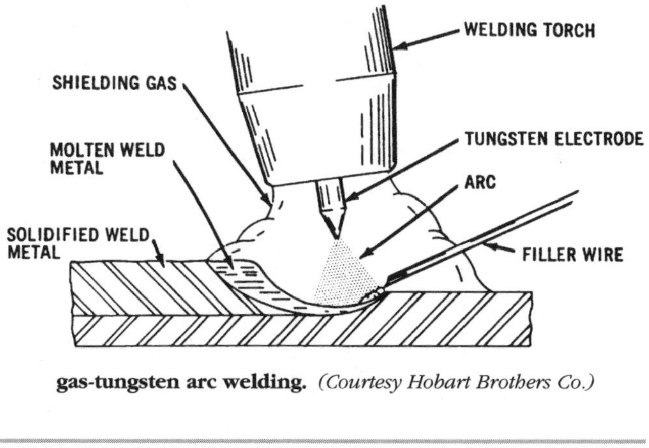

gas-tungsten arc welding. *(Courtesy Hobart Brothers Co.)*

□ The process, sometimes called *tungsten inert-gas (TIG) welding,* is used with shielding gas and without the application of pressure.

□ *See also* HOT-WIRE WELDING *and* PULSED GAS-TUNGSTEN ARC WELDING.

goggles Protective glasses equipped with filter plates set in a frame that fits snugly against the face and used primarily with oxyfuel gas processes.

graduated thermal spray deposit A composite thermal spray deposit composed of mixed materials in successive layers that progressively change in composition from the material adjacent to the substrate to the material at the surface of the thermal spray deposit.

gravity feed welding A shielded-metal arc-welding process for making a fillet weld where a long electrode slides down a tripod-mounted electrode holder as the electrode is consumed.

ground connection An electrical connection of the welding machine frame to the earth for safety.

□ Also called *workpiece connection* and *workpiece lead.*

hand shield A protective device used in arc welding, arc cutting, and thermal spraying for shielding the eyes, face, and neck.

□ It is equipped with a filter plate and is designed to be held by hand.

hardfacing The process of obtaining the correct dimensions and properties by applying an integral layer of metal onto a surface of a base metal or composition to produce an abrasion- and wear-resistant surface.

□ Hardfacing metals have an iron, nickel, copper, or cobalt base.

□ Alloying elements such as carbon, chromium, manganese, nitrogen, silicon, titanium, and vanadium tend to form carbides.

□ Hardfacing can be performed by oxyfuel welding and arc welding.

heat-affected zone The portion of the base metal whose mechanical properties or microstructure has been altered by the heat of welding, brazing, soldering, or thermal cutting.

high-frequency resistance welding A group of resistance welding processes that use high-frequency welding current to concentrate the welding heat at the desired location.

high-frequency upset welding (UW–HF) An upset welding process that transfers high-frequency welding current through the electrode into the workpiece.

high-vacuum electron-beam welding (EBW-HV) An electron-beam welding process in which welding is accomplished at a pressure of 10^{-4} to 10^{-1} Pa (approximately 10^{-6} to 10^{-3} torr).

horizontal welding position (fillet weld) The welding position in which the weld is on the upper side of an approximately horizontal surface and against an approximately vertical surface. *(See figure.)*

hot isostatic pressure welding A diffusion welding process that produces coalescence of metals by heating and applying hot inert gas under pressure.

hot-pressure welding (HPW) A solid-state welding process that produces a weld with heat and application of pressure sufficient to produce macrodeformation of the workpieces.

□ *See also* COLD WELDING, DIFFUSION WELDING, *and* FORGE WELDING.

hot-wire welding A fusion welding process in which a filler-metal wire is resistance-heated by current flowing through the wire as it is fed into the weld pool.

inclusion Entrapped foreign solid material, such as slag, flux, tungsten, or oxide.

incomplete fusion A weld discontinuity in which fusion did not occur between weld metal and fusion faces or adjoining weld beads.

incomplete joint penetration A joint root condition in a groove weld in which weld metal does not extend through the joint thickness.

□ *See also* COMPLETE JOINT PENETRATION WELD, PARTIAL JOINT PENETRATION WELD, *and* JOINT PENETRATION.

indirect welding A resistance welding secondary circuit in which the welding current flows through the workpieces in locations away from, as well as at, the welds for resistance spot, seam, or projection welding.

induction brazing (IB) A brazing process that uses heat from the resistance of the workpieces to induce electric current.

induction seam welding (RSEW-1) A resistance seam welding process in which high-frequency welding current is induced in the workpieces.

□ *See also* HIGH-FREQUENCY RESISTANCE WELDING.

induction upset welding (UW-I) An upset welding process where high-frequency welding current is induced in the workpieces.

□ *See also* HIGH-FREQUENCY RESISTANCE WELDING *and* HIGH-FREQUENCY UPSET WELDING.

induction welding (IW) A welding process that produces coalescence of metals by the heat from the resistance of the workpieces to the flow of induced

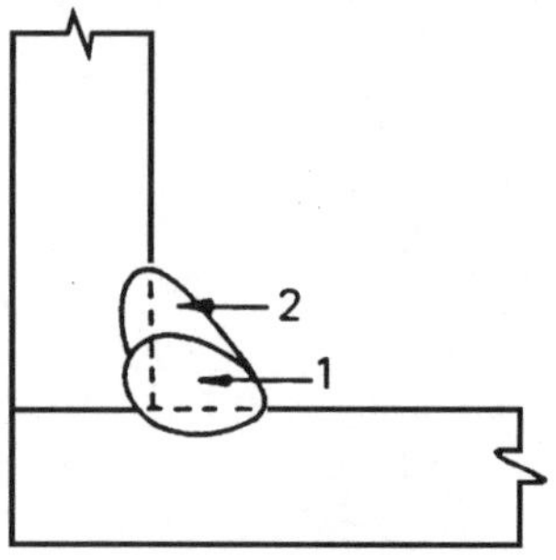

horizontal welding position (fillet weld). (1) First weld bead, (2) second weld bead. (*Courtesy Lincoln Electric Co.)*

high-frequency welding current with or without the application of pressure.

□ The effect of the high-frequency welding current is to concentrate the welding heat at the desired location.

ionic bond A primary bond arising from the electrostatic attraction between two oppositely charged ions.

joint The junction of members or the edges of members that are to be joined or have been joined.

joint penetration The distance that the weld metal extends from the weld face into a joint, exclusive of weld reinforcement.

joint tracking A function of an adaptive control that determines changes in joint location during welding and directs the welding machine to take appropriate action.

kerf The width of the cut produced during a cutting process.

keyhole welding A welding technique in which a concentrated heat source penetrates partially or completely through a workpiece, forming a hole (keyhole) at the leading edge of the weld pool.

□ As the heat source progresses, the molten metal fills in behind the hole to form the weld bead.

laser A device that produces a concentrated coherent light beam by stimulated electronic or molecular transitions to lower energy levels.

□ Laser is an acronym for *l*ight *a*mplification by *s*timulated *e*mission of *r*adiation.

laser-beam cutting (LBC) A thermal cutting process that cuts metal by locally melting or vaporizing with the heat from a laser beam. *(See figure.)*

□ The process is used with or without an assist gas to aid the removal of molten and vaporized material.

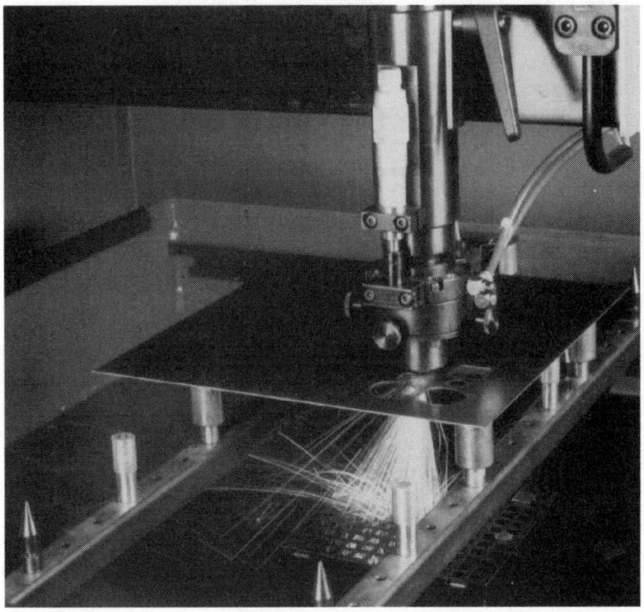

laser-beam cutting. *(Courtesy Emco Maier Corp.)*

□ *See also* LASER-BEAM EVAPORATIVE CUTTING *and* LASER-BEAM OXYGEN CUTTING.

laser-beam evaporative cutting (LBC-EV) A laser-beam cutting process that vaporizes the workpiece, with or without an assist gas, typically an inert gas, to aid the removal of vaporized material.

laser-beam oxygen cutting (LBC-O) A laser-beam cutting process that uses the heat from the chemical reaction between oxygen and the base metal at elevated temperatures.

□ The necessary temperature is maintained with a laser beam.

laser-beam welding (LBW) A welding process that produces coalescence with the heat from a laser beam striking the joint.

□ The process is used without a shielding gas and without the application of pressure.

laser welding This process obtains fusion by directing a highly concentrated beam of coherent light on a very small spot.

□ Welds through transparent materials are possible because there is no contact between the material and the welding equipment.

□ This type of welding is used for connecting leads in integrated circuits, welding heat-treated alloys, and treating difficult-to-weld metals such as tungsten, titanium alloys, aluminum, and nickel alloys.

lasing medium A material that emits coherent radiation by stimulation of electronic or molecular transitions to lower energy.

lightly coated electrode A filler-metal electrode consisting of a metal wire with a light coating applied after the drawing operation, primarily for stabilizing the arc.

liquidus The lowest temperature at which a metal or an alloy is completely liquid.

longitudinal sequence The order in which the weld passes of a continuous weld are made with respect to its length.

□ *See also* BLOCK SEQUENCE *and* CASCADE SEQUENCE.

manual Referring to the control of a process with the torch, gun, or electrode holder held and manipulated by hand.

□ Accessory equipment, such as part motion devices and manually controlled material feeders, may be used.

□ *See also* ADAPTIVE CONTROL WELDING, AUTOMATIC WELDING, MECHANIZED WELDING, ROBOTIC WELDING, *and* SEMIAUTOMATIC WELDING.

manual welding Welding with a torch, gun, or electrode holder held and manipulated by hand.

□ Accessory equipment, such as part motion devices and manually controlled filler material feeders, may be used.

□ Variations of this term are *manual brazing, manual soldering, manual thermal cutting,* and *manual thermal spraying*.

□ *See also* ADAPTIVE CONTROL WELDING, AUTOMATIC WELDING, MECHANIZED WELDING, ROBOTIC WELDING, *and* SEMIAUTOMATIC WELDING.

mechanized welding Welding with equipment that requires manual adjustment of the equipment controls in response to visual observation of the welding, with the torch, gun, or electrode holder held by a mechanical device. *(See figure.)*

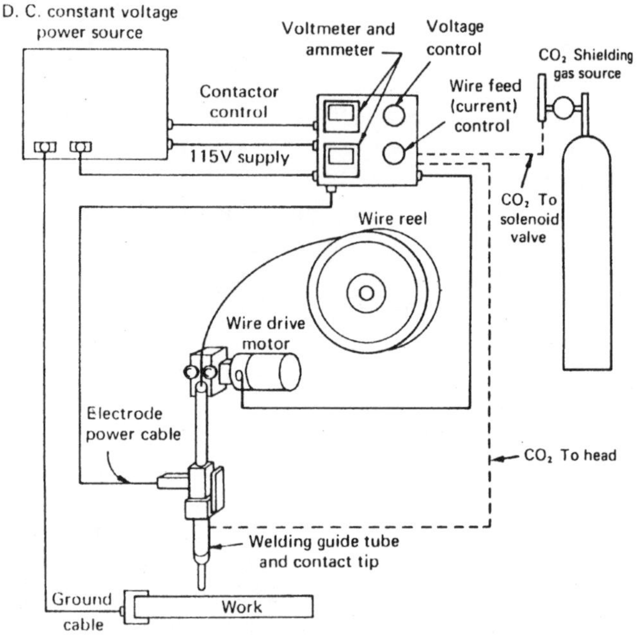

mechanized welding. *(Courtesy Lincoln Electric Co.)*

□ Variations of this term are *mechanized brazing, mechanized soldering, mechanized thermal cutting,* and *mechanized thermal spraying.*

□ *See also* ADAPTIVE CONTROL WELDING, AUTOMATIC WELDING, MANUAL WELDING, ROBOTIC WELDING, *and* SEMIAUTOMATIC WELDING.

medium-vacuum electron-beam welding (EBW-MV) An electron-beam welding process in which welding is done at a pressure of 10^{-1} to 3×10^3 Pa (approximately 10^{-3} to 25 torr).

metal-cored electrode A composite tubular filler metal electrode consisting of a metal sheath and a core of various powdered materials, producing no more than slag islands on the face of a weld bead.

□ External shielding may be required.

metal electrode A filler or nonfiller metal electrode used in arc welding and cutting consisting of a bare or covered metal wire or rod.

metallic bond The principal bond that holds metals together. It is a primary bond arising from the increased spatial extension of the valence electron wavefunctions when an aggregate of metal atoms is brought close together.

□ *See also* BONDING FORCE, COVALENT BOND, *and* IONIC BOND.

metallizing A nonstandard term used for thermal spraying or the application of a metal coating.

metal powder cutting (POC) An oxyfuel cutting process that uses heat from an oxyfuel gas flame, with iron or other metal powder to aid cutting.

mig welding A nonstandard term for gas-metal arc welding and flux-cored arc welding.

multiple-impulse welding A resistance welding process in which welds are made by more than one impulse.

multiple welding position An orientation for a nonrotated circumferential joint requiring welding in more than one welding position.

multiport nozzle A constricting nozzle of the plasma arc torch that contains two or more orifices located in a pattern to provide some control over the arc shape.

narrow-groove welding A welding process that uses multiple-pass welding with filler metal.

□ The use of a small root opening, with either a square or a V groove and a small groove angle, yields a weld with a high depth:width ratio.

neutral flame An oxyfuel gas inner flame that has nonozidizing or nonreducing characteristics. *(See figure.)*

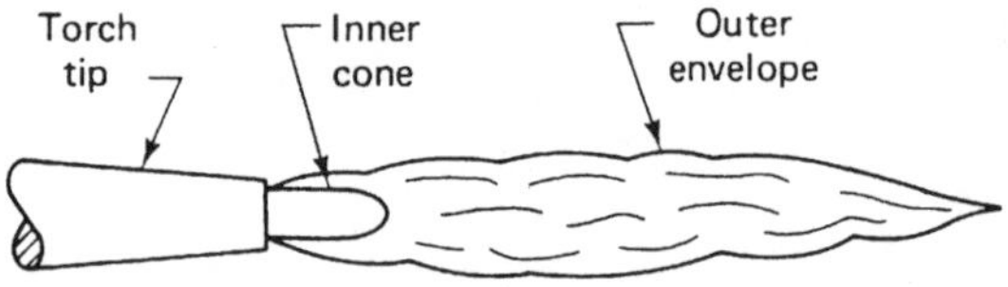

neutral flame. *(Courtesy Lincoln Electric Co.)*

□ Oxygen from the air is used to complete the combustion of CO_2 and H_2 in the inner flame.

□ *See also* CARBURIZING FLAME, OXIDIZING FLAME, *and* REDUCING FLAME.

neutral flux (submerged arc welding) A flux that will not cause a significant change in the weld metal composition when there is a marked change in the arc voltage.

□ *See also* ACTIVE FLUX.

noncorrosive flux A soldering flux that in either its original or residual form does not chemically attack the base metal.

□ Usually composed of rosin-base materials.

nonvacuum electron-beam welding (EBW-NV) An electron-beam welding process in which welding is performed at atmospheric pressure.

1F (pipe) A welding test position designation for a circumferential fillet weld applied to a joint in pipe, with its axis approximately 45° from horizontal, where the weld is made in the flat welding position by rotating the pipe about its axis.

1F (plate) A welding test position designation for a linear fillet weld applied to a joint where the weld is made in the flat welding position.

1G (pipe) A welding test position designation for a

circumferential groove weld applied to a joint in a pipe, where the weld is made in the flat welding position by rotating the pipe about its axis.

1G (plate) A welding test position designation for a linear groove weld applied to a joint where the weld is made in the flat welding position.

orifice gas The gas that is directed into the plasma arc torch or thermal spraying gun to surround the electrode.

□ It becomes ionized in the arc to form the arc plasma, and issues from the constricting orifice of the nozzle as a plasma jet.

oxidizing flame An oxyfuel gas flame in which there is an excess of oxygen, resulting in an oxygen-rich zone extending around and beyond the cone. *(See figure.)*

oxidizing flame. *(Courtesy Lincoln Electric Co.)*

□ *See also* CARBURIZING FLAME, NEUTRAL FLAME, *and* REDUCING FLAME.

oxyacetylene cutting (OFC-A) An oxyfuel gas cutting process that uses acetylene as the fuel gas to cut metal at elevated temperatures through the chemical reaction of oxygen with the base metal.

oxyacetylene welding (OAW) An oxyfuel gas welding process that uses acetylene as the fuel gas.

□ The process may be used without the application of pressure and filler metal.

oxyfuel gas cutting (OFC) A group of oxygen cutting processes that use heat from an oxyfuel (oxyacetylene, oxyhydrogen, oxynatural gas, and oxypropane) gas flame.

oxyfuel gas welding (OFW) A group of welding processes that produce coalescence of workpieces by heating them with an oxyfuel gas flame.

□ The processes are used with or without the application of pressure and with or without filler metal.

oxygen arc cutting (AOC) An oxygen cutting process that uses an arc between the workpiece and a consumable tubular electrode through which oxygen is directed to the workpiece.

□ The necessary temperature is maintained by an arc between the electrode and the base metal.

oxygen cutting (OC) A group of thermal cutting processes that cut or remove metal by means of the chemical reaction between oxygen and the base metal at elevated temperature.

□ The necessary temperature is maintained by the heat from an arc, an oxyfuel gas flame, or other source.

oxygen lance cutting (LOC) An oxygen cutting process that uses oxygen supplied through a consumable lance.

□ The preheat to start the cutting is obtained by other means.

oxyhydrogen cutting (OFC-H) An oxyfuel gas cutting process that uses hydrogen as the fuel gas.

□ The necessary temperature is maintained by the flame resulting from the combustion of hydrogen and oxygen.

oxyhydrogen welding (OHW) An oxyfuel gas welding process that uses hydrogen as the fuel gas.

□ The process is used without the application of pressure.

parallel welding A resistance welding secondary circuit in which the secondary current is divided and conducted through the workpieces and electrodes in parallel electrical paths to simultaneously form multiple resistance spot, seam, or projection welds.

partial joint penetration weld A joint root condition in a groove weld in which an incomplete joint penetration exists.

□ *See also* COMPLETE JOINT PENETRATION, COMPLETE JOINT PENETRATION WELD, INCOMPLETE JOINT PENETRATION, *and* JOINT PENETRATION.

penetration A nonstandard term used for depth of fusion, joint penetration, or root penetration.

percussion welding (PEW) A type of flash welding in which fusion occurs simultaneously by an arc resulting from a rapid discharge of electrical energy.

□ An electric arc, created by the rapid discharge of stored energy, produces the heat that causes fusion.

□ During or immediately after the electrical discharge, pressure is applied as a hammer blow.

plasma arc cutting (PAC) An arc cutting process that uses a constricted arc and removes the molten metal with a high-velocity jet of ionized gas from the constricting orifice.

plasma arc welding (PAW) An arc-welding process that uses a constricted arc between a nonconsumable electrode and the weld pool (transferred arc) or between the electrode and the constricting nozzle (nontransferred arc). *(See figure.)*

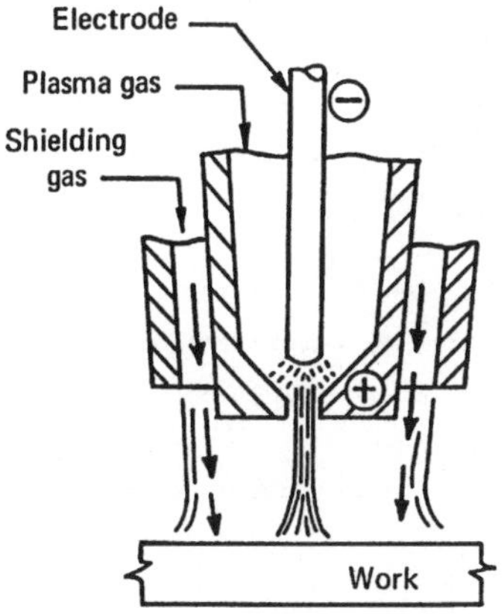

plasma arc welding. *(Courtesy Lincoln Electric Co.)*

□ Shielding is obtained from the ionized gas coming from the torch, which may be supplemented by an auxiliary source of shielding gas.
□ The process is used without the application of pressure.
□ *See also* HOT-WIRE WELDING.

plasma spraying (PSP) A thermal spraying process in which the nontransferred arc of a gun is used to create an arc plasma for melting and propelling the surfacing material to the substrate.

postheating The application of heat to an assembly after welding, brazing, soldering, thermal spraying, or thermal cutting.

powder flame spraying A flame-spraying process in which the surfacing material is in powder form.
□ *See also* FLAME SPRAYING.

preheat The heat applied to the base metal or substrate to attain and maintain preheat temperature.

prequalified welding procedure specification A welding procedure specification that complies with the stated conditions of a particular welding code or specification and is therefore acceptable for use under that code or specification without a requirement for testing.

pressure-controlled welding A resistance welding process in which a number of spot or projection welds are made with several electrodes functioning progressively under the control of a pressure-sequencing device.

pressure gas welding (PGW) An oxyfuel gas welding process that produces a weld simultaneously over the entire faying surfaces.
□ This process is used with the application of pressure and without filler metal.

projection welding (PW) A resistance welding process that produces a weld by the heat obtained from the resistance to the flow of the welding current.
□ The resulting welds are localized at predetermined points by projections, embossments, or intersections.

protective atmosphere A gas or vacuum envelope surrounding the workpieces, used to prevent or reduce the formation of oxides and other detrimental surface substances, and to aid in their removal.

puddle A nonstandard term used for weld pool.

pulsed gas-metal arc welding (GMAW-P) A gas-metal arc-welding process in which the current is pulsed.
□ *See also* PULSED POWER WELDING.

pulsed gas-tungsten arc welding (GTAW-P) A gas-tungsten arc-welding process in which the current is pulsed.
□ *See also* PULSED POWER WELDING.

pulsed laser A laser whose output is controlled to produce a pulse whose duration is 25 ms or less.

pulsed power welding An arc-welding process in which the power is cyclically programmed to pulse so that effective but short-duration values of power can be used.
□ Such short-duration values are significantly different from the average value of power.
□ Equivalent terms are *pulsed voltage* or *pulsed current welding*.
□ *See also* PULSED SPRAY WELDING.

pulsed spray welding An arc-welding process in which the current is pulsed to utilize the advantages of the spray mode of metal transfer at average currents equal to or less than the globular-to-spray transition current.

push angle The angle between the electrode and the work when the electrode is pointing in the direction of the weld travel.

reaction flux (soldering) A flux composition in which one or more of the ingredients react with a base metal on heating to deposit one or more metals.

residual stress The stress present in a joint member or material that is free of external forces or thermal gradients.

resistance projection welding (RPW) A resistance welding process similar to spot welding that uses an embossment, intersection, or projection to localize and direct the flow of electric current.
□ The projections, or small raised areas, can be circular, diamond, oblong, or round.
□ This method is used to join galvanized sheets, stainless steels, and steel plate.

resistance seam welding (RSEW) A resistance welding process that produces a weld at the faying surfaces of overlapped parts progressively along a length of a joint. *(See figure.)*

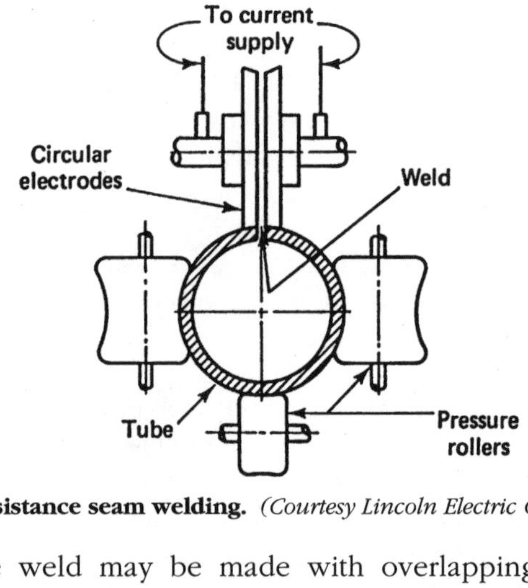

resistance seam welding. *(Courtesy Lincoln Electric Co.)*

□ The weld may be made with overlapping weld nuggets, with a continuous weld nugget, or by forging the joint as it is heated to the welding temperature by resistance to the flow of the welding current.
□ Rotating electrode wheels are used instead of cylindrical or shaped electrodes.

□ *See also* HIGH-FREQUENCY RESISTANCE WELDING *and* INDUCTION SEAM WELDING.

resistance spot welding (RSW) A resistance welding process that produces a weld at the faying surfaces of a joint by the heat obtained from resistance to welding current flow through the workpieces from electrodes that concentrate the welding current and pressure at the weld area.

resistance welding (RW) A group of welding processes in which coalescence is produced by the heat obtained from the resistance of the workpiece to the flow of low-voltage, high-density electric current.

□ During the welding cycle, mating surfaces are usually joined as the result of heat and pressure; they do not have to melt for the weld to occur.

□ Processes include spot (RSW), seam (RSEW), flash (FW), high-frequency seam (RSEW-HF), percussion (PEW), projection (PW), and upset (UW) welding.

resistance welding electrode The part of a resistance welding machine through which the welding current and, in most cases, force are applied directly to the workpiece.

□ The electrode may be a rotating wheel, roll, bar, cylinder, plate, clamp, chuck, or modification thereof.

robotic welding Welding that is automatically performed and controlled by robotic equipment. *(See figure.)*

robotic welding. *(Courtesy Cincinnati Milacron, Inc.)*

□ Variations of this term are *robotic brazing, robotic soldering, robotic thermal cutting,* and *robotic thermal spraying.*

□ *See also* ADAPTIVE CONTROL WELDING, AUTOMATIC WELDING, MANUAL WELDING, MECHANIZED WELDING, *and* SEMIAUTOMATIC WELDING.

rotary roughening (thermal spraying) A method of surface roughening in which a revolving tool is pressed against the surface being prepared, while either the work or the tool, or both, move.

seal-bonding material (thermal spraying) A material that partially forms, in the as-sprayed condition, a metallic bond with the substrate.

seam weld A continuous weld made between or on overlapping members, where coalescence may start and occur on the faying surfaces, or may have proceeded from the outer surface of one member.

□ The continuous weld may consist of a single weld bead or a series of overlapping spot welds.

□ *See also* ARC SEAM WELD *and* RESISTANCE SEAM WELDING.

seam welding (RSEW) This process is similar to spot welding except that the spots are spaced so close that they actually overlap each weld to make a continuous seam.

□ Roller-type electrodes in the form of wheels are used, and the workpieces are cooled by a constant stream of water.

self-fluxing alloy (thermal spraying) A surfacing material that wets the substrate and coalesces when heated to its melting point, with no flux other than the boron and silicon contained in the alloy.

semiautomatic welding Manual welding with equipment that automatically controls one or more of the welding conditions.

□ Variations of this term are *semiautomatic brazing, semiautomatic soldering, semiautomatic thermal cutting,* and *semiautomatic thermal spraying.*

□ *See also* ADAPTIVE CONTROL WELDING, AUTOMATIC WELDING, MANUAL WELDING, MECHANIZED WELDING, *and* ROBOTIC WELDING.

shadow mask (thermal spraying) A device that partially shields an area of the workpiece, producing a feathered edge of the thermal spray deposit.

shielded carbon arc welding (CAW-S) A carbon arc-welding process that uses shielding from the combustion of solid material fed into the arc, or from a blanket of flux on the workpieces, or both.

□ Pressure and filler metal may or may not be used.

shielded-metal arc cutting (SMAC) An arc-cutting process that uses a covered electrode to cut metal by the heat of the arc between the electrode and the base metal.

□ Assist gases are not used; metal is removed by gravity.

shielded-metal arc welding (SMAW) An arc-welding

process with an arc between a covered electrode and the weld pool. *(See figure.)*

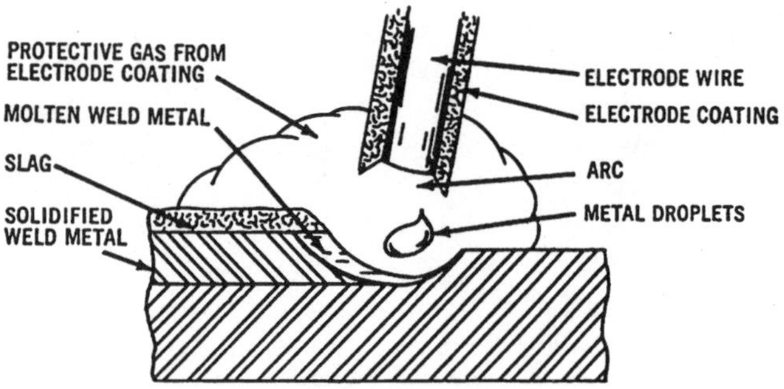

shielded-metal arc welding. *(Courtesy Hobart Brothers Co.)*

□ The process is used with shielding from the decomposition of the electrode covering, without the application of pressure, and with filler metal from the electrode.

□ Also called "firecracker welding."

silver soldering A nonstandard term for brazing with a silver-base filler metal.

6F A welding test position designation for a circumferential fillet weld applied to a joint in pipe, with its axis approximately 45° from horizontal, where the weld is made in flat, vertical, and overhead welding positions.

□ The pipe remains fixed until welding is complete.

6G A welding test position designation for a circumferential groove weld applied to a joint in pipe, with its axis approximately 45° from horizontal, where the weld is made in the flat, vertical, and overhead welding positions.

□ The pipe remains fixed until welding is complete.

6GR A welding test position designation for a circumferential groove weld applied to a joint in a pipe, with its axis approximately 45° from horizontal, where the weld is made in a flat, vertical, and overhead welding position.

□ A restriction ring is added, adjacent to the joint, to restrict access to the weld.

□ The pipe remains fixed until welding is complete.

slot weld A weld made in an elongated hole in one member of a joint, fusing that member to another member.

□ The hole may be open at one end.

□ A fillet welded slot is not to be construed as conforming to this definition.

solder The metal or alloy used as a filler metal in soldering, that has a liquidus not exceeding 450°C (840°F) and below the solidus of the base metal.

soldering (S) A group of welding processes that produce coalescence of materials by heating them to the soldering temperature and by using a filler metal having a liquidus not exceeding 450°C (840°F) and below the solidus of the base metals.

□ The filler metal is distributed between closely fitted faying surfaces of the joint by capillary action.

□ Flux is generally used to ensure wetting by the molten solder.

solid-state welding (SSW) A group of welding processes that produce coalescence by the application of pressure at a welding temperature below the melting temperatures of the base metal and the filler metal.

solidus The highest temperature at which a metal or an alloy is completely solid.

spot weld A weld made between or on overlapping members in which coalescence may start and occur on the faying surfaces or may proceed from the outer surface of one member.

□ The weld cross section is approximately circular.

□ *See also* SPOT WELDING *and* RESISTANCE SPOT WELDING.

spot welding The process in which relatively thin sheets of metal are joined together as in the production of automobiles and similar products. *(See figure.)*

spot welding. *(Courtesy Cincinnati Milacron, Inc.)*

□ The materials are held together by the pressure of the two electrodes.

□ The size and shape of the electrodes control the weld shape.

□ The heat created by the resistance to the electric current flow joins the parts together.

spray transfer (arc welding) Metal transfer in which molten metal from a consumable electrode is propelled axially across the arc in small droplets.

static electrode force The force exerted by electrodes on the workpieces in making spot, seam, or projection welds by resistance welding under welding conditions, but with no current flowing and no movement in the welding machine.

stored-energy welding A resistance welding process in which welds are made with electrical energy accumulated electrostatically, electromagnetically, or electrochemically at a relatively low rate and made available at the required welding rate.

stranded electrode A composite filler-metal electrode consisting of stranded wires that may mechanically enclose materials to improve properties, stabilize the arc, or provide shielding.

stress-relief cracking Intergranular cracking in the heat-affected zone or weld metal as a result of the combined action of residual stresses and postweld exposure to an elevated temperature.

stress-relief heat treatment Uniform heating of a structure or a portion thereof to a sufficient temperature to relieve the major portion of the residual stresses, followed by uniform cooling.

stud welding (SW) A fusion arc stud welding process in which a stud or other fastener acts as an electrode. *(See figure.)*

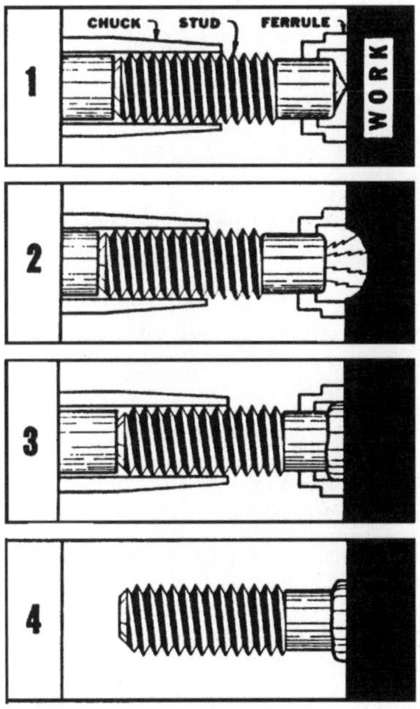

stud arc welding. *(Courtesy Lincoln Electric Co.)*

□ A high-current dc discharge creates the weld puddle, and the stud is plunged into the puddle to complete the joint.

submerged arc welding (SAW) An arc-welding process that uses an arc or arcs between a bare metal electrode or electrodes and the weld pool. *(See figure.)*

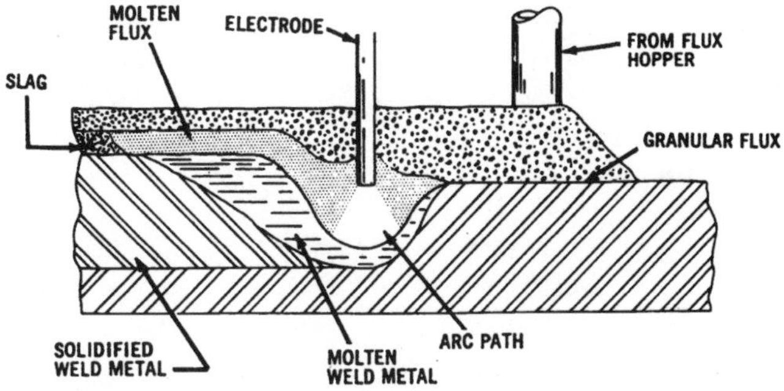

submerged arc welding. *(Courtesy Hobart Brothers Co.)*

□ The arc and the molten metal are shielded by a blanket of granular flux on the workpieces.

□ The process is used without pressure and with filler metal from the electrode and sometimes from a welding rod, flux, or metal granules.

substrate Any material to which a thermal spray deposit is applied.

surface roughening A group of methods for producing irregularities on a surface.

□ *See also* DOVETAILING *and* ROTARY ROUGHENING.

surfacing The application by welding, brazing, or thermal spraying of a layer of material to a surface in order to obtain desired properties or dimensions.

□ *See also* BUILDUP, BUTTERING, CLADDING, *and* HARDFACING.

sweat soldering A soldering process in which workpieces that have been precoated with solder are reheated and assembled into a joint without the use of additional solder.

tack weld A weld made to hold the parts of a weldment in proper alignment until the final welds are made.

tension test A test in which a specimen is loaded in tension until failure occurs.

thermal cutting (TC) A group of cutting processes that cut or remove metal by localized melting, burning, or vaporizing of the workpieces.

□ *See also* ARC CUTTING, ELECTRON-BEAM CUTTING, LASER-BEAM CUTTING, *and* OXYGEN CUTTING.

thermal spraying (THSP) A group of processes in which finely divided metallic or nonmetallic surfacing materials are deposited in a molten or semimolten condition on a substrate to form a thermal spray deposit.

□ The surfacing material may be powder, rod, cord, or wire.

□ Three main methods of applying thermal spray are with wire, powder, and crucible guns.

□ *See also* ARC SPRAYING, FLAME SPRAYING, *and* PLASMA SPRAYING.

thermite welding (TW) A welding process that produces coalescence of metals by heating them with superheated liquid metal from a chemical reaction between a metal oxide and aluminum, with or without the application of pressure.

□ Filler metal is obtained from the liquid metal.

3F A welding test position designation for a linear fillet weld applied to a joint in which the weld is made in the vertical welding position.

3G A welding test position designation for a linear groove weld applied to a joint in which the weld is made in the vertical welding position.

throat area The area bounded by the physical parts of the secondary circuit in a resistance spot, seam, or projection welding machine.

□ Used to determine the dimensions of a part that can be welded and determine, in part, the secondary impedance of the equipment.

travel angle The angle less than 90° between the electrode axis and a line perpendicular to the weld axis, in a plane determined by the electrode and weld axes.

□ This angle can also be used to partially define the position of guns, torches, rods, and beams.

□ *See also* DRAG ANGLE.

tungsten electrode A nonfiller metal electrode, consisting primarily of tungsten, which is used in arc cutting, arc welding, and plasma spraying.

2F (pipe) A welding test position designation for a circumferential fillet weld applied to a pipe joint, with its axis approximately vertical, in which the weld is made in the horizontal welding position.

2F (plate) A welding test position designation for a linear fillet weld applied to a joint in which the weld is made in the horizontal welding position.

2FR A welding test position designation for a circumferential fillet weld applied to a pipe joint, with its axis approximately horizontal, in which the weld is made in the horizontal welding position by rotating the pipe about its axis.

2G (pipe) A welding test position designation for a circumferential groove weld applied to a pipe joint, with its axis approximately vertical, in which the weld is made in the horizontal welding position.

2G (plate) A welding test position designation for a linear groove weld applied to a joint in which the weld is made in the horizontal welding position.

ultrasonic welding (USW) A solid-state welding process that uses high-frequency vibrations in overlapping areas of metals to join similar and dissimilar metals.

□ Fluxes and filler metals are not required, and only localized heating occurs.

□ No melting occurs because temperatures are below the melting point of the materials.

upset A bulk deformation that results from the application of pressure during a welding operation.

□ This upset may be measured as a percent increase in interface area, a reduction in length, a percent reduction in lap joint thickness, or a reduction in cross-wire, weld-stack height.

upset welding (UW) A resistance welding process that produces coalescence over the entire faying surface area or along a butt joint by the heat obtained from the resistance to the welding current flow through the area where those surfaces are in contact.

□ Pressure is applied as a hammer blow during or immediately after the electrical discharge.

□ *See also* INDUCTION UPSET WELDING.

vacuum plasma-spraying (VPSP) A thermal spraying process using a plasma spraying gun confined to a stable enclosure that is partially evacuated.

vertical welding position The welding position in which the weld axis, at the point of welding, is approximately vertical and the weld face lies in an approximately vertical plane. *(See figure.)*

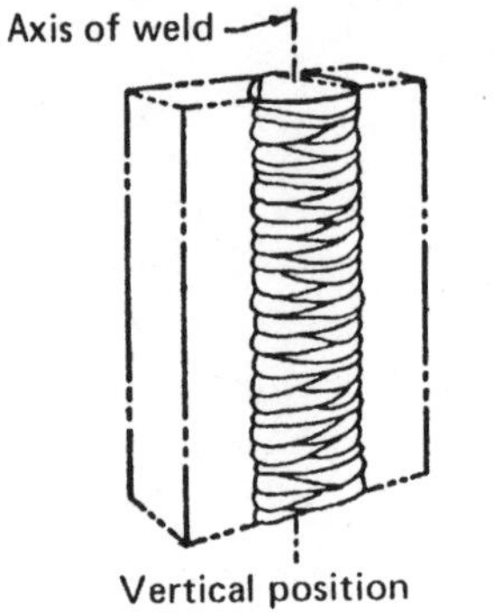

vertical welding position. *(Courtesy Lincoln Electric Co.)*

weld A localized coalescence of metals or nonmetals produced either by heating the materials to the welding temperature, with or without the application of pressure, or by the application of pressure alone, with or without the use of filler material.

weld symbol A graphical character connected to the welding symbol indicating the type of weld.

weldability The properties of material that makes it suitable to be welded into a specific structure and to perform as intended in service.

welding A joining process that produces coalescence of materials by heating them to the welding temperature, with or without the application of pressure or by the application of pressure alone, and with or without the use of filler metal.

□ *See also* ARC WELDING, GTAW, RESISTANCE WELDING, *and* FORGE WELDING.

welding arc A controlled electrical discharge between the electrode and the workpiece that is formed and maintained by a gaseous conductive medium, called an *arc plasma*.

welding electrode A component of the welding circuit through which current is conducted and terminates at the arc, molten conductive slag, or base metal.

□ *See also* ARC WELDING ELECTRODE, BARE ELECTRODE, CARBON ELECTRODE, COMPOSITE ELECTRODE, COVERED ELECTRODE, ELECTROSLAG WELDING ELECTRODE, FLUX-CORED ELECTRODE, METAL-CORED ELECTRODE, METAL ELECTRODE, RESISTANCE WELDING ELECTRODE, STRANDED ELECTRODE, *and* TUNGSTEN ELECTRODE.

welding rod *See* WELDING ELECTRODE *and* WELDING WIRE.

welding wire A form of welding filler metal, normally packaged as coils or spools, that may or may not conduct electrical current depending on the welding process used.

□ *See also* WELDING ELECTRODE.

work angle The angle made by the electrode with a line at 90° (perpendicular) to the weld axis.

Common Abbreviations

Computers

ABC	Atanasoff-Berry Computer
ALU	Arithmetic/Logic Unit
APL	A Programming Language
ATM	Asynchronous Transfer Mode
BCD	Binary-Coded Decimal
BIOS	Basic Input/Output System
BIT	Binary Digit
CAI	Computer-Assisted Instruction
CD-I	Compact Disk—Interactive
CD-ROM	Compact Disk—Read Only Memory
CGA	Color Graphics Adapter
CISC	Complex Instruction Set Computer
CMOS	Complementary Metal Oxide Semiconductor
COBOL	Common Business-Oriented Language
CP/M	Control Program for Microcomputers
CPS	Characters per Second
CPU	Central Processing Unit
CRT	Cathode-Ray Tube
DASD	Direct-Access Storage Device
DDP	Distributed Data Processing
DMA	Direct Memory Access
DRAM	Dynamic Random-Access Memory
DVI	Digital-Video Interactive
EGA	Enhanced Graphics Adapter
EIA	Electronic Industries Association
EISA	Extended Industry Standard Architecture
ENIAC	Electronic Numerical Integrator and Calculator
EPROM	Erasable Programmable Read-Only Memory
EPS	Encapsulated PostScript
FAT	File Allocation Table
FAX	Facsimile Machine
GB	Gigabyte
GUI	Graphical User Interface
HEX	Hexadecimal
HMA	High-Memory Area
IBG	Interblock Gap
IC	Integrated Circuit
I/O	Input/Output
IPX	Internetwork Packet Exchange
ISA	Industry Standard Architecture
JCL	Job Control Language
K	Kilobyte
LAN	Local Area Network
LCD	Liquid Crystal Display
LISP	List Processor
LSI	Large-Scale Integration
MAN	Metropolitan Area Network
Mbits/s	Megabits per Second
Mbyte	Megabyte
MHz	Megahertz
MICR	Magnetic-Ink Character Recognition
MIDI	Music Instrument Digital Interface
mips	Million Instructions per Second
MIS	Management Information System
MNP	Microcom Networking Protocol
MODEM	*Mo*dulator-*Dem*odulator
MS-DOS	Microsoft Disk Operating System
MTBF	Mean Time between Failures
NCP	Network Control Program
NIC	Network Interface Card
NLQ	Near–Letter Quality
OCR	Optical Character Reader
OLE	Object Linking and Embedding
OMR	Optical-Mark Reader
OS	Operating System
PC	Personal Computer
PL/1	Programming Language 1
PNP	Plug and Play
POST	Power-On Self-Test
PPM	Pages per Minute
PROM	Programmable Read-Only Memory
RAM	Random-Access Memory
RFM	Radio-Frequency Modulator
RGB	Red-Green-Blue Monitors
RISC	Reduced Instruction Set Computing
RJE	Remote Job Entry
ROM	Read-Only Memory
RPG	Report Program Generator
SCSI	Small Computer System Interface
SIMM	Single In-line Memory Module
SPA	Software Publishers Association

SQL	Structured Query Language	VGA	Video Graphics Array
SVGA	Super Video Graphics Array	VL-BUS	VESA Local Bus
TSR	Terminate and Stay Resident	VLSI	Very Large-Scale Integration
UMA	Upper-Memory Area	WAN	Wide Area Network
UPS	Uninterruptible Power Supply	WORM	Write Once, Read Many
VESA	Video Electronics Standards Association	WYSIWYG	What You See Is What You Get

The Internet

ACK	Acknowledgment
AD	Administrative Domain
ADN	Advanced Digital Network
AOL	America OnLine
API	Application Programming Interface
ARP	Address Resolution Protocol
AUP	Acceptable-Use Policy
BBS	Bulletin Board System
BIND	Berkeley Internet Name Domain
BITNET	Because It's Time Network
BPS	Bits per Second
BTA	But Then Again
BTW	By the Way
CGI	Common Gateway Interface
CNDIR	Clearinghouse for Networked Information Discovery and Retrieval
COM	Commercial Domain
CWIS	Campus-Wide Information System
DNS	Domain Name System
e-mail	Electronic Mail
FAQ	Frequently Asked Question
FDDI	Fiber Distributed Data Interface
FTP	File Transfer Protocol
FYA	For Your Amusement
FYI	For Your Information
GIF	Graphics Interchange Format
HTML	HyperText Markup Language
HTTP	HyperText Transport Protocol
IAB	Internet Architecture Board
IETF	Internet Engineering Task Force
INIC	Internet Network Information Center
IP	Internet Protocol
IPng	Internet Protocol: The Next Generation
IRC	Internet Relay Chat
ISDN	Integrated Services Digital Network
ISOC	Internet Society
ISP	Internet Service Provider
JPEG	Joint Photographics Expert Group
LOL	Laughing Out Loud
MBONE	Multicast Backbone
MIME	Multipurpose Internet Mail Extensions
MOO	Mud, Object-Oriented
MPEG	Motion Picture Experts Group
MUSE	Multiuser Simulated Environment
Net	The Internet
NIC	Network Information Center
NNTP	Network News Transmission Protocol
NTP	Network Time Protocol
OCLC	Online Computer Library Catalog
PEM	Privacy-Enhanced Mail
PING	Packet Internet Groper
POP	Post Office Protocol
PPP	Point-to-Point Protocol
RFC	Request for Comments
RTFM	Read the Fine Manual
RTT	Round-Trip Time
SGML	Standard Generalized Markup Language
SLIP	Serial Line Internet Protocol
SMDS	Switched Multimegabit Data Service
SMTP	Simple Mail Transfer Protocol
SNMP	Simple Network Management Protocol
SYSOP	System Operator
TCP	Transmission Control Protocol
TCP/IP	Transmission Control Protocol/Internet Protocol
URL	Uniform Resource Locator
UUCP	Unix-to-Unix Copy Program
VMS	Virtual Memory System
VRML	Virtual Reality Modeling Language
WAIS	Wide Area Information Servers
WWW	World Wide Web
YP	Yellow Pages

Metalworking and Manufacturing

AA	Arithmetic Average
AEM	Assembly Evaluation Method
AFM	Abrasive Flow Machining
AGV	Automated Guided Vehicle
AI	Artificial Intelligence
AISI	American Iron and Steel Institute
AJM	Abrasive Jet Machining
AL	Arm Language
ANOVA	Analysis of Variance
AOQL	Average Outgoing Quality Limit
APC	Automatic Pallet Changer
APT	Automatically Programmed Tools
AQL	Acceptable Quality Level
ASCII	American Standard Code for Information Exchange
ASME	American Society of Mechanical Engineers

ASQ	American Society for Quality
ASRS/W	Automatic Storage and Retrieval System/Warehouse
ASTM	American Society for Testing and Materials
BTR	Behind-the-Tape Reader
BUE	Built-up Edge
CAD	Computer-Aided Design
CAE	Computer-Aided Engineering
CAM	Computer-Assisted Machining
CAPP	Computer-Aided Process Planning
CAR	Computer-Aided Robotics
CBN	Cubic Boron Nitride
CBORE	Counterbore
CD	Continuous Dressing
CHM	Chemical Milling
CIM	Computer-Integrated Manufacturing
CIRP	International Institution for Production Engineering Research
CLA	Centre Line Average (British Standard)
CMM	Coordinate Measuring Machine
CNC	Computer Numerical Control
CPU	Central Processing Unit
CRT	Cathode-Ray Tube
CS	Cutting Speed
CSG	Constructive Solid Geometry
CSK	Countersink
CSS	Constant Surface Speed
CVD	Chemical Vapor Deposition
D	Diamond
DAC	Digital-to-Analog Converter
DCC	Direct Computer Control
DFA	Design for Assembly
DIA	Diameter
DNC	Direct Numerical Control
DOD	Department of Defense
DOE	Design of Experiments
DTC	Difficult to Cut
DTG	Difficult to Grind
DWMI	Diamond Wheel Manufacturers Institute
EBM	Electron-Beam Machining
ECDG	Electrochemical Discharge Grinding
ECG	Electrochemical Grinding
ECH	Electrochemical Honing
ECM	Electrochemical Machining
ECT	Electrochemical Turning
EDG	Electrical Discharge Grinding
EDM	Electrical Discharge Machining
EDWC	Electrical Discharge Wire Cut
ELG	Electrolytic Grinding
EMI	Electromagnetic Interface
EP	Extreme Pressure
ETG	Easy to Grind
FAO	Finish All Over
FAS	Flexible Assembly System
FDM	Fused Deposition Modeling
FEPA	(Federation Européene des Fabricants de Produits Abrasifs) European Federation of Abrasive Manufacturers
FMC	Flexible Manufacturing Cell
FMS	Flexible Manufacturing System
ft/min	Feet per Minute
GT	Group Technology
HAZ	Heat-Affected Zone
HBM	Horizontal Boring Machine
HDM	Hydromatic Machining
HDN	Harden
HERF	High-Energy-Rate Forming
HMC	Horizontal Machining Center
hp	Horsepower
HSLA	High-Strength, Low-Alloy Steel
HSS	High-Speed Steel
ID	Inside Diameter
IDA	Industrial Diamond Association of America, Inc.
in	Inch
ISO	International Standards Organization
IVD	Ion Vapor Deposition
JIT	Just-in-Time Manufacturing
L	Lead
laser	Light Amplification by Stimulated Emission of Radiation
LBM	Laser-Beam Machining
LBT	Laser-Beam Torch
LERT	Linear, Extensional, Rotational, and Twist
LH	Left Hand
LMC	Least Material Condition
LOM	Laminated-Object Manufacturing
LSI	Large-Scale Integration
m	Meter
MAP	Manufacturing Automation Protocol
MBG	Metal-Bond Diamond Grinding Wheel
MBS	Metal-Bond Diamond Sawing Wheel
MCL	Manufacturing Control Language
MDI	Manual Data Input
MIC	Minimum Inscribed Circle
mm	Millimeter
MMC	Maximum Material Condition
MRP	Material Requirements Planning
MRR	Material-Removal Rate
NC	Numerical Control
NPT	National Pipe Thread
OC	Operating Characteristic Curve
OD	Outside Diameter
P	Pitch
PAM	Plasma Arc Machining
PAU	Position Analog Unit
PCA	Postcuring Apparatus
PCBN	Polycrystalline Cubic Boron Nitride
PCD	Polycrystalline Diamond
PH	Precipitation Hardening
PLC	Programmable Logic Controller
PM	Powder Metallurgy

psi	Pounds per Square Inch (lb/in^2)
PTP	Point-to-Point Control/Positioning
PVD	Physical Vapor Deposition
QC	Quality Control [Rockwell Hardness Test Scale (Rc)]
R&D	Research and Development
RH	Right Hand
r/min	Revolutions per Minute
RMS	Root Mean Square
RP	Rapid Response
RPL	Robot Programming Language
RP&M	Rapid Prototyping and Manufacturing
RUM	Rotary Ultrasonic-Assisted Machining
SAE	Society of Automotive Engineers
SCARA	Selective Compliance Articulated Robot Arm
SiC	Silicon Carbide
sf/min	Surface (Linear) Feed per Minute
SGC	Solid-Ground Curing
SL	Stereolithography
SLA	Stereolithography Apparatus
SLS	Selective Laser Sintering
SPC	Statistical Process Control
SQC	Statistical Quality Control
STEM	Shaped-Tube Electrolytic Machining
TCP	Tool Center Point
TDS	Tap Drill Size
TEM	Thermal Energy Finishing
t/ft	Taper per Foot
THD	Thread
TIR	Total Indicated Runout
tpi	Threads per Inch
TQC	Total Quality Control
TQM	Total Quality Management
TTT	Time-Temperature Transformation
UAM	Ultrasonic-Assisted Machining
UMC	Universal Machining Center
UNC	Unified National Coarse
UNF	Unified National Fine
UNS	Unified Numbering System
USM	Ultrasonic Machining
UT	Ultrasonic Testing
VMC	Vertical Machining Center
WCS	World Coordinate System
WJC	Waterjet Cutting

Plastics

ACS	American Chemical Society
APC	Automated Process Control
ASTM	American Society for Testing and Materials
BOPP	Biaxially Oriented Polypropylene
Btu	British Thermal Unit
CAMM	Computer-Assisted Moldmaking
CM	Compression Molding
CR	Polychloroprene Rubber
DB	Dinkelberry (Caused by Leaking Die Endplates, Deckles, or Other Extruder Matting Surfaces)
deg	Degree (Angle)
E	Modulus of Elasticity
EAA	Ethylene Acrylic Acid (Copolymer)
EBM	Extrusion Blow Molding
EEA	Ethylene Ethyl Acrylate (Copolymer)
elong	Elongation
EMAA	Ethylene Methyl Acrylic Acid (Copolymer)
EMI	Electromagnetic Interference
EnBA	Ethylene-*n*-Butyl Acrylate (Copolymer)
ESCR	Environmental Stress Cracking Resistance
EVA	Ethylene Vinyl Acetate (Copolymer)
EVOH	Ethylene Vinyl Alcohol (Copolymer)
FDA	Food and Drug Administration, the U.S.
flex	Flexural
FPA	Flexible Packaging Association
FR	Flame Retardant
FRP	Fiber-Reinforced Plastic
g	Gram
GP	General-Purpose
HALS	Hindered Amine Light Stabilizer
H/C	Hopper Car
HDPE	High-Density Polyethylene
HMW	High Molecular Weight
IBM	Injection Blow Molding
imp	Impact
IPN	Interpenetrating Polymer Network
IR	Infrared
J	Joule
K	Kelvin
kpsi	1000 pounds per square inch (1000 lb/in^2)
LCP	Liquid Crystal Polymer Plastic
lbf	Pound-force
L/D	Length-to-Diameter Ratio of Screw
LDPE	Low-Density Polyethylene
LLDPE	Linear Low-Density Polyethylene
MD	Machine Direction
MDPE	Medium-Density Polyethylene
MI	Melt Index
MIL	Military, as in Military Standard (MIL-STD)
mod	Modulus
mol%	Mole Percent
MVTR	Moisture Vapor Transmission Rate
MW	Molecular Weight
N	Newton
NR	Natural Rubber
OPP	Oriented Polypropylene
PA	Polyamide
PAN	Polycrylonitrile Plastic
PAR	Polyarylite Plastic
PBT	Polybutylene Terephthalate
PE	Polyethylene

PET	Polyethylene Terephthalate
PI	Polyimide Plastic
PP	Polypropylene
pphr	Parts per Hundred Resin, Parts per Hour
ppm	Parts per Million
PS	Polystyrene Plastic
psi	Pounds per Square Inch (lb/in^2)
PUR	Polyurethane Plastic
PVB	Polyvinyl Plastic
PVDC	Polyvinylidene Chloride
PVDF	Polyvinylidene Flouride Plastic
RH	Relative Humidity
RIM	Reaction Injection Molding
r/min	Revolutions per Minute
RP	Reinforced Plastic
RTM	Resin Transfer Molding
SAN	Styrene-Acrylonitrile (Copolymer)
SBM	Stretch Blow Molding
SCR	Silicon Controlled Rectifier
SCT	Soluble-Core Technology
SPC	Statistical Process Control
SPE	Society of Plastics Engineers
sp gr	Specific Gravity
SPI	Society of the Plastics Industry, The
SQC	Statistical Quality Control
TAPPI	Technical Association of the Pulp and Paper Industry
TD	Transdirectional
ten.	Tensile
Tg	Glass Transition Temperature (Crystalline Polymers)
T/L	Truckload
Tm	Melt Temperature (Amorphous Polymers)
TM	Transfer Molding
TP	Thermoplastic
TPO	Thermoplastic Olefin
UHMW	Ultra-High Molecular Weight HDPE
UF	Urea Formaldehyde Plastic
ult	Ultimate
UV	Ultraviolet
VAM	Vinyl Acetate Monomer
WVTR	Water Vapor Transmission Rate
yld	Yield

Welding

AAW	Air-Acetylene Welding
ABW	Arc Braze Welding
AC	Arc Cutting
AHW	Atomic Hydrogen Welding
AOC	Oxygen Arc Cutting
ASP	Arc Spraying
AW	Arc Welding
B	Brazing
BMAW	Bare-Metal Arc Welding
BW	Braze Welding
CABW	Carbon Arc Braze Welding
CAC	Carbon Arc Cutting
CAC-A	Air Carbon Arc Cutting
CAW	Carbon Arc Welding
CAW-S	Shielded-Carbon Arc Welding
CEW	Coextrusion Welding
CW	Cold Welding
DB	Dip Brazing
DFB	Diffusion Brazing
DFW	Diffusion Welding
DS	Dip Soldering
EBBW	Electron-Beam Braze Welding
EBC	Electron-Beam Cutting
EBW	Electron-Beam Welding
EBW-HV	High-Vacuum Electron-Beam Welding
EBW-MV	Medium-Vacuum Electron-Beam Welding
EBW-NV	Nonvacuum Electron-Beam Welding
EGW	Electrogas Welding
ESW	Electroslag Welding
EXB	Exothermic Brazing
EXBW	Exothermic Braze Welding
EXW	Explosion Welding
FB	Furnace Brazing
FCAW	Flux-Cored Arc Welding
FCAW-G	Gas-Shielded Flux-Cored Arc Welding
FLOW	Flow Welding
FLSP	Flame Spraying
FOW	Forge Welding
FRW	Friction Welding
FW	Flash Welding
GMAC	Gas-Metal Arc Cutting
GMAW	Gas-Metal Arc Welding
GMAW-P	Pulsed Gas-Metal Arc Welding
GTAC	Gas-Tungsten Arc Cutting
GTAW	Gas-Tungsten Arc Welding
GTAW-P	Pulsed Gas-Tungsten Arc Welding
HPW	Hot-Pressure Welding
IB	Induction Brazing
IW	Induction Welding
LBC	Laser-Beam Cutting
LBC-EV	Laser-Beam Evaporative Cutting
LBC-O	Laser-Beam Oxygen Cutting
LBW	Laser-Beam Welding
LOC	Oxygen Lance Cutting
LW	Laser Welding
OAW	Oxyacetylene Welding
OC	Oxygen Cutting
OFC	Oxyfuel Gas Cutting
OFC-A	Oxyacetylene Cutting
OFC-H	Oxyhydrogen Cutting
OFW	Oxyfuel Gas Welding
OHW	Oxyhydrogen Welding

PAC	Plasma Arc Cutting
PAW	Plasma Arc Welding
PEW	Percussion Welding
PGW	Pressure Gas Welding
POC	Metal Powder Cutting
PSP	Plasma Spraying
PW	Projection Welding
RPW	Resistance Projection Welding
RSEW	Resistance Seam Welding
RSEW-I	Induction Seam Welding
RSW	Resistance Spot Welding
RW	Resistance Welding
S	Soldering
SAW	Submerged Arc Welding
SMAC	Shielded-Metal Arc Cutting
SMAW	Shielded-Metal Arc Welding
SSW	Solid-State Welding
SW	Stud Welding
TC	Thermal Cutting
THSP	Thermal Spraying
TW	Thermite Welding
USW	Ultrasonic Welding
UW	Upset Welding
UW-I	Induction Upset Welding
VPSP	Vacuum Plasma Spraying

APPENDIX A

Geometric Symbols, Characteristics, and Definitions

The latest published standard ASME Y14.5M—1994, Dimensioning and Tolerancing, a revision of the ANSI Y14.5M—1982 standard, has adopted many of the ISO standards to improve national and international drawing communications. The 1994 standard contains additions and modifications to the 1982 standard; some of the more important changes are

- ☐ The adoption of the universal (ISO) datum feature symbol

  now replaces 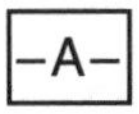

  The △ in the symbol can be open or closed ▲.
- ☐ The RFS symbol Ⓢ has been discontinued.
- ☐ The placement of the *projected tolerance zone* symbol and its height in the *feature control zone* has been modified.
- ☐ The symmetry symbol ⌯ has been reinstated to be applied only on a RFS basis.

The following tables on Geometric Symbols, Characteristics, and Definitions use the 1994 standard in some of the more industrial applications.

Table A1. Geometric Symbols and Definitions

▱ | .002	⌖ | Ø .005 Ⓜ | A
Individual Features	Related Features

Feature-Control Frame

A specification box that shows a particular geometric characteristic (flatness, straightness, etc.) applied to a part feature and states the allowable tolerance. The feature's tolerance may be individual, or related to one or more datums. Any datum references and tolerance modifiers are also shown.

[A] (datum flag) **Datum Feature** A flag which designates a physical feature of the part to be used as a reference to measure geometric characteristics of other part features.	A1 A2 Ø.10 A3 **Datum Targets** Callouts occasionally needed to designate specific points, lines, or areas on an actual part to be used to establish a theoretical datum feature.
[1.500] **Basic Dimension** A box around any drawing dimension makes it a "basic" dimension, a theoretically exact value used as a reference for measuring geometric characteristics and tolerance of other part features.	Ø **Cylindrical Tolerance Zone** This symbol, commonly used to indicate a diameter dimension, also specifies a cylindrically shaped tolerance zone in a feature-control frame.
Ⓜ **Maximum Material Condition** Abbreviation: MMC. A tolerance modifier that applies the stated tight tolerance zone only while the part theoretically contains the maximum amount of material permitted within its dimensional limits (e.g. minimum hole diameters and maximum shaft diameters), allowing more variation under normal conditions.	Ⓛ **Least Material Condition** Abbreviation: LMC. A tolerance modifier that applies the stated tight tolerance zone only while the part theoretically contains the minimum amount of material permitted within its dimensional limits (e.g. maximum hole diameters and minimum shaft diameters), allowing more variation under normal conditions.
RFS **(Regardless of Feature Size)** For all applicable geometric tolerances, "regardless of feature size" (RFS) applies to the individual tolerance, datum reference, or both where no modifying symbol is specified.	Ⓟ ⌖ | Ø .008 Ⓜ Ⓟ XX | A | B | C **Projected Tolerance Zone** The Ⓟ symbol followed by a dimension in the feature control frame indicates the minimum height of the tolerance zone.

Table A2. Geometric Characteristics

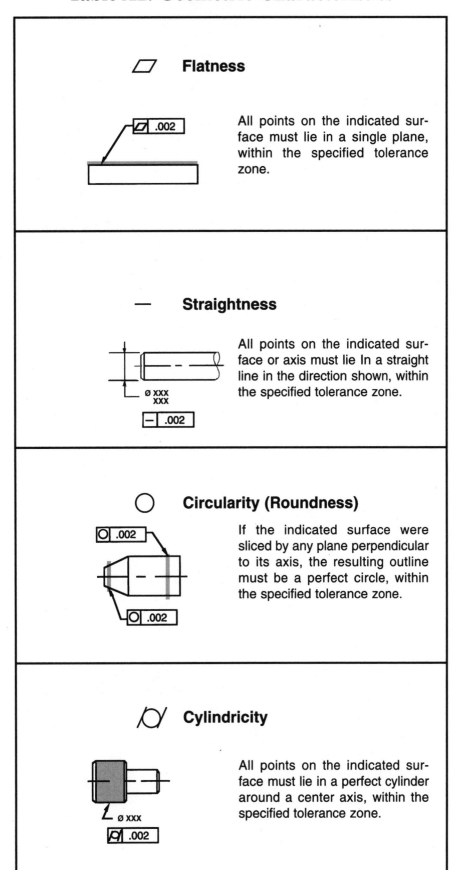

⏥ Flatness

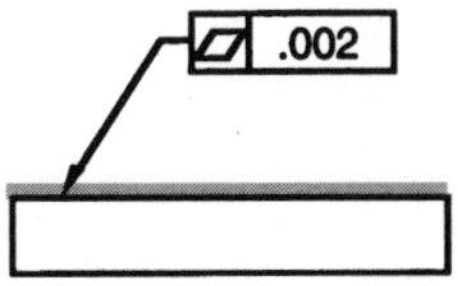

All points on the indicated surface must lie in a single plane, within the specified tolerance zone.

— Straightness

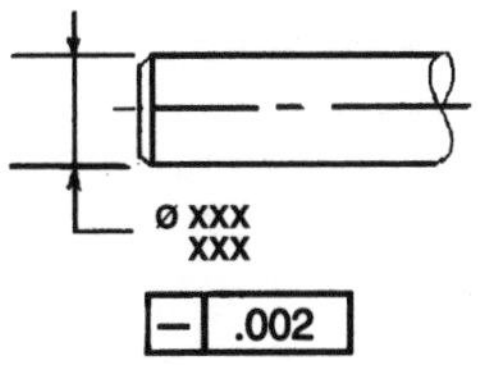

All points on the indicated surface or axis must lie In a straight line in the direction shown, within the specified tolerance zone.

○ Circularity (Roundness)

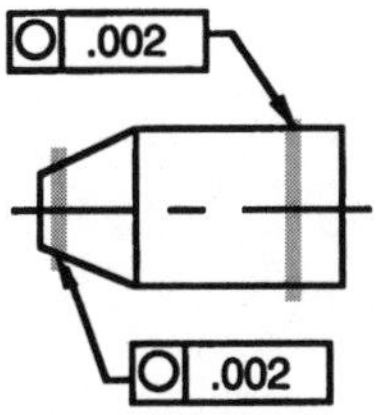

If the indicated surface were sliced by any plane perpendicular to its axis, the resulting outline must be a perfect circle, within the specified tolerance zone.

⌭ Cylindricity

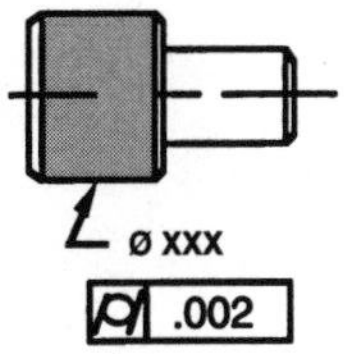

All points on the indicated surface must lie in a perfect cylinder around a center axis, within the specified tolerance zone.

Table A2 (Cont.)

⌒ Linear Profile

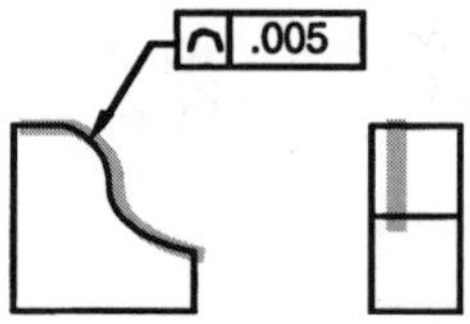

All points on any full slice of the indicated surface must lie on its theoretical two-dimensional profile, as defined by basic dimensions, within the specified tolerance zone. The profile may or may not be oriented with respect to datums.

⌓ Surface Profile

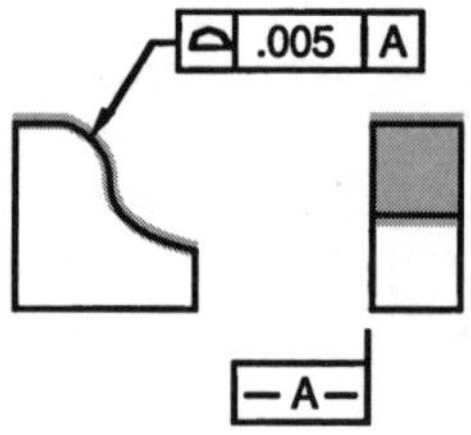

All points on the indicated surface must lie on its theoretical three-dimensional profile, as defined by basic dimensions, within the specified tolerance zone. The profile may or may not be oriented with respect to datums.

Table A2 (Cont.)

⊥ Perpendicularity (Squareness)

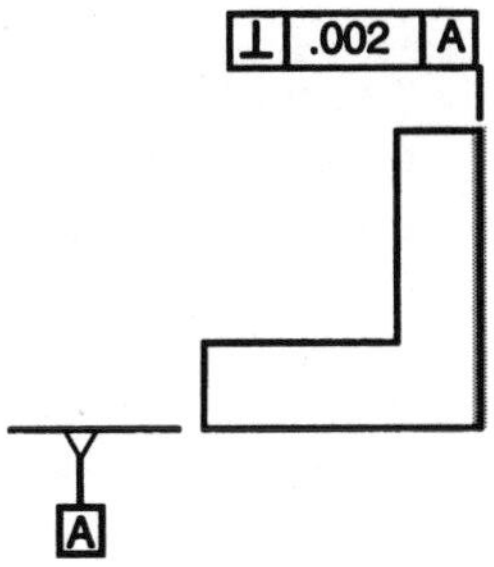

All points on the indicated surface, axis, or line must lie in a single plane exactly 90° from the designated datum plane or axis, within the specified tolerance zone.

∠ Angularity

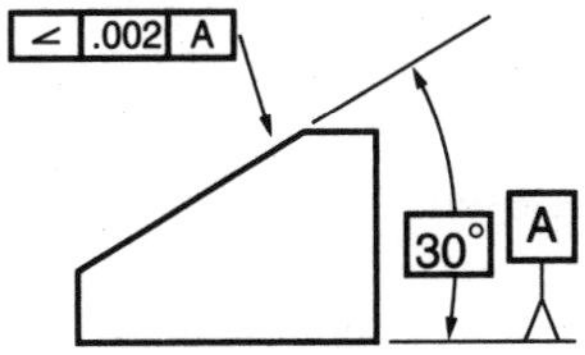

All points on the indicated surface or axis must lie in a single plane at exactly the specified angle from the designated datum plane or axis, within the specified tolerance zone.

// Parallelism

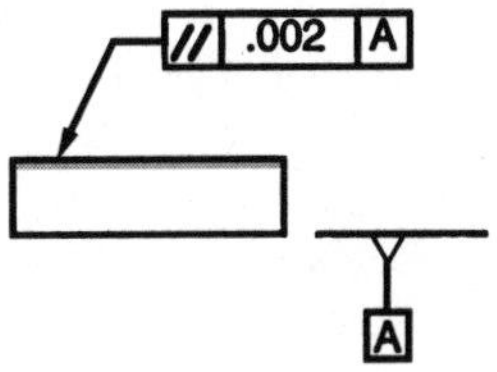

All points on the indicated surface or axis must lie in a single plane parallel to the designated datum plane or axis, within the specified tolerance zone.

Table A2 (Cont.)

Circular Runout

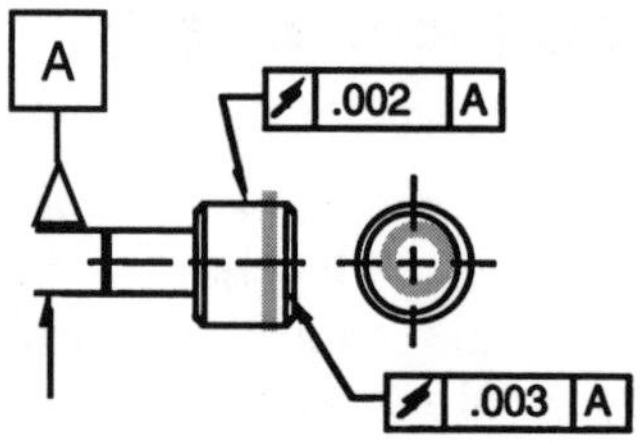

Each circular element of the indicated surface is allowed to deviate only the specified amount from its theoretical form and orientation during 360° rotation about the designated datum axis.

Total Runout

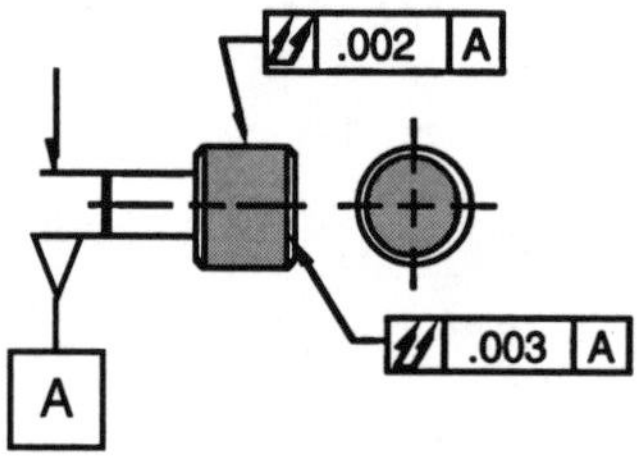

The entire indicated surface is allowed to deviate only the specified amount from its theoretical form and during 360° rotation about the designated datum axis.

Concentricity

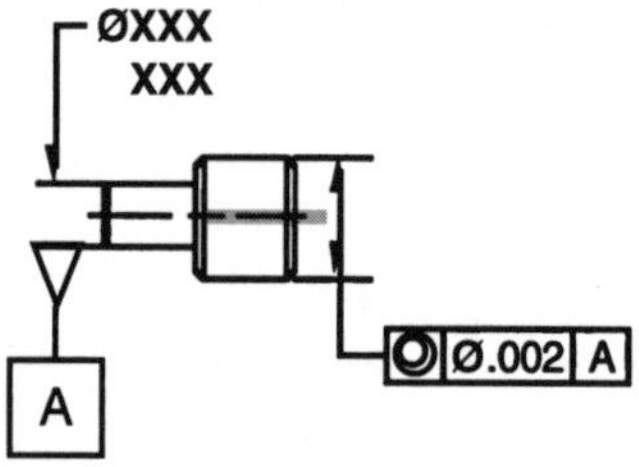

If the indicated surface were sliced by any plane perpendicular to the designated datum axis, every slice's center of area must lie on the datum axis, within the specified cylindrical tolerance zone (controls rotational balance).

Symmetry

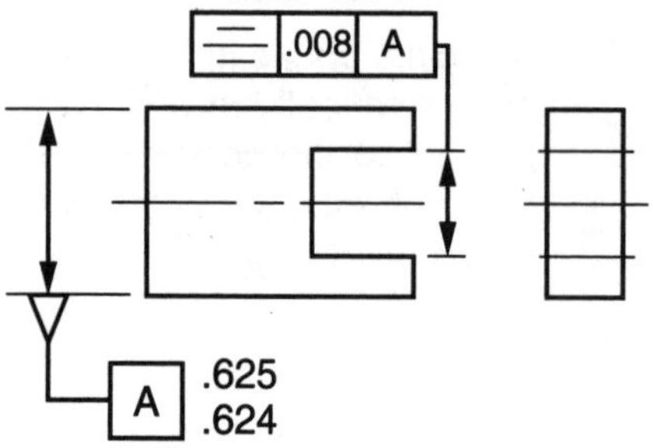

The indicated quality where features on each side of the center line are identical in size, shape, and location. This may be dimensioned on one side of the enter line of symmetry.

APPENDIX B

Tables

Table B1. Decimal Inch, Fractional Inch, and Millimeter Equivalents 290

Table B2. Metric-English Conversion Table 291

Table B3. Formulas for Circles and Squares 292

Table B4. Tap Drill Sizes (Inch) 293

Table B5. Tapers and Angles 294

Table B6. Morse Tapers 295

Table B7. Milling Machine Tapers 296

Table B8. ISO Metric Pitch and Diameter Combinations 297

Table B9. Metric Tap Drill Sizes 297

Table B10. Three-Wire Thread Measurement 298

Table B11. Surface Finishes (Production) 299

Table B12. Hardness Conversion Chart 300

Table B13. AISI/SAE Numbering System for Carbon and Alloy Steels 301

Table B14. Right-Angle Triangles 302

Table B15. Oblique Triangles 303

Table B16. Coordinate Factors and Angles 304

Table B17. Sine Bar Constants 309

Table B18. Natural Trigonometric Functions 312

Table B1.
Decimal Inch, Fractional Inch, and Millimeter Equivalents

Decimal inch	Fractional inch	Millimeter	Decimal inch	Fractional inch	Millimeter
.015625	1/64	0.397	.515625	33/64	13.097
.03125	1/32	0.794	.53125	17/32	13.494
.046875	3/64	1.191	.546875	35/64	13.891
.0625	1/16	1.588	.5625	9/16	14.288
.078125	5/64	1.984	.578125	37/64	14.684
.09375	3/32	2.381	.59375	19/32	15.081
.109375	7/64	2.778	.609375	39/64	15.478
.125	1/8	3.175	.625	5/8	15.875
.140625	9/64	3.572	.640625	41/64	16.272
.15625	5/32	3.969	.65625	21/32	16.669
.171875	11/64	4.366	.671875	43/64	17.066
.1875	3/16	4.762	.6875	11/16	17.462
.203125	13/64	5.159	.703125	45/64	17.859
.21875	7/32	5.556	.71875	23/32	18.256
.234375	15/64	5.953	.734375	47/64	18.653
.25	1/4	6.35	.75	3/4	19.05
.265625	17/64	6.747	.765625	49/64	19.447
.28125	9/32	7.144	.78125	25/32	19.844
.296875	19/64	7.541	.796875	51/64	20.241
.3125	5/16	7.938	.8125	13/16	20.638
.328125	21/64	8.334	.828125	53/64	21.034
.34375	11/32	8.731	.84375	27/32	21.431
.359375	23/64	9.128	.859375	55/64	21.828
.375	3/8	9.525	.875	7/8	22.225
.390625	25/64	9.922	.890625	57/64	22.622
.40625	13/32	10.319	.90625	29/32	23.019
.421875	27/64	10.716	.921875	59/64	23.416
.4375	7/16	11.112	.9375	15/16	23.812
.453125	29/64	11.509	.953125	61/64	24.209
.46875	15/32	11.906	.96875	31/32	24.606
.484375	31/64	12.303	.984375	63/64	25.003
.5	1/2	12.7	1.	1	25.4

Table B2.
Metric-English Conversion Table

Multiply	By	To get equivalent number of
	Length	
Inch	25.4	Millimeters (mm)
Foot	0.304 8	Meters (m)
Yard	0.914 4	Meters
Mile	1.609	Kilometers (km)
	Area	
$Inch^2$	645.2	$Millimeters^2$ (mm^2)
	6.45	$Centimeters^2$ (cm^2)
$Foot^2$	0.092 9	$Meters^2$ (m^2)
$Yard^2$	0.836 1	$Meters^2$
	Volume	
$Inch^3$	16 387.	mm^3
	16.387	cm^3
	0.016 4	Liters (l)
Quart (U.S.)	0.946 4	Liters
Quart (Imperial)	1.136	Liters
Gallon (U.S.)	3.785 4	Liters
Gallon (Imperial)	4.459	Liters
$Yard^3$	0.764 6	$Meters^3$ (m^3)
	Mass	
Pound	0.453 6	Kilograms (kg)
Ton	907.18	Kilograms (kg)
Ton	0.907	Tonne (t)
	Force	
Kilogram	9.807	Newtons (N)
Ounce	0.278 0	Newtons
Pound	4.448	Newtons
	Temperature	
Degree Fahrenheit	(°F − 32) ÷ 1.8	Degree Celsius (C)
Degree Celsius	(°C × 1.8) +32	Degree Fahrenheit (F)

Multiply	By	To get equivalent number of
	Acceleration	
$Foot/second^2$	0.304 8	Meter per $second^2$ (m/s^2)
$Inch/second^2$	0.025 4	Meter per $second^2$
	Torque	
Pound-inch	0.112 98	Newton-meters (N-m)
Pound-foot	1.355 8	Newton-meters
	Power	
Horsepower	0.746	Kilowatts (kW)
	Pressure or stress	
Inches of water	0.249 1	Kilopascals (kPa)
Pounds/square inch	6.895	Kilopascals
	Energy or work	
BTU	1 055.	Joules (J)
Foot-pound	1.355 8	Joules
Kilowatthour	3 600 000. or 3.6×10^6	Joules (J = one Ws)
	Light	
Footcandle	1.076 4	Lumens per $meter^2$ (lm/m^2)
	Fuel performance	
Miles per gallon	0.425 1	Kilometers per liter (km/l)
Gallons per mile	2.352 7	Liters per kilometers (l/km)
	Velocity	
Miles per hour	1.609 3	Kilometers per hr (km/h)

Table B3.
Formulas for Circles and Squares

D is diameter of stock necessary to turn shape desired.
E is distance "across flats," or diameter of inscribed circle.
C is depth of cut into stock turned to correct diameter.

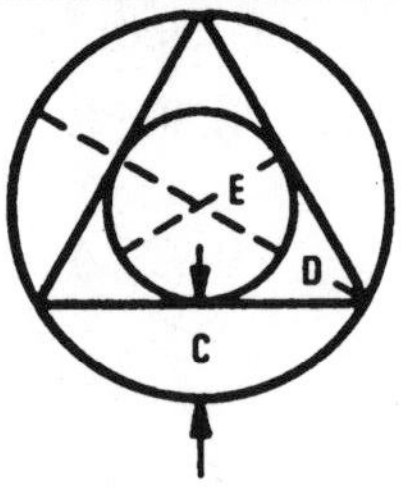

Triangle

$E = \text{side} \times .57735$
$D = \text{side} \times 1.1547 = 2E$
$\text{Side} = D \times .866$
$C = E \times .5 = D \times .25$

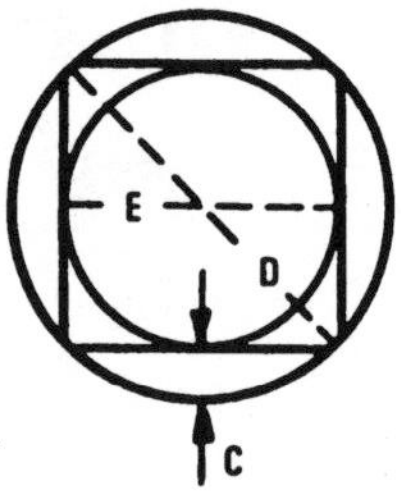

Square

$E = \text{side} = D \times .7071$
$D = \text{side} \times 1.4142 = \text{diagonal}$
$\text{Side} = D \times .7071$
$C = D \times .14645$

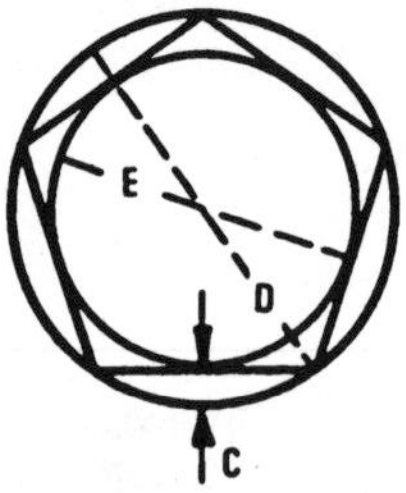

Pentagon

$E = \text{side} \times 1.3764 = D \times .809$
$D = \text{side} \times 1.7013 = E \times 1.2361$
$\text{Side} = D \times .5878$
$C = D \times .0955$

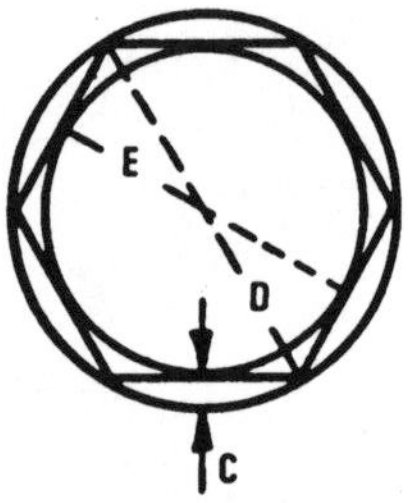

Hexagon

$E = \text{side} \times 1.7321 = D \times .866$
$D = \text{side} \times 2 = E \times 1.1547$
$\text{Side} = D \times .5$
$C = D \times .067$

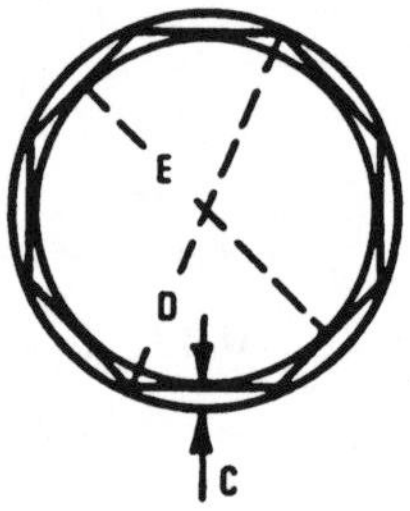

Octagon

$E = \text{side} \times 2.4142 = D \times .9239$
$D = \text{side} \times 2.6131 = E \times 1.0824$
$\text{Side} = D \times .3827$
$C = D \times .038$

Table B4.
Tap Drill Sizes (Inch)

NC National Coarse			NF National Fine		
Tap Size	Threads per inch	Tap Drill Size	Tap Size	Threads per inch	Tap Drill Size
# 5	40	#38	# 5	44	#37
# 6	32	#36	# 6	40	#33
# 8	32	#29	# 8	36	#29
#10	24	#25	#10	32	#21
#12	24	#16	#12	28	#14
1/4	20	# 7	1/4	28	# 3
5/16	18	F	5/16	24	I
3/8	16	5/16	3/8	24	Q
7/16	14	U	7/16	20	25/64
1/2	13	27/64	1/2	20	29/64
9/16	12	31/64	9/16	18	33/64
5/8	11	17/32	5/8	18	37/64
3/4	10	21/32	3/4	16	11/16
7/8	9	49/64	7/8	14	13/16
1	8	7/8	1	14	15/16
1-1/8	7	63/64	1-1/8	12	1-3/64
1-1/4	7	1-7/64	1-1/4	12	1-11/64
1-3/8	6	1-7/32	1-3/8	12	1-19/64
1-1/2	6	1-11/32	1-1/2	12	1-27/64
1-3/4	5	1-9/16			
2	4-1/2	1-25/32			
NPT NATIONAL PIPE THREAD					
1/8	27	11/32	1	11-1/2	1-5/32
1/4	18	7/16	1-1/4	11-1/2	1-1/2
3/8	18	19/32	1-1/2	11-1/2	1-23/32
1/2	14	23/32	2	11-1/2	2-3/16
3/4	14	15/16	2-1/2	8	2-5/8

The major diameter of an NC or NF number size tap or screw= (N × .013) + .060
EXAMPLE: The major diameter of a #5 tap equals
(5 × .013) + .060 = .125 diameter

Table B5.
Tapers and Angles

Taper per Foot	Included Angle		With Center Line		Taper per Inch	Taper per Inch from Center Line
	Degree	Minute	Degree	Minute		
1/8	0	36	0	18	.010416	.005208
3/16	0	54	0	27	.015625	.007812
1/4	1	12	0	36	.020833	.010416
5/16	1	30	0	45	.026042	.013021
3/8	1	47	0	53	.031250	.015625
7/16	2	05	1	02	.036458	.018229
1/2	2	23	1	11	.041667	.020833
9/16	2	42	1	21	.046875	.023438
5/8	3	00	1	30	.052084	.026042
11/16	3	18	1	39	.057292	.028646
3/4	3	35	1	48	.062500	.031250
13/16	3	52	1	56	.067708	.033854
7/8	4	12	2	06	.072917	.036458
15/16	4	28	2	14	.078125	.039063
1	4	45	2	23	.083330	.041667
1-1/4	5	58	2	59	.104166	.052083
1-1/2	7	08	3	34	.125000	.062500
1-3/4	8	20	4	10	.145833	.072917
2	9	32	4	46	.166666	.083333
2-1/2	11	54	5	57	.208333	.104166
3	14	16	7	08	.250000	.125000
3-1/2	16	36	8	18	.291666	.145833
4	18	56	9	28	.333333	.166666
4-1/2	21	14	10	37	.375000	.187500
5	23	32	11	46	.416666	.208333
6	28	04	14	02	.500000	.250000

Courtesy Morse Twist Drill & Machine Co.

Table B6.
Morse Tapers

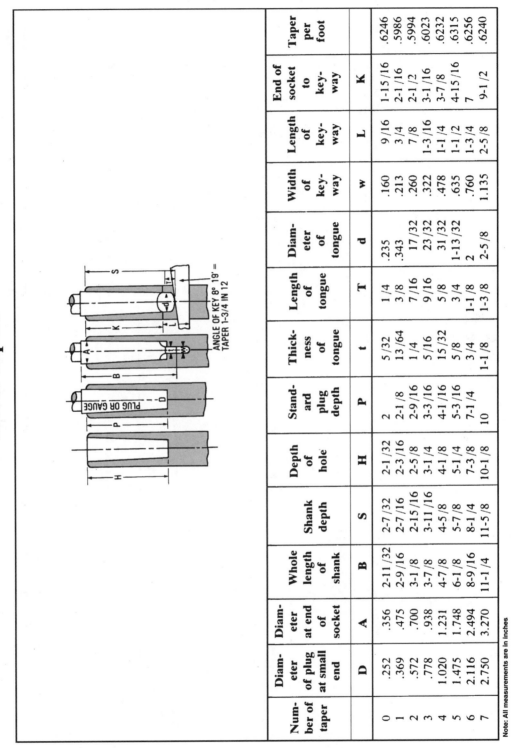

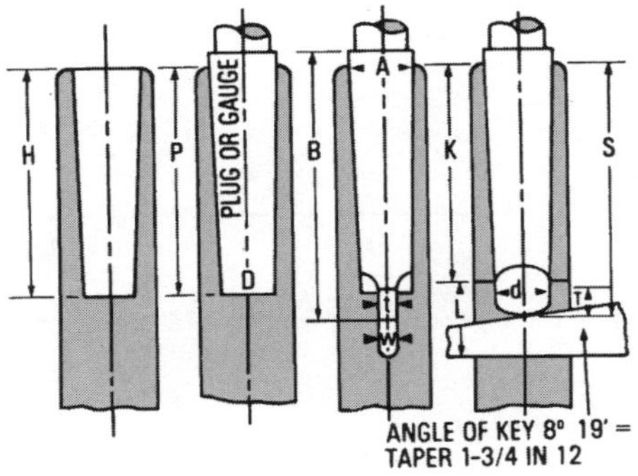

Number of taper	Diameter of plug at small end	Diameter at end of socket	Whole length of shank	Shank depth	Depth of hole	Standard plug depth	Thickness of tongue	Length of tongue	Diameter of tongue	Width of keyway	Length of keyway	End of socket to keyway	Taper per foot
	D	A	B	S	H	P	t	T	d	w	L	K	
0	.252	.356	2-11/32	2-7/32	2-1/32	2	5/32	1/4	.235	.160	9/16	1-15/16	.6246
1	.369	.475	2-9/16	2-7/16	2-3/16	2-1/8	13/64	3/8	.343	.213	3/4	2-1/16	.5986
2	.572	.700	3-1/8	2-15/16	2-5/8	2-9/16	1/4	7/16	17/32	.260	7/8	2-1/2	.5994
3	.778	.938	3-7/8	3-11/16	3-1/4	3-3/16	5/16	9/16	23/32	.322	1-3/16	3-1/16	.6023
4	1.020	1.231	4-7/8	4-5/8	4-1/8	4-1/16	15/32	5/8	31/32	.478	1-1/4	3-7/8	.6232
5	1.475	1.748	6-1/8	5-7/8	5-1/4	5-3/16	5/8	3/4	1-13/32	.635	1-1/2	4-15/16	.6315
6	2.116	2.494	8-9/16	8-1/4	7-3/8	7-1/4	3/4	1-1/8	2	.760	1-3/4	7	.6256
7	2.750	3.270	11-1/4	11-5/8	10-1/8	10	1-1/8	1-3/8	2-5/8	1.135	2-5/8	9-1/2	.6240

Note: All measurements are in inches

Table B7.
Milling Machine Tapers

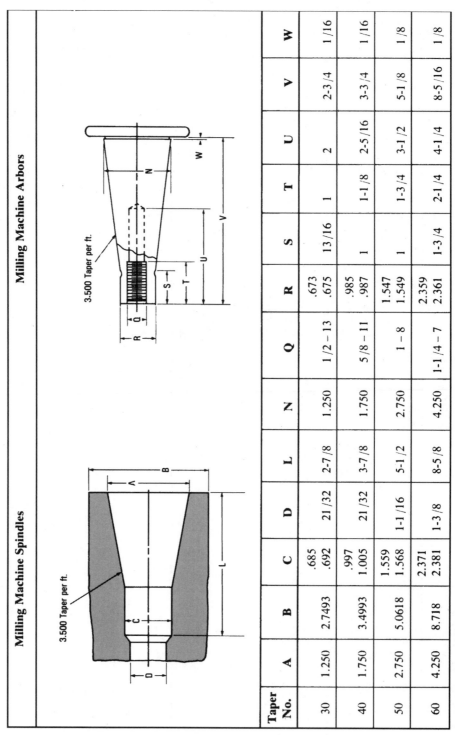

Milling Machine Spindles	Milling Machine Arbors

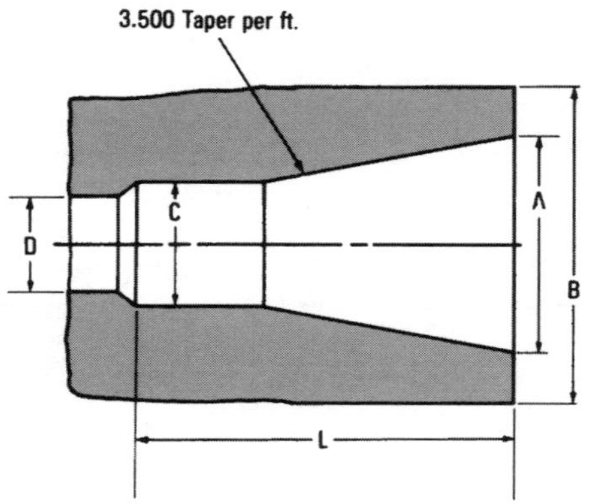

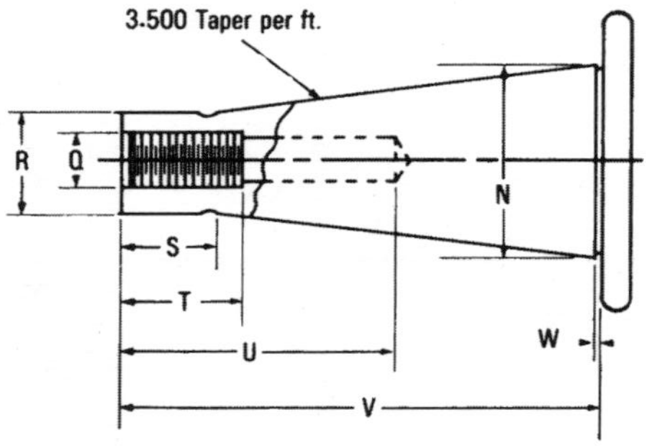

Taper No.	A	B	C	D	L	N	Q	R	S	T	U	V	W
30	1.250	2.7493	.685 .692	21/32	2-7/8	1.250	1/2 – 13	.673 .675	13/16	1	2	2-3/4	1/16
40	1.750	3.4993	.997 1.005	21/32	3-7/8	1.750	5/8 – 11	.985 .987	1	1-1/8	2-5/16	3-3/4	1/16
50	2.750	5.0618	1.559 1.568	1-1/16	5-1/2	2.750	1 – 8	1.547 1.549	1	1-3/4	3-1/2	5-1/8	1/8
60	4.250	8.718	2.371 2.381	1-3/8	8-5/8	4.250	1-1/4 – 7	2.359 2.361	1-3/4	2-1/4	4-1/4	8-5/16	1/8

Note: All measurements are in inches

Table B8.
ISO Metric Pitch and Diameter Combinations

Nominal diameter, mm	Thread pitch, mm	Nominal diameter, mm	Thread pitch, mm
1.6	0.35	20	2.5
2	0.4	24	3
2.5	0.45	30	3.5
3	0.5	36	4
3.5	0.6	42	4.5
4	0.7	48	5
5	0.8	56	5.5
6	1	64	6
8	1.25	72	6
10	1.5	80	6
12	1.75	90	6
14	2	100	6
16	2		

Table B9.
Metric Tap Drill Sizes

Nominal diameter, mm	Thread pitch, mm	Tap drill size, mm	Nominal diameter, mm	Thread pitch, mm	Tap drill size, mm
1.6	0.35	1.2	20	2.5	17.5
2	0.4	1.6	24	3	21
2.5	0.45	2.05	30	3.5	26.5
3	0.5	2.5	36	4	32
3.5	0.6	2.9	42	4.5	37.5
4	0.7	3.3	48	5	43
5	0.8	4.2	56	5.5	50.5
6	1	5.3	64	6	58
8	1.25	6.8	72	6	66
10	1.5	8.5	80	6	74
12	1.75	10.2	90	6	84
14	2	12	100	6	94
16	2	14			

Table B10.
Three-Wire Thread Measurement

THREE WIRE THREAD MEASUREMENT
(60° Metric Thread)

M = PD + C **PD = M − C**

M = Measurement over wires
PD = Pitch diameter
C = Constant

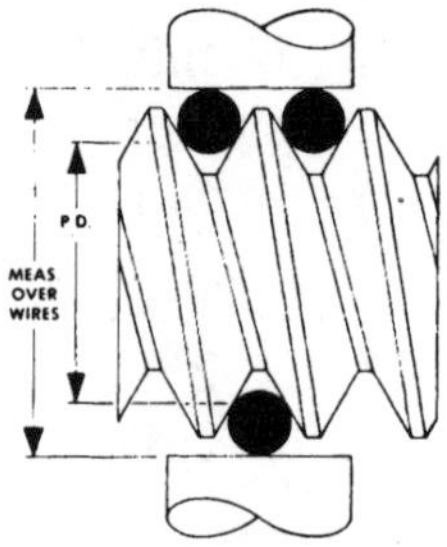

Pitch		Best Wire Size		Constant	
mm	Inches	mm	Inches	mm	Inches
0.2	.00787	0.1155	.00455	0.1732	.00682
0.225	.00886	0.1299	.00511	0.1949	.00767
0.25	.00934	0.1443	.00568	0.2165	.00852
0.3	.01181	0.1732	.00682	0.2598	.01023
0.35	.01378	0.2021	.00796	0.3031	.01193
0.4	.01575	0.2309	.00909	0.3464	.01364
0.45	.01772	0.2598	.01023	0.3897	.01534
0.5	.01969	0.2887	.01137	0.433	.01705
0.6	.02362	0.3464	.01364	0.5196	.02046
0.7	.02756	0.4041	.01591	0.6062	.02387
0.75	.02953	0.433	.01705	0.6495	.02557
0.8	.0315	0.4619	.01818	0.6928	.02728
0.9	.03543	0.5196	.02046	0.7794	.03069
1	.03937	0.5774	.02273	0.866	.0341
1.25	.04921	0.7217	.02841	1.0825	.04262
1.5	.05906	0.866	.0341	1.299	.05114
1.75	.0689	1.0104	.03978	1.5155	.05967
2	.07874	1.1547	.04546	1.7321	.06819
2.5	.09843	1.4434	.05683	2.1651	.08524
3	.11811	1.7321	.06819	2.5981	.10229
3.5	.1378	2.0207	.07956	3.0311	.11933
4	.15748	2.3094	.09092	3.4641	.13638
4.5	.17717	2.5981	.10229	3.8971	.15343
5	.19685	2.8868	.11365	4.3301	.17048
5.5	.21654	3.1754	.12502	4.7631	.18753
6	.23622	3.4641	.13638	5.1962	.20457
7	.27559	4.0415	.15911	6.0622	.23867
8	.31496	4.6188	.18184	6.9282	.27276
9	.35433	5.1962	.20457	7.7942	.30686
10	.3937	5.7735	.2273	8.6603	.34095

Table B11. Surface Finishes (Production)

Roughness Average

Microinches	2000	1000	500	250	125	63	32	16	8	4	2	1
Micrometers	50	25	12.5	6.3	3.2	1.6	.8	.4	.2	.1	.05	.025
Roughness Grade (N)	12	11	10	9	8	7	6	5	4	3	2	1
Sand Casting		■	■									
Hot Rolling		■	■									
Forging			■	■	■							
PermanentMold Casting					■	■						
Investment Casting					■	■						
Extruding					■	■	■					
Cold Rolling					■	■	■					
Die Casting						■	■					
Flame Cutting		■	■									
Sawing		■	■	■	■	■						
Planing			■	■	■	■						
Drilling				■	■	■						
EDM				■	■	■						
Milling				■	■	■	■					
Laser Cutting				■	■	■	■					
Reaming					■	■	■					
Broaching					■	■	■					
Turning and Boring				■	■	■	■	■				
Tumbling							■	■	■			
Roller Burnishing								■	■			
Grinding						■	■	■	■	■		
Honing							■	■	■	■		
Electro-polishing							■	■	■	■		
Polishing								■	■	■		
Lapping'								■	■	■	■	
Superfinishing									■	■	■	■
Metric Roughness Symbol		▽			▽▽			▽▽▽			▽▽▽▽	

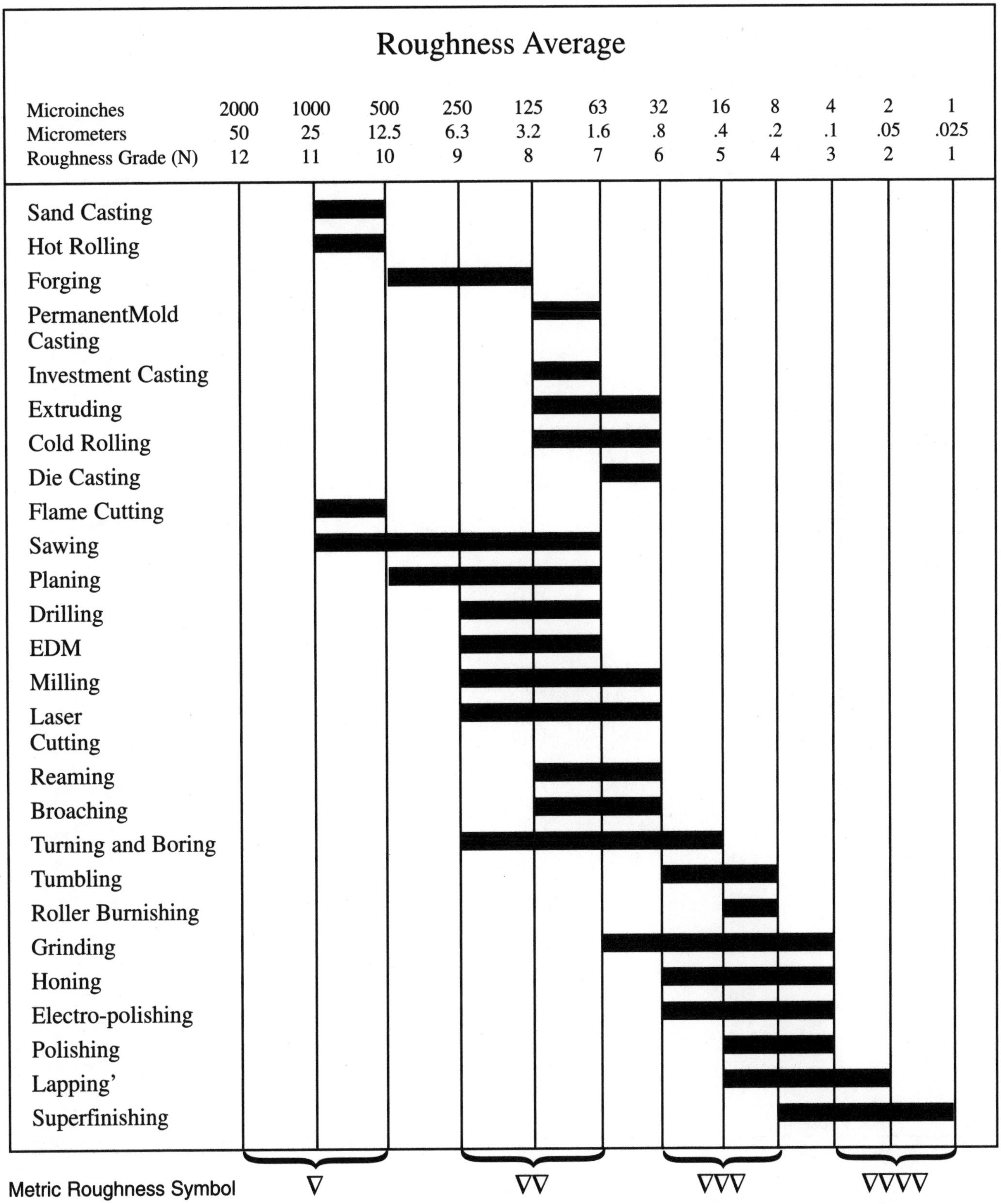

Table B12.
Hardness Conversion Chart

10 mm Ball 3000 kg	120° Cone 150 kg	1/16 in. Ball 100 kg	Model C	Mpa	10 mm Ball 3000 kg	120° Cone 150 kg	1/16 in. Ball 100 kg	Model C	Mpa
Brinell	**Rockwell C**	**Rockwell B**	**Shore Scleroscope**	**Tensile Strength**	**Brinell**	**Rockwell C**	**Rockwell B**	**Shore Scleroscope**	**Tensile Strength**
800	72	...	100	...	276	30	105	42	938
780	71	...	99	...	269	29	104	41	910
760	70	...	98	...	261	28	103	40	889
745	68	...	97	2530	258	27	102	39	876
725	67	...	96	2460	255	26	102	39	862
712	66	...	95	2413	249	25	101	38	848
682	65	...	93	2324	245	24	100	37	820
668	64	...	91	2248	240	23	99	36	807
652	63	...	89	2193	237	23	99	35	793
626	62	...	87	2110	229	22	98	34	779
614	61	...	85	2062	224	21	97	33	758
601	60	...	83	2013	217	20	96	33	738
590	59	...	81	2000	211	19	95	32	717
576	57	...	79	1937	206	18	94	32	703
552	56	...	76	1862	203	17	94	31	689
545	55	...	75	1848	200	16	93	31	676
529	54	...	74	1786	196	15	92	30	662
514	53	120	72	1751	191	14	92	30	648
502	52	119	70	1703	187	13	91	29	634
495	51	119	69	1682	185	12	91	29	627
477	49	118	67	1606	183	11	90	28	621
461	48	117	66	1565	180	10	89	28	614
451	47	117	65	1538	175	9	88	27	593
444	46	116	64	1510	170	7	87	27	579
427	45	115	62	1441	167	6	87	27	565
415	44	115	60	1407	165	5	86	26	558
401	43	114	58	1351	163	4	85	26	552
388	42	114	57	1317	160	3	84	25	538
375	41	113	55	1269	156	2	83	25	524
370	40	112	54	1255	154	1	82	25	517
362	39	111	53	1234	152	...	82	24	510
351	38	111	51	1193	150	...	81	24	510
346	37	110	50	1172	147	...	80	24	496
341	37	110	49	1158	145	...	79	23	490
331	36	109	47	1124	143	...	79	23	483
323	35	109	46	1089	141	...	78	23	476
311	34	108	46	1055	140	...	77	22	476
301	33	107	45	1020	135	...	75	22	462
293	32	106	44	993	130	...	72	22	448
285	31	105	43	965	...	...	...		

Table B13. AISI/SAE Numbering System for Carbon and Alloy Steels

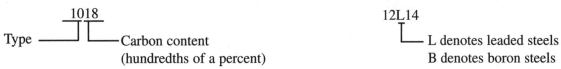

1018: Type (first two digits); Carbon content (hundredths of a percent) (last two digits)

12L14: L denotes leaded steels; B denotes boron steels

Carbon steels	10XX	Plain carbon, Mn 1.00% max
	11XX	Resulfurized free machining
	12XX	Resulfurized/rephosphorized free machining
	15XX	Plain carbon, Mn 1.00–1.65%
Manganese steels	13XX	Mn 1.75%
Nickel steels	23XX	Ni 3.50%
	25XX	Ni 5.00%
Nickel-chromium steels	31XX	Ni 1.25%, Cr 0.65–0.80%
	32XX	Ni 1.75%, Cr 1.07%
	33XX	Ni 3.50%, Cr 1.50–1.57%
	34XX	Ni 3.00%, Cr 0.77%
Molybdenum steels	40XX	Mo 0.20–0.25%
	44XX	Mo 0.40–0.52%
Chromium-molybdenum steels	41XX	Cr 0.50–0.95%, Mo 0.12–0.30%
Nickel-chromium-molybdenum steels	43XX	Ni 1.82%, Cr 0.50–0.80%, Mo 0.25%
	47XX	Ni 1.05%, Cr 0.45%, Mo 0.20–0.35%
Nickel-molybdenum steels	46XX	Ni 0.85–1.82%, Mo 0.20–0.25%
	48XX	Ni 3.50%, Mo 0.25%
Chromium steels	50XX	Cr 0.27–0.65%
	51XX	Cr 0.80–1.05%
	50XXX	Cr 0.50%, C 1.00% min
	51XXX	Cr 1.02%, C 1.00% min
	52XXX	Cr 1.45%, C 1.00% min
Chromium-vanadium steels	61XX	Cr 0.60–0.95%, V 0.10–0.15%
Tungsten-chromium steels	72XX	W 1.75%, Cr 0.75%
Nickel-chromium-molybdenum steels	81XX	Ni 0.30%, Cr 0.40%, Mo 0.12%
	86XX	Ni 0.55%, Cr 0.50%, Mo 0.20%
	87XX	Ni 0.55%, Cr 0.50%, Mo 0.25%
	88XX	Ni 0.55%, Cr 0.50%, Mo 0.35%
Silicon-manganese steels	92XX	Si 1.40–2.00%, Mn 0.65–0.85%, Cr 0–0.65%
Nickel-chromium-molybdenum steels	93XX	Ni 3.25%, Cr 1.20%, Mo 0.12%
	94XX	Ni 0.45%, Cr 0.40%, Mo 0.12%
	97XX	Ni 0.55%, Cr 0.20%, Mo 0.20%
	98XX	Ni 1.00%, Cr 0.80%, Mo 0.25%

Table B14.
Right-Angle Triangles

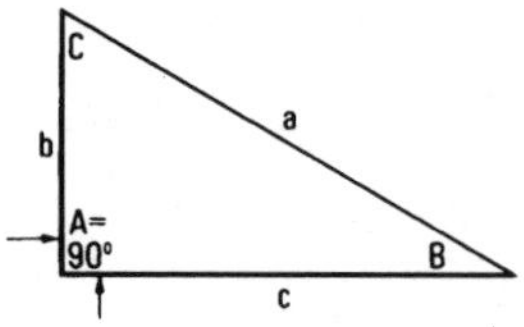	Sine ∠ = $\frac{\text{Side opposite}}{\text{Hypotenuse}}$ Cosine ∠ = $\frac{\text{Side adjacent}}{\text{Hypotenuse}}$ Tangent ∠ = $\frac{\text{Side opposite}}{\text{Side adjacent}}$	Cosecant ∠ = $\frac{\text{Hypotenuse}}{\text{Side opposite}}$ Secant ∠ = $\frac{\text{Hypotenuse}}{\text{Side adjacent}}$ Cotangent ∠ = $\frac{\text{Side adjacent}}{\text{Side opposite}}$

Knowing		**Formulas to find**	
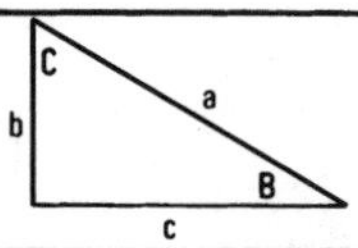	Sides a & b	$c = \sqrt{a^2 - b^2}$	$\sin B = \frac{b}{a}$
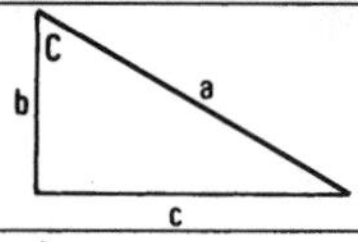	Side a & angle B	$b = a \times \sin B$	$c = a \times \cos B$
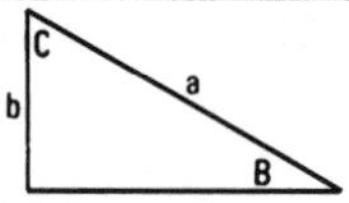	Sides a & c	$b = \sqrt{a^2 - c^2}$	$\sin C = \frac{c}{a}$
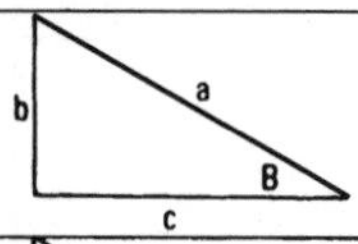	Side a & angle C	$b = a \times \cos C$	$c = a \times \sin C$
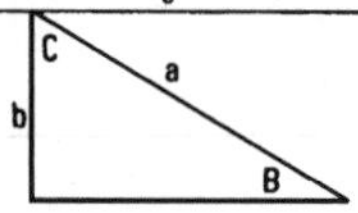	Sides b & c	$a = \sqrt{b^2 + c^2}$	$\tan B = \frac{b}{c}$
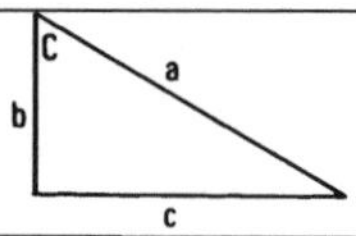	Side b & angle B	$a = \frac{b}{\sin B}$	$c = b \times \cot B$
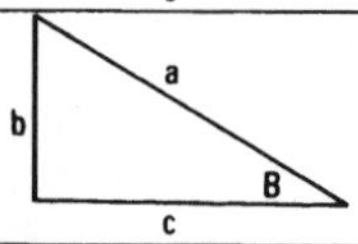	Side b & angle C	$a = \frac{b}{\cos C}$	$c = b \times \tan C$
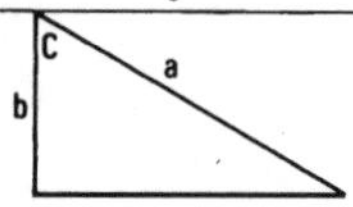	Side c & angle B	$a = \frac{c}{\cos B}$	$b = c \times \tan B$
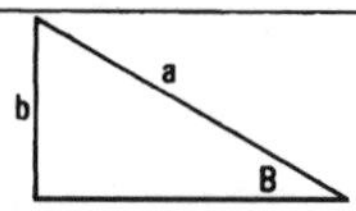	Side c & angle C	$a = \frac{c}{\sin C}$	$b = c \times \cot C$

Table B15. Oblique Triangles

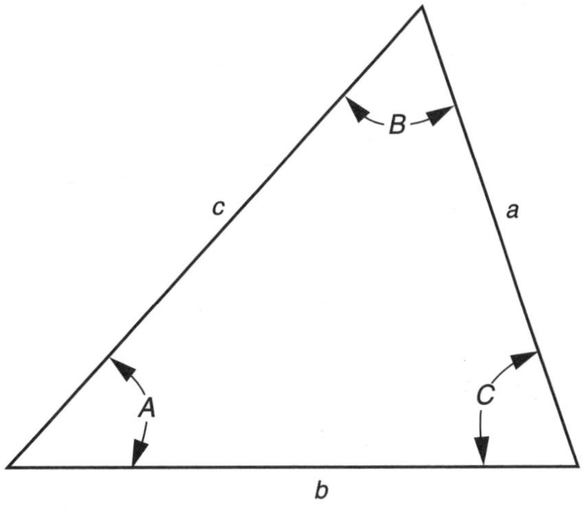

Known sides and angles	Unknown Sides and Angles			Area
All three sides a, b, c	$A = \arc\cos \frac{b^2+c^2-a^2}{2bc}$	$B = \arc\sin \frac{b \times \sin A}{a}$	$C = 180°\text{-A-B}$	$\frac{a \times b \times \sin C}{2}$
Two sides and the angle between them a, b, $\angle C$	$c = \sqrt{a^2+b^2-(2ab \times \cos C)}$	$A = \arc\tan \frac{a \times \sin C}{b-(a \times \cos C)}$	$B = 180°\text{-A-C}$	$\frac{a \times b \times \sin C}{2}$
Two sides and the angle opposite one of the sides a, b, $\angle A$ ($\angle B$ less than 90°)	$B = \arc\sin \frac{b \times \sin A}{a}$	$C = 180°\text{-A-B}$	$c = \frac{a \times \sin C}{\sin A}$	$\frac{a \times b \times \sin C}{2}$
Two sides and the angle opposite one of the sides a, b, $\angle A$ ($\angle B$ greater than 90°)	$B = 180° - \arc\sin \frac{b \times \sin A}{a}$	$C = 180°\text{-A-B}$	$c = \frac{a \times \sin C}{\sin A}$	$\frac{a \times b \times \sin C}{2}$
One side and two angles a, $\angle A$, $\angle B$	$b = \frac{a \times \sin B}{\sin A}$	$C = 180°\text{-A-B}$	$c = \frac{a \times \sin C}{\sin A}$	$\frac{a \times b \times \sin C}{2}$

Table B16. Coordinate Factors and Angles

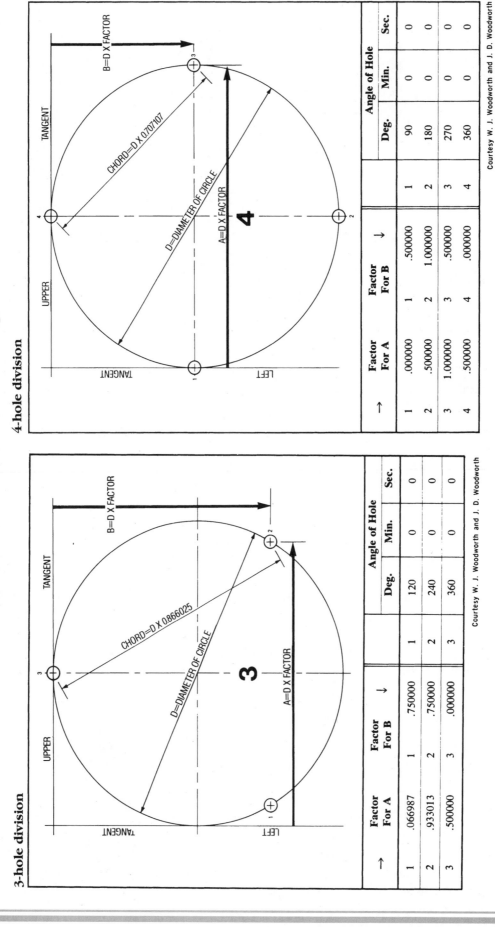

3-hole division

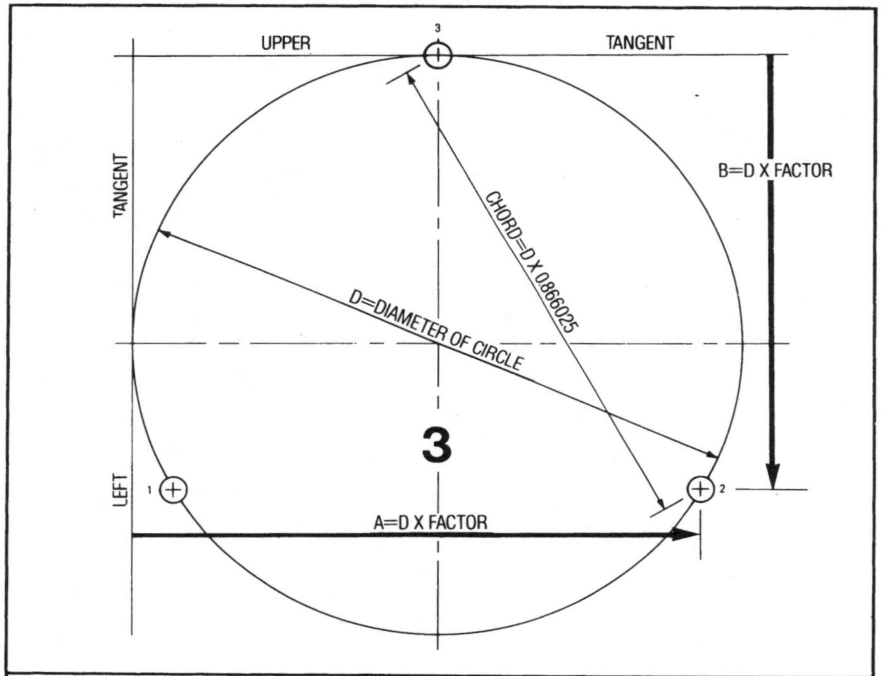

→	Factor For A	Factor For B	↓		Angle of Hole Deg.	Min.	Sec.
1	.066987	1	.750000	1	120	0	0
2	.933013	2	.750000	2	240	0	0
3	.500000	3	.000000	3	360	0	0

Courtesy W. J. Woodworth and J. D. Woodworth

4-hole division

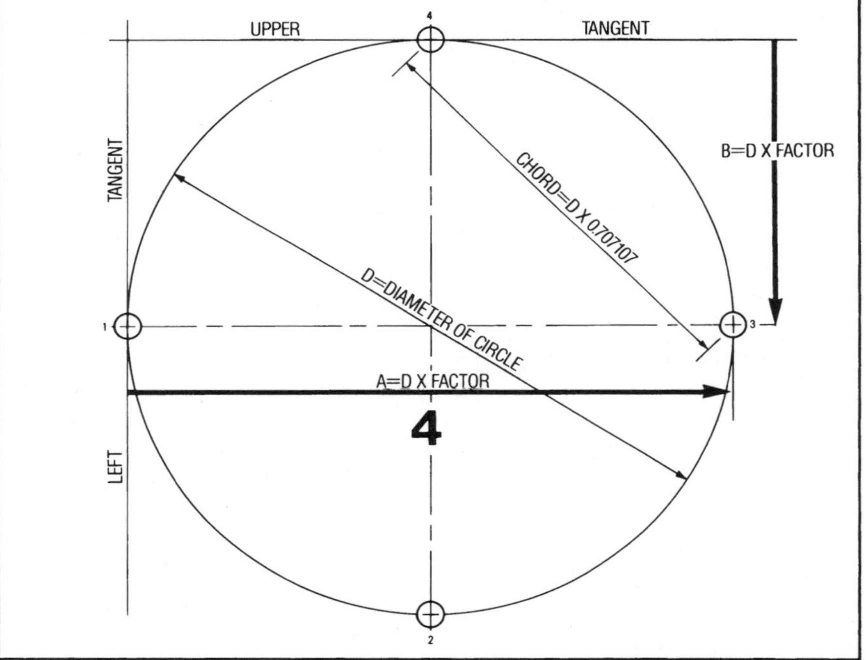

→	Factor For A	Factor For B	↓		Angle of Hole Deg.	Min.	Sec.
1	.000000	1	.500000	1	90	0	0
2	.500000	2	1.000000	2	180	0	0
3	1.000000	3	.500000	3	270	0	0
4	.500000	4	.000000	4	360	0	0

Courtesy W. J. Woodworth and J. D. Woodworth

5-hole division

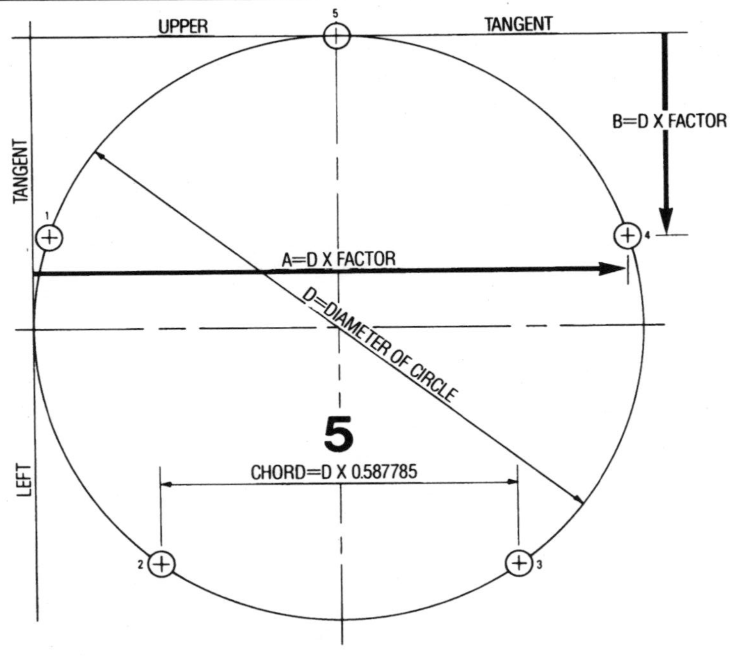

→	Factor For A	Factor For B	↓		Angle of Hole Deg.	Min.	Sec.
1	.024472	1	.345492	1	72	0	0
2	.206107	2	.904508	2	144	0	0
3	.793893	3	.904508	3	216	0	0
4	.975528	4	.345492	4	288	0	0
5	.500000	5	.000000	5	360	0	0

Courtesy W. J. Woodworth and J. D. Woodworth

6-hole division

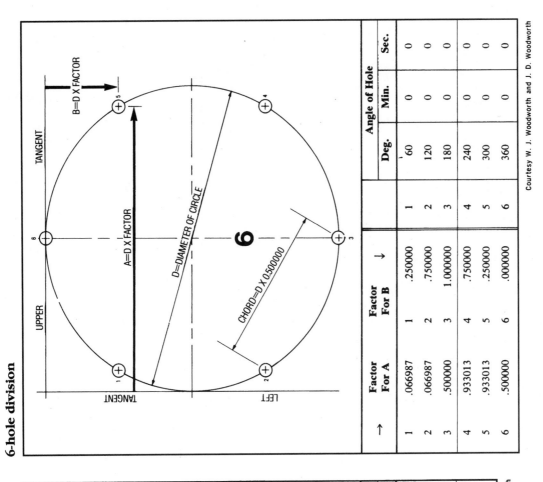

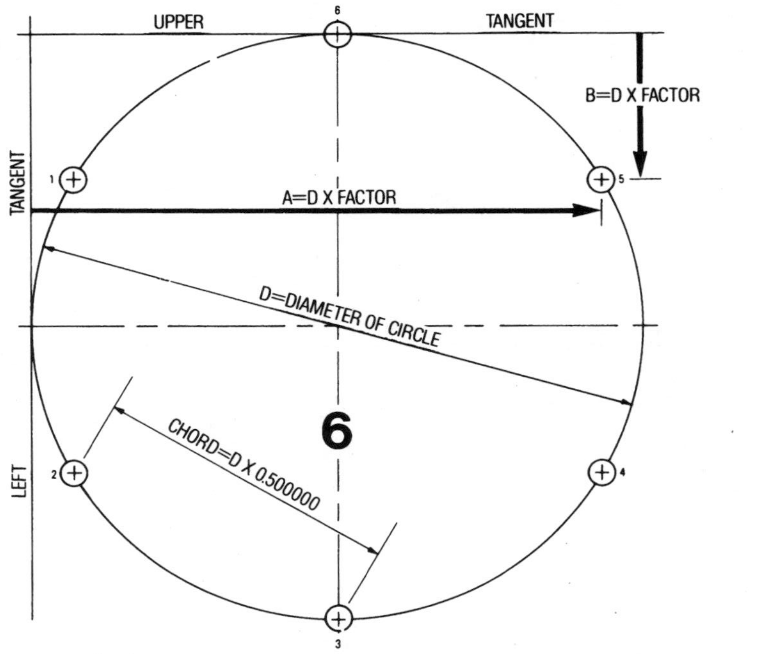

→	Factor For A	Factor For B	↓		Angle of Hole Deg.	Min.	Sec.
1	.066987	1	.250000	1	60	0	0
2	.066987	2	.750000	2	120	0	0
3	.500000	3	1.000000	3	180	0	0
4	.933013	4	.750000	4	240	0	0
5	.933013	5	.250000	5	300	0	0
6	.500000	6	.000000	6	360	0	0

Courtesy W. J. Woodworth and J. D. Woodworth

(Continued)

Table B16. (Cont.)

7-hole division

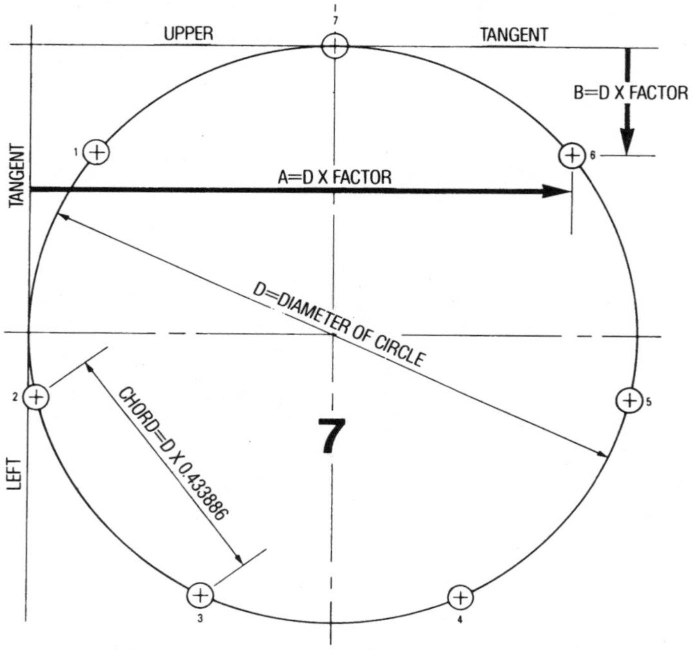

→	Factor For A	Factor For B	↓		Angle of Hole Deg.	Min.	Sec.
1	.109084	1	.188255	1	51	25	42-6/7
2	.012536	2	.611261	2	102	51	23-5/7
3	.283058	3	.950484	3	154	17	8-4/7
4	.716942	4	.950484	4	205	42	51-3/7
5	.987464	5	.611261	5	257	8	34-2/7
6	.890916	6	.188255	6	308	34	17-1/7
7	.500000	7	.000000	7	360	0	0

Courtesy W. J. Woodworth and J. D. Woodworth

8-hole division

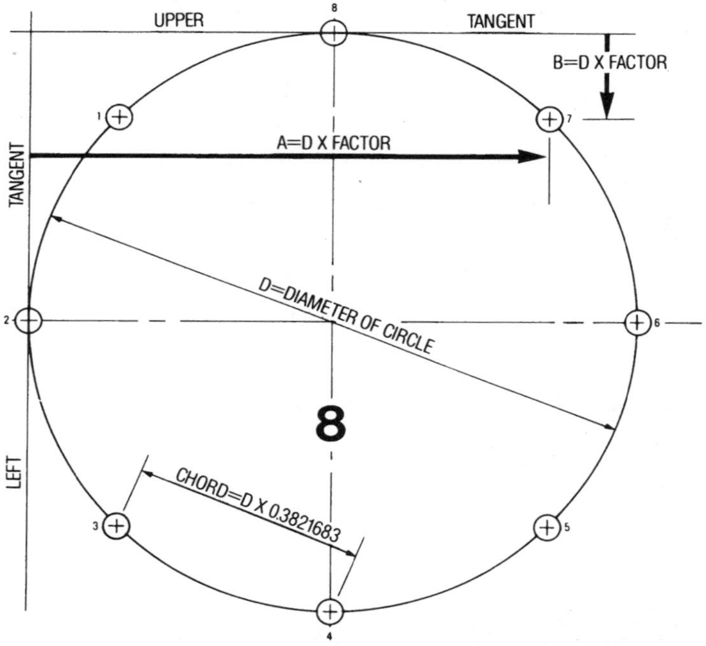

→	Factor For A	Factor For B	↓		Angle of Hole Deg.	Min.	Sec.
1	.146447	1	.146447	1	45	0	0
2	.000000	2	.500000	2	90	0	0
3	.146447	3	.853553	3	135	0	0
4	.500000	4	1.000000	4	180	0	0
5	.853553	5	.853553	5	225	0	0
6	1.000000	6	.500000	6	270	0	0
7	.853553	7	.146447	7	315	0	0
8	.500000	8	.000000	8	360	0	0

Courtesy W. J. Woodworth and J. D. Woodworth

9-hole division

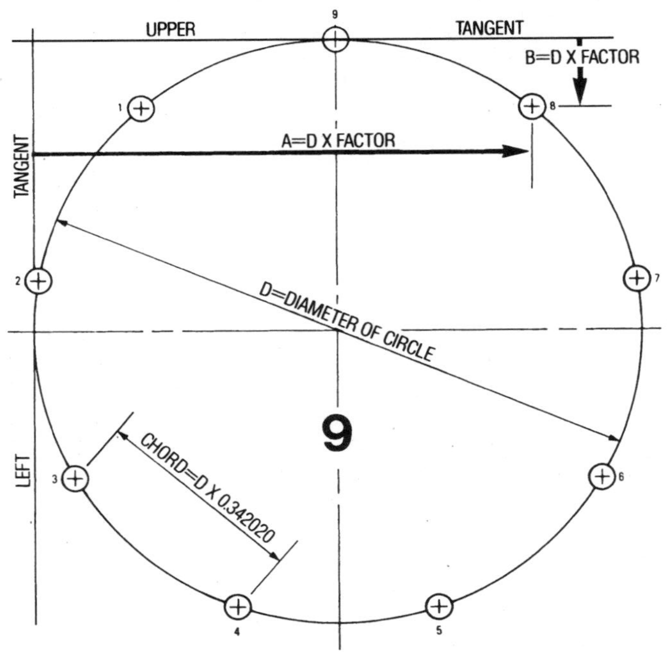

→	Factor For A	Factor For B	↓		Angle of Hole Deg.	Min.	Sec.
1	.178606	1	.116978	1	40	0	0
2	.007596	2	.413176	2	80	0	0
3	.066987	3	.750000	3	120	0	0
4	.328990	4	.969846	4	160	0	0
5	.671010	5	.969846	5	200	0	0
6	.933013	6	.750000	6	240	0	0
7	.992404	7	.413176	7	280	0	0
8	.821394	8	.116978	8	320	0	0
9	.500000	9	.000000	9	360	0	0

Courtesy W. J. Woodworth and J. D. Woodworth

10-hole division

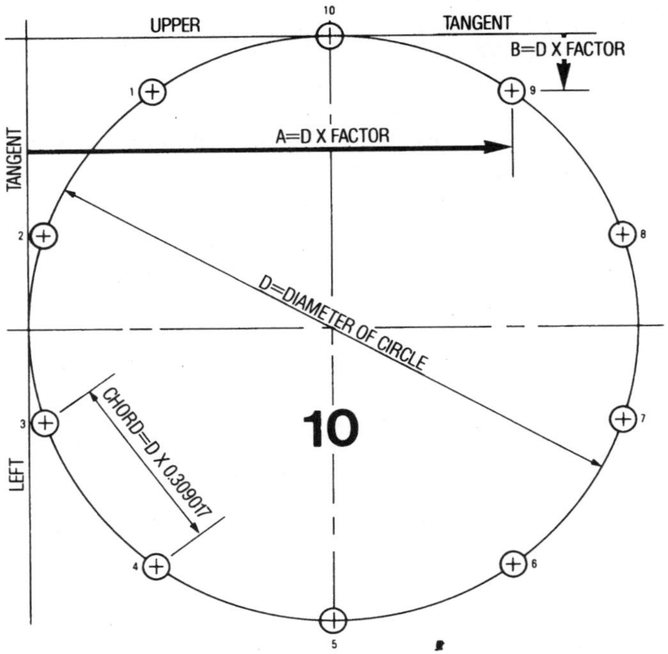

→	Factor For A	Factor For B	↓		Angle of Hole Deg.	Min.	Sec.
1	.206107	1	.095492	1	36	0	0
2	.024472	2	.345492	2	72	0	0
3	.024472	3	.654508	3	108	0	0
4	.206107	4	.904508	4	144	0	0
5	.500000	5	1.000000	5	180	0	0
6	.793893	6	.904508	6	216	0	0
7	.975528	7	.654508	7	252	0	0
8	.975528	8	.345492	8	288	0	0
9	.793893	9	.095492	9	324	0	0
10	.500000	10	.000000	10	360	0	0

Courtesy W. J. Woodworth and J. D. Woodworth

(Continued)

Table B16. (Cont.)

11-hole division

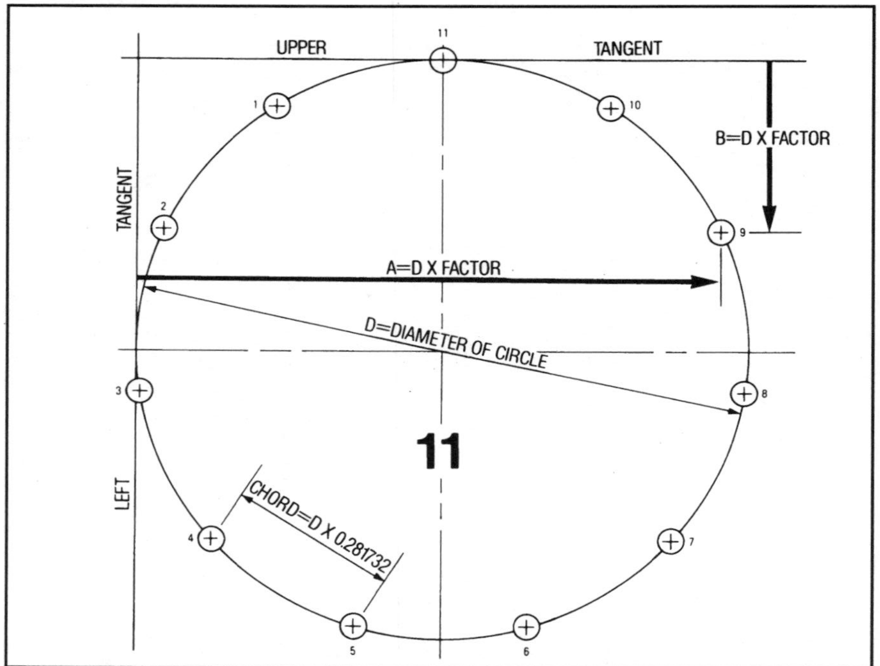

→	Factor For A	Factor For B	↓		Angle of Hole Deg.	Min.	Sec.
1	.229680	1	.079373	1	32	43	38-2/11
2	.045184	2	.292293	2	65	27	16-4/11
3	.005089	3	.571157	3	98	10	54-6/11
4	.122125	4	.827430	4	130	54	32-8/11
5	.359134	5	.979746	5	163	38	10-10/11
6	.640866	6	.979746	6	196	21	49-1/11
7	.877875	7	.827430	7	229	5	27-3/11
8	.994911	8	.571157	8	261	49	5-5/11
9	.954816	9	.292293	9	294	32	43-7/11
10	.770320	10	.079373	10	327	16	21-9/11
11	.500000	11	.000000	11	360	0	0

Courtesy W. J. Woodworth and J. D. Woodworth

Table B17. Sine Bar Constants (5-in Bar)

(Multiply Constants by Two for a 10-in Sine Bar)

Min.	0°	1°	2°	3°	4°	5°	6°	7°	8°	9°	10°	11°	12°	13°	14°	15°	16°	17°	18°	19°	Min.
0	.00000	.08725	.17450	.26170	.34880	.43580	.52265	.60935	.69585	.78215	.86825	.95405	1.0395	1.1247	1.2096	1.2941	1.3782	1.4618	1.5451	1.6278	0
2	.00290	.09015	.17740	.26460	.35170	.43870	.52555	.61225	.69875	.78505	.87110	.95690	.0424	.1276	.2124	.2969	.3810	.4646	.5478	.6306	2
4	.00580	.09310	.18030	.26750	.35460	.44155	.52845	.61510	.70165	.78790	.87395	.95975	.0452	.1304	.2152	.2997	.3838	.4674	.5506	.6333	4
6	.00875	.09600	.18320	.27040	.35750	.44445	.53130	.61800	.70450	.79080	.87685	.96260	.0481	.1332	.2181	.3025	.3865	.4702	.5534	.6361	6
8	.01165	.09890	.18615	.27330	.36040	.44735	.53420	.62090	.70740	.79365	.87970	.96545	.0509	.1361	.2209	.3053	.3893	.4730	.5561	.6388	8
10	.01455	.10180	.18905	.27620	.36330	.45025	.53710	.62380	.71025	.79655	.88255	.96830	1.0538	1.1389	1.2237	1.3081	1.3921	1.4757	1.5589	1.6416	10
12	.01745	.10470	.19195	.27910	.36620	.45315	.54000	.62665	.71315	.79940	.88540	.97115	.0566	.1417	.2265	.3109	.3949	.4785	.5616	.6443	12
14	.02035	.10760	.19485	.28200	.36910	.45605	.54290	.62955	.71600	.80230	.88830	.97405	.0594	.1446	.2293	.3137	.3977	.4813	.5644	.6471	14
16	.02325	.11055	.19775	.28490	.37200	.45895	.54580	.63245	.71890	.80515	.89115	.97690	.0623	.1474	.2322	.3165	.4005	.4841	.5672	.6498	16
18	.02620	.11345	.20065	.28780	.37490	.46185	.54865	.63530	.72180	.80800	.89400	.97975	.0651	.1502	.2350	.3193	.4033	.4868	.5699	.6525	18
20	.02910	.11635	.20355	.29070	.37780	.46475	.55155	.63820	.72465	.81090	.89685	.98260	1.0680	1.1531	1.2378	1.3221	1.4061	1.4896	1.5727	1.6553	20
22	.03200	.11925	.20645	.29365	.38070	.46765	.55445	.64110	.72755	.81375	.89975	.98545	.0708	.1559	.2406	.3250	.4089	.4924	.5755	.6580	22
24	.03490	.12215	.20940	.29655	.38360	.47055	.55735	.64400	.73040	.81665	.90260	.98830	.0737	.1587	.2434	.3278	.4117	.4952	.5782	.6608	24
26	.03780	.12505	.21230	.29945	.38650	.47345	.56025	.64685	.73330	.81950	.90545	.99115	.0765	.1615	.2462	.3306	.4145	.4980	.5810	.6635	26
28	.04070	.12800	.21520	.30235	.38940	.47635	.56315	.64975	.73615	.82235	.90830	.99400	.0793	.1644	.2491	.3334	.4173	.5007	.5837	.6663	28
30	.04365	.13090	.21810	.30525	.39230	.47925	.56600	.65265	.73905	.82525	.91120	.99685	1.0822	1.1672	1.2519	1.3362	1.4201	1.5035	1.5865	1.6690	30
32	.04655	.13380	.22100	.30815	.39520	.48210	.56890	.65550	.74190	.82810	.91405	.99970	.0850	.1700	.2547	.3390	.4228	.5063	.5893	.6718	32
34	.04945	.13670	.22390	.31105	.39810	.48500	.57180	.65840	.74480	.83100	.91690	1.0016	.0879	.1729	.2575	.3418	.4256	.5091	.5920	.6745	34
36	.05235	.13960	.22680	.31395	.40100	.48790	.57470	.66130	.74770	.83385	.91975	.0054	.0907	.1757	.2603	.3446	.4284	.5118	.5948	.6772	36
38	.05525	.14250	.22970	.31685	.40390	.49080	.57760	.66415	.75055	.83670	.92260	.0082	.0935	.1785	.2631	.3474	.4312	.5146	.5975	.6800	38
40	.05820	.14540	.23265	.31975	.40680	.49370	.58045	.66705	.75345	.83960	.92545	1.0110	1.0964	1.1813	1.2660	1.3502	1.4340	1.5174	1.6003	1.6827	40
42	.06110	.14835	.23555	.32265	.40970	.49660	.58335	.66995	.75630	.84245	.92835	.0139	.0992	.1842	.2688	.3530	.4368	.5201	.6030	.6855	42
44	.06400	.15125	.23845	.32555	.41260	.49950	.58625	.67280	.75920	.84530	.93120	.0168	.1020	.1870	.2716	.3558	.4396	.5229	.6058	.6882	44
46	.06690	.15415	.24135	.32845	.41550	.50240	.58915	.67570	.76205	.84820	.93405	.0196	.1049	.1898	.2744	.3586	.4423	.5257	.6085	.6909	46
48	.06980	.15705	.24425	.33135	.41840	.50530	.59200	.67860	.76495	.85105	.93690	.0225	.1077	.1926	.2772	.3614	.4451	.5285	.6113	.6937	48
50	.07270	.15995	.24715	.33425	.42130	.50820	.59490	.68145	.76780	.85390	.93975	1.0253	1.1106	1.1955	1.2800	1.3642	1.4479	1.5312	1.6141	1.6964	50
52	.07565	.16285	.25005	.33715	.42420	.51105	.59780	.68435	.77070	.85680	.94260	.0281	.1134	.1983	.2828	.3670	.4507	.5340	.6168	.6991	52
54	.07855	.16580	.25295	.34010	.42710	.51395	.60070	.68720	.77355	.85965	.94550	.0310	.1162	.2011	.2856	.3698	.4535	.5368	.6196	.7019	54
56	.08145	.16870	.25585	.34300	.43000	.51685	.60355	.69010	.77645	.86250	.94835	.0338	.1191	.2039	.2884	.3726	.4563	.5395	.6223	.7046	56
58	.08435	.17160	.25875	.34590	.43290	.51975	.60645	.69300	.77930	.86540	.95120	.0367	.1219	.2068	.2913	.3754	.4591	.5423	.6251	.7073	58
60	.08725	.17450	.26170	.34880	.43580	.52265	.60935	.69585	.78215	.86825	.95405	1.0395	1.1247	1.2096	1.2941	1.3782	1.4618	1.5451	1.6278	1.7101	60

(Continued)

Table B17. Sine Bar Constants (Cont.)

Min.	20°	21°	22°	23°	24°	25°	26°	27°	28°	29°	30°	31°	32°	33°	34°	35°	36°	37°	38°	39°	Min.
0	1.7101	1.7918	1.8730	1.9536	2.0337	2.1131	2.1918	2.2699	2.3473	2.4240	2.5000	2.5752	2.6496	2.7232	2.7959	2.8679	2.9389	3.0091	3.0783	3.1466	0
2	.7128	.7945	.8757	.9563	.0363	.1157	.1944	.2725	.3499	.4266	.5025	.5777	.6520	.7256	.7984	.8702	.9413	.0114	.0806	.1488	2
4	.7155	.7972	.8784	.9590	.0390	.1183	.1971	.2751	.3525	.4291	.5050	.5802	.6545	.7280	.8008	.8726	.9436	.0137	.0829	.1511	4
6	.7183	.8000	.8811	.9617	.0416	.1210	.1997	.2777	.3550	.4317	.5075	.5826	.6570	.7305	.8032	.8750	.9460	.0160	.0852	.1534	6
8	.7210	.8027	.8838	.9643	.0443	.1236	.2023	.2803	.3576	.4342	.5100	.5851	.6594	.7329	.8056	.8774	.9483	.0183	.0874	.1556	8
10	1.7237	1.8054	1.8865	1.9670	2.0469	2.1262	2.2049	2.2829	2.3602	2.4367	2.5126	2.5876	2.6619	2.7354	2.8080	2.8798	2.9507	3.0207	3.0897	3.1579	10
12	.7265	.8081	.8892	.9697	.0496	.1289	.2075	.2855	.3627	.4393	.5151	.5901	.6644	.7378	.8104	.8821	.9530	.0230	.0920	.1601	12
14	.7292	.8108	.8919	.9724	.0522	.1315	.2101	.2881	.3653	.4418	.5176	.5926	.6668	.7402	.8128	.8845	.9554	.0253	.0943	.1624	14
16	.7319	.8135	.8946	.9750	.0549	.1341	.2127	.2906	.3679	.4444	.5201	.5951	.6693	.7427	.8152	.8869	.9577	.0276	.0966	.1646	16
18	.7347	.8162	.8973	.9777	.0575	.1368	.2153	.2932	.3704	.4469	.5226	.5976	.6717	.7451	.8176	.8893	.9600	.0299	.0989	.1669	18
20	1.7374	1.8189	1.8999	1.9804	2.0602	2.1394	2.2179	2.2958	2.3730	2.4494	2.5251	2.6001	2.6742	2.7475	2.8200	2.8916	2.9624	3.0322	3.1012	3.1691	20
22	.7401	.8217	.9026	.9830	.0628	.1420	.2205	.2984	.3755	.4520	.5276	.6025	.6767	.7499	.8224	.8940	.9647	.0345	.1034	.1714	22
24	.7428	.8244	.9053	.9857	.0655	.1447	.2232	.3010	.3781	.4545	.5301	.6050	.6791	.7524	.8248	.8964	.9671	.0369	.1057	.1736	24
26	.7456	.8271	.9080	.9884	.0681	.1473	.2258	.3036	.3807	.4570	.5327	.6075	.6816	.7548	.8272	.8988	.9694	.0392	.1080	.1759	26
28	.7483	.8298	.9107	.9911	.0708	.1499	.2284	.3061	.3832	.4596	.5352	.6100	.6840	.7572	.8296	.9011	.9718	.0415	.1103	.1781	28
30	1.7510	1.8325	1.9134	1.9937	2.0734	2.1525	2.2310	2.3087	2.3858	2.4621	2.5377	2.6125	2.6865	2.7597	2.8320	2.9035	2.9741	3.0438	3.1125	3.1804	30
32	.7537	.8352	.9161	.9964	.0761	.1552	.2336	.3113	.3883	.4646	.5402	.6149	.6889	.7621	.8344	.9059	.9764	.0461	.1148	.1826	32
34	.7565	.8379	.9188	.9991	.0787	.1578	.2362	.3139	.3909	.4672	.5427	.6174	.6914	.7645	.8368	.9082	.9788	.0484	.1171	.1849	34
36	.7592	.8406	.9215	2.0017	.0814	.1604	.2388	.3165	.3934	.4697	.5452	.6199	.6938	.7669	.8392	.9106	.9811	.0507	.1194	.1871	36
38	.7619	.8433	.9241	.0044	.0840	.1630	.2414	.3190	.3960	.4722	.5477	.6224	.6963	.7694	.8416	.9130	.9834	.0530	.1216	.1893	38
40	1.7646	1.8460	1.9268	2.0070	2.0867	2.1656	2.2440	2.3216	2.3985	2.4747	2.5502	2.6249	2.6987	2.7718	2.8440	2.9153	2.9858	3.0553	3.1239	3.1916	40
42	.7673	.8487	.9295	.0097	.0893	.1683	.2466	.3242	.4011	.4773	.5527	.6273	.7012	.7742	.8464	.9177	.9881	.0576	.1262	.1938	42
44	.7701	.8514	.9322	.0124	.0920	.1709	.2492	.3268	.4036	.4798	.5552	.6298	.7036	.7766	.8488	.9200	.9904	.0599	.1285	.1961	44
46	.7728	.8541	.9349	.0150	.0946	.1735	.2518	.3293	.4062	.4823	.5577	.6323	.7061	.7790	.8512	.9224	.9928	.0622	.1307	.1983	46
48	.7755	.8568	.9376	.0177	.0972	.1761	.2544	.3319	.4087	.4848	.5602	.6348	.7085	.7815	.8535	.9248	.9951	.0645	.1330	.2005	48
50	1.7782	1.8595	1.9402	2.0204	2.0999	2.1787	2.2570	2.3345	2.4113	2.4874	2.5627	2.6372	2.7110	2.7839	2.8559	2.9271	2.9974	3.0668	3.1353	3.2028	50
52	.7809	.8622	.9429	.0230	.1025	.1814	.2596	.3371	.4138	.4899	.5652	.6397	.7134	.7863	.8583	.9295	.9997	.0691	.1375	.2050	52
54	.7837	.8649	.9456	.0257	.1052	.1840	.2621	.3396	.4164	.4924	.5677	.6422	.7158	.7887	.8607	.9318	3.0021	.0714	.1398	.2072	54
56	.7864	.8676	.9483	.0283	.1078	.1866	.2647	.3422	.4189	.4949	.5702	.6446	.7183	.7911	.8631	.9342	.0044	.0737	.1421	.2095	56
58	.7891	.8703	.9510	.0310	.1104	.1892	.2673	.3448	.4215	.4975	.5727	.6471	.7207	.7935	.8655	.9365	.0067	.0760	.1443	.2117	58
60	1.7918	1.8730	1.9536	2.0337	2.1131	2.1918	2.2699	2.3473	2.4240	2.5000	2.5752	2.6496	2.7232	2.7959	2.8679	2.9389	3.0091	3.0783	3.1466	3.2139	60

Min.	40°	41°	42°	43°	44°	45°	46°	47°	48°	49°	50°	51°	52°	53°	54°	55°	56°	57°	58°	59°	Min.
0	3.2139	3.2803	3.3456	3.4100	3.4733	3.5355	3.5967	3.6567	3.7157	3.7735	3.8302	3.8857	3.9400	3.9932	4.0451	4.0957	4.1452	4.1933	4.2402	4.2858	0
2	.2161	.2825	.3478	.4121	.4754	.5376	.5987	.6587	.7176	.7754	.8321	.8875	.9418	.9949	.0468	.0974	.1468	.1949	.2418	.2873	2
4	.2184	.2847	.3499	.4142	.4774	.5396	.6007	.6607	.7196	.7773	.8339	.8894	.9436	.9967	.0485	.0991	.1484	.1965	.2433	.2888	4
6	.2206	.2869	.3521	.4163	.4795	.5417	.6027	.6627	.7215	.7792	.8358	.8912	.9454	.9984	.0502	.1007	.1500	.1981	.2448	.2903	6
8	.2228	.2890	.3543	.4185	.4816	.5437	.6047	.6647	.7235	.7811	.8377	.8930	.9472	4.0001	.0519	.1024	.1517	.1997	.2464	.2918	8
10	3.2250	3.2912	3.3564	3.4206	3.4837	3.5458	3.6068	3.6666	3.7254	3.7830	3.8395	3.8948	3.9490	4.0019	4.0536	4.1041	4.1533	4.2012	4.2479	4.2933	10
12	.2273	.2934	.3586	.4227	.4858	.5478	.6088	.6686	.7274	.7850	.8414	.8967	.9508	.0036	.0553	.1057	.1549	.2028	.2494	.2948	12
14	.2295	.2956	.3607	.4248	.4879	.5499	.6108	.6706	.7293	.7869	.8433	.8985	.9525	.0054	.0570	.1074	.1565	.2044	.2510	.2963	14
16	.2317	.2978	.3629	.4269	.4900	.5519	.6128	.6726	.7312	.7887	.8451	.9003	.9543	.0071	.0587	.1090	.1581	.2060	.2525	.2978	16
18	.2339	.3000	.3650	.4291	.4921	.5540	.6148	.6745	.7332	.7906	.8470	.9021	.9561	.0089	.0604	.1107	.1597	.2075	.2540	.2992	18
20	3.2361	3.3022	3.3672	3.4312	3.4941	3.5560	3.6168	3.6765	3.7351	3.7925	3.8488	3.9039	3.9579	4.0106	4.0621	4.1124	4.1614	4.2091	4.2556	4.3007	20
22	.2384	.3044	.3693	.4333	.4962	.5581	.6188	.6785	.7370	.7944	.8507	.9058	.9596	.0123	.0638	.1140	.1630	.2107	.2571	.3022	22
24	.2406	.3065	.3715	.4354	.4983	.5601	.6208	.6805	.7390	.7963	.8525	.9076	.9614	.0141	.0655	.1157	.1646	.2122	.2586	.3037	24
26	.2428	.3087	.3736	.4375	.5004	.5621	.6228	.6824	.7409	.7982	.8544	.9094	.9632	.0158	.0672	.1173	.1662	.2138	.2601	.3052	26
28	.2450	.3109	.3758	.4396	.5024	.5642	.6248	.6844	.7428	.8001	.8562	.9112	.9650	.0175	.0689	.1190	.1678	.2154	.2617	.3066	28
30	3.2472	3.3131	3.3779	3.4417	3.5045	3.5662	3.6268	3.6864	3.7448	3.8020	3.8581	3.9130	3.9667	4.0193	4.0706	4.1206	4.1694	4.2169	4.2632	4.3081	30
32	.2494	.3153	.3801	.4439	.5066	.5683	.6288	.6883	.7467	.8039	.8599	.9148	.9685	.0210	.0722	.1223	.1710	.2185	.2647	.3096	32
34	.2516	.3174	.3822	.4460	.5087	.5703	.6308	.6903	.7486	.8058	.8618	.9166	.9703	.0227	.0739	.1239	.1726	.2201	.2662	.3111	34
36	.2538	.3196	.3844	.4481	.5107	.5723	.6328	.6923	.7505	.8077	.8636	.9184	.9720	.0244	.0756	.1255	.1742	.2216	.2677	.3125	36
38	.2561	.3218	.3865	.4502	.5128	.5744	.6348	.6942	.7525	.8096	.8655	.9202	.9738	.0262	.0773	.1272	.1758	.2232	.2692	.3140	38
40	3.2583	3.3240	3.3886	3.4523	3.5149	3.5764	3.6368	3.6962	3.7544	3.8114	3.8673	3.9221	3.9756	4.0279	4.0790	4.1288	4.1774	4.2247	4.2708	4.3155	40
42	.2605	.3261	.3908	.4544	.5169	.5784	.6388	.6981	.7563	.8133	.8692	.9239	.9773	.0296	.0807	.1305	.1790	.2263	.2723	.3170	42
44	.2627	.3283	.3929	.4565	.5190	.5805	.6408	.7001	.7582	.8152	.8710	.9257	.9791	.0313	.0823	.1321	.1806	.2278	.2738	.3184	44
46	.2649	.3305	.3950	.4586	.5211	.5825	.6428	.7020	.7601	.8171	.8729	.9275	.9809	.0331	.0840	.1337	.1822	.2294	.2753	.3199	46
48	.2671	.3326	.3972	.4607	.5231	.5845	.6448	.7040	.7620	.8190	.8747	.9293	.9826	.0348	.0857	.1354	.1838	.2309	.2768	.3213	48
50	3.2693	3.3348	3.3993	3.4628	3.5252	3.5866	3.6468	3.7060	3.7640	3.8208	3.8765	3.9311	3.9844	4.0365	4.0874	4.1370	4.1854	4.2325	4.2783	4.3228	50
52	.2715	.3370	.4014	.4649	.5273	.5886	.6488	.7079	.7659	.8227	.8784	.9329	.9861	.0382	.0891	.1386	.1870	.2340	.2798	.3243	52
54	.2737	.3391	.4036	.4670	.5293	.5906	.6508	.7099	.7678	.8246	.8802	.9347	.9879	.0399	.0907	.1403	.1886	.2356	.2813	.3257	54
56	.2759	.3413	.4057	.4691	.5314	.5926	.6528	.7118	.7697	.8265	.8820	.9364	.9896	.0416	.0924	.1419	.1902	.2371	.2828	.3272	56
58	.2781	.3435	.4078	.4712	.5335	.5947	.6548	.7138	.7716	.8283	.8839	.9382	.9914	.0433	.0941	.1435	.1917	.2387	.2843	.3286	58
60	3.2803	3.3456	3.4100	3.4733	3.5355	3.5967	3.6567	3.7157	3.7735	3.8302	3.8857	3.9400	3.9932	4.0451	4.0957	4.1452	4.1933	4.2402	4.2858	4.3301	60

(Continued)

Table B18. Natural Trigonometric Functions

	0°						1°					
′	sin	cos	tan	cot	sec	cosec	sin	cos	tan	cot	sec	cosec
0	.00000	1.0000	.00000	Infinite	1.0000	Infinite	.01745	.99985	.01745	57.290	1.0001	57.299
1	.00029	.0000	.00029	3437.7	.0000	3437.7	.01774	.99984	.01775	56.350	.0001	56.359
2	.00058	.0000	.00058	1718.9	.0000	1718.9	.01803	.99984	.01804	55.441	.0001	55.450
3	.00087	.0000	.00087	1145.9	.0000	1145.9	.01832	.99983	.01833	54.561	.0002	54.570
4	.00116	.0000	.00116	859.44	.0000	859.44	.01861	.99983	.01862	53.708	.0002	53.718
5	.00145	1.0000	.00145	687.55	.0000	687.55	.01891	.99982	.01891	52.882	.0002	52.891
6	.00174	.0000	.00174	572.96	.0000	572.96	.01920	.99981	.01920	52.081	.0002	52.090
7	.00204	.0000	.00204	491.11	.0000	491.11	.01949	.99981	.01949	51.303	.0002	51.313
8	.00233	.0000	.00233	429.72	.0000	429.72	.01978	.99980	.01978	50.548	.0002	50.558
9	.00262	.0000	.00262	381.97	.0000	381.97	.02007	.99980	.02007	49.816	.0002	49.826
10	.00291	0.99999	.00291	343.77	.0000	343.77	.02036	.99979	.02036	49.104	.0002	49.114
11	.00320	.99999	.00320	312.52	.0000	312.52	.02065	.99979	.02066	48.412	.0002	48.422
12	.00349	.99999	.00349	286.48	.0000	286.48	.02094	.99978	.02095	47.739	.0002	47.750
13	.00378	.99999	.00378	264.44	.0000	264.44	.02123	.99977	.02124	47.085	.0002	47.096
14	.00407	.99999	.00407	245.55	.0000	245.55	.02152	.99977	.02153	46.449	.0002	46.460
15	.00436	.99999	.00436	229.18	.0000	229.18	.02181	.99976	.02182	45.829	.0002	45.840
16	.00465	.99999	.00465	214.86	.0000	214.86	.02210	.99975	.02211	45.226	.0002	45.237
17	.00494	.99999	.00494	202.22	.0000	202.22	.02240	.99975	.02240	44.638	.0002	44.650
18	.00524	.99999	.00524	190.98	.0000	190.99	.02269	.99974	.02269	44.066	.0002	44.077
19	.00553	.99998	.00553	180.93	.0000	180.93	.02298	.99974	.02298	43.508	.0003	43.520
20	.00582	.99998	.00582	171.88	.0000	171.89	.02326	.99973	.02327	42.964	.0003	42.976
21	.00611	.99998	.00611	163.70	.0000	163.70	.02356	.99972	.02357	42.433	.0003	42.445
22	.00640	.99998	.00640	156.26	.0000	156.26	.02385	.99971	.02386	41.916	.0003	41.928
23	.00669	.99998	.00669	149.46	.0000	149.47	.02414	.99971	.02415	41.410	.0003	41.423
24	.00698	.99997	.00698	143.24	.0000	143.24	.02443	.99970	.02444	40.917	.0003	40.930
25	.00727	.99997	.00727	137.51	.0000	137.51	.02472	.99969	.02473	40.436	.0003	40.448
26	.00756	.99997	.00756	132.22	.0000	132.22	.02501	.99969	.02502	39.965	.0003	39.978
27	.00785	.99997	.00785	127.32	.0000	127.32	.02530	.99968	.02531	39.506	.0003	39.518
28	.00814	.99997	.00814	122.77	.0000	122.78	.02559	.99967	.02560	39.057	.0003	39.069
29	.00843	.99996	.00844	118.54	.0000	118.54	.02589	.99966	.02589	38.618	.0003	38.631
30	.00873	.99996	.00873	114.59	.0000	114.59	.02618	.99966	.02618	38.188	.0003	38.201
31	.00902	.99996	.00902	110.89	.0000	110.90	.02647	.99965	.02648	37.769	.0003	37.782
32	.00931	.99996	.00931	107.43	.0000	107.43	.02676	.99964	.02677	37.358	.0003	37.371
33	.00960	.99995	.00960	104.17	.0000	104.17	.02705	.99963	.02706	36.956	.0004	36.969
34	.00989	.99995	.00989	101.11	.0000	101.11	.02734	.99963	.02735	36.563	.0004	36.576
35	.01018	.99995	.01018	98.218	.0000	98.223	.02763	.99962	.02764	36.177	.0004	36.191
36	.01047	.99994	.01047	95.489	.0000	95.495	.02792	.99961	.02793	35.800	.0004	35.814
37	.01076	.99994	.01076	92.908	.0000	92.914	.02821	.99960	.02822	35.431	.0004	35.445
38	.01105	.99994	.01105	90.463	.0001	90.469	.02850	.99959	.02851	35.069	.0004	35.084
39	.01134	.99993	.01134	88.143	.0001	88.149	.02879	.99958	.02880	34.715	.0004	34.729
40	.01163	.99993	.01164	85.940	.0001	85.946	.02908	.99958	.02910	34.368	.0004	34.382
41	.01193	.99993	.01193	83.843	.0001	83.849	.02937	.99957	.02939	34.027	.0004	34.042
42	.01222	.99992	.01222	81.847	.0001	81.853	.02967	.99956	.02968	33.693	.0004	33.708
43	.01251	.99992	.01251	79.943	.0001	79.950	.02996	.99955	.02997	33.366	.0004	33.381
44	.01280	.99992	.01280	78.126	.0001	78.133	.03025	.99954	.03026	33.045	.0004	33.060
45	.01309	.99991	.01309	76.390	.0001	76.396	.03054	.99953	.03055	32.730	.0005	32.745
46	.01338	.99991	.01338	74.729	.0001	74.736	.03083	.99952	.03084	32.421	.0005	32.437
47	.01367	.99991	.01367	73.139	.0001	73.146	.03112	.99951	.03113	32.118	.0005	32.134
48	.01396	.99990	.01396	71.615	.0001	71.622	.03141	.99951	.03143	31.820	.0005	31.836
49	.01425	.99990	.01425	70.153	.0001	70.160	.03170	.99950	.03172	31.528	.0005	31.544
50	.01454	.99989	.01454	68.750	.0001	68.757	.03199	.99949	.03201	31.241	.0005	31.257
51	.01483	.99989	.01484	67.402	.0001	67.409	.03228	.99948	.03230	30.960	.0005	30.976
52	.01512	.99988	.01513	66.105	.0001	66.113	.03257	.99947	.03259	30.683	.0005	30.699
53	.01542	.99988	.01542	64.858	.0001	64.866	.03286	.99946	.03288	30.411	.0005	30.428
54	.01571	.99988	.01571	63.657	.0001	63.664	.03315	.99945	.03317	30.145	.0005	30.161
55	.01600	.99987	.01600	62.499	.0001	62.507	.03344	.99944	.03346	29.882	.0005	29.899
56	.01629	.99987	.01629	61.383	.0001	61.391	.03374	.99943	.03375	29.624	.0006	29.641
57	.01658	.99987	.01658	60.306	.0001	60.314	.03403	.99942	.03405	29.371	.0006	29.388
58	.01687	.99986	.01687	59.266	.0001	59.274	.03432	.99941	.03434	29.122	.0006	29.139
59	.01716	.99985	.01716	58.261	.0001	58.270	.03461	.99940	.03463	28.877	.0006	28.894
60	.01745	.99985	.01745	57.290	.0001	57.299	.03490	.99939	.03492	28.636	.0006	28.654
′	cos	sin	cot	tan	cosec	sec	cos	sin	cot	tan	cosec	sec
	89°						88°					

	2°						3°						
	sin	cos	tan	cot	sec	cosec	sin	cos	tan	cot	sec	cosec	′
	.03490	.99939	.03492	28.636	1.0006	28.654	.05234	.99863	.05241	19.081	1.0014	19.107	60
	.03519	.99938	.03521	28.399	.0006	28.417	.05263	.99861	.05270	18.975	.0014	19.002	59
	.03548	.99937	.03550	28.166	.0006	28.184	.05292	.99860	.05299	18.871	.0014	18.897	58
	.03577	.99936	.03579	27.937	.0006	27.955	.05321	.99858	.05328	18.768	.0014	18.794	57
	.03606	.99935	.03608	27.712	.0006	27.730	.05350	.99857	.05357	18.665	.0014	18.692	56
	.03635	.99934	.03638	27.490	.0007	27.508	.05379	.99855	.05387	18.564	.0014	18.591	55
	.03664	.99933	.03667	27.271	.0007	27.290	.05408	.99854	.05416	18.464	.0015	18.491	54
	.03693	.99932	.03696	27.056	.0007	27.075	.05437	.99852	.05445	18.365	.0015	18.393	53
	.03722	.99931	.03725	26.845	.0007	26.864	.05466	.99850	.05474	18.268	.0015	18.295	52
	.03751	.99930	.03754	26.637	.0007	26.655	.05495	.99849	.05503	18.171	.0015	18.198	51
	.03781	.99928	.03783	26.432	.0007	26.450	.05524	.99847	.05532	18.075	.0015	18.103	50
	.03810	.99927	.03812	26.230	.0007	26.249	.05553	.99846	.05562	17.980	.0015	18.008	49
	.03839	.99926	.03842	26.031	.0007	26.050	.05582	.99844	.05591	17.886	.0016	17.914	48
	.03868	.99925	.03871	25.835	.0007	25.854	.05611	.99842	.05620	17.793	.0016	17.821	47
	.03897	.99924	.03900	25.642	.0008	25.661	.05640	.99841	.05649	17.701	.0016	17.730	46
	.03926	.99923	.03929	25.452	.0008	25.471	.05669	.99839	.05678	17.610	.0016	17.639	45
	.03955	.99922	.03958	25.264	.0008	25.284	.05698	.99837	.05707	17.520	.0016	17.549	44
	.03984	.99921	.03987	25.080	.0008	25.100	.05727	.99836	.05737	17.431	.0016	17.460	43
	.04013	.99919	.04016	24.898	.0008	24.918	.05756	.99834	.05766	17.343	.0017	17.372	42
	.04042	.99918	.04045	24.718	.0008	24.739	.05785	.99832	.05795	17.256	.0017	17.285	41
	.04071	.99917	.04075	24.542	.0008	24.562	.05814	.99831	.05824	17.169	.0017	17.198	40
	.04100	.99916	.04104	24.367	.0008	24.388	.05843	.99829	.05853	17.084	.0017	17.113	39
	.04129	.99915	.04133	24.196	.0008	24.216	.05872	.99827	.05883	16.999	.0017	17.028	38
	.04158	.99913	.04162	24.026	.0009	24.047	.05902	.99826	.05912	16.915	.0017	16.944	37
	.04187	.99912	.04191	23.859	.0009	23.880	.05931	.99824	.05941	16.832	.0018	16.861	36
	.04217	.99911	.04220	23.694	.0009	23.716	.05960	.99822	.05970	16.750	.0018	16.779	35
	.04246	.99910	.04249	23.532	.0009	23.553	.05989	.99820	.05999	16.668	.0018	16.698	34
	.04275	.99908	.04279	23.372	.0009	23.393	.06018	.99819	.06029	16.587	.0018	16.617	33
	.04304	.99907	.04308	23.214	.0009	23.235	.06047	.99817	.06058	16.507	.0018	16.538	32
	.04333	.99906	.04337	23.058	.0009	23.079	.06076	.99815	.06087	16.428	.0018	16.459	31
	.04362	.99905	.04366	22.904	.0009	22.925	.06105	.99813	.06116	16.350	.0019	16.380	30
	.04391	.99903	.04395	22.752	.0010	22.774	.06134	.99812	.06145	16.272	.0019	16.303	29
	.04420	.99902	.04424	22.602	.0010	22.624	.06163	.99810	.06175	16.195	.0019	16.226	28
	.04449	.99901	.04453	22.454	.0010	22.476	.06192	.99808	.06204	16.119	.0019	16.150	27
	.04478	.99900	.04483	22.308	.0010	22.330	.06221	.99806	.06233	16.043	.0019	16.075	26
	.04507	.99898	.04512	22.164	.0010	22.186	.06250	.99804	.06262	15.969	.0019	16.000	25
	.04536	.99897	.04541	22.022	.0010	22.044	.06279	.99803	.06291	15.894	.0020	15.926	24
	.04565	.99896	.04570	21.881	.0010	21.904	.06308	.99801	.06321	15.821	.0020	15.853	23
	.04594	.99894	.04599	21.742	.0010	21.765	.06337	.99799	.06350	15.748	.0020	15.780	22
	.04623	.99893	.04628	21.606	.0011	21.629	.06366	.99797	.06379	15.676	.0020	15.708	21
	.04652	.99892	.04657	21.470	.0011	21.494	.06395	.99795	.06408	15.605	.0020	15.637	20
	.04681	.99890	.04687	21.337	.0011	21.360	.06424	.99793	.06437	15.534	.0021	15.566	19
	.04711	.99889	.04716	21.205	.0011	21.228	.06453	.99791	.06467	15.464	.0021	15.496	18
	.04740	.99888	.04745	21.075	.0011	21.098	.06482	.99790	.06496	15.394	.0021	15.427	17
	.04769	.99886	.94774	20.946	.0011	20.970	.06511	.99788	.06525	15.325	.0021	15.358	16
	.04798	.99885	.04803	20.819	.0011	20.843	.06540	.99786	.06554	15.257	.0021	15.290	15
	.04827	.99883	.04832	20.693	.0012	20.717	.06569	.99784	.06583	15.189	.0022	15.222	14
	.04856	.99882	.04862	20.569	.0012	20.593	.06598	.99782	.06613	15.122	.0022	15.155	13
	.04885	.99881	.04891	20.446	.0012	20.471	.06627	.99780	.06642	15.056	.0022	15.089	12
	.04914	.99879	.04920	20.325	.0012	20.350	.06656	.99778	.06671	14.990	.0022	15.023	11
	.04943	.99878	.04949	20.205	.0012	20.230	.06685	.99776	.06700	14.924	.0022	14.958	10
	.04972	.99876	.04978	20.087	.0012	20.112	.06714	.99774	.06730	14.860	.0023	14.893	9
	.05001	.99875	.05007	19.970	.0012	19.995	.06743	.99772	.06759	14.795	.0023	14.829	8
	.05030	.99873	.05037	19.854	.0013	19.880	.06772	.99770	.06788	14.732	.0023	14.765	7
	.05059	.99872	.05066	19.740	.0013	19.766	.06801	.99768	.06817	14.668	.0023	14.702	6
	.05088	.99870	.05095	19.627	.0013	19.653	.06830	.99766	.06846	14.606	.0023	14.640	5
	.05117	.99869	.05124	19.515	.0013	19.541	.06859	.99764	.06876	14.544	.0024	14.578	4
	.05146	.99867	.05153	19.405	.0013	19.431	.06888	.99762	.06905	14.482	.0024	14.517	3
	.05175	.99866	.05182	19.296	.0013	19.322	.06918	.99760	.06934	14.421	.0024	14.456	2
	.05204	.99864	.05212	19.188	.0013	19.214	.06947	.99758	.06963	14.361	.0024	14.395	1
	.05234	.99863	.05241	19.081	.0014	19.107	.06976	.99756	06993	14.301	.0024	14.335	0
	cos	sin	cot	tan	cosec	sec	cos	sin	cot	tan	cosec	sec	′
	87°						86°						

	4°						5°						6°						7°						
′	sin	cos	tan	cot	sec	cosec	sin	cos	tan	cot	sec	cosec	sin	cos	tan	cot	sec	cosec	sin	cos	tan	cot	sec	cosec	′
0	.06976	.99756	.06993	14.301	1.0024	14.335	.08715	.99619	.08749	11.430	1.0038	11.474	.10453	.99452	.10510	9.5144	1.0055	9.5668	.12187	.99255	.12278	8.1443	1.0075	8.2055	60
1	.07005	.99754	.07022	14.241	.0025	14.276	.08744	.99617	.08778	11.392	.0038	11.436	.10482	.99449	.10540	.4878	.0055	.5404	.12216	.99251	.12308	.1248	.0075	.1861	59
2	.07034	.99752	.07051	14.182	.0025	14.217	.08773	.99614	.08807	11.354	.0039	11.398	.10511	.99446	.10569	.4614	.0056	.5141	.12245	.99247	.12337	.1053	.0076	.1668	58
3	.07063	.99750	.07080	14.123	.0025	14.159	.08802	.99612	.08837	11.316	.0039	11.360	.10540	.99443	.10599	.4351	.0056	.4880	.12273	.99244	.12367	.0860	.0076	.1476	57
4	.07092	.99748	.07110	14.065	.0025	14.101	.08831	.99609	.08866	11.279	.0039	11.323	.10568	.99440	.10628	.4090	.0056	.4620	.12302	.99240	.12396	.0667	.0076	.1285	56
5	.07121	.99746	.07139	14.008	.0025	14.043	.08860	.99607	.08895	11.242	.0039	11.286	.10597	.99437	.10657	9.3831	.0057	9.4362	.12331	.99237	.12426	8.0476	.0077	8.1094	55
6	.07150	.99744	.07168	13.951	.0026	13.986	.08889	.99604	.08925	11.205	.0040	11.249	.10626	.99434	.10687	.3572	.0057	.4105	.12360	.99233	.12456	.0285	.0077	.0905	54
7	.07179	.99742	.07197	13.894	.0026	13.930	.08918	.99601	.08954	11.168	.0040	11.213	.10655	.99431	.10716	.3315	.0057	.3850	.12389	.99229	.12485	.0095	.0078	.0717	53
8	.07208	.99740	.07226	13.838	.0026	13.874	.08947	.99599	.08983	11.132	.0040	11.176	.10684	.99428	.10746	.3060	.0057	.3596	.12418	.99226	.12515	7.9906	.0078	.0529	52
9	.07237	.99738	.07256	13.782	.0026	13.818	.08976	.99596	.09013	11.095	.0040	11.140	.10713	.99424	.10775	.2806	.0058	.3343	.12447	.99222	.12544	.9717	.0078	.0342	51
10	.07266	.99736	.07285	13.727	.0026	13.763	.09005	.99594	.09042	11.059	.0041	11.104	.10742	.99421	.10805	9.2553	.0058	9.3092	.12476	.99219	.12574	7.9530	.0079	8.0156	50
11	.07295	.99733	.07314	13.672	.0027	13.708	.09034	.99591	.09071	11.024	.0041	11.069	.10771	.99418	.10834	.2302	.0058	.2842	.12504	.99215	.12603	.9344	.0079	7.9971	49
12	.07324	.99731	.07343	13.617	.0027	13.654	.09063	.99588	.09101	10.988	.0041	11.033	.10800	.99415	.10863	.2051	.0059	.2593	.12533	.99211	.12633	.9158	.0079	.9787	48
13	.07353	.99729	.07373	13.563	.0027	13.600	.09092	.99586	.09130	10.953	.0041	10.998	.10829	.99412	.10893	.1803	.0059	.2346	.12562	.99208	.12662	.8973	.0080	.9604	47
14	.07382	.99727	.07402	13.510	.0027	13.547	.09121	.99583	.09159	10.918	.0042	10.963	.10858	.99409	.10922	.1555	.0059	.2100	.12591	.99204	.12692	.8789	.0080	.9421	46
15	.07411	.99725	.07431	13.457	.0027	13.494	.09150	.99580	.09189	10.883	.0042	10.929	.10887	.99406	.10952	9.1309	.0060	9.1855	.12620	.99200	.12722	7.8606	.0080	7.9240	45
16	.07440	.99723	.07460	13.404	.0028	13.441	.09179	.99578	.09218	10.848	.0042	10.894	.10916	.99402	.10981	.1064	.0060	.1612	.12649	.99197	.12751	.8424	.0081	.9059	44
17	.07469	.99721	.07490	13.351	.0028	13.389	.09208	.99575	.09247	10.814	.0043	10.860	.10944	.99399	.11011	.0821	.0060	.1370	.12678	.99193	.12781	.8243	.0081	.8879	43
18	.07498	.99718	.07519	13.299	.0028	13.337	.09237	.99572	.09277	10.780	.0043	10.826	.10973	.99396	.11040	.0579	.0061	.1129	.12706	.99189	.12810	.8062	.0082	.8700	42
19	.07527	.99716	.07548	13.248	.0028	13.286	.09266	.99570	.09306	10.746	.0043	10.792	.11002	.99393	.11069	.0338	.0061	.0890	.12735	.99186	.12840	.7882	.0082	.8522	41
20	.07556	.99714	.07577	13.917	.0029	13.235	.09295	.99567	.09335	10.712	.0043	10.758	.11031	.99390	.11099	9.0098	.0061	9.0651	.12764	.99182	.12869	7.7703	.0082	7.8344	40
21	.07585	.99712	.07607	13.146	.0029	13.184	.09324	.99564	.09365	10.678	.0044	10.725	.11060	.99386	.11128	8.9860	.0062	.0414	.12793	.99178	.12899	.7525	.0083	.8168	39
22	.07614	.99710	.07636	13.096	.0029	13.134	.09353	.99562	.09394	10.645	.0044	10.692	.11089	.99383	.11158	.9623	.0062	.0179	.12822	.99174	.12928	.7348	.0083	.7992	38
23	.07643	.99707	.07665	13.046	.0029	13.084	.09382	.99559	.09423	10.612	.0044	10.659	.11118	.99380	.11187	.9387	.0062	8.9944	.12851	.99171	.12958	.7171	.0084	.7817	37
24	.07672	.99705	.07694	12.996	.0029	13.034	.09411	.99556	.09453	10.579	.0044	10.626	.11147	.99377	.11217	.9152	.0063	.9711	.12879	.99167	.12988	.6996	.0084	.7642	36
25	.07701	.99703	.07724	12.947	.0030	12.985	.09440	.99553	.09482	10.546	.0045	10.593	.11176	.99373	.11246	8.8918	.0063	8.9479	.12908	.99163	.13017	7.6821	.0084	7.7469	35
26	.07730	.99701	.07753	12.898	.0030	12.937	.09469	.99551	.09511	10.514	.0045	10.561	.11205	.99370	.11276	.8686	.0063	.9248	.12937	.99160	.13047	.6646	.0085	.7296	34
27	.07759	.99698	.07782	12.849	.0030	12.888	.09498	.99548	.09541	10.481	.0045	10.529	.11234	.99367	.11305	.8455	.0064	.9018	.12966	.99156	.13076	.6473	.0085	.7124	33
28	.07788	.99696	.07812	12.801	.0030	12.840	.09527	.99545	.09570	10.449	.0046	10.497	.11262	.99364	.11335	.8225	.0064	.8790	.12995	.99152	.13106	.6300	.0085	.6953	32
29	.07817	.99694	.07841	12.754	.0031	12.793	.09556	.99542	.09599	10.417	.0046	10.465	.11291	.99360	.11364	.7996	.0064	.8563	.13024	.99148	.13136	.6129	.0086	.6783	31
30	.07846	.99692	.07870	12.706	.0031	12.745	.09584	.99540	.09629	10.385	.0046	10.433	.11320	.99357	.11393	8.7769	.0065	8.8337	.13053	.99144	.13165	7.5957	.0086	7.6613	30
31	.07875	.99689	.07899	12.659	.0031	12.698	.09613	.99537	.09658	10.354	.0046	10.402	.11349	.99354	.11423	.7542	.0065	.8112	.13081	.99141	.13195	.5787	.0087	.6444	29
32	.07904	.99687	.07929	12.612	.0031	12.652	.09642	.99534	.09688	10.322	.0047	10.371	.11378	.99350	.11452	.7317	.0065	.7888	.13110	.99137	.13224	.5617	.0087	.6276	28
33	.07933	.99685	.07958	12.566	.0032	12.606	.09671	.99531	.09717	10.291	.0047	10.340	.11407	.99347	.11482	.7093	.0066	.7665	.13139	.99133	.13254	.5449	.0087	.6108	27
34	.07962	.99682	.07987	12.520	.0032	12.560	.09700	.99528	.09746	10.260	.0047	10.309	.11436	.99344	.11511	.6870	.0066	.7444	.13168	.99129	.13284	.5280	.0088	.5942	26
35	.07991	.99680	.08016	12.474	.0032	12.514	.09729	.99525	.09776	10.229	.0048	10.278	.11465	.99341	.11541	8.6648	.0066	8.7223	.13197	.99125	.13313	7.5113	.0088	7.5776	25
36	.08020	.99678	.08046	12.429	.0032	12.469	.09758	.99523	.09805	10.199	.0048	10.248	.11494	.99337	.11570	.6427	.0067	.7004	.13226	.99121	.13343	.4946	.0089	.5611	24
37	.08049	.99675	.08075	12.384	.0032	12.424	.09787	.99520	.09834	10.168	.0048	10.217	.11523	.99334	.11600	.6208	.0067	.6786	.13254	.99118	.13372	.4780	.0089	.5446	23
38	.08078	.99673	.08104	12.339	.0033	12.379	.09816	.99517	.09864	10.138	.0048	10.187	.11551	.99330	.11629	.5989	.0067	.6569	.13283	.99114	.13402	.4615	.0089	.5282	22
39	.08107	.99671	.08134	12.295	.0033	12.335	.09845	.99514	.09893	10.108	.0049	10.157	.11580	.99327	.11659	.5772	.0068	.6353	.13312	.99110	.13432	.4451	.0090	.5119	21
40	.08136	.99668	.08163	12.250	.0033	12.291	.09874	.99511	.09922	10.078	.0049	10.127	.11609	.99324	.11688	8.5555	.0068	8.6138	.13341	.99106	.13461	7.4287	.0090	7.4957	20
41	.08165	.99666	.08192	12.207	.0033	12.248	.09903	.99508	.09952	10.048	.0049	10.098	.11638	.99320	.11718	.5340	.0068	.5924	.13370	.99102	.13491	.4124	.0090	.4795	19
42	.08194	.99664	.08221	12.163	.0034	12.204	.09932	.99505	.09981	10.019	.0050	10.068	.11667	.99317	.11747	.5126	.0069	.5711	.13399	.99098	.13520	.3961	.0091	.4634	18
43	.08223	.99661	.08251	12.120	.0034	12.161	.09961	.99503	.10011	9.9893	.0050	10.039	.11696	.99314	.11777	.4913	.0069	.5499	.13427	.99094	.13550	.3800	.0091	.4474	17
44	.08252	.99659	.08280	12.077	.0034	12.118	.09990	.99500	.10040	9.9601	.0050	10.010	.11725	.99310	.11806	.4701	.0069	.5289	.13456	.99090	.13580	.3639	.0092	.4315	16
45	.08281	.99656	.08309	12.035	.0034	12.076	.10019	.99497	.10069	9.9310	.0050	9.9812	.11754	.99307	.11836	8.4489	.0070	8.5079	.13485	.99086	.13609	7.3479	.0092	7.4156	15
46	.08310	.99654	.08339	11.992	.0035	12.034	.10048	.99494	.10099	9.9021	.0051	9.9525	.11783	.99303	.11865	.4279	.0070	.4871	.13514	.99083	.13639	.3319	.0092	.3998	14
47	.08339	.99652	.08368	11.950	.0035	11.992	.10077	.99491	.10128	9.8734	.0051	9.9239	.11811	.99300	.11895	.4070	.0070	.4663	.13543	.99079	.13669	.3160	.0093	.3840	13
48	.08368	.99649	.08397	11.909	.0035	11.950	.10106	.99488	.10158	9.8448	.0051	9.8955	.11840	.99296	.11924	.3862	.0071	.4457	.13571	.99075	.13698	.3002	.0093	.3683	12
49	.08397	.99647	.08426	11.867	.0035	11.909	.10134	.99485	.10187	9.8164	.0052	9.8672	.11869	.99293	.11954	.3655	.0071	.4251	.13600	.99071	.13728	.2844	.0094	.3527	11
50	.08426	.99644	.08456	11.826	.0036	11.868	.10163	.99482	.10216	9.7882	.0052	9.8391	.11898	.99290	.11983	8.3449	.0071	8.4046	.13629	.99067	.13757	7.2687	.0094	7.3372	10
51	.08455	.99642	.08485	11.785	.0036	11.828	.10192	.99479	.10246	9.7601	.0052	9.8112	.11927	.99286	.12013	.3244	.0072	.3843	.13658	.99063	.13787	.2531	.0094	.3217	9
52	.08484	.99639	.08514	11.745	.0036	11.787	.10221	.99476	.10275	9.7322	.0053	9.7834	.11956	.99283	.12042	.3040	.0072	.3640	.13687	.99059	.13817	.2375	.0095	.3063	8
53	.08513	.99637	.08544	11.704	.0036	11.747	.10250	.99473	.10305	9.7044	.0053	9.7558	.11985	.99279	.12072	.2837	.0073	.3439	.13716	.99055	.13846	.2220	.0095	.2909	7
54	.08542	.99634	.08573	11.664	.0037	11.707	.10279	.99470	.10334	9.6768	.0053	9.7283	.12014	.99276	.12101	.2635	.0073	.3238	.13744	.99051	.13876	.2066	.0096	.2757	6
55	.08571	.99632	.08602	11.625	.0037	11.668	.10308	.99467	.10363	9.6493	.0053	9.7010	.12042	.99272	.12131	8.2434	.0073	8.3039	.13773	.99047	.13906	7.1912	.0096	7.2604	5
56	.08600	.99629	.08632	11.585	.0037	11.628	.10337	.99464	.10393	9.6220	.0054	9.6739	.12071	.99269	.12160	.2234	.0074	.2840	.13802	.99043	.13935	.1759	.0097	.2453	4
57	.08629	.99627	.08661	11.546	.0037	11.589	.10366	.99461	.10422	9.5949	.0054	9.6469	.12100	.99265	.12190	.2035	.0074	.2642	.13831	.99039	.13965	.1607	.0097	.2302	3
58	.08658	.99624	.08690	11.507	.0038	11.550	.10395	.99458	.10452	9.5679	.0054	9.6200	.12129	.99262	.12219	.1837	.0074	.2446	.13860	.99035	.13995	.1455	.0097	.2152	2
59	.08687	.99622	.08719	11.468	.0038	11.512	.10424	.99455	.10481	9.5411	.0055	9.5933	.12158	.99258	.12249	.1640	.0075	.2250	.13888	.99031	.14024	.1304	.0098	.2002	1
60	.08715	.99619	.08749	11.430	.0038	11.474	.10453	.99452	.10510	9.5144	.0055	9.5668	.12187	.99255	.12278	8.1443	.0075	8.2055	.13917	.99027	.14054	7.1154	.0098	7.1853	0
′	cos	sin	cot	tan	cosec	sec	cos	sin	cot	tan	cosec	sec	cos	sin	cot	tan	cosec	sec	cos	sin	cot	tan	cosec	sec	′
	85°						84°						83°						82°						

(Continued)

Table B18. (Cont.)

	8°						9°						10°						11°						
′	sin	cos	tan	cot	sec	cosec	sin	cos	tan	cot	sec	cosec	sin	cos	tan	cot	sec	cosec	sin	cos	tan	cot	sec	cosec	′
0	.13917	.99027	.14054	7.1154	1.0098	7.1853	.15643	.98769	.15838	6.3137	1.0215	6.3924	.17365	.98481	.17633	5.6713	1.0154	5.7588	.19081	.98163	.19438	5.1445	1.0187	5.2408	60
1	.13946	.99023	.14084	.1004	.0099	.1704	.15672	.98764	.15868	.3019	.0125	.3807	.17393	.98476	.17663	.6616	.0155	.7493	.19109	.98157	.19468	.1366	.0188	.2330	59
2	.13975	.99019	.14113	.0854	.0099	.1557	.15701	.98760	.15898	.2901	.0125	.3690	.17422	.98471	.17693	.6520	.0155	.7398	.19138	.98152	.19498	.1286	.0188	.2252	58
3	.14004	.99015	.14143	.0706	.0099	.1409	.15730	.98755	.15928	.2783	.0126	.3574	.17451	.98465	.17723	.6425	.0156	.7304	.19166	.98146	.19529	.1207	.0189	.2174	57
4	.14032	.99010	.14173	.0558	.0100	.1263	.15758	.98750	.15958	.2665	.0126	.3458	.17479	.98460	.17753	.6329	.0156	.7210	.19195	.98140	.19559	.1128	.0189	.2097	56
5	.14061	.99006	.14202	7.0410	.0100	7.1117	.15787	.98746	.15987	6.2548	.0127	6.3343	.17508	.98455	.17783	5.6234	.0157	5.7117	.19224	.98135	.19589	5.1049	.0190	5.2019	55
6	.14090	.99002	.14232	.0264	.0101	.0972	.15816	.98741	.16017	.2432	.0217	.3228	.17537	.98450	.17813	.6140	.0157	.7023	.19252	.98129	.19619	.0970	.0191	.1942	54
7	.14119	.98998	.14262	.0117	.0101	.0827	.15844	.98737	.16047	.2316	.0128	.3113	.17565	.98445	.17843	.6045	.0158	.6930	.19281	.98124	.19649	.0892	.0191	.1865	53
8	.14148	.98994	.14291	6.9972	.0102	.0683	.15873	.98732	.16077	.2200	.0128	.2999	.17594	.98440	.17873	.5951	.0158	.6838	.19309	.98118	.19680	.0814	.0192	.1788	52
9	.14176	.98990	.14321	.9827	.0102	.0539	.15902	.98727	.16107	.2085	.0129	.2885	.17622	.98435	.17903	.5857	.0159	.6745	.19338	.98112	.19710	.0736	.0192	.1712	51
10	.14205	.98986	.14351	6.9682	.0102	7.0396	.15931	.98723	.16137	6.1970	.0129	6.2772	.17651	.98430	.17933	5.5764	.0159	5.6653	.19366	.98107	.19740	5.0658	.0193	5.1636	50
11	.14234	.98982	.14380	.9538	.0103	.0254	.15959	.98718	.16167	.1856	.0130	.2659	.17680	.98425	.17963	.5670	.0160	.6561	.19395	.98101	.19770	.0581	.0193	.1560	49
12	.14263	.98978	.14410	.9395	.0103	.0112	.15988	.98714	.16196	.1742	.0130	.2546	.17708	.98419	.17993	.5578	.0160	.6470	.19423	.98095	.19800	.0504	.0194	.1484	48
13	.14292	.98973	.14440	.9252	.0104	6.9971	.16017	.98709	.16226	.1628	.0131	.2434	.17737	.98414	.18023	.5485	.0161	.6379	.19452	.98090	.19831	.0427	.0195	.1409	47
14	.14320	.98969	.14470	.9110	.0104	.9830	.16045	.98704	.16256	.1515	.0131	.2322	.17766	.98409	.18053	.5393	.0162	.6288	.19480	.98084	.19861	.0350	.0195	.1333	46
15	.14349	.98965	.14499	6.8969	.0104	6.9690	.16074	.98700	.16286	6.1402	.0132	6.2211	.17794	.98404	.18083	5.5301	.0162	5.6197	.19509	.98078	.19891	5.0273	.0196	5.1258	45
16	.14378	.98961	.14529	.8828	.0105	.9550	.16103	.98695	.16316	.1290	.0132	.2100	.17823	.98399	.18113	.5209	.0163	.6107	.19537	.98073	.19921	.0197	.0196	.1183	44
17	.14407	.98957	.14559	.8687	.0105	.9411	.16132	.98690	.16346	.1178	.0133	.1990	.17852	.98394	.18143	.5117	.0163	.6017	.19566	.98067	.19952	.0121	.0197	.1109	43
18	.14436	.98952	.14588	.8547	.0106	.9273	.16160	.98685	.16376	.1066	.0133	.1880	.17880	.98388	.18173	.5026	.0164	.5928	.19595	.98061	.19982	.0045	.0198	.1034	42
19	.14464	.98948	.14618	.8408	.0106	.9135	.16189	.98681	.16405	.0955	.0134	.1770	.17909	.98383	.18203	.4936	.0164	.5838	.19623	.98056	.20012	4.9969	.0198	.0960	41
20	.14493	.98944	.14648	6.8269	.0107	6.8998	.16218	.98676	.16435	6.0844	.0134	6.1661	.17937	.98378	.18233	5.4845	.0165	5.5749	.19652	.98050	.20042	4.9894	.0199	5.0886	40
21	.14522	.98940	.14677	.8131	.0107	.8861	.16246	.98671	.16465	.0734	.0135	.1552	.17966	.98373	.18263	.4755	.0165	.5660	.19680	.98044	.20073	.9819	.0199	.0812	39
22	.14551	.98936	.14707	.7993	.0107	.8725	.16275	.98667	.16495	.0624	.0135	.1443	.17995	.98368	.18293	.4665	.0166	.5572	.19709	.98039	.20103	.9744	.0200	.0739	38
23	.14579	.98931	.14737	.7856	.0108	.8589	.16304	.98662	.16525	.0514	.0136	.1335	.18023	.98362	.18323	.4575	.0166	.5484	.19737	.98033	.20133	.9669	.0201	.0666	37
24	.14608	.98927	.14767	.7720	.0108	.8454	.16333	.98657	.16555	.0405	.0136	.1227	.18052	.98357	.18353	.4486	.0167	.5396	.19766	.98027	.20163	.9594	.0201	.0593	36
25	.14637	.98923	.14796	6.7584	.0109	6.8320	.16361	.98652	.16585	6.0296	.0136	6.1120	.18080	.98352	.18383	5.4396	.0167	5.5308	.19794	.98021	.20194	4.9520	.0202	5.0520	35
26	.14666	.98919	.14826	.7448	.0109	.8185	.16390	.98648	.16615	.0188	.0137	.1013	.18109	.98347	.18413	.4308	.0168	.5221	.19823	.98016	.20224	.9446	.0202	.0447	34
27	.14695	.98914	.14856	.7313	.0110	.8052	.16419	.98643	.16644	.0080	.0137	.0906	.18138	.98341	.18444	.4219	.0169	.5134	.19851	.98010	.20254	.9372	.0203	.0375	33
28	.14723	.98910	.14886	.7179	.0110	.7919	.16447	.98638	.16674	5.9972	.0138	.0800	.18166	.98336	.18474	.4131	.0169	.5047	.19880	.98004	.20285	.9298	.0204	.0302	32
29	.14752	.98906	.14915	.7045	.0111	.7787	.16476	.98633	.16704	.9865	.0138	.0694	.18195	.98331	.18504	.4043	.0170	.4960	.19908	.97998	.20315	.9225	.0204	.0230	31
30	.14781	.98901	.14945	6.6911	.0111	6.7655	.16505	.98628	.16734	5.9758	.0139	6.0588	.18223	.98325	.18534	5.3955	.0170	5.4874	.19937	.97992	.20345	4.9151	.0205	5.0158	30
31	.14810	.98897	.14975	.6779	.0111	.7523	.16533	.98624	.16764	.9651	.9139	.0483	.18252	.98320	.18564	.3868	.0171	.4788	.19965	.97987	.20375	.9078	.0205	.0087	29
32	.14838	.98893	.15004	.6646	.0112	.7392	.16562	.98619	.16794	.9545	.9140	.0379	.18281	.98315	.18594	.3780	.0171	.4702	.19994	.97981	.20406	.9006	.0206	.0015	28
33	.14867	.98889	.15034	.6514	.0112	.7262	.16591	.98614	.16824	.9439	.0140	.0274	.18309	.98309	.18624	.3694	.0172	.4617	.20022	.97975	.20436	.8933	.0207	4.9944	27
34	.14896	.98884	.15064	.6383	.0113	.7132	.16619	.98609	.16854	.9333	.0141	.0170	.18338	.98304	.18654	.3607	.0172	.4532	.20051	.97969	.20466	.8860	.0207	.9873	26
35	.14925	.98880	.15094	6.6252	.0113	6.7003	.16648	.98604	.16884	5.9228	.0141	6.0066	.18366	.98299	.18684	5.3521	.0173	5.4447	.20079	.97963	.20497	4.8788	0208	4.9802	25
36	.14953	.98876	.15123	.6122	.0114	.6874	.16677	.98600	.16914	.9123	.0142	5.9963	.18395	.98293	.18714	.3434	.0174	.4362	.20108	.97957	.20527	.8716	.0208	.9732	24
37	.14982	.98871	.15153	.5992	.0114	.6745	.16705	.98595	.16944	.9019	.0142	.9860	.18424	.98288	.18745	.3349	.0174	.4278	.20136	.97952	.20557	.8644	.0209	.9661	23
38	.15011	.98867	.15183	.5863	.0115	.6617	.16734	.98590	.16973	.8915	.0143	.9758	.18452	.98283	.18775	.3263	.0175	.4194	.20165	.97946	.20588	.8573	.0210	.9591	22
39	.15050	.98862	.15213	.5734	.0115	.6490	.16763	.98585	.17003	.8811	.0143	.9655	.18481	.98277	.18805	.3178	.0175	.4110	.20193	.97940	.20618	.8501	.0210	.9521	21
40	.15068	.98858	.15243	6.5605	.0115	6.6363	.16791	.98580	.17033	5.8708	.0144	5.9554	.18509	.98272	.18835	5.3093	.0176	5.4026	.20222	.97934	.20648	4.8430	.0211	4.9452	20
41	.15097	.98854	.15272	.5478	.0116	.6237	.16820	.98575	.17063	.8605	.0144	.9452	.18538	.98267	.18865	.3008	.0176	.3943	.20250	.97928	.20679	.8359	.0211	.9382	19
42	.15126	.98849	.15302	.5350	.0116	.6111	.16849	.98570	.17093	.8502	.0145	.9351	.18567	.98261	.18895	.2923	.0177	.3860	.20279	.97922	.20709	.8288	.0212	.9313	18
43	.15155	.98845	.15332	.5223	.0117	.5985	.16878	.98565	.17123	.8400	.0145	.9250	.18595	.98256	.18925	.2839	.0177	.3777	.20307	.97916	.20739	.8217	.0213	.9243	17
44	.15183	.98840	.15362	.5097	.0117	.5860	.16906	.98560	.17153	.8298	.0146	.9150	.18624	.98250	.18955	.2755	.0178	.3695	.20336	.97910	.20770	.8147	.0213	.9175	16
45	.15212	.98836	.15391	6.4971	.0118	6.5736	.16935	.98556	.17183	5.8196	.0146	5.9049	.18652	.98245	.18985	5.2671	.0179	5.3612	.20364	.97904	.20800	4.8077	.0214	4.9106	15
46	.15241	.98832	.15421	.4845	.0118	.5612	.16964	.98551	.17213	.8095	.0147	.8950	.18681	.98240	.19016	.2588	.0179	.3530	.20393	.97899	.20830	.8007	.0215	.9037	14
47	.15270	.98827	.15451	.4720	.0119	.5488	.16992	.98546	.17243	.7994	.0147	.8850	.18709	.98234	.19046	.2505	.0180	.3449	.20421	.97893	.20861	.7937	.0215	.8969	13
48	.15298	.98823	.15481	.4596	.0119	.5365	.17021	.98541	.17273	.7894	.0148	.8751	.18738	.98229	.19076	.2422	.0180	.3367	.20450	.97887	.20891	.7867	.0216	.8901	12
49	.15328	.98818	.15511	.4472	.0119	.5243	.17050	.98536	.17303	.7794	.0148	.8652	.18767	.98223	.19106	.2339	.0181	.3286	.20478	.97881	.20921	.7798	.0216	.8833	11
50	.15356	.98814	.15540	6.4348	.0120	6.5121	.17078	.98531	.17333	5.7694	.0149	5.8554	.18795	.98218	.19136	5.2257	.0181	5.3205	.20506	.97875	.20952	4.7728	.0217	4.8765	10
51	.15385	.98809	.15570	.4225	.0120	.4999	.17107	.98526	.17363	.7594	.0150	.8456	.18824	.98212	.19166	.2174	.0182	.3124	.20535	.97869	.20982	.7659	.0218	.8967	9
52	.15413	.98805	.15600	.4103	.0121	.4878	.17136	.98521	.17393	.7495	.0150	.8358	.18852	.98207	.19197	.2092	.0182	.3044	.20563	.97863	.21012	.7591	.0218	.8630	8
53	.15442	.98800	.15630	.3980	.0121	.4757	.17164	.98516	.17423	.7396	.0151	.8261	.18881	.98201	.19227	.2011	.0183	.2963	.20592	.97857	.21043	.7522	.0219	.8563	7
54	.15471	.98796	.15659	.3859	.0122	.4637	.17193	.98511	.17453	.7297	.0151	.8163	.18909	.98196	.19257	.1929	.0184	.2883	.20620	.97851	.21073	.7453	.0220	.8496	6
55	.15500	.98791	.15689	6.3737	.0122	6.4517	.17221	.98506	.17483	5.7199	.0152	5.8067	.18938	.98190	.19287	5.1848	.0184	5.2803	.20649	.97845	.21104	4.7385	.0220	4.8429	5
56	.15528	.98787	.15719	.3616	.0123	.4398	.17250	.98501	.17513	.7101	.0152	.7970	.18967	.98185	.19317	.1767	.0185	.2724	.20677	.97839	.21134	.7317	.0221	.8362	4
57	.15557	.98782	.15749	.3496	.0123	.4279	.17279	.98496	.17543	.7004	.0153	.7874	.18995	.98179	.19347	.1686	.0185	.2645	.20706	.97833	.21164	.7249	.0221	.8296	3
58	.15586	.98778	.15779	.3376	.0124	.4160	.17307	.98491	.17573	.6906	.0153	.7778	.19024	.98174	.19378	.1606	.0186	.2566	.20734	.97827	.21195	.7181	.0222	.8229	2
59	.15615	.98773	.15809	.3257	.0124	.4042	.17336	.98486	.17603	.6809	.0154	.7683	.19052	.98168	.19408	.1525	.0186	.2487	.20763	.97821	.21225	.7114	.0223	.8163	1
60	.15643	.98769	.15838	6.3137	.0125	6.3924	.17365	.98481	.17633	5.6713	.0154	5.7588	.19081	.98163	.19438	5.1445	.0187	5.2408	.20791	.97815	.21256	4.7046	.0223	4.8097	0
′	cos	sin	cot	tan	cosec	sec	cos	sin	cot	tan	cosec	sec	cos	sin	cot	tan	cosec	sec	cos	sin	cot	tan	cosec	sec	′
	81°						80°						79°						78°						

′	12° sin	12° cos	12° tan	12° cot	12° sec	12° cosec	13° sin	13° cos	13° tan	13° cot	13° sec	13° cosec
0	.20791	.97815	.21256	4.7046	1.0223	4.8097	.22495	.97437	.23087	4.3315	1.0263	4.4454
1	.20820	.97809	.21286	.6979	.0224	.8032	.22523	.97430	.23117	.3257	.0264	.4398
2	.20848	.97803	.21316	.6912	.0225	.7966	.22552	.97424	.23148	.3200	.0264	.4342
3	.20876	.97797	.21347	.6845	.0225	.7901	.22580	.97417	.23179	.3143	.0265	.4287
4	.20905	.97790	.21377	.6778	.0226	.7835	.22608	.97411	.23209	.3086	.0266	.4231
5	.20933	.97784	.21408	4.6712	.0226	4.7770	.22637	.97404	.23240	4.3029	.0266	4.4176
6	.20962	.97778	.21438	.6646	.0227	.7706	.22665	.97398	.23270	.2972	.0267	.4121
7	.20990	.97772	.21468	.6580	.0228	.7641	.22693	.97391	.23301	.2916	.0268	.4065
8	.21019	.97766	.21499	.6514	.0228	.7576	.22722	.97384	.23332	.2859	.0268	.4011
9	.21047	.97760	.21529	.6448	.0229	.7512	.22750	.97378	.23363	.2803	.0269	.3956
10	.21076	.97754	.21560	4.6382	.0230	4.7448	.22778	.97371	.23393	4.2747	.0270	4.3901
11	.21104	.97748	.21590	.6317	.0230	.7384	.22807	.97364	.23424	.2691	.0271	.3847
12	.21132	.97741	.21621	.6252	.0231	.7320	.22835	.97358	.23455	.2635	.0271	.3792
13	.21161	.97735	.21651	.6187	.0232	.7257	.22863	.97351	.23485	.2579	.0272	.3738
14	.21189	.97729	.21682	.6122	.0232	.7193	.22892	.97344	.23516	.2524	.0273	.3684
15	.21218	.97723	.21712	4.6057	.0233	4.7130	.22920	.97338	.23547	4.2468	.0273	4.3630
16	.21246	.97717	.21742	.5993	.0234	.7067	.22948	.97331	.23577	.2413	.0274	.3576
17	.21275	.97711	.21773	.5928	.0234	.7004	.22977	.97324	.23608	.2358	.0275	.3522
18	.21303	.97704	.21803	.5864	.0235	.6942	.23005	.97318	.23639	.2303	.0276	.3469
19	.21331	.97698	.21834	.5800	.0235	.6879	.23033	.97311	.23670	.2248	.0276	.3415
20	.21360	.97692	.21864	4.5736	.0236	4.6817	.23061	.97304	.23700	4.2193	.0277	4.3362
21	.21388	.97686	.21895	.5673	.0237	.6754	.23090	.97298	.23731	.2139	.0278	.3309
22	.21417	.97680	.21925	.5609	.0237	.6692	.23118	.97291	.23762	.2084	.0278	.3256
23	.21445	.97673	.21956	.5546	.0238	.6631	.23146	.97284	.23793	.2030	.0279	.3203
24	.21473	.97667	.21986	.5483	.0239	.6569	.23175	.97277	.23823	.1976	.0280	.3150
25	.21502	.97661	.22017	4.5420	.0239	4.6507	.23202	.97271	.23854	4.1921	.0280	4.3098
26	.21530	.97655	.22047	.5357	.0240	.6446	.23231	.97264	.23885	.1867	.0281	.3045
27	.21559	.97648	.22078	.5294	.0241	.6385	.23260	.97257	.23916	.1814	.0282	.2993
28	.21587	.97642	.22108	.5232	.0241	.6324	.23288	.97250	.23946	.1760	.0283	.2941
29	.21615	.97636	.22139	.5169	.0242	.6263	.23316	.97244	.23977	.1706	.0283	.2838
30	.21644	.97630	.22169	4.5107	.0243	4.6201	.23344	.97237	.24008	4.1653	.0284	4.2836
31	.21672	.97623	.22200	.5045	.0243	.6142	.23373	.97230	.24039	.1600	.0285	.2785
32	.21701	.97617	.22230	.4983	.0244	.6081	.23401	.97223	.24069	.1546	.0285	.2733
33	.21729	.97611	.22261	.4921	.0245	.6021	.23429	.97216	.24100	.1493	.0286	.2681
34	.21757	.97604	.22291	.4860	.0245	.5961	.23458	.97210	.24131	.1440	.0287	.2630
35	.21786	.97598	.22322	4.4799	.0246	4.5901	.23486	.97203	.24162	4.1388	.0288	4.2579
36	.21814	.97592	.22353	.4737	.0247	.5841	.23514	.97196	.24192	.1335	.0288	.2527
37	.21843	.97575	.22383	.4676	.0247	.5782	.23542	.97189	.24223	.1282	.0289	.2476
38	.21871	.97579	.22414	.4615	.0248	.5722	.23571	.97182	.24254	.1230	.0290	.2425
39	.21899	.97573	.22444	.4555	.0249	.5663	.23599	.97175	.24285	.1178	.0291	.2375
40	.21928	.97566	.22475	4.4494	.0249	4.5604	.23627	.97169	.24316	4.1126	.0291	4.2324
41	.21956	.97560	.22505	.4434	.0250	.5545	.23655	.97162	.24346	.1073	.0292	.2273
42	.21985	.97553	.22536	.4373	.0251	.5486	.23684	.97155	.24377	.1022	.0293	.2223
43	.22013	.97547	.22566	.4313	.0251	.5428	.23712	.97148	.24408	.0970	.0293	.2173
44	.22041	.97541	.22597	.4253	.0252	.5369	.23740	.97141	.24439	.0918	.0294	.2122
45	.22070	.97534	.22628	4.4194	.0253	4.5311	.23768	.97134	.24470	4.0867	.0295	4.2072
46	.22098	.97528	.22658	.4134	.0253	.5253	.23797	.97127	.24501	.0815	.0296	.2022
47	.22126	.97521	.22689	.4074	.0254	.5195	.23825	.97120	.24531	.0764	.0296	.1972
48	.22155	.97515	.22719	.4015	.0255	.5137	.23853	.97113	.24562	.0713	.0297	.1923
49	.22183	.97508	.22750	.3956	.0255	.5079	.23881	.97106	.24593	.0662	.0298	.1873
50	.22211	.97502	.22781	4.3897	.0256	4.5021	.23910	.97099	.24624	4.0611	.0299	4.1824
51	.22240	.97495	.22811	.3838	.0257	.4964	.23938	.97092	.24655	.0560	.0299	.1774
52	.22268	.97489	.22842	.3779	.0257	.4907	.23966	.97086	.24686	.0509	.0300	.1725
53	.22297	.97483	.22872	.3721	.0258	.4850	.23994	.97079	.24717	.0458	.0301	.1676
54	.22325	.97476	.22903	.3662	.0259	.4793	.24023	.97072	.24747	.0408	.0302	.1627
55	.22353	.97470	.22934	4.3604	.0260	4.4736	.24051	.97065	.24778	4.0358	.0302	4.1578
56	.22382	.97463	.22964	.3546	.0260	.4679	.24079	.97058	.24809	.0307	.0303	.1529
57	.22410	.97457	.22995	.3488	.0261	.4623	.24107	.97051	.24840	.0257	.0304	.1481
58	.22438	.97450	.23025	.3430	.0262	.4566	.24136	.97044	.24871	.0207	.0305	.1432
59	.22467	.97443	.23056	.3372	.0262	.4510	.24164	.97037	.24902	.0157	.0305	.1384
60	.22495	.97437	.23087	4.3315	.0263	4.4454	.24192	.97029	.24933	4.0108	.0306	4.1336
′	cos	sin	cot	tan	cosec	sec	cos	sin	cot	tan	cosec	sec
	77°						76°					

14° sin	14° cos	14° tan	14° cot	14° sec	14° cosec	15° sin	15° cos	15° tan	15° cot	15° sec	15° cosec	′
.24192	.97029	.24933	4.0108	1.0306	4.1336	.25882	.96592	.26795	3.7320	1.0353	3.8637	60
.24220	.97022	.24964	.0058	.0307	.1287	.25910	.96585	.26826	.7277	.0353	.8595	59
.24249	.97015	.24995	.0009	.0308	.1239	.25938	.96577	.26857	.7234	.0354	.8553	58
.24277	.97008	.25025	3.9959	.0308	.1191	.25966	.96570	.26888	.7191	.0355	.8512	57
.24305	.97001	.25056	.9910	.0309	.1144	.25994	.96562	.26920	.7147	.0356	.8470	56
.24333	.96994	.25087	3.9861	.0310	4.1096	.26022	.96555	.26951	3.7104	.0357	3.8428	55
.24361	.96987	.25118	.9812	.0311	.1048	.26050	.96547	.26982	.7062	.0358	.8387	54
.24390	.96980	.25149	.9763	.0311	.1001	.26078	.96540	.27013	.7019	.0358	.8346	53
.24418	.96973	.25180	.9714	.0312	.0953	.26107	.96532	.27044	.6976	.0359	.8304	52
.24446	.96966	.25211	.9665	.0313	.0906	.26135	.96524	.27076	.6933	.0360	.8263	51
.24474	.96959	.25242	3.9616	.0314	4.0859	.26163	.96517	.27107	3.6891	.0361	3.8222	50
.24502	.96952	.25273	.9568	.0314	.0812	.26191	.96509	.27138	.6848	.0362	.8181	49
.24531	.96944	.25304	.9520	.0315	.0765	.26219	.96502	.27169	.6806	.0362	.8140	48
.24559	.96937	.25335	.9471	.0316	.0718	.26247	.96494	.27201	.6764	.0363	.8100	47
.24587	.96930	.25366	.9423	.0317	.0672	.26275	.96486	.27232	.6722	.0364	.8059	46
.24615	.96923	.25397	3.9375	.0317	4.0625	.26303	.96479	.27263	3.6679	.0365	3.8018	45
.24643	.96916	.25428	.9327	.0318	.0579	.26331	.96471	.27294	.6637	.0366	.7978	44
.24672	.96909	.25459	.9279	.0319	.0532	.26359	.96463	.27326	.6596	.0367	.7937	43
.24700	.96901	.25490	.9231	.0320	.0486	.26387	.96456	.27357	.6554	.0367	.7897	42
.24728	.96894	.25521	.9184	.0320	.0440	.26415	.96448	.27388	.6512	.0368	.7857	41
.24756	.96887	.25552	3.9136	.0321	4.0394	.26443	.96440	.27419	3.6470	.0369	3.7816	40
.24784	.96880	.25583	.9089	.0322	.0348	.26471	.96433	.27451	.6429	.0370	.7776	39
.24813	.96873	.25614	.9042	.0323	.0302	.26499	.96425	.27482	.6387	.0371	.7736	38
.24841	.96865	.25645	.8994	.0323	.0256	.26527	.96417	.27513	.6346	.0371	.7697	37
.24869	.96858	.25676	.8947	.0324	.0211	.26556	.96409	.27544	.6305	.0372	.7657	36
.24897	.96851	.25707	3.8900	.0325	4.0165	.26584	.96402	.27576	3.6263	.0373	3.7617	35
.24925	.96844	.25738	.8853	.0326	.0120	.26612	.96394	.27607	.6222	.0374	.7577	34
.24953	.96836	.25769	.8807	.0327	.0074	.26640	.96386	.27638	.6181	.0375	.7538	33
.24982	.96829	.25800	.8760	.0327	.0029	.26668	.96378	.27670	.6140	.0376	.7498	32
.25010	.96822	.25831	.8713	.0328	3.9984	.26696	.96371	.27701	.6100	.0376	.7459	31
.25038	.96815	.25862	3.8667	.0329	3.9939	.26724	.96363	.27732	3.6059	.0377	3.7420	30
.25066	.96807	.25893	.8621	.0330	.9894	.26752	.96355	.27764	.6018	.0378	.7380	29
.25094	.96800	.25924	.8574	.0330	.9850	.26780	.96347	.27795	.5977	.0379	.7341	28
.25122	.96793	.25955	.8528	.0331	.9805	.26808	.96340	.27826	.5937	.0380	.7302	27
.25151	.96785	.25986	.8482	.0332	.97( )	.26836	.96332	.27858	.5896	.0381	.7263	26
.25179	.96778	.26017	3.8436	.0333	3.9716	.26864	.96324	.27889	3.5856	.0382	3.7224	25
.25207	.96771	.26048	.8390	.0334	.9672	.26892	.96316	.27920	.5816	.0382	.7186	24
.25235	.96763	.26079	.8345	.0334	.9627	.26920	.96308	.27952	.5776	.0383	.7147	23
.25263	.96756	.26110	.8299	.0335	.9583	.26948	.96301	.27983	.5736	.0384	.7108	22
.25291	.96749	.26141	.8254	.0336	.9539	.26976	.96293	.28014	.5696	.0385	.7070	21
.25319	.96741	.26172	3.8208	.0337	3.9495	.27004	.96285	.28046	3.5656	.0386	3.7031	20
.25348	.96734	.26203	.8163	.0338	.9451	.27032	.96277	.28077	.5616	.0387	.6993	19
.25376	.96727	.26234	.8118	.0338	.9408	.27060	.96269	.28109	.5576	.0387	.6955	18
.25404	.96719	.26266	.8073	.0339	.9364	.27088	.96261	.28140	.5536	.0388	.6917	17
.25432	.96712	.26297	.8027	.0340	.9320	.27116	.96253	.28171	.5497	.0389	.6878	16
.25460	.96704	.26328	3.7983	.0341	3.9277	.27144	.96245	.28203	3.5457	.0390	3.6840	15
.25488	.96697	.26359	.7938	.0341	.9234	.27172	.96238	.28234	.5418	.0391	.6802	14
.25516	.96690	.26390	.7893	.0342	.9190	.27200	.96230	.28266	.5378	.0392	.6765	13
.25544	.96682	.26421	.7848	.0343	.9147	.27228	.96222	.28297	.5339	.0393	.6727	12
.25573	.96675	.26452	.7804	.0344	.9104	.27256	.96214	.28328	.5300	.0393	.6689	11
.25601	.96667	.26483	3.7759	.0345	3.9061	.27284	.96206	.28360	3.5261	.0394	3.6651	10
.25629	.96660	.26514	.7715	.0345	.9018	.27312	.96198	.28391	.5222	.0395	.6614	9
.25657	.96652	.26546	.7671	.0346	.8976	.27340	.96190	.28423	.5183	.0396	.6576	8
.25685	.96645	.26577	.7627	.0347	.8933	.27368	.96182	.28454	.5144	.0397	.6539	7
.25713	.96638	.26608	.7583	.0348	.8890	.27396	.96174	.28486	.5105	.0398	.6502	6
.25741	.96630	.26639	3.7539	.0349	3.8848	.27424	.96166	.28517	3.5066	.0399	3.6464	5
.25769	.96623	.26670	.7495	.0349	.8805	.27452	.96158	.28549	.5028	.0399	.6427	4
.25798	.96615	.26701	.7451	.0350	.8763	.27480	.96150	.28580	.4989	.0400	.6390	3
.25826	.96608	.26732	.7407	.0351	.8721	.27508	.96142	.28611	.4951	.0401	.6353	2
.25854	.96600	.26764	.7364	.0352	.8679	.27536	.96134	.28643	.4912	.0402	.6316	1
.25882	.96592	.26795	3.7320	.0353	3.8637	.27564	.96126	.28674	3.4874	.0403	3.6279	0
cos	sin	cot	tan	cosec	sec	cos	sin	cot	tan	cosec	sec	′
75°						74°						

(Continued)

Table B18. (Cont.)

′	16°						17°						18°						19°						′
′	sin	cos	tan	cot	sec	cosec	sin	cos	tan	cot	sec	cosec	sin	cos	tan	cot	sec	cosec	sin	cos	tan	cot	sec	cosec	′
0	.27564	.96126	.28674	3.4874	1.0403	3.6279	.29237	.95630	.30573	3.2708	1.0457	3.4203	.30902	.95106	.32492	3.0777	1.0515	3.2361	.32557	.94552	.34433	2.9042	1.0576	3.0715	60
1	.27592	.96118	.28706	.4836	.0404	.6243	.29265	.95622	.30605	.2674	.0458	.4170	.30929	.95097	.32524	.0746	.0516	.2332	.32584	.94542	.34465	.9015	.0577	.0690	59
2	.27620	.96110	.28737	.4798	.0405	.6206	.29293	.95613	.30637	.2640	.0459	.4138	.30957	.95088	.32556	.0716	.0517	.2303	.32612	.94533	.34498	.8987	.0578	.0664	58
3	.27648	.96102	.28769	.4760	.0406	.6169	.29321	.95605	.30668	.2607	.0460	.4106	.30985	.95079	.32588	.0686	.0518	.2274	.32639	.94523	.34530	.8960	.0579	.0638	57
4	.27675	.96094	.28800	.4722	.0406	.6133	.29348	.95596	.30700	.2573	.0461	.4073	.31012	.95070	.32621	.0655	.0519	.2245	.32667	.94514	.34563	.8933	.0580	.0612	56
5	.27703	.96086	.28832	3.4684	.0407	3.6096	.29376	.95588	.30732	3.2539	.0461	3.4041	.31040	.95061	.32653	3.0625	.0520	3.2216	.32694	.94504	.34595	2.8905	.0581	3.0586	55
6	.27731	.96078	.28863	.4646	.0408	.6060	.29404	.95579	.30764	.2505	.0462	.4009	.31068	.95051	.32685	.0595	.0521	.2188	.32722	.94495	.34628	.8878	.0582	.0561	54
7	.27759	.96070	.28895	.4608	.0409	.6024	.29432	.95571	.30796	.2472	.0463	.3977	.31095	.95042	.32717	.0565	.0522	.2159	.32749	.94485	.34661	.8851	.0584	.0535	53
8	.27787	.96062	.28926	.4570	.0410	.5987	.29460	.95562	.30828	.2438	.0464	.3945	.31123	.95033	.32749	.0535	.0523	.2131	.32777	.94476	.34693	.8824	.0585	.0509	52
9	.27815	.96054	.28958	.4533	.0411	.5951	.29487	.95554	.30859	.2405	.0465	.3913	.31150	.95024	.32782	.0505	.0524	.2102	.32804	.94466	.34726	.8797	.0586	.0484	51
10	.27843	.96045	.28990	3.4495	.0412	3.5915	.29515	.95545	.30891	3.2371	.0466	3.3881	.31178	.95015	.32814	3.0475	.0525	3.2074	.32832	.94457	.34758	2.8770	.0587	3.0458	50
11	.27871	.96037	.29021	.4458	.0413	.5879	.29543	.95536	.30923	.2338	.0467	.3849	.31206	.95006	.32846	.0445	.0526	.2045	.32859	.94447	.34791	.8743	.0588	.0433	49
12	.27899	.96029	.29053	.4420	.0413	.5843	.29571	.95528	.30955	.2305	.0468	.3817	.31233	.94997	.32878	.0415	.0527	.2017	.32887	.94438	.34824	.8716	.0589	.0407	48
13	.27927	.96021	.29084	.4383	.0414	.5807	.29598	.95519	.30987	.2271	.0469	.3785	.31261	.94988	.32910	.0385	.0528	.1989	.32914	.94428	.34856	.8689	.0590	.0382	47
14	.27955	.96013	.29116	.4346	.0415	.5772	.29626	.95511	.31019	.2238	.0470	.3754	.31289	.94979	.32943	.0356	.0529	.1960	.32942	.94418	.34889	.8662	.0591	.0357	46
15	.27983	.96005	.29147	3.4308	.0416	3.5736	.29654	.95502	.31051	3.2205	.0471	3.3722	.31316	.94970	.32975	3.0326	.0530	3.1932	.32969	.94409	.34921	2.8636	.0592	3.0331	45
16	.28011	.95997	.29179	.4271	.0417	.5700	.29682	.95493	.31083	.2172	.0472	.3690	.31344	.94961	.33007	.0296	.0531	.1904	.32996	.94399	.34954	.8609	.0593	.0306	44
17	.28039	.95989	.29210	.4234	.0418	.5665	.29710	.95485	.31115	.2139	.0473	.3659	.31372	.94952	.33039	.0267	.0532	.1876	.33024	.94390	.34987	.8582	.0594	.0281	43
18	.28067	.95980	.29242	.4197	.0419	.5629	.29737	.95476	.31146	.2106	.0474	.3627	.31399	.94942	.33072	.0237	.0533	.1848	.33051	.94380	.35019	.8555	.0595	.0256	42
19	.28094	.95972	.29274	.4160	.0420	.5594	.29765	.95467	.31178	.2073	.0475	.3596	.31427	.94933	.33104	.0208	.0534	.1820	.33079	.94370	.35052	.8529	.0596	.0231	41
20	.28122	.95964	.29305	3.4124	.0420	3.5559	.29793	.95459	.31210	3.2041	.0476	3.3565	.31454	.94924	.33136	3.0178	.0535	3.1792	.33106	.94361	.35085	2.8502	.0598	3.0206	40
21	.28150	.95956	.29337	.4087	.0421	.5523	.29821	.95450	.31242	.2008	.0477	.3534	.31482	.94915	.33169	.0149	.0536	.1764	.33134	.94351	.35117	.8476	.0599	.0181	39
22	.28178	.95948	.29368	.4050	.0422	.5488	.29848	.95441	.31274	.1975	.0478	.3502	.31510	.94906	.33201	.0120	.0537	.1736	.33161	.94341	.35150	.8449	.0600	.0156	38
23	.28206	.95940	.29400	.4014	.0423	.5453	.29876	.95433	.31306	.1942	.0478	.3471	.31537	.94897	.33233	.0090	.0528	.1708	.33189	.94332	.35183	.8423	.0601	.0131	37
24	.28234	.95931	.29432	.3977	.0424	.5418	.29904	.95424	.31338	.1910	.0479	.3440	.31565	.94888	.33265	.0061	.0539	.1681	.33216	.94322	.35215	.8396	.0602	.0106	36
25	.28262	.95923	.29463	3.3941	.0425	3.5383	.29932	.95415	.31370	3.1877	.0480	3.3409	.31592	.94878	.33298	3.0032	.0540	3.1653	.33243	.94313	.35248	2.8370	.0603	3.0081	35
26	.28290	.95915	.29495	.3904	.0426	.5348	.29959	.95407	.31402	.1845	.0481	.3378	.31620	.94869	.33330	.0003	.0541	.1625	.33271	.94303	.35281	.8344	.0604	.0056	34
27	.28318	.95907	.29526	.3868	.0427	.5313	.29987	.95398	.31434	.1813	.0482	.3347	.31648	.94860	.33362	2.9974	.0542	.1598	.33298	.94293	.35314	.8318	.0605	.0031	33
28	.28346	.95898	.29558	.3832	.0428	.5279	.30015	.95389	.31466	.1780	.0483	.3316	.31675	.94851	.33395	.9945	.0543	.1570	.33326	.94283	.35346	.8291	.0606	.0007	32
29	.28374	.95890	.29590	.3795	.0428	.5244	.30043	.95380	.31498	.1748	.0484	.3286	.31703	.94841	.33427	.9916	.0544	.1543	.33353	.94274	.35379	.8265	.0607	2.9982	31
30	.28401	.95882	.29621	3.3759	.0429	3.5209	.30070	.95372	.31530	3.1716	.0485	3.3255	.31730	.94832	.33459	2.9887	.0545	3.1515	.33381	.94264	.35412	2.8239	.0608	2.9957	30
31	.28429	.95874	.29653	.3723	.0430	.5175	.30098	.95363	.31562	.1684	.0486	.3224	.31758	.94823	.33492	.9858	.0546	.1488	.33408	.94254	.35445	.8213	.0609	.9933	29
32	.28457	.95865	.29685	.3687	.0431	.5140	.30126	.95354	.31594	.1652	.0487	.3194	.31786	.94814	.33524	.9829	.0547	.1461	.33435	.94245	.35477	.8187	.0611	.9908	28
33	.28485	.95857	.29716	.3651	.0432	.5106	.30154	.95345	.31626	.1620	.0488	.3163	.31813	.94805	.33557	.9800	.0548	.1433	.33463	.94235	.35510	.8161	.0612	.9884	27
34	.28513	.95849	.29748	.3616	.0433	.5072	.30181	.95337	.31658	.1588	.0489	.3133	.31841	.94795	.33589	.9772	.0549	.1406	.33490	.94225	.35543	.8135	.0613	.9859	26
35	.28541	.95840	.29780	3.3580	.0434	3.5037	.30209	.95328	.31690	3.1556	.0490	3.3102	.31868	.94786	.33621	2.9743	.0550	3.1379	.33518	.94215	.35576	2.8109	.0614	2.9835	25
36	.28569	.95832	.29811	.3544	.0435	.5003	.30237	.95319	.31722	.1524	.0491	.3072	.31896	.94777	.33654	.9714	.0551	.1352	.33545	.94206	.35608	.8083	.0615	.9810	24
37	.28597	.95824	.29843	.3509	.0436	.4969	.30265	.95310	.31754	.1492	.0492	.3042	.31923	.94767	.33686	.9686	.0552	.1325	.33572	.94196	.35641	.8057	.0616	.9786	23
38	.28624	.95816	.29875	.3473	.0437	.4935	.30292	.95301	.31786	.1460	.0493	.3011	.31951	.94758	.33718	.9657	.0553	.1298	.33600	.94186	.35674	.8032	.0617	.9762	22
39	.28652	.95807	.29906	.3438	.0438	.4901	.30320	.95293	.31818	.1429	.0494	.2981	.31978	.94749	.33751	.9629	.0554	.1271	.33627	.94176	.35707	.8006	.0618	.9738	21
40	.28680	.95799	.29938	3.3402	.0438	3.4867	.30348	.95284	.31850	3.1397	.0495	3.2951	.32006	.94740	.33783	2.9600	.0555	3.1244	.33655	.94167	.35739	2.7980	.0619	2.9713	20
41	.28708	.95791	.29970	.3367	.0439	.4833	.30375	.95275	.31882	.1366	.0496	.2921	.32034	.94730	.33816	.9572	.0556	.1217	.33682	.94157	.35772	.7954	.0620	.9689	19
42	.28736	.95782	.30001	.3332	.0440	.4799	.30403	.95266	.31914	.1334	.0497	.2891	.32061	.94721	.33848	.9544	.0557	.1190	.33709	.94147	.35805	.7929	.0622	.9665	18
43	.28764	.95774	.30033	.3296	.0441	.4766	.30431	.95257	.31946	.1303	.0498	.2861	.32089	.94712	.33880	.9515	.0558	.1163	.33737	.94137	.35838	.7903	.0623	.9641	17
44	.28792	.95765	.30065	.3261	.0442	.4732	.30459	.95248	.31978	.1271	.0499	.2831	.32116	.94702	.33913	.9487	.0559	.1137	.33764	.94127	.35871	.7878	.0624	.9617	16
45	.28820	.95757	.30096	3.3226	.0443	3.4698	.30486	.95239	.32010	3.1240	.0500	3.2801	.32144	.94693	.33945	2.9459	.0560	3.1110	.33792	.94118	.35904	2.7852	.0625	2.9593	15
46	.28847	.95749	.30128	.3191	.0444	.4665	.30514	.95231	.32042	.1209	.0501	.2772	.32171	.94684	.33978	.9431	.0561	.1083	.33819	.94108	.35936	.7827	.0626	.9569	14
47	.28875	.95740	.30160	.3156	.0445	.4632	.30542	.95222	.32074	.1177	.0502	.2742	.32199	.94674	.34010	.9403	.0562	.1057	.33846	.94098	.35969	.7801	.0627	.9545	13
48	.28903	.95732	.30192	.3121	.0446	.4598	.30569	.95213	.32106	.1146	.0503	.2712	.32226	.94665	.34043	.9375	.0563	.1030	.33874	.94088	.36002	.7776	.0628	.9521	12
49	.28931	.95723	.30223	.3087	.0447	.4565	.30597	.95204	.32138	.1115	.0504	.2683	.32254	.94655	.34075	.9347	.0565	.1004	.33901	.94078	.36035	.7751	.0629	.9497	11
50	.28959	.95715	.30255	3.3052	.0448	3.4532	.30625	.95195	.32171	3.1084	.0505	3.2653	.32282	.94646	.34108	2.9319	.0566	3.0977	.33928	.94068	.36068	2.7725	.0630	2.9474	10
51	.28987	.95707	.30287	.3017	.0448	.4498	.30653	.95186	.32203	.1053	.0506	.2624	.32309	.94637	.34140	.9291	.0567	.0951	.33956	.94058	.36101	.7700	.0632	.9450	9
52	.29014	.95698	.30319	.2983	.0449	.4465	.30680	.95177	.32235	.1022	.0507	.2594	.32337	.94627	.34173	.9263	.0568	.0925	.33983	.94049	.36134	.7675	.0633	.9426	8
53	.29042	.95690	.30350	.2948	.0450	.4432	.30708	.95168	.32267	.0991	.0508	.2565	.32364	.94618	.34205	.9235	.0569	.0898	.34011	.94039	.36167	.7650	.0634	.9402	7
54	.29070	.95681	.30382	.2914	.0451	.4399	.30736	.95159	.32299	.0960	.0509	.2535	.32392	.94608	.34238	.9208	.0570	.0872	.34038	.94029	.36199	.7625	.0635	.9379	6
55	.29098	.95673	.30414	3.2879	.0452	3.4366	.30763	.95150	.32331	3.0930	.0510	3.2506	.32419	.94599	.34270	2.9180	.0571	3.0846	.34065	.94019	.36232	2.7600	.0636	2.9355	5
56	.29126	.95664	.30446	.2845	.0453	.4334	.30791	.95141	.32363	.0899	.0511	.2477	.32447	.94590	.34303	.9152	.0572	.0820	.34093	.94009	.36265	.7575	.0637	.9332	4
57	.29154	.95656	.30478	.2811	.0454	.4301	.30819	.95132	.32395	.0868	.0512	.2448	.32474	.94580	.34335	.9125	.0573	.0793	.34120	.93999	.36298	.7550	.0638	.9308	3
58	.29181	.95647	.30509	.2777	.0455	.4268	.30846	.95124	.32428	.0838	.0513	.2419	.32502	.94571	.34368	.9097	.0574	.0767	.34147	.93989	.36331	.7525	.0639	.9285	2
59	.29209	.95639	.30541	.2742	.0456	.4236	.30874	.95115	.32460	.0807	.0514	.2390	.32529	.94561	.34400	.9069	.0575	.0741	.34175	.93979	.36364	.7500	.0641	.9261	1
60	.29237	.95630	.30573	3.2708	.0457	3.4203	.30902	.95106	.32492	3.0777	.0515	3.2361	.32557	.94552	.34433	2.9042	.0576	3.0715	.34202	.93969	.36397	2.7475	.0642	2.9238	0
′	cos	sin	cot	tan	cosec	sec	cos	sin	cot	tan	cosec	sec	cos	sin	cot	tan	cosec	sec	cos	sin	cot	tan	cosec	sec	′
	73°						72°						71°						70°						

′	20° sin	20° cos	20° tan	20° cot	20° sec	20° cosec	21° sin	21° cos	21° tan	21° cot	21° sec	21° cosec	22° sin	22° cos	22° tan	22° cot	22° sec	22° cosec	23° sin	23° cos	23° tan	23° cot	23° sec	23° cosec	′
0	.34202	.93969	.36397	2.7475	1.0642	2.9238	.35837	.93358	.38386	2.6051	1.0711	2.7904	.37461	.92718	.40403	2.4751	1.0785	2.6695	.39073	.92050	.42447	2.3558	1.0864	2.5593	60
1	.34229	.93959	.36430	.7450	.0643	.9215	.35864	.93348	.38420	.6028	.0713	.7883	.37488	.92707	.40436	.4730	.0787	.6675	.39100	.92039	.42482	.3539	.0865	.5575	59
2	.34257	.93949	.36463	.7425	.0644	.9191	.35891	.93337	.38453	.6006	.0714	.7862	.37514	.92696	.40470	.4709	.0788	.6656	.39126	.92028	.42516	.3520	.0866	.5558	58
3	.34284	.93939	.36496	.7400	.0645	.9168	.35918	.93327	.38486	.5983	.0715	.7841	.37541	.92686	.40504	.4689	.0789	.6637	.39153	.92016	.42550	.3501	.0868	.5540	57
4	.34311	.93929	.36529	.7376	.0646	.9145	.35945	.93316	.38520	.5960	.0716	.7820	.37568	.92675	.40538	.4668	.0790	.6618	.39180	.92005	.42585	.3482	.0869	.5523	56
5	.34339	.93919	.36562	2.7351	.0647	2.9122	.35972	.93306	.38553	2.5938	.0717	2.7799	.37595	.92664	.40572	2.4647	.0792	2.6599	.39207	.91993	.42619	2.3463	.0870	2.5506	55
6	.34366	.93909	.36595	.7326	.0648	.9098	.36000	.93295	.38587	.5916	.0719	.7778	.37622	.92653	.40606	.4627	.0793	.6580	.39234	.91982	.42654	.3445	.0872	.5488	54
7	.34393	.93899	.36628	.7302	.0650	.9075	.36027	.93285	.38620	.5893	.0720	.7757	.37649	.92642	.40640	.4606	.0794	.6561	.39260	.91971	.42688	.3426	.0873	.5471	53
8	.34421	.93889	.36661	.7277	.0651	.9052	.36054	.93274	.38654	.5871	.0721	.7736	.37676	.92631	.40673	.4586	.0795	.6542	.39287	.91959	.42722	.3407	.0874	.5453	52
9	.34448	.93879	.36694	.7252	.0652	.9029	.36081	.93264	.38687	.5848	.0722	.7715	.37703	.92620	.40707	.4565	.0797	.6523	.39314	.91948	.42757	.3388	.0876	.5436	51
10	.34475	.93869	.36727	2.7228	.0653	2.9006	.36108	.93253	.38720	2.5826	.0723	2.7694	.37730	.92609	.40741	2.4545	.0798	2.6504	.39341	.91936	.42791	2.3369	.0877	2.5419	50
11	.34502	.93859	.36760	.7204	.0654	.8983	.36135	.93243	.38754	.5804	.0725	.7674	.37757	.92598	.40775	.4525	.0799	.6485	.39367	.91925	.42826	.3350	.0878	.5402	49
12	.34530	.93849	.36793	.7179	.0655	.8960	.36162	.93232	.38787	.5781	.0726	.7653	.37784	.92587	.40809	.4504	.0801	.6466	.39394	.91913	.42860	.3332	.0880	.5384	48
13	.34557	.93839	.36826	.7155	.0656	.8937	.36189	.93222	.38821	.5759	.0727	.7632	.37811	.92576	.40843	.4484	.0802	.6447	.39421	.91902	.42894	.3313	.0881	.5367	47
14	.34584	.93829	.36859	.7130	.0658	.8915	.36217	.93211	.38854	.5737	.0728	.7611	.37838	.92565	.40877	.4463	.0803	.6428	.39448	.91891	.42929	.3294	.0882	.5350	46
15	.34612	.93819	.36892	2.7106	.0659	2.8892	.36244	.93201	.38888	2.5715	.0729	2.7591	.37865	.92554	.40911	2.4443	.0804	2.6410	.39474	.91879	.42963	2.3276	.0884	2.5333	45
16	.34639	.93809	.36925	.7082	.0660	.8869	.36271	.93190	.38921	.5693	.0731	.7570	.37892	.92543	.40945	.4423	.0806	.6391	.39501	.91868	.42998	.3257	.0885	.5316	44
17	.34666	.93799	.36958	.7058	.0661	.8846	.36298	.93180	.38955	.5671	.0732	.7550	.37919	.92532	.40979	.4403	.0807	.6372	.39528	.91856	.43032	.3238	.0886	.5299	43
18	.34693	.93789	.36991	.7033	.0662	.8824	.36325	.93169	.38988	.5649	.0733	.7529	.37946	.92521	.41013	.4382	.0808	.6353	.39554	.91845	.43067	.3220	.0888	.5281	42
19	.34721	.93779	.37024	.7009	.0663	.8801	.36352	.93158	.39022	.5627	.0734	.7509	.37972	.92510	.41047	.4362	.0810	.6335	.39581	.91833	.43101	.3201	.0889	.5264	41
20	.34748	.93769	.37057	2.6985	.0664	2.8778	.36379	.93148	.39055	2.5605	.0736	2.7488	.37999	.92499	.41081	2.4342	.0811	2.6316	.39608	.91822	.43136	2.3183	.0891	2.5247	40
21	.34775	.93758	.37090	.6961	.0666	.8756	.36406	.93137	.39089	.5583	.0737	.7468	.38026	.92488	.41115	.4322	.0812	.6297	.39635	.91810	.43170	.3164	.0892	.5230	39
22	.34803	.93748	.37123	.6937	.0667	.8733	.36433	.93127	.39122	.5561	.0738	.7447	.38053	.92477	.41149	.4302	.0813	.6279	.39661	.91798	.43205	.3145	.0893	.5213	38
23	.34830	.93738	.37156	.6913	.0668	.8711	.36460	.93116	.39156	.5539	.0739	.7427	.38080	.92466	.41183	.4282	.0815	.6260	.39688	.91787	.43239	.3127	.0895	.5196	37
24	.34857	.93728	.37190	.6889	.0669	.8688	.36488	.93105	.39189	.5517	.0740	.7406	.38107	.92455	.41217	.4262	.0816	.6242	.39715	.91775	.43274	.3109	.0896	.5179	36
25	.34884	.93718	.37223	2.6865	.0670	2.8666	.36515	.93095	.39223	2.5495	.0742	2.7386	.38134	.92443	.41251	2.4242	.0817	2.6223	.39741	.91764	.43308	2.3090	.0897	2.5163	35
26	.34912	.93708	.37256	.6841	.0671	.8644	.36542	.93084	.39257	.5473	.0743	.7366	.38161	.92432	.41285	.4222	.0819	.6205	.39768	.91752	.43343	.3072	.0899	.5146	34
27	.34939	.93698	.37289	.6817	.0673	.8621	.36569	.93074	.39290	.5451	.0744	.7346	.38188	.92421	.41319	.4202	.0820	.6186	.39795	.91741	.43377	.3053	.0900	.5129	33
28	.34966	.93687	.37322	.6794	.0674	.8599	.36596	.93063	.39324	.5430	.0745	.7325	.38214	.92410	.41353	.4182	.0821	.6168	.39821	.91729	.43412	.3035	.0902	.5112	32
29	.34993	.93677	.37355	.6770	.0675	.8577	.36623	.93052	.39357	.5408	.0747	.7305	.38241	.92399	.41387	.4162	.0823	.6150	.39848	.91718	.43447	.3017	.0903	.5095	31
30	.35021	.93667	.37388	2.6746	.0676	2.8554	.36650	.93042	.39391	2.5386	.0748	2.7285	.38268	.92388	.41421	2.4142	.0824	2.6131	.39875	.91706	.43481	2.2998	.0904	2.5078	30
31	.35048	.93657	.37422	.6722	.0677	.8532	.36677	.93031	.39425	.5365	.0749	.7265	.38295	.92377	.41455	.4122	.0825	.6113	.39901	.91694	.43516	.2980	.0906	.5062	29
32	.35075	.93647	.37455	.6699	.0678	.8510	.36704	.93020	.39458	.5343	.0750	.7245	.38322	.92366	.41489	.4102	.0826	.6095	.39928	.91683	.43550	.2962	.0907	.5045	28
33	.35102	.93637	.37488	.6675	.0679	.8488	.36731	.93010	.39492	.5322	.0751	.7225	.38349	.92354	.41524	.4083	.0828	.6076	.39955	.91671	.43585	.2944	.0908	.5028	27
34	.35130	.93626	.37521	.6652	.0681	.8466	.36758	.92999	.39525	.5300	.0753	.7205	.38376	.92343	.41558	.4063	.0829	.6058	.39981	.91659	.43620	.2925	.0910	.5011	26
35	.35157	.93616	.37554	2.6628	.0682	2.8444	.36785	.92988	.39559	2.5278	.0754	2.7185	.38403	.92332	.41592	2.4043	.0830	2.6040	.40008	.91648	.43654	2.2907	.0911	2.4995	25
36	.35184	.93606	.37587	.6604	.0683	.8422	.36812	.92978	.39593	.5257	.0755	.7165	.38429	.92321	.41626	.4023	.0832	.6022	.40035	.91636	.43689	.2889	.0913	.4978	24
37	.35211	.93596	.37621	.6581	.0684	.8400	.36839	.92967	.39626	.5236	.0756	.7145	.38456	.92310	.41660	.4004	.0833	.6003	.40061	.91625	.43723	.2871	.0914	.4961	23
38	.35239	.93585	.37654	.6558	.0685	.8378	.36866	.92956	.39660	.5214	.0758	.7125	.38483	.92299	.41694	.3984	.0834	.5985	.40088	.91613	.43758	.2853	.0915	.4945	22
39	.35266	.93575	.37687	.6534	.0686	.8356	.36893	.92945	.39694	.5193	.0759	.7105	.38510	.92287	.41728	.3964	.0836	.5967	.40115	.91601	.43793	.2835	.0917	.4928	21
40	.35293	.93565	.37720	2.6511	.0688	2.8334	.36921	.92935	.39727	2.5171	.0760	2.7085	.38537	.92276	.41762	2.3945	.0837	2.5949	.40141	.91590	.43827	2.2817	.0918	2.4912	20
41	.35320	.93555	.37754	.6487	.0689	.8312	.36948	.92924	.39761	.5150	.0761	.7065	.38564	.92265	.41797	.3925	.0838	.5931	.40168	.91578	.43862	.2799	.0920	.4895	19
42	.35347	.93544	.37787	.6464	.0690	.8290	.36975	.92913	.39795	.5129	.0763	.7045	.38591	.92254	.41831	.3906	.0840	.5913	.40195	.91566	.43897	.2781	.0921	.4879	18
43	.35375	.93534	.37820	.6441	.0691	.8269	.37002	.92902	.39828	.5108	.0764	.7026	.38617	.92242	.41865	.3886	.0841	.5895	.40221	.91554	.43932	.2763	.0922	.4862	17
44	.35402	.93524	.37853	.6418	.0692	.8247	.37029	.92892	.39862	.5086	.0765	.7006	.38644	.92231	.41899	.3867	.0842	.5877	.40248	.91543	.43966	.2745	.0924	.4846	16
45	.35429	.93513	.37887	2.6394	.0694	2.8225	.37056	.92881	.39896	2.5065	.0766	2.6986	.38671	.92220	.41933	2.3847	.0844	2.5958	.40205	.91531	.44001	2.2727	.0925	2.4829	15
46	.35456	.93503	.37920	.6371	.0695	.8204	.37083	.92870	.39930	.5044	.0768	.6967	.38698	.92209	.41968	.3828	.0845	.5841	.40301	.91519	.44036	.2709	.0927	.4813	14
47	.35483	.93493	.37953	.6348	.0696	.8182	.37110	.92859	.39963	.5023	.0769	.6947	.38725	.92197	.42002	.3808	.0846	.5823	.40328	.91508	.44070	.2691	.0928	.4797	13
48	.35511	.93482	.37986	.6325	.0697	.8160	.37137	.92848	.39997	.5002	.0770	.6927	.38751	.92186	.42036	.3789	.0847	.5805	.40354	.91496	.44105	.2673	.0929	.4780	12
49	.35538	.93472	.38020	.6302	.0698	.8139	.37164	.92838	.40031	.4981	.0771	.6908	.38778	.92105	.42070	.3770	.0849	.5787	.40381	.91484	.44140	.2655	.0931	.4764	11
50	.35565	.93462	.38053	2.6279	.0699	2.8117	.37191	.92827	.40065	2.4960	.0773	2.6888	.38805	.92164	.42105	2.3750	.0850	2.5770	.40408	.91472	.44175	2.2637	.0932	2.4748	10
51	.35592	.93451	.38086	.6256	.0701	.8096	.37218	.92816	.40098	.4939	.0774	.6869	.38832	.92152	.42139	.3731	.0851	.5752	.40434	.91461	.44209	.2619	.0934	.4731	9
52	.35619	.93441	.38120	.6233	.0702	.8074	.37245	.92805	.40132	.4918	.0775	.6849	.38859	.92141	.42173	.3712	.0853	.5734	.40461	.91449	.44244	.2602	.0935	.4715	8
53	.35647	.93431	.38153	.6210	.0703	.8053	.37272	.92794	.40166	.4897	.0776	.6830	.38886	.92130	.42207	.3692	.0854	.5716	.40487	.91437	.44279	.2584	.0936	.4699	7
54	.35674	.93420	.38186	.6187	.0704	.8032	.37299	.92784	.40200	.4876	.0778	.6810	.38912	.92118	.42242	.3673	.0855	.5699	.40514	.91425	.44314	.2566	.0938	.4683	6
55	.35701	.93410	.38220	2.6164	.0705	2.8010	.37326	.92773	.40233	2.4855	.0779	2.6791	.38939	.92107	.42276	2.3654	.0857	2.5681	.40541	.91414	.44349	2.2548	.0939	2.4666	5
56	.35728	.93400	.38253	.6142	.0707	.7989	.37353	.92762	.40267	.4834	.0780	.6772	.38966	.92096	.42310	.3635	.0858	.5663	.40567	.91402	.44383	.2531	.0941	.4650	4
57	.35755	.93389	.38286	.6119	.0708	.7968	.37380	.92751	.40301	.4813	.0781	.6752	.38993	.92084	.42344	.3616	.0859	.5646	.40594	.91390	.44418	.2513	.0942	.4634	3
58	.35782	.93379	.38320	.6096	.0709	.7947	.37407	.92740	.40335	.4792	.0783	.6733	.39019	.92073	.42379	.3597	.0861	.5628	.40620	.91378	.44453	.2495	.0943	.4618	2
59	.35810	.93368	.38353	.6073	.0710	.7925	.37434	.92729	.40369	.4772	.0784	.6714	.39046	.92062	.42413	.3577	.0862	.5610	.40647	.91366	.44488	.2478	.0945	.4602	1
60	.35837	.93358	.38386	2.6051	.0711	2.7904	.37461	.92718	.40403	2.4751	.0785	2.6695	.39073	.92050	.42447	2.3558	.0864	2.5593	.40674	.91354	.44523	2.2460	.0946	2.4586	0
′	69° cos	69° sin	69° cot	69° tan	69° cosec	69° sec	68° cos	68° sin	68° cot	68° tan	68° cosec	68° sec	67° cos	67° sin	67° cot	67° tan	67° cosec	67° sec	66° cos	66° sin	66° cot	66° tan	66° cosec	66° sec	′

(Continued)

Table B18. (Cont.)

	24°						25°						26°						27°						
′	sin	cos	tan	cot	sec	cosec	sin	cos	tan	cot	sec	cosec	sin	cos	tan	cot	sec	cosec	sin	cos	tan	cot	sec	cosec	′
0	.40674	.91354	.44523	2.2460	1.0946	2.4586	.42262	.90631	.46631	2.1445	1.1034	2.3662	.43837	.89879	.48773	2.0503	1.1126	2.2812	.45399	.89101	.50952	1.9626	1.1223	2.2027	60
1	.40700	.91343	.44558	.2443	.0948	.4570	.42288	.90618	.46666	.1429	.1035	.3647	.43863	.89867	.48809	.0488	.1127	.2798	.45425	.89087	.50989	.9612	.1225	.2014	59
2	.40727	.91331	.44593	.2425	.0949	.4554	.42314	.90606	.46702	.1412	.1037	.3632	.43889	.89854	.48845	.0473	.1129	.2784	.45451	.89074	.51026	.9598	.1226	.2002	58
3	.40753	.91319	.44627	.2408	.0951	.4538	.42341	.90594	.46737	.1396	.1038	.3618	.43915	.89841	.48881	.0458	.1131	.2771	.45477	.89061	.51062	.9584	.1228	.1989	57
4	.40780	.91307	.44662	.2390	.0952	.4522	.42367	.90581	.46772	.1380	.1040	.3603	.43942	.89828	.48917	.0443	.1132	.2757	.45503	.89048	.51099	.9570	.1230	.1977	56
5	.40806	.91295	.44697	2.2373	.0953	2.4506	.42394	.90569	.46808	2.1364	.1041	2.3588	.43968	.89815	.48953	2.0427	.1134	2.2744	.45528	.89034	.51136	1.9556	.1231	2.1964	55
6	.40833	.91283	.44732	.2355	.0955	.4490	.42420	.90557	.46843	.1348	.1043	.3574	.43994	.89803	.48989	.0412	.1135	.2730	.45554	.89021	.51172	.9542	.1233	.1952	54
7	.40860	.91271	.44767	.2338	.0956	.4474	.42446	.90544	.46879	.1331	.1044	.3559	.44020	.89790	.49025	.0397	.1137	.2717	.45580	.89008	.51209	.9528	.1235	.1939	53
8	.40886	.91260	.44802	.2320	.0958	.4458	.42473	.90532	.46914	.1315	.1046	.3544	.44046	.89777	.49062	.0382	.1139	.2703	.45606	.88995	.51246	.9514	.1237	.1927	52
9	.40913	.91248	.44837	.2303	.0959	.4442	.42499	.90520	.46950	.1299	.1047	.3530	.44072	.89764	.49098	.0367	.1140	.2690	.45632	.88981	.51283	.9500	.1238	.1914	51
10	.40939	.91236	.44872	2.2286	.0961	2.4426	.42525	.90507	.46985	2.1283	.1049	2.3515	.44098	.89751	.49134	2.0352	.1142	2.2676	.45658	.88968	.51319	1.9486	.1240	2.1902	50
11	.40966	.91224	.44907	.2268	.0962	.4418	.42552	.90495	.47021	.1267	.1050	.3501	.44124	.89739	.49170	.0338	.1143	.2663	.45684	.88955	.51356	.9472	.1242	.1889	49
12	.40992	.91212	.44942	.2251	.0963	.4395	.42578	.90483	.47056	.1251	.1052	.3486	.44150	.89726	.49206	.0323	.1145	.2650	.45710	.88942	.51393	.9458	.1243	.1877	48
13	.41019	.91200	.44977	.2234	.0965	.4379	.42604	.90470	.47092	.1235	.1053	.3472	.44177	.89713	.49242	.0308	.1147	.2636	.45736	.88928	.51430	.9444	.1245	.1865	47
14	.41045	.91188	.45012	.2216	.0966	.4363	.42630	.90458	.47127	.1219	.1055	.3457	.44203	.89700	.49278	.0293	.1148	.2623	.45761	.88915	.51466	.9430	.1247	.1852	46
15	.41072	.91176	.45047	2.2199	.0968	2.4347	.42657	.90445	.47163	2.1203	.1056	2.3443	.44229	.89687	.49314	2.0278	.1150	2.2610	.45787	.88902	.51503	1.9416	.1248	2.1840	45
16	.41098	.91164	.45082	.2182	.0969	.4332	.42683	.90433	.47199	.1187	.1058	.3428	.44255	.89674	.49351	.0263	.1151	.2596	.45813	.88888	.51540	.9402	.1250	.1828	44
17	.41125	.91152	.45117	.2165	.0971	.4316	.42709	.90421	.47234	.1171	.1059	.3414	.44281	.89661	.49387	.0248	.1153	.2583	.45839	.88875	.51577	.9388	.1252	.1815	43
18	.41151	.91140	.45152	.2147	.0972	.4300	.42736	.90408	.47270	.1155	.1061	.3399	.44307	.89649	.49423	.0233	.1155	.2570	.45865	.88862	.51614	.9375	.1253	.1803	42
19	.41178	.91128	.45187	.2130	.0973	.4285	.42762	.90396	.47305	.1139	.1062	.3385	.44333	.89636	.49459	.0219	.1156	.2556	.45891	.88848	.51651	.9361	.1255	.1791	41
20	.41204	.91116	.45222	2.2113	.0975	2.4269	.42788	.90383	.47341	2.1123	.1064	2.3371	.44359	.89623	.49495	2.0204	.1158	2.2543	.45917	.88835	.51687	1.9347	.1257	2.1778	40
21	.41231	.91104	.45257	.2096	.0976	.4254	.42815	.90371	.47376	.1107	.1065	.3356	.44385	.89610	.49532	.0189	.1159	.2530	.45942	.88822	.51724	.9333	.1258	.1766	39
22	.41257	.91092	.45292	.2079	.0978	.4238	.42841	.90358	.47412	.1092	.1067	.3342	.44411	.89597	.49568	.0174	.1161	.2517	.45968	.88808	.51761	.9319	.1260	.1754	38
23	.41284	.91080	.45327	.2062	.0979	.4222	.42867	.90346	.47448	.1076	.1068	.3328	.44437	.89584	.49604	.0159	.1163	.2503	.45994	.88795	.51798	.9306	.1262	.1742	37
24	.41310	.91068	.45362	.2045	.0981	.4207	.42893	.90333	.47483	.1060	.1070	.3313	.44463	.89571	.49640	.0145	.1164	.2490	.46020	.88781	.51835	.9292	.1264	.1730	36
25	.41337	.91056	.45397	2.2028	.0982	2.4191	42920	.90321	.47519	2.1044	.1072	2.3299	.44489	.89558	.49677	2.0130	.1166	2.2477	.46046	.88768	.51872	1.9278	.1265	2.1717	35
26	.41363	.91044	.45432	.2011	.0984	.4176	.42946	.90308	.47555	.1028	.1073	.3285	.44516	.89545	.49713	.0115	.1167	.2464	.46072	.88755	.51909	.9264	.1267	.1705	34
27	.41390	.91032	.45467	.1994	.0985	.4160	.42972	.90296	.47590	.1013	.1075	.3271	.44542	.89532	.49749	.0101	.1169	.2451	.46097	.88741	.51946	.9251	.1269	.1693	33
28	.41416	.91020	.45502	.1977	.0986	.4145	.42998	.90283	.47626	.0997	.1076	.3256	.44568	.89519	.49785	.0086	.1171	.2438	.46123	.88728	.51983	.9237	.1270	.1681	32
29	.41443	.91008	.45537	.1960	.0988	.4130	.43025	.90271	.47662	.0981	.1078	.3242	.44594	.89506	.49822	.0071	.1172	.2425	.46149	.88714	.52020	.9223	.1272	.1669	31
30	.41469	.90996	.45573	2.1943	.0989	2.4114	.43051	.90258	.47697	2.0965	.1079	2.3228	.44620	.89493	.49858	2.0057	.1174	2.2411	.46175	.88701	.52057	1.9210	.1274	2.1657	30
31	.41496	.90984	.45608	.1926	.0991	.4099	.43077	.90246	.47733	.0950	.1081	.3214	.44646	.89480	.49894	.0042	.1176	.2398	.46201	.88688	.52094	.9196	.1275	.1645	29
32	.41522	.90972	.45643	.1909	.0992	.4083	.43104	.90233	.47769	.0934	.1082	.3200	.44672	.89467	.49931	.0028	.1177	.2385	.46226	.88674	.52131	.9182	.1277	.1633	28
33	.41549	.90960	.45678	.1892	.0994	.4068	.43130	.90221	.47805	.0918	.1084	.3186	.44698	.89454	.49967	.0013	.1179	.2372	.46252	.88661	.52168	.9169	.1279	.1620	27
34	.41575	.90948	.45713	.1875	.0995	.4053	.43156	.90208	.47840	.0903	.1085	.3172	.44724	.89441	.50003	1.9998	.1180	.2359	.46278	.88647	.52205	.9155	.1281	.1608	26
35	.41602	.90936	.45748	2.1859	.0997	2.4037	.43182	.90196	.47876	2.0887	.1087	2.3158	.44750	.89428	.50040	1.9984	.1182	2.2346	.46304	.88634	.52242	1.9142	.1282	2.1596	25
36	.41628	.90924	.45783	.1842	.0998	.4022	.43208	.90183	.47912	.0872	.1088	.3143	.44776	.89415	.50076	.9969	.1184	.2333	.46330	.88620	.52279	.9128	.1284	.1584	24
37	.41654	.90911	.45819	.1825	.1000	.4007	.43235	.90171	.47948	.0856	.1090	.3129	.44802	.89402	.50113	.9955	.1185	.2320	.46355	.88607	.52316	.9115	.1286	.1572	23
38	.41681	.90899	.45854	.1808	.1001	.3992	.43261	.90158	.47983	.0840	.1092	.3115	.44828	.89389	.50149	.9940	.1187	.2307	.46381	.88593	.52353	.9101	.1287	.1560	22
39	.41707	.90887	.45889	.1792	.1003	.3976	.43287	.90145	.48019	.0825	.1093	.3101	.44854	.89376	.50185	.9926	.1189	.2294	.46407	.88580	.52390	.9088	.1289	.1548	21
40	.41734	.90875	.45924	2.1775	.1004	2.3961	.43313	.90133	.48055	2.0809	.1095	2.3087	.44880	.89363	.50222	1.9912	.1190	2.2282	.46433	.88566	.52427	1.9074	.1291	2.1536	20
41	.41760	.90863	.45960	.1758	.1005	.3946	.43340	.90120	.48091	.0794	.1096	.3073	.44906	.89350	.50258	.9897	.1192	.2269	.46458	.88553	.52464	.9061	.1293	.1525	19
42	.41787	.90851	.45995	.1741	.1007	.3931	.43366	.90108	.48127	.0778	.1098	.3059	.44932	.89337	.50295	.9883	.1193	.2256	.46484	.88539	.52501	.9047	.1294	.1513	18
43	.41813	.90839	.46030	.1725	.1008	.3916	.43392	.90095	.48162	.0763	.1099	.3046	.44958	.89324	.50331	.9868	.1195	.2243	.46510	.88526	.52538	.9034	.1296	.1501	17
44	.41839	.90826	.46065	.1708	.1010	.3901	.43418	.90082	.48198	.0747	.1101	.3032	.44984	.89311	.50368	.9854	.1197	.2230	.46536	.88512	.52575	.9020	.1298	.1489	16
45	.41866	.90814	.46101	2.1692	.1011	2.3886	.43444	.90070	.48234	2.0732	.1102	2.3018	.45010	.89298	.50404	1.9840	.1198	2.2217	.46561	.88499	.52612	1.9007	.1299	2.1477	15
46	.41892	.90812	.46136	.1675	.1013	.3871	.43471	.90057	.48270	.0717	.1104	.3004	.45036	.89285	.50441	.9825	.1200	.2204	.46587	.88485	.52650	.8993	.1301	.1465	14
47	.41919	.90790	.46171	.1658	.1014	.3856	.43497	.90044	.48306	.0701	.1106	.2990	.45062	.89272	.50477	.9811	.1202	.2192	.46613	.88472	.52687	.8980	.1303	.1453	13
48	.41945	.90778	.46206	.1642	.1016	.3841	.43523	.90032	.48342	.0686	.1107	.2976	.45088	.89258	.50514	.9797	.1203	.2179	.46639	.88458	.52724	.8967	.1305	.1441	12
49	.41972	.90765	.46242	.1625	.1017	.3826	.43549	.90019	.48378	.0671	.1109	.2962	.45114	.89245	.50550	.9782	.1205	.2166	.46664	.88444	.52761	.8953	.1306	.1430	11
50	.41998	.90753	.46277	2.1609	.1019	2.3811	.43575	.90006	.48414	2.0655	.1110	2.2949	.45140	.89232	.50587	1.9768	.1207	2.2153	.46690	.88431	.52798	1.8940	.1308	2.1418	10
51	.42024	.90741	.46312	.1592	.1020	.3796	.43602	.89994	.48449	.0640	.1112	.2935	.45166	.89219	.50623	.9754	.1208	.2141	.46716	.88417	.52836	.8927	.1310	.1406	9
52	.42051	.90729	.46348	.1576	.1022	.3781	.43628	.89981	.48485	.0625	.1113	.2921	.45191	.89206	.50660	.9739	.1210	.2128	.46741	.88404	.52873	.8913	.1312	.1394	8
53	.42077	.90717	.46383	.1559	.1023	.3766	.43654	.89968	.48521	.0609	.1115	.2907	.45217	.89193	.50696	.9725	.1212	.2115	.46767	.88390	.52910	.8900	.1313	.1382	7
54	.42103	.90704	.46418	.1543	.1025	.3751	.43680	.89956	.48557	.0594	.1116	.2894	.45243	.89180	.50733	.9711	.1213	.2103	.46793	.88376	.52947	.8887	.1315	.1371	6
55	.42130	.90692	.46454	2.1527	.1026	2.3736	.43706	.89943	.48593	2.0579	.1118	2.2880	.45269	.89166	.50769	1.9697	.1215	2.2090	.46819	.88363	.52984	1.8873	.1317	2.1359	5
56	.42156	.90680	.46489	.1510	.1028	.3721	.43732	.89930	.48629	.0564	.1120	.2866	.45295	.89153	.50806	.9683	.1217	.2077	.46844	.88349	.53022	.8860	.1319	.1347	4
57	.42183	.90668	.46524	.1494	.1029	.3706	.43759	.89918	.48665	.0548	.1121	.2853	.45321	.89140	.50843	.9668	.1218	.2065	.46870	.88336	.53059	.8847	.1320	.1335	3
58	.42209	.90655	.46560	.1478	.1031	.3691	.43785	.89905	.48701	.0533	.1123	.2839	.45347	.89127	.50879	.9654	.1220	.2052	.46896	.88322	.53096	.8834	.1322	.1324	2
59	.42235	.90643	.46595	.1461	.1032	.3677	.43811	.89892	.48737	.0518	.1124	.2825	.45373	.89114	.50916	.9640	.1222	.2039	.46921	.88308	.53134	.8820	.1324	.1312	1
60	.42262	.90631	.46631	2.1445	.1034	2.3662	.43837	.89879	.48773	2.0503	.1126	2.2812	.45399	.89101	.50952	1.9626	.1223	2.2027	.46947	.88295	.53171	1.8807	.1326	2.1300	0
′	cos	sin	cot	tan	cosec	sec	cos	sin	cot	tan	cosec	sec	cos	sin	cot	tan	cosec	sec	cos	sin	cot	tan	cosec	sec	′
	65°						64°						63°						62°						

′	28°						29°						30°						31°						′
	sin	cos	tan	cot	sec	cosec	sin	cos	tan	cot	sec	cosec	sin	cos	tan	cot	sec	cosec	sin	cos	tan	cot	sec	cosec	
0	.46947	.88295	.53171	1.8807	1.1326	2.1300	.48481	.87462	.55431	1.8040	1.1433	2.0627	.50000	.86603	.57735	1.7320	1.1547	2.0000	.51504	.85717	.60086	1.6643	1.1666	1.9416	60
1	.46973	.88281	.53208	.8794	.1327	.1289	.48506	.87448	.55469	.8028	.1435	.0616	.50025	.86588	.57774	.7309	.1549	1.9990	.51529	.85702	.60126	.6632	.1668	.9407	59
2	.46998	.88267	.53245	.8781	.1329	.1277	.48532	.87434	.55507	.8016	.1437	.0605	.50050	.86573	.57813	.7297	.1551	.9980	.51554	.85687	.60165	.6621	.1670	.9397	58
3	.47024	.88254	.53283	.8768	.1331	.1266	.48557	.87420	.55545	.8003	.1439	.0594	.50075	.86559	.57851	.7286	.1553	.9970	.51578	.85672	.60205	.6610	.1672	.9388	57
4	.47050	.88240	.53320	.8754	.1333	.1254	.48583	.87405	.55583	.7991	.1441	.0583	.50101	.86544	.57890	.7274	.1555	.9960	.51603	.85657	.60244	.6599	.1674	.9378	56
5	.47075	.88226	.53358	1.8741	.1334	2.1242	.48608	.87391	.55621	1.7979	.1443	2.0573	.50126	.86530	.57929	1.7262	.1557	1.9950	.51628	.85642	.60284	1.6588	.1676	1.9369	55
6	.47101	.88213	.53395	.8728	.1336	.1231	.48633	.87377	.55659	.7966	.1445	.0562	.50151	.86515	.57968	.7251	.1559	.9940	.51653	.85627	.60324	.6577	.1678	.9360	54
7	.47127	.88199	.53432	.8715	.1338	.1219	.48659	.87363	.55697	.7954	.1446	.0551	.50176	.86500	.58007	.7239	.1561	.9930	.51678	.85612	.60363	.6566	.1681	.9350	53
8	.47152	.88185	.53470	.8702	.1340	.1208	.48684	.87349	.55735	.7942	.1448	.0540	.50201	.86486	.58046	.7228	.1562	.9920	.51703	.85597	.60403	.6555	.1683	.9341	52
9	.47178	.88171	.53507	.8689	.1341	.1196	.48710	.87335	.55774	.7930	.1450	.0530	.50226	.86471	.58085	.7216	.1564	.9910	.51728	.85582	.60443	.6544	.1685	.9332	51
10	.47204	.88158	.53545	1.8676	.1343	2.1185	.48735	.87320	.55812	1.7917	.1452	2.0519	.50252	.86457	.58123	1.7205	.1566	1.9900	.51753	.85566	.60483	1.6534	.1687	1.9322	50
11	.47229	.88144	.53582	.8663	.1345	.1173	.48760	.87306	.55850	.7905	.1454	.0508	.50277	.86442	.58162	.7193	.1568	.9890	.51778	.85551	.60522	.6523	.1689	.9313	49
12	.47255	.88130	.53619	.8650	.1347	.1162	.48786	.87292	.55888	.7893	.1456	.0498	.50302	.86427	.58201	.7182	.1570	.9880	.51803	.85536	.60562	.6512	.1691	.9304	48
13	.47281	.88117	.53657	.8637	.1349	.1150	.48811	.87278	.55926	.7881	.1458	.0487	.50327	.86413	.58240	.7170	.1572	.9870	.51827	.85521	.60602	.6501	.1693	.9295	47
14	.47306	.88103	.53694	.8624	.1350	.1139	.48837	.87264	.55964	.7868	.1459	.0476	.50352	.86398	.58279	.7159	.1574	.9860	.51852	.85506	.60642	.6490	.1695	.9285	46
15	.47332	.88089	.53732	1.8611	.1352	2.1127	.48862	.87250	.56003	1.7856	.1461	2.0466	.50377	.86383	.58318	1.7147	.1576	1.9850	.51877	.85491	.60681	1.6479	.1697	1.9276	45
16	.47357	.88075	.53769	.8598	.1354	.1116	.48887	.87235	.56041	.7844	.1463	.0455	.50402	.86369	.58357	.7136	.1578	.9840	.51902	.85476	.60721	.6469	.1699	.9267	44
17	.47383	.88061	.53807	.8585	.1356	.1104	.48913	.87221	.56079	.7832	.1465	.0444	.50428	.86354	.58396	.7124	.1580	.9830	.51927	.85461	.60761	.6458	.1701	.9258	43
18	.47409	.88048	.53844	.8572	.1357	.1093	.48938	.87207	.56117	.7820	.1467	.0434	.50453	.86339	.58435	.7113	.1582	.9820	.51952	.85446	.60801	.6447	.1703	.9248	42
19	.47434	.88034	.53882	.8559	.1359	.1082	.48964	.87193	.56156	.7808	.1469	.0423	.50478	.86325	.58474	.7101	.1584	.9811	.51977	.85431	.60841	.6436	.1705	.9239	41
20	.47460	.88020	.53919	1.8546	.1361	2.1070	.48989	.87178	.56194	1.7795	.1471	2.0413	.50503	.86310	.58513	1.7090	.1586	1.9801	.52002	.85416	.60881	1.6425	.1707	1.9230	40
21	.47486	.88006	.53957	.8533	.1363	.1059	.49014	.87164	.56232	.7783	.1473	.0402	.50528	.86295	.58552	.7079	.1588	.9791	.52026	.85400	.60920	.6415	.1709	.9221	39
22	.47511	.87992	.53995	.8520	.1365	.1048	.49040	.87150	.56270	.7771	.1474	.0392	.50553	.86281	.58591	.7067	.1590	.9781	.52051	.85385	.60960	.6404	.1712	.9212	38
23	.47537	.87979	.54032	.8507	.1366	.1036	.49065	.87136	.56309	.7759	.1476	.0381	.50578	.86266	.58630	.7056	.1592	.9771	.52076	.85370	.61000	.6393	.1714	.9203	37
24	.47562	.87965	.54070	.8495	.1368	.1025	.49090	.87121	.56347	.7747	.1478	.0370	.50603	.86251	.58670	.7044	.1594	.9761	.52101	.85355	.61040	.6383	.1716	.9193	36
25	.47588	.87951	.54107	1.8482	.1370	2.1014	.49116	.87107	.56385	1.7735	.1480	2.0360	.50628	.86237	.58709	1.7033	.1596	1.9752	.52126	.85340	.61080	1.6372	.1718	1.9184	35
26	.47613	.87937	.54145	.8469	.1372	.1002	.49141	.87093	.56424	.7723	.1482	.0349	.50653	.86222	.58748	.7022	.1598	.9742	.52151	.85325	.61120	.6361	.1720	.9175	34
27	.47639	.87923	.54183	.8456	.1373	.0991	.49166	.87078	.56462	.7711	.1484	.0339	.50679	.86207	.58787	.7010	.1600	.9732	.52175	.85309	.61160	.6350	.1722	.9166	33
28	.47665	.87909	.54220	.8443	.1375	.0980	.49192	.87064	.56500	.7699	.1486	.0329	.50704	.86192	.58826	.6999	.1602	.9722	.52200	.85294	.61200	.6340	.1724	.9157	32
29	.47690	.87895	.54258	.8430	.1377	.0969	.49217	.87050	.56539	.7687	.1488	.0318	.50729	.86178	.58865	.6988	.1604	.9713	.52225	.85279	.61240	.6329	.1726	.9148	31
30	.47716	.87882	.54295	1.8418	.1379	2.0957	.49242	.87035	.56577	1.7675	.1489	2.0308	.50754	.86163	.58904	1.6977	.1606	1.9703	.52250	.85264	.61280	1.6318	.1728	1.9139	30
31	.47741	.87868	.54333	.8405	.1381	.0946	.49268	.87021	.56616	.7663	.1491	.0297	.50779	.86148	.58944	.6965	.1608	.9693	.52275	.85249	.61320	.6308	.1730	.9130	29
32	.47767	.87854	.54371	.8392	.1382	.0935	.49293	.87007	.56654	.7651	.1493	.0287	.50804	.86133	.58983	.6954	.1610	.9683	.52299	.85234	.61360	.6297	.1732	.9121	28
33	.47792	.87840	.54409	.8379	.1384	.0924	.49318	.86992	.56692	.7639	.1495	.0276	.50829	.86118	.59022	.6943	.1612	.9674	.52324	.85218	.61400	.6286	.1734	.9112	27
34	.47818	.87826	.54446	.8367	.1386	.0912	.49343	.86978	.56731	.7627	.1497	.0266	.50854	.86104	.59061	.6931	.1614	.9664	.52349	.85203	.61440	.6276	.1737	.9102	26
35	.47844	.87812	.54484	1.8354	.1388	2.0901	.49369	.86964	.56769	1.7615	.1499	2.0256	.50879	.86089	.59100	1.6920	.1616	1.9654	.52374	.85188	.61480	1.6265	.1739	1.9093	25
36	.47869	.87798	.54522	.8341	.1390	.0890	.49394	.86949	.56808	.7603	.1501	.0245	.50904	.86074	.59140	.6909	.1618	.9645	.52398	.85173	.61520	.6255	.1741	.9084	24
37	.47895	.87784	.54559	.8329	.1391	.0879	.49419	.86935	.56846	.7591	.1503	.0235	.50929	.86059	.59179	.6898	.1620	.9635	.52423	.85157	.61560	.6244	.1743	.9075	23
38	.47920	.87770	.54597	.8316	.1393	.0868	.49445	.86921	.56885	.7579	.1505	.0224	.50954	.86044	.59218	.6887	.1622	.9625	.52448	.85142	.61601	.6233	.1745	.9066	22
39	.47946	.87756	.54635	.8303	.1395	.0857	.49470	.86906	.56923	.7567	.1507	.0214	.50979	.86030	.59258	.6875	.1624	.9616	.52473	.85127	.61641	.6223	.1747	.9057	21
40	.47971	.87742	.54673	1.8291	.1397	2.0846	.49495	.86892	.56962	1.7555	.1508	2.0204	.51004	.86015	.59297	1.6864	.1626	1.9606	.52498	.85112	.61681	1.6212	.1749	1.9048	20
41	.47997	.87728	.54711	.8278	.1399	.0835	.49521	.86877	.57000	.7544	.1510	.0194	.51029	.86000	.59336	.6853	.1628	.9596	.52522	.85096	.61721	.6202	.1751	.9039	19
42	.48022	.87715	.54748	.8265	.1401	.0824	.49546	.86863	.57039	.7532	.1512	.0183	.51054	.85985	.59376	.6842	.1630	.9587	.52547	.85081	.61761	.6191	.1753	.9030	18
43	.48048	.87701	.54786	.8253	.1402	.0812	.49571	.86849	.57077	.7520	.1514	.0173	.51079	.85970	.59415	.6831	.1632	.9577	.52572	.85066	.61801	.6181	.1756	.9021	17
44	.48073	.87687	.54824	.8240	.1404	.0801	.49596	.86834	.57116	.7508	.1516	.0163	.51104	.85955	.59454	.6820	.1634	.9568	.52597	.85050	.61842	.6170	.1758	.9013	16
45	.48099	.87673	.54862	1.8227	.1406	2.0790	.49622	.86820	.57155	1.7496	.1518	2.0152	.51129	.85941	.59494	1.6808	.1636	1.9558	.52621	.85735	.61882	1.6160	.1760	1.9004	15
46	.48124	.87659	.54900	.8215	.1408	.0779	.49647	.86805	.57193	.7484	.1520	.0142	.51154	.85926	.59533	.6797	.1638	.9549	.52646	.85020	.61922	.6149	.1762	.8995	14
47	.48150	.87645	.54937	.8202	.1410	.0768	.49672	.86791	.57232	.7473	.1522	.0132	.51179	.85911	.59572	.6786	.1640	.9539	.52671	.85004	.61962	.6139	.1764	.8986	13
48	.48175	.87631	.54975	.8190	.1411	.0757	.49697	.86776	.57270	.7461	.1524	.0122	.51204	.85896	.59612	.6775	.1642	.9530	.52695	.84989	.62003	.6128	.1766	.8977	12
49	.48201	.87617	.55013	.8177	.1413	.0746	.49723	.86762	.57309	.7449	.1526	.0111	.51229	.85881	.59651	.6764	.1644	.9520	.52720	.84974	.62043	.6118	.1768	.8968	11
50	.48226	.87603	.55051	1.8165	.1415	2.0735	.49748	.86748	.57348	1.7437	.1528	2.0101	.51254	.85866	.59691	1.6753	.1646	1.9510	.52745	.84959	.62083	1.6107	.1770	1.8959	10
51	.48252	.87588	.55089	.8152	.1417	.0725	.49773	.86733	.57386	.7426	.1530	.0091	.51279	.85851	.59730	.6742	.1648	.9501	.52770	.84943	.62123	.6097	.1772	.8950	9
52	.48277	.87574	.55127	.8140	.1419	.0714	.49798	.86719	.57425	.7414	.1531	.0081	.51304	.85836	.59770	.6731	.1650	.9491	.52794	.84928	.62164	.6086	.1775	.8941	8
53	.48303	.87560	.55165	.8127	.1421	.0703	.49823	.86704	.57464	.7402	.1533	.0071	.51329	.85821	.59809	.6720	.1652	.9482	.52819	.84912	.62204	.6076	.1777	.8932	7
54	.48328	.87546	.55203	.8115	.1422	.0692	.49849	.86690	.57502	.7390	.1535	.0061	.51354	.85806	.59849	.6709	.1654	.9473	.52844	.84897	.62244	.6066	.1779	.8924	6
55	.48354	.87532	.55241	1.8102	.1424	2.0681	.49874	.86675	.57541	1.7379	.1537	2.0050	.51379	.85791	.59888	1.6698	.1656	1.9463	.52868	.84882	.62285	1.6055	.1781	1.8915	5
56	.48379	.87518	.55279	.8090	.1426	.0670	.49899	.86661	.57580	.7367	.1539	.0040	.51404	.85777	.59928	.6687	.1658	.9454	.52893	.84866	.62325	.6045	.1783	.8906	4
57	.48405	.87504	.55317	.8078	.1428	.0659	.49924	.86646	.57619	.7355	.1541	.0030	.51429	.85762	.59967	.6676	.1660	.9444	.52918	.84851	.62366	.6034	.1785	.8897	3
58	.48430	.87490	.55355	.8065	.1430	.0648	.49950	.86632	.57657	.7344	.1543	.0020	.51454	.85747	.60007	.6665	.1662	.9435	.52942	.84836	.62406	.6024	.1787	.8888	2
59	.48455	.87476	.55393	.8053	.1432	.0637	.49975	.86617	.57696	.7332	.1545	.0010	.51479	.85732	.60046	.6654	.1664	.9425	.52967	.84820	.62446	.6014	.1790	.8879	1
60	.48481	.87462	.55431	1.8040	.1433	2.0627	.50000	.86603	.57735	1.7320	.1547	2.0000	.51504	.85717	.60086	1.6643	.1666	1.9416	.52992	.84805	.62487	1.6003	.1792	1.8871	0
′	cos	sin	cot	tan	cosec	sec	cos	sin	cot	tan	cosec	sec	cos	sin	cot	tan	cosec	sec	cos	sin	cot	tan	cosec	sec	′
	61°						60°						59°						58°						

(Continued)

Table B18. (Cont.)

′	32°						33°					
	sin	cos	tan	cot	sec	cosec	sin	cos	tan	cot	sec	cosec
0	.52992	.84805	.62487	1.6003	1.1792	1.8871	.54464	.83867	.64941	1.5399	1.1924	1.8361
1	.53016	.84789	.62527	.5993	.1794	.8862	.54488	.83851	.64982	.5389	.1926	.8352
2	.53041	.84774	.62568	.5983	.1796	.8853	.54513	.83835	.65023	.5379	.1928	.8344
3	.53066	.84758	.62608	.5972	.1798	.8844	.54537	.83819	.65065	.5369	.1930	.8336
4	.53090	.84743	.62649	.5962	.1800	.8836	.54561	.83804	.65106	.5359	.1933	.8328
5	.53115	.84728	.62689	1.5952	.1802	1.8827	.54586	.83788	.65148	1.5350	.1935	1.8320
6	.53140	.84712	.62730	.5941	.1805	.8818	.54610	.83772	.65189	.5340	.1937	.8311
7	.53164	.84697	.62770	.5931	.1807	.8809	.54634	.83756	.65231	.5330	.1939	.8303
8	.53189	.84681	.62811	.5921	.1809	.8801	.54659	.83740	.65272	.5320	.1942	.8295
9	.53214	.84666	.62851	.5910	.1811	.8792	.54683	.83724	.65314	.5311	.1944	.8287
10	.53238	.84650	.62892	1.5900	.1813	1.8783	.54708	.83708	.65355	1.5301	.1946	1.8279
11	.53263	.84635	.62933	.5890	.1815	.8775	.54732	.83692	.65397	.5291	.1948	.8271
12	.53288	.84619	.62973	.5880	.1818	.8766	.54756	.83676	.65438	.5282	.1951	.8263
13	.53312	.84604	.63014	.5869	.1820	.8757	.54781	.83660	.65480	.5272	.1953	.8255
14	.53337	.84588	.63055	.5859	.1822	.8749	.54805	.83644	.65521	.5262	.1955	.8246
15	.53361	.84573	.63095	1.5849	.1824	1.8740	.54829	.83629	.65563	1.5252	.1958	1.8238
16	.53386	.84557	.63136	.5839	.1826	.8731	.54854	.83613	.65604	.5234	.1960	.8230
17	.53411	.84542	.63177	.5829	.1828	.8723	.54878	.83597	.65646	.5233	.1962	.8222
18	.53435	.84526	.63217	.5818	.1831	.8714	.54902	.83581	.65688	.5223	.1964	.8214
19	.53460	.84511	.63258	.5808	.1833	.8706	.54926	.83565	.65729	.5214	.1967	.8206
20	.53484	.84495	.63299	1.5798	.1835	1.8697	.54951	.83549	.65771	1.5204	.1969	1.8198
21	.53509	.84479	.63339	.5788	.1837	.8688	.54975	.83533	.65813	.5195	.1971	.8190
22	.53533	.84464	.63380	.5778	.1839	.8680	.54999	.83517	.65854	.5185	.1974	.8182
23	.53558	.84448	.63421	.5768	.1841	.8671	.55024	.83501	.65896	.5175	.1976	.8174
24	.53583	.84433	.63462	.5757	.1844	.8663	.55048	.83485	.65938	.5166	.1978	.8166
25	.53607	.84417	.63503	1.5747	.1846	1.8654	.55002	.83469	.65980	1.5156	.1980	1.8158
26	.53632	.84402	.63543	.5737	.1848	.8646	.55090	.83453	.66021	.5147	.1983	.8150
27	.53656	.84386	.63584	.5727	.1850	.8637	.55121	.83437	.66063	.5137	.1985	.8142
28	.53681	.84370	.63625	.5717	.1852	.8629	.55145	.83421	.66105	.5127	.1987	.8134
29	.53705	.84355	.63666	.5707	.1855	.8620	.55169	.83405	.66147	.5118	.1990	.8126
30	.53730	.84339	.63707	1.5697	.1857	1.8611	.55194	.83388	.56188	1.5108	.1992	1.8118
31	.53754	.84323	.63748	.5687	.1859	.8603	.55218	.83372	.66230	.5099	.1994	.8110
32	.53779	.84308	.63789	.5677	.1861	.8595	.55242	.83356	.66272	.5089	.1997	.8102
33	.53803	.84292	.63830	.5667	.1863	.8586	.55266	.83340	.66314	.5080	.1999	.8094
34	.53828	.84276	.63871	.5657	.1866	.8578	.55291	.83324	.66356	.5070	.2001	.8086
35	.53852	.84261	.63912	1.5646	.1868	1.8569	.55315	.83308	.66398	1.5061	.2004	1.8078
36	.53877	.84245	.63953	.5636	.1870	.8561	.55339	.83292	.66440	.5051	.2006	.8070
37	.53901	.84229	.63994	.5626	.1872	.8552	.55363	.83276	.66482	.5042	.2008	.8062
38	.53926	.84214	.64035	.5616	.1874	.8544	.55388	.83260	.66524	.5032	.2010	.8054
39	.53950	.84198	.64076	.5606	.1877	.8535	.55412	.83244	.66566	.5023	.2013	.8047
40	.53975	.84182	.64117	1.5596	.1879	1.8527	.55436	.83228	.66608	1.5013	.2015	1.8039
41	.53999	.84167	.64158	.5586	.1881	.8519	.55460	.83211	.66650	.5004	.2017	.8031
42	.54024	.84151	.64199	.5577	.1883	.8510	.55484	.83195	.66692	.4994	.2020	.8023
43	.54048	.84135	.64240	.5567	.1886	.8502	.55509	.83179	.66734	.4985	.2022	.8015
44	.54073	.84120	.64281	.5557	.1888	.8493	.55533	.83163	.66776	.4975	.2024	.8007
45	.54097	.84104	.64322	1.5547	.1890	1.8485	.55557	.83147	.66818	1.4966	.2027	1.7999
46	.54122	.84088	.64363	.5537	.1892	.8477	.55581	.83131	.66860	.4957	.2029	.7992
47	.54146	.84072	.64404	.5527	.1894	.8468	.55605	.83115	.66902	.4947	.2031	.7984
48	.54171	.84057	.64446	.5517	.1897	.8460	.55629	.83098	.66944	.4938	.2034	.7976
49	.54195	.84041	.64487	.5507	.1899	.8452	.55654	.83082	.66986	.4928	.2036	.7968
50	.54220	.84025	.64528	1.5497	.1901	1.8443	.55678	.83066	.67028	1.4919	.2039	1.7960
51	.54244	.84009	.64569	.5487	.1903	.8435	.55702	.83050	.67071	.4910	.2041	.7953
52	.54268	.83993	.64610	.5477	.1906	.8427	.55726	.83034	.67113	.4900	.2043	.7945
53	.54293	.83978	.64652	.5467	.1908	.8418	.55750	.83017	.67155	.4891	.2046	.7937
54	.54317	.83962	.64693	.5458	.1910	.8410	.55774	.83001	.67197	.4881	.2048	.7929
55	.54342	.83946	.64734	1.5448	.1912	1.8402	.55799	.82985	.67239	1.4872	.2050	1.7921
56	.54366	.83930	.64775	.5438	.1915	.8394	.55823	.82969	.67282	.4863	.2053	.7914
57	.54391	.83914	.64817	.5428	.1917	.8385	.55847	.82952	.67324	.4853	.2055	.7906
58	.54415	.83899	.64858	.5418	.1919	.8377	.55871	.82936	.67366	.4844	.2057	.7898
59	.54439	.83883	.64899	.5408	.1921	.8369	.55895	.82920	.67408	.4835	.2060	.7891
60	.54464	.83867	.64941	1.5399	.1922	1.8361	.55919	.82904	.67451	1.4826	.2062	1.7883
′	cos	sin	cot	tan	cosec	sec	cos	sin	cot	tan	cosec	sec
	57°						56°					

34°						35°						′
sin	cos	tan	cot	sec	cosec	sin	cos	tan	cot	sec	cosec	
.55919	.82904	.67451	1.4826	1.2062	1.7883	.57358	.81915	.70021	1.4281	1.2208	1.7434	60
.55943	.82887	.67493	.4816	.2064	.7875	.57381	.81898	.70064	.4273	.2210	.7427	59
.55967	.82871	.67535	.4807	.2067	.7867	.57405	.81882	.70107	.4264	.2213	.7420	58
.55992	.82855	.67578	.4798	.2069	.7860	.57429	.81865	.70151	.4255	.2215	.7413	57
.56016	.82839	.67620	.4788	.2072	.7852	.57453	.81848	.70194	.4246	.2218	.7405	56
.56040	.82822	.67663	1.4779	.2074	1.7844	.57477	.81832	.70238	1.4237	.2220	1.7398	55
.56064	.82806	.67705	.4770	.2076	.7837	.57500	.81815	.70281	.4228	.2223	.7391	54
.56088	.82790	.67747	.4761	.2079	.7829	.57524	.81798	.70325	.4220	.2225	.7384	53
.56112	.82773	.67790	.4751	.2081	.7821	.57548	.81781	.70368	.4211	.2228	.7377	52
.56136	.82757	.67832	.4742	.2083	.7814	.57572	.81765	.70412	.4202	.2230	.7369	51
.56160	.82741	.67875	1.4733	.2086	1.7806	.57596	.81748	.70455	1.4193	.2233	1.7362	50
.56184	.82724	.67917	.4724	.2088	.7798	.57619	.81731	.70499	.4185	.2235	.7355	49
.56208	.82708	.67960	.4714	.2091	.7791	.57643	.81714	.70542	.4176	.2238	.7348	48
.56232	.82692	.68002	.4705	.2093	.7783	.57667	.81698	.70586	.4167	.2240	.7341	47
.56256	.82675	.68045	.4696	.2095	.7776	.57691	.81681	.70629	.4158	.2243	.7334	46
.56280	.82659	.68087	1.4687	.2098	1.7768	.57714	.81664	.70673	1.4150	.2245	1.7327	45
.56304	.82643	.68130	.4678	.2100	.7760	.57738	.81647	.70717	.4141	.2248	.7319	44
.56328	.82626	.68173	.4669	.2103	.7753	.57762	.81630	.70760	.4132	.2250	.7312	43
.56353	.82610	.68215	.4659	.2105	.7745	.57786	.81614	.70804	.4123	.2253	.7305	42
.56377	.82593	.68258	.4650	.2107	.7738	.57809	.81597	.70848	.4115	.2255	.7298	41
.56401	.82577	.68301	1.4641	.2110	1.7730	.57833	.81580	.70891	1.4106	.2258	1.7291	40
.56425	.82561	.68343	.4632	.2112	.7723	.57857	.81563	.70935	.4097	.2260	.7284	39
.56449	.82544	.68386	.4623	.2115	.7715	.57881	.81546	.70979	.4089	.2263	.7277	38
.56473	.82528	.68429	.4614	.2117	.7708	.57904	.81530	.71022	.4080	.2265	.7270	37
.56497	.82511	.68471	.4605	.2119	.7700	.57928	.81513	.71066	.4071	.2268	.7263	36
.56521	.82495	.68514	1.4595	.2122	1.7693	.57952	.81496	.71110	1.4063	.2270	1.7256	35
.56545	.82478	.68557	.4586	.2124	.7685	.57975	.81479	.71154	.4054	.2273	.7249	34
.56569	.82462	.68600	.4577	.2127	.7678	.57999	.81462	.71198	.4045	.2276	.7242	33
.56593	.82445	.68642	.4568	.2129	.7670	.58023	.81445	.71241	.4037	.2278	.7234	32
.56617	.82429	.68685	.4559	.2132	.7663	.58047	.81428	.71285	.4028	.2281	.7227	31
.56641	.82413	.68728	1.4550	.2134	1.7655	.58070	.81411	.71329	1.4019	.2283	1.7220	30
.56664	.82396	.68771	.4541	.2136	.7648	.58094	.81395	.71373	.4011	.2286	.7213	29
.56688	.82380	.68814	.4532	.2139	.7640	.58118	.81378	.71417	.4002	.2288	.7206	28
.56712	.82363	.68857	.4523	.2141	.7633	.58141	.81361	.71461	.3994	.2291	.7199	27
.56736	.82347	.68899	.4514	.2144	.7625	.58165	.81344	.71505	.3985	.2293	.7192	26
.56760	.82330	.68942	1.4505	.2146	1.7618	.58189	.81327	.71549	1.3976	.2296	1.7185	25
.56784	.82314	.68985	.4496	.2149	.7610	.58212	.81310	.71593	.3968	.2298	.7178	24
.56808	.82297	.69028	.4487	.2151	.7603	.58236	.81293	.71637	.3959	.2301	.7171	23
.56832	.82280	.69071	.4478	.2153	.7596	.58259	.81276	.71681	.3951	.2304	.7164	22
.56856	.82264	.69114	.4469	.2156	.7588	.58283	.81259	.71725	.3942	.2306	.7157	21
.56880	.82247	.69157	1.4460	.2158	1.7581	.58307	.81242	.71769	1.3933	.2309	1.7151	20
.56904	.82231	.69200	.4451	.2161	.7573	.58330	.81225	.71813	.3925	.2311	.7144	19
.56928	.82214	.69243	.4442	.2163	.7566	.58354	.81208	.71857	.3916	.2314	.7137	18
.56952	.82198	.69286	.4433	.2166	.7559	.58378	.81191	.71901	.3908	.2316	.7130	17
.56976	.82181	.69329	.4424	.2168	.7551	.58401	.81174	.71945	.3899	.2319	.7123	16
.57000	.82165	.69372	1.4415	.2171	1.7544	.58425	.81157	.71990	1.3891	.2322	1.7116	15
.57023	.82148	.69415	.4406	.2173	.7537	.58448	.81140	.72034	.3882	.2324	.7109	14
.57047	.82131	.69459	.4397	.2175	.7529	.58472	.81123	.72078	.3874	.2327	.7102	13
.57071	.82115	.69502	.4388	.2178	.7522	.58496	.81106	.72122	.3865	.2329	.7095	12
.57095	.82098	.69545	.4379	.2180	.7514	.58519	.81089	.72166	.3857	.2332	.7088	11
.57119	.82082	.69588	1.4370	.2183	1.7507	.58543	.81072	.72211	1.3848	.2335	1.7081	10
.57143	.82065	.69631	.4361	.2185	.7500	.58566	.81055	.72255	.3840	.2337	.7075	9
.57167	.82048	.69674	.4352	.2188	.7493	.58590	.81038	.72299	.3831	.2340	.7068	8
.57191	.82032	.69718	.4343	.2190	.7485	.58614	.81021	.72344	.3823	.2342	.7061	7
.57214	.82015	.69761	.4335	.2193	.7478	.58637	.81004	.72388	.3814	.2345	.7054	6
.57238	.81998	.69804	1.4326	.2195	1.7471	.58661	.80987	.72432	1.3806	.2348	1.7047	5
.57262	.81982	.69847	.4317	.2198	.7463	.58684	.80970	.72477	.3797	.2350	.7040	4
.57286	.81965	.69891	.4308	.2200	.7456	.58708	.80953	.72521	.3789	.2353	.7033	3
.57310	.81948	.69934	.4299	.2203	.7449	.58731	.80936	.72565	.3781	.2355	.7027	2
.57334	.81932	.69977	.4290	.2205	.7442	.58755	.80919	.72610	.3772	.2358	.7020	1
.57358	.81915	.70021	1.4281	.2208	1.7434	.58778	.80902	.72654	1.3764	.2361	1.7013	0
cos	sin	cot	tan	cosec	sec	cos	sin	cot	tan	cosec	sec	′
55°						54°						

′	36°						37°						38°						39°						′
′	sin	cos	tan	cot	sec	cosec	sin	cos	tan	cot	sec	cosec	sin	cos	tan	cot	sec	cosec	sin	cos	tan	cot	sec	cosec	′
0	.58778	.80902	.72654	1.3764	1.2361	1.7013	.60181	.79863	.75355	1.3270	1.2521	1.6616	.61566	.78801	.78128	1.2799	1.2690	1.6243	.62932	.77715	.80978	1.2349	1.2867	1.5890	60
1	.58802	.80885	.72699	.3755	.2363	.7006	.60205	.79846	.75401	.3262	.2524	.6610	.61589	.78783	.78175	.2792	.2693	.6237	.62955	.77696	.81026	.2342	.2871	.5884	59
2	.58825	.80867	.72743	.3747	.2366	.6999	.60228	.79828	.75447	.3254	.2527	.6603	.61612	.78765	.78222	.2784	.2696	.6231	.62977	.77678	.81075	.2334	.2874	.5879	58
3	.58849	.80850	.72788	.3738	.2368	.6993	.60251	.79811	.75492	.3246	.2530	.6597	.61635	.78747	.78269	.2776	.2699	.6224	.63000	.77660	.81123	.2327	.2877	.5873	57
4	.58873	.80833	.72832	.3730	.2371	.6986	.60274	.79793	.75538	.3238	.2532	.6591	.61658	.78729	.78316	.2769	.2702	.6218	.63022	.77641	.81171	.2320	.2880	.5867	56
5	.58896	.80816	.72877	1.3722	.2374	1.6979	.60298	.79776	.75584	1.3230	.2535	1.6584	.61681	.78711	.78363	1.2761	.2705	1.6212	.63045	.77623	.81219	1.2312	.2883	1.5862	55
6	.58920	.80799	.72921	.3713	.2376	.6972	.60320	.79758	.75629	.3222	.2538	.6578	.61703	.78693	.78410	.2753	.2707	.6206	.63067	.77605	.81268	.2305	.2886	.5856	54
7	.58943	.80782	.72966	.3705	.2379	.6965	.60344	.79741	.75675	.3214	.2541	.6572	.61726	.78675	.78457	.2746	.2710	.6200	.63090	.77586	.81316	.2297	.2889	.5850	53
8	.58967	.80765	.73010	.3697	.2382	.6959	.60367	.79723	.75721	.3206	.2543	.6565	.61749	.78657	.78504	.2738	.2713	.6194	.63113	.77568	.81364	.2290	.2892	.5854	52
9	.58990	.80747	.73055	.3688	.2384	.6952	.60390	.79706	.75767	.3198	.2546	.6559	.61772	.78640	.78551	.2730	.2716	.6188	.63135	.77549	.81413	.2283	.2895	.5839	51
10	.59014	.80730	.73100	1.3680	.2387	1.6945	.60413	.79688	.75812	1.3190	.2549	1.6552	.61795	.78622	.78598	1.2723	.2719	1.6182	.63158	.77531	.81461	1.2276	.2898	1.5833	50
11	.59037	.80713	.73144	.3672	.2389	.6938	.60437	.79670	.75858	.3182	.2552	.6546	.61818	.78604	.78645	.2715	.2722	.6176	.63180	.77513	.81509	.2268	.2901	.5828	49
12	.59060	.80696	.73189	.3663	.2392	.6932	.60460	.79653	.75904	.3174	.2554	.6540	.61841	.78586	.78692	.2708	.2725	.6170	.63203	.77494	.81558	.2261	.2904	.5822	48
13	.59084	.80679	.73234	.3655	.2395	.6925	.60483	.79635	.75950	.3166	.2557	.6533	.61864	.78568	.78739	.2700	.2728	.6164	.63225	.77476	.81606	.2254	.2907	.5816	47
14	.59107	.80662	.73278	.3647	.2397	.6918	.60506	.79618	.75996	.3159	.2560	.6527	.61886	.78550	.78786	.2692	.2731	.6159	.63248	.77458	.81655	.2247	.2910	.5811	46
15	.59131	.80644	.73323	1.3638	.2400	1.6912	.60529	.79600	.76042	1.3151	.2563	1.6521	.61909	.78532	.78834	1.2685	.2734	1.6153	.63270	.77439	.81703	1.2239	.2913	1.5805	45
16	.59154	.80627	.73368	.3630	.2403	.6905	.60552	.79582	.76088	.3143	.2565	.6514	.61932	.78514	.78881	.2677	.2737	.6147	.63293	.77421	.81752	.2232	.2916	.5799	44
17	.59178	.80610	.73412	.3622	.2405	.6898	.60576	.79565	.76134	.3135	.2568	.6508	.61955	.78496	.78928	.2670	.2739	.6141	.63315	.77402	.81800	.2225	.2919	.5794	43
18	.59201	.80593	.73457	.3613	.2408	.6891	.60599	.79547	.76179	.3127	.2571	.6502	.61978	.78478	.78975	.2662	.2742	.6135	.63338	.77384	.81849	.2218	.2922	.5788	42
19	.59225	.80576	.73502	.3605	.2411	.6885	.60622	.79530	.76225	.3119	.2574	.6496	.62001	.78460	.79022	.2655	.2745	.6129	.63360	.77365	.81898	.2210	.2926	.5783	41
20	.59248	.80558	.73547	1.3597	.2413	1.6878	.60645	.79512	.76271	1.3111	.2577	1.6489	.62023	.78441	.79070	1.2647	.2748	1.6123	.63383	.77347	.81946	1.2203	.2929	1.5777	40
21	.59272	.80541	.73592	.3588	.2416	.6871	.60668	.79494	.76317	.3103	.2579	.6483	.62046	.78423	.79117	.2639	.2751	.6117	.63405	.77329	.81995	.2196	.2932	.5771	39
22	.59295	.80524	.73637	.3580	.2419	.6865	.60691	.79477	.76364	.3095	.2582	.6477	.62069	.78405	.79164	.2632	.2754	.6111	.63428	.77310	.82043	.2189	.2935	.5766	38
23	.59318	.80507	.73681	.3572	.2421	.6858	.60714	.79459	.76410	.3087	.2585	.6470	.62092	.78387	.79212	.2624	.2757	.6105	.63450	.77292	.82092	.2181	.2938	.5760	37
24	.59342	.80489	.73726	.3564	.2424	.6851	.60737	.79441	.76456	.3079	.2588	.6464	.62115	.78369	.79259	.2617	.2760	.6099	.63473	.77273	.82141	.2174	.2941	.5755	36
25	.59365	.80472	.73771	1.3555	.2427	1.6845	.60761	.79424	.76502	1.3071	.2591	1.6458	.62137	.78351	.79306	1.2609	.2763	1.6093	.63495	.77255	.82190	1.2167	.2944	1.5749	35
26	.59389	.80455	.73816	.3547	.2429	.6838	.60784	.79406	.76548	.3064	.2593	.6452	.62160	.78333	.79354	.2602	.2766	.6087	.63518	.77236	.82238	.2160	.2947	.5743	34
27	.59412	.80437	.73861	.3539	.2432	.6831	.60807	.79388	.76594	.3056	.2596	.6445	.62183	.78315	.79401	.2594	.2769	.6081	.63540	.77218	.82287	.2152	.2950	.5738	33
28	.59435	.80420	.73906	.3531	.2435	.6825	.60830	.79371	.76640	.3048	.2599	.6439	.62206	.78297	.79449	.2587	.2772	.6077	.63563	.77199	.82336	.2145	.2953	.5732	32
29	.59459	.80403	.73951	.3522	.2437	.6818	.60853	.79353	.76686	.3040	.2602	.6433	.62229	.78279	.79496	.2579	.2775	.6070	.63585	.77181	.82385	.2138	.2956	.5727	31
30	.59482	.80386	.73996	1.3514	.2440	1.6812	.60876	.79335	.76733	1.3032	.2605	1.6427	.62251	.78261	.79543	1.2572	.2778	1.6064	.63608	.77162	.82434	1.2131	.2960	1.5721	30
31	.59506	.80368	.74041	.3506	.2443	.6805	.60899	.79318	.76779	.3024	.2607	.6420	.62274	.78243	.79591	.2564	.2781	.6058	.63630	.77144	.82482	.2124	.2963	.5716	29
32	.59529	.80351	.74086	.3498	.2445	.6798	.60922	.79300	.76825	.3016	.2610	.6414	.62297	.78224	.79639	.2557	.2784	.6052	.63653	.77125	.82531	.2117	.2966	.5710	28
33	.59552	.80334	.74131	.3489	.2448	.6792	.60945	.79282	.76871	.3009	.2613	.6408	.62320	.78206	.79686	.2549	.2787	.6046	.63675	.77107	.82580	.2109	.2969	.5705	27
34	.59576	.80316	.74176	.3481	.2451	.6785	.60968	.79264	.76918	.3001	.2616	.6402	.62342	.78188	.79734	.2542	.2790	.6040	.63697	.77088	.82629	.2102	.2972	.5699	26
35	.59599	.80299	.74221	1.3473	.2453	1.6779	.60991	.79247	.76964	1.2993	.2619	1.6396	.62365	.78170	.79781	1.2534	.2793	1.6034	.63720	.77070	.82678	1.2095	.2975	1.5694	25
36	.59622	.80282	.74266	.3465	.2456	.6772	.61014	.79229	.77010	.2985	.2622	.6389	.62388	.78152	.79829	.2527	.2795	.6029	.63742	.77051	.82727	.2088	.2978	.5688	24
37	.59646	.80264	.74312	.3457	.2459	.6766	.61037	.79211	.77057	.2977	.2624	.6383	.62411	.78134	.79876	.2519	.2798	.6023	.63765	.77033	.82776	.2081	.2981	.5683	23
38	.59669	.80247	.74357	.3449	.2461	.6759	.61061	.79193	.77103	.2970	.2627	.6377	.62433	.78116	.79924	.2512	.2801	.6017	.63787	.77014	.82825	.2074	.2985	.5677	22
39	.59692	.80230	.74402	.3440	.2464	.6752	.61084	.79176	.77149	.2962	.2630	.6371	.62456	.78097	.79972	.2504	.2804	.6011	.63810	.76996	.82874	.2066	.2988	.5672	21
40	.59716	.80212	.74447	1.3432	.2467	1.6746	.61107	.79158	.77196	1.2954	.2633	1.6365	.62479	.78079	.80020	1.2497	.2807	1.6005	.63832	.76977	.82923	1.2059	.2991	1.5666	20
41	.59739	.80195	.74492	.3424	.2470	.6739	.61130	.79140	.77242	.2946	.2636	.6359	.62501	.78061	.80067	.2489	.2810	.6000	.63854	.76958	.82972	.2052	.2994	.5661	19
42	.59762	.80177	.74538	.3416	.2472	.6733	.61153	.79122	.77289	.2938	.2639	.6352	.62524	.78043	.80115	.2482	.2813	.5994	.63877	.76940	.83022	.2045	.2997	.5655	18
43	.59786	.80160	.74583	.3408	.2475	.6726	.61176	.79104	.77335	.2931	.2641	.6346	.62547	.78025	.80163	.2475	.2816	.5988	.63899	.76921	.83071	.2038	.3000	.5650	17
44	.59809	.80143	.74628	.3400	.2478	.6720	.61199	.79087	.77382	.2923	.2644	.6340	.62570	.78007	.80211	.2467	.2819	.5982	.63921	.76903	.83120	.2031	.3003	.5644	16
45	.59832	.80125	.74673	1.3392	.2480	1.6713	.61222	.79069	.77428	1.2915	.2647	1.6334	.62592	.77988	.80258	1.2460	.2822	1.5976	.63944	.76884	.83169	1.2024	.3006	1.5639	15
46	.59856	.80108	.74019	.3383	.2483	.6707	.61245	.79051	.77475	.2907	.2650	.6328	.62615	.77970	.80306	.2452	.2825	.5971	.63966	.76865	.83218	.2016	.3010	.5633	14
47	.59879	.80090	.74064	.3375	.2486	.6700	.61268	.79033	.77521	.2900	.2653	.6322	.62638	.77952	.80354	.2445	.2828	.5965	.63989	.76847	.83267	.2009	.3013	.5628	13
48	.59902	.80073	.74809	.3367	.2488	.6694	.61290	.79015	.77568	.2892	.2656	.6316	.62660	.77934	.80402	.2437	.2831	.5959	.64011	.76828	.83317	.2002	.3016	.5622	12
49	.59926	.80056	.74855	.3359	.2491	.6687	.61314	.78998	.77614	.2884	.2659	.6309	.62683	.77915	.80450	.2430	.2834	.5953	.64033	.76810	.83366	.1995	.3019	.5617	11
50	.59949	.80038	.74900	1.3351	.2494	1.6681	.61337	.78980	.77661	1.2876	.2661	1.6303	.62706	.77897	.80498	1.2423	.2837	1.5947	.64056	.76791	.83415	1.1988	.3022	1.5611	10
51	.59972	.80021	.74946	.3343	.2497	.6674	.61360	.78962	.77708	.2869	.2664	.6297	.62728	.77879	.80546	.2415	.2840	.5942	.64078	.76772	.83465	.1981	.3025	.5606	9
52	.59995	.80003	.74991	.3335	.2499	.6668	.61383	.78944	.77754	.2861	.2667	.6291	.62751	.77861	.80594	.2408	.2843	.5936	.64100	.76754	.83514	.1974	.3029	.5600	8
53	.60019	.79986	.75037	.3327	.2502	.6661	.61405	.78926	.77801	.2853	.2670	.6285	.62774	.77842	.80642	.2400	.2846	.5930	.64123	.76735	.83563	.1967	.3032	.5595	7
54	.60042	.79968	.75082	.3319	.2505	.6655	.61428	.78908	.77848	.2845	.2673	.6279	.62796	.77824	.80690	.2393	.2849	.5924	.64145	.76716	.83613	.1960	.3035	.5590	6
55	.60065	.79951	.75128	1.3311	.2508	1.6648	.61451	.78890	.77895	1.2838	.2676	1.6273	.62819	.77806	.80738	1.2386	.2852	1.5919	.64160	.76698	.83662	1.1953	.3038	1.5584	5
56	.60088	.79933	.75173	.3303	.2510	.6642	.61474	.78873	.77941	.2830	.2679	.6267	.62841	.77788	.80786	.2378	.2855	.5913	.64189	.76679	.83712	.1946	.3041	.5579	4
57	.60112	.79916	.75219	.3294	.2513	.6636	.61497	.78855	.77988	.2822	.2681	.6261	.62864	.77769	.80834	.2371	.2858	.5907	.64212	.76660	.83761	.1939	.3044	.5573	3
58	.60135	.79898	.75264	.3286	.2516	.6629	.61520	.78837	.78035	.2815	.2684	.6255	.62887	.77751	.80882	.2364	.2861	.5901	.64234	.76642	.83811	.1932	.3048	.5568	2
59	.60158	.79881	.75310	.3278	.2519	.6623	.61543	.78819	.78082	.2807	.2687	.6249	.62909	.77733	.80930	.2356	.2864	.5896	.64256	.76623	.83860	.1924	.3051	.5563	1
60	.60181	.79863	.75355	1.3270	.2521	1.6616	.61566	.78801	.78128	1.2799	.2690	1.6243	.62932	.77715	.80978	1.2349	.2867	1.5890	.64279	.76604	.83910	1.1917	.3054	1.5557	0
′	cos	sin	cot	tan	cosec	sec	cos	sin	cot	tan	cosec	sec	cos	sin	cot	tan	cosec	sec	cos	sin	cot	tan	cosec	sec	′
	53°						52°						51°						50°						

(Continued)

Table B18. (Cont.)

′	40° sin	40° cos	40° tan	40° cot	40° sec	40° cosec	41° sin	41° cos	41° tan	41° cot	41° sec	41° cosec
0	.64279	.76604	.83910	1.1917	1.3054	1.5557	.65606	.75471	.86929	1.1504	1.3250	1.5242
1	.64301	.76586	.83959	.1910	.3057	.5552	.65628	.75452	.86980	.1497	.3253	.5237
2	.64323	.76567	.84009	.1903	.3060	.5546	.65650	.75433	.87031	.1490	.3257	.5232
3	.64345	.76548	.84059	.1896	.3064	.5541	.65672	.75414	.87082	.1483	.3260	.5227
4	.64368	.76530	.84108	.1889	.3067	.5536	.65694	.75394	.87133	.1477	.3263	.5222
5	.64390	.76511	.84158	1.1882	.3070	1.5530	.65716	.75375	.87184	1.1470	.3267	1.5217
6	.64412	.76492	.84208	.1875	.3073	.5525	.65037	.75356	.87235	.1463	.3270	.5212
7	.64435	.76473	.84257	.1868	.3076	.5520	.65759	.75337	.87287	.1456	.3274	.5207
8	.64457	.76455	.84307	.1861	.3080	.5514	.65781	.75318	.87338	.1450	.3277	.5202
9	.64479	.76436	.84357	.1854	.3083	.5509	.65803	.75299	.87389	.1443	.3280	.5197
10	.64501	.76417	.84407	1.1847	.3086	1.5503	.65285	.75280	.87441	1.1436	.3284	1.5192
11	.64523	.76398	.84457	.1840	.3089	.5498	.65847	.75261	.87492	.1430	.3287	.5187
12	.64546	.76380	.84506	.1833	.3092	.5493	.65869	.75241	.87543	.1423	.3290	.5182
13	.64568	.76361	.84556	.1826	.3096	.5487	.65891	.75222	.87595	.1416	.3294	.5177
14	.64590	.76342	.84606	.1819	.3099	.5482	.65913	.75203	.87646	.1409	.3297	.5171
15	.64612	.76323	.84656	1.1812	.3102	1.5477	.65934	.75184	.87698	1.1403	.3301	1.5166
16	.64635	.76304	.84706	.1805	.3105	.5471	.65956	.75165	.87749	.1396	.3304	.5161
17	.64657	.76286	.84056	.1798	.3109	.5466	.65978	.75146	.87801	.1389	.3307	.5156
18	.64679	.76267	.84806	.1791	.3112	.5461	.66000	.75126	.87852	.1383	.3311	.5151
19	.64701	.76248	.84856	.1785	.3115	.5456	.66022	.75107	.87904	.1376	.3314	.5146
20	.64723	.76229	.84906	1.1778	.3118	1.5450	.66044	.75088	.87955	1.1369	.3318	1.5141
21	.64745	.76210	.84956	.1771	.3121	.5445	.66066	.75069	.88007	.1363	.3321	.5136
22	.64768	.76191	.85006	.1764	.3125	.5440	.66087	.75049	.88058	.1356	.3324	.5131
23	.64790	.76173	.85056	.1757	.3128	.5434	.66109	.75030	.88110	.1349	.3328	.5126
24	.64812	.76154	.85107	.1750	.3131	.5429	.66131	.75011	.88162	.1343	.3331	.5121
25	.64834	.76135	.85157	1.1743	.3134	1.5424	.66153	.74992	.88213	1.1336	.3335	1.5116
26	.64856	.76116	.85207	.1736	.3138	.5419	.66175	.74973	.88265	.1329	.3338	.5111
27	.64878	.76097	.85257	.1729	.3141	.5413	.66197	.74953	.88317	.1323	.3342	.5106
28	.64900	.76078	.85307	.1722	.3144	.5408	.66218	.74934	.88369	.1316	.3345	.5101
29	.64923	.76059	.85358	.1715	.3148	.5403	.66240	.74915	.88421	.1309	.3348	.5096
30	.64945	.76041	.85408	1.1708	.3151	1.5398	.66262	.74895	.88472	1.1303	.3352	1.5092
31	.64967	.76022	.85458	.1702	.3154	.5392	.66284	.74876	.88524	.1296	.3355	.5087
32	.64989	.76003	.85509	.1695	.3157	.5387	.66305	.74857	.88576	.1290	.3359	.5082
33	.65011	.75984	.85559	.1688	.3161	.5382	.66327	.74838	.88628	.1283	.3362	.5077
34	.65033	.75965	.85609	.1681	.3164	.5377	.66349	.74818	.88680	.1276	.3366	.5072
35	.65055	.75946	.85660	1.1674	.3167	1.5371	.66371	.74799	.88732	1.1270	.3369	1.5067
36	.65077	.75927	.85710	.1667	.3170	.5366	.66393	.74780	.88784	.1263	.3372	.5062
37	.65100	.75908	.85761	.1660	.3174	.5361	.66414	.74760	.88836	.1257	.3376	.5057
38	.65121	.75889	.85811	.1653	.3177	.5356	.66436	.74741	.88888	.1250	.3379	.5052
39	.65144	.75870	.85862	.1647	.3180	.5351	.66458	.74722	.88940	.1243	.3383	.5047
40	.65166	.75851	.85912	1.1640	.3184	1.5345	.66479	.74702	.88992	1.1237	.3386	1.5042
41	.65188	.75832	.85963	.1633	.3187	.5340	.66501	.74683	.89044	.1230	.3390	.5037
42	.65210	.75813	.86013	.1626	.3190	.5335	.66523	.74664	.89097	.1224	.3393	.5032
43	.65232	.75794	.86064	.1619	.3193	.5330	.66545	.74644	.89149	.1217	.3397	.5027
44	.65254	.75775	.86115	.1612	.3197	.5325	.66566	.74625	.89201	.1211	.3400	.5022
45	.65276	.75756	.86165	1.1605	.3200	1.5319	.66588	.74606	.89253	1.1204	1.3404	1.5018
46	.65298	.75737	.86216	.1599	.3203	.5314	.66610	.74586	.89306	.1197	.3407	.5013
47	.65320	.75718	.86267	.1592	.3207	.5309	.66631	.74567	.89358	.1191	.3411	.5008
48	.65342	.75700	.86318	.1585	.3210	.5304	.66653	.74548	.89410	.1184	.3414	.5003
49	.65364	.75680	.86368	.1578	.3213	.5299	.66675	.74528	.89463	.1178	.3418	.4998
50	.65386	.75661	.86419	1.1571	.3217	1.5294	.66697	.74509	.89515	1.1171	.3421	1.4993
51	.65408	.75642	.86470	.1565	.3220	.5289	.66718	.74489	.89567	.1165	.3425	.4988
52	.65430	.75623	.86521	.1558	.3223	.5283	.66740	.74470	.89620	.1158	.3428	.4983
53	.65452	.75604	.86572	.1551	.3227	.5278	.66762	.74450	.89672	.1152	.3432	.4979
54	.65474	.75585	.86623	.1544	.3230	.5273	.66783	.74431	.89725	.1145	.3435	.4974
55	.65496	.75566	.86674	1.1537	.3233	1.5268	.66805	.74412	.89777	1.1139	.3439	1.4969
56	.65518	.75547	.86725	.1531	.3237	.5263	.66826	.74392	.89830	.1132	.3442	.4964
57	.65540	.75528	.86775	.1524	.3240	.5258	.66848	.74373	.89882	.1126	.3446	.4959
58	.65562	.75509	.86826	.1517	.3243	.5253	.66870	.74353	.89935	.1119	.3449	.4954
59	.65584	.75490	.86878	.1510	.3247	.5248	.66891	.74334	.89988	.1113	.3453	.4949
60	.65606	.75401	.86929	1.1504	.3250	1.5242	.66913	.74314	.90040	1.1106	.3456	1.4945
′	49° cos	49° sin	49° cot	49° tan	49° cosec	49° sec	48° cos	48° sin	48° cot	48° tan	48° cosec	48° sec

′	42° sin	42° cos	42° tan	42° cot	42° sec	42° cosec	43° sin	43° cos	43° tan	43° cot	43° sec	43° cosec	′
0	.66913	.74314	.90040	1.1106	1.3456	1.4945	.68200	.73135	.93251	1.0724	1.3673	1.4663	60
1	.66935	.74295	.90093	.1100	.3460	.4940	.68221	.73115	.93306	.0717	.3677	.4658	59
2	.66956	.74276	.90146	.1093	.3463	.4935	.68242	.73096	.93360	.0711	.3681	.4654	58
3	.66978	.74256	.90198	.1086	.3467	.4930	.68264	.73076	.93415	.0705	.3684	.4649	57
4	.66999	.74236	.90251	.1080	.3470	.4925	.68285	.73056	.93469	.0699	.3688	.4644	56
5	.67021	.74217	.90304	1.1074	.3474	1.4921	.68306	.73036	.93524	1.0692	.3692	1.4640	55
6	.67043	.74197	.90357	.1067	.3477	.4916	.68327	.73016	.93578	.0686	.3695	.4635	54
7	.67064	.74178	.90410	.1061	.3481	.4911	.68349	.72996	.93633	.0680	.3699	.4631	53
8	.67086	.74158	.90463	.1054	.3485	.4906	.68370	.72976	.93687	.0674	.3703	.4626	52
9	.67107	.74139	.90515	.1048	.3488	.4901	.68391	.72956	.93742	.0667	.3707	.4622	51
10	.67129	.74119	.90568	1.1041	.3492	1.4897	.68412	.72937	.93797	1.0661	.3710	1.4617	50
11	.67150	.74100	.90621	.1035	.3495	.4892	.68433	.72917	.93851	.0655	.3714	.4613	49
12	.67172	.74080	.90674	.1028	.3499	.4887	.68455	.72897	.93906	.0649	.3718	.4608	48
13	.67194	.74061	.90727	.1022	.3502	.4882	.68476	.72877	.93961	.0643	.3722	.4604	47
14	.67215	.74041	.90780	.1015	.3506	.4877	.68497	.72857	.94016	.0636	.3725	.4599	46
15	.67237	.74022	.90834	1.1009	.3509	1.4873	.68518	.72837	.94071	1.0630	.3729	1.4595	45
16	.67258	.74002	.90887	.1003	.3513	.4868	.68539	.72817	.94125	.0624	.3733	.4590	44
17	.67280	.73983	.90940	.0996	.3517	.4863	.68561	.72797	.94180	.0618	.3737	.4586	43
18	.67301	.73963	.90993	.0990	.3520	.4858	.68582	.72777	.94235	.0612	.3740	.4581	42
19	.67323	.73943	.91046	.0983	.3524	.4854	.68603	.72757	.94290	.0605	.3744	.4577	41
20	.67344	.73924	.91099	1.0977	.3527	1.4849	.68624	.72737	.94345	1.0599	.3748	1.4572	40
21	.67366	.73904	.91153	.0971	.3531	.4844	.68645	.72717	.94400	.0593	.3752	.4568	39
22	.67387	.73885	.91206	.0964	.3534	.4839	.68666	.72697	.94455	.0587	.3756	.4563	38
23	.67409	.73865	.91259	.0958	.3538	.4835	.68688	.72677	.94510	.0581	.3759	.4559	37
24	.67430	.73845	.91312	.0951	.3542	.4830	.68709	.72657	.94565	.0575	.3763	.4554	36
25	.67452	.73826	.91366	1.0945	.3545	1.4825	.68730	.72637	.94620	1.0568	.3767	1.4550	35
26	.67473	.73806	.91419	.0939	.3549	.4821	.68751	.72617	.94675	.0562	.3771	.4545	34
27	.67495	.73787	.91473	.0932	.3552	.4816	.68772	.72597	.94731	.0556	.3774	.4541	33
28	.67516	.73767	.91526	.0926	.3556	.4811	.68793	.72577	.94786	.0550	.3778	.4536	32
29	.67537	.73747	.91580	.0919	.3560	.4806	.68814	.72557	.94841	.0544	.3782	.4532	31
30	.67559	.73728	.91633	1.0913	.3563	1.4802	.68835	.72537	.94896	1.0538	.3786	1.4527	30
31	.67580	.73708	.91687	.0907	.3567	.4797	.68856	.72517	.94952	.0532	.3790	.4523	29
32	.67602	.73688	.91740	.0900	.3571	.4792	.68878	.72497	.95007	.0525	.3794	.4518	28
33	.67623	.73669	.91794	.0894	.3574	.4788	.68899	.72477	.95062	.0519	.3797	.4514	27
34	.67645	.73649	.91847	.0888	.3578	.4783	.68920	.72457	.95118	.0513	.3801	.4510	26
35	.67666	.73629	.91901	1.0881	.3581	1.4778	.68941	.72437	.95173	1.0507	.3805	1.4505	25
36	.67688	.73610	.91955	.0875	.3585	.4774	.68962	.72417	.95229	.0501	.3809	.4501	24
37	.67709	.73590	.92008	.0868	.3589	.4769	.68983	.72397	.95284	.0495	.3813	.4496	23
38	.67730	.73570	.92062	.0862	.3592	.4764	.69004	.72377	.95340	.0489	.3816	.4492	22
39	.67752	.73551	.92116	.0856	.3596	.4760	.69025	.72357	.95395	.0483	.3820	.4487	21
40	.67773	.73531	.92170	1.0849	.3600	1.4755	.69046	.72337	.95451	1.0476	.3824	1.4483	20
41	.67794	.73511	.92223	.0843	.3603	.4750	.69067	.72317	.95506	.0470	.3828	.4479	19
42	.67816	.73491	.92277	.0837	.3607	.4746	.69088	.72297	.95562	.0464	.3832	.4474	18
43	.67837	.73472	.92331	.0830	.3611	.4741	.69109	.72277	.95618	.0458	.3836	.4470	17
44	.67859	.73452	.92385	.0824	.3614	.4736	.69130	.72256	.95673	.0452	.3839	.4465	16
45	.67880	.73432	.92439	1.0818	.3618	1.4732	.69151	.72236	.95729	1.0446	.3843	1.4461	15
46	.67901	.73412	.92493	.0812	.3622	.4727	.69172	.72216	.95785	.0440	.3847	.4457	14
47	.67923	.73393	.92547	.0805	.3625	.4723	.69193	.72196	.95841	.0434	.3851	.4452	13
48	.67944	.73373	.92601	.0799	.3629	.4718	.69214	.72176	.95896	.0428	.3855	.4448	12
49	.67965	.73353	.92655	.0793	.3633	.4713	.69235	.72156	.95952	.0422	.3859	.4443	11
50	.67987	.73333	.92709	1.0786	.3636	1.4709	.69256	.72136	.96008	1.0416	.3863	1.4439	10
51	.68008	.73314	.92763	.0780	.3640	.4704	.69277	.72115	.96064	.0410	.3867	.4435	9
52	.68029	.73294	.92817	.0774	.3644	.4699	.69298	.72095	.96120	.0404	.3870	.4430	8
53	.68051	.73274	.92871	.0767	.3647	.4695	.69319	.72075	.96176	.0397	.3874	.4426	7
54	.68072	.73254	.92926	.0761	.3651	.4690	.69340	.72055	.96232	.0391	.3878	.4422	6
55	.68093	.73234	.92980	1.0755	.3655	1.4686	.69361	.72035	.96288	1.0385	.3882	1.4417	5
56	.68115	.73215	.93034	.0749	.3658	.4681	.69382	.72015	.96344	.0379	.3886	.4413	4
57	.68136	.73195	.93088	.0742	.3662	.4676	.69403	.71994	.96400	.0373	.3890	.4408	3
58	.68157	.73175	.93143	.0736	.3666	.4672	.69424	.71974	.96456	.0367	.3894	.4404	2
59	.68178	.73155	.93197	.0730	.3669	.4667	.69445	.71954	.96513	.0361	.3898	.4400	1
60	.68200	.73135	.93251	1.0724	.3673	1.4663	.69466	.71934	.96569	1.0355	.3902	1.4395	0
′	47° cos	47° sin	47° cot	47° tan	47° cosec	47° sec	46° cos	46° sin	46° cot	46° tan	46° cosec	46° sec	′

44°

′	sin	cos	tan	cot	sec	cosec	′
0	.69466	.71934	.96569	1.0355	1.3902	1.4395	60
1	.69487	.71914	.96625	.0349	.3905	.4391	59
2	.69508	.71893	.96681	.0343	.3909	.4387	58
3	.69528	.71873	.96738	.0337	.3913	.4382	57
4	.69549	.71853	.96794	.0331	.3917	.4378	56
5	.69570	.71833	.96850	1.0325	.3921	1.4374	55
6	.69591	.71813	.96907	.0319	.3925	.4370	54
7	.69612	.71792	.96963	.0313	.3929	.4365	53
8	.69633	.71772	.97020	.0307	.3933	.4361	52
9	.69654	.71752	.97076	.0301	.3937	.4357	51
10	.69675	.71732	.97133	1.0295	.3941	1.4352	50
11	.69696	.71711	.97189	.0289	.3945	.4348	49
12	.69716	.71691	.97246	.0283	.3949	.4344	48
13	.69737	.71671	.97302	.0277	.3953	.4339	47
14	.69758	.71650	.97359	.0271	.3957	.4335	46
15	.69779	.71630	.97416	1.0265	.3960	1.4331	45
16	.69800	.71610	.97472	.0259	.3964	.4327	44
17	.69821	.71589	.97529	.0253	.3968	.4322	43
18	.69841	.71569	.97586	.0247	.3972	.4318	42
19	.69862	.71549	.97643	.0241	.3976	.4314	41
20	.69883	.71529	.97700	1.0235	.3980	1.4310	40
21	.69904	.71508	.97756	.0229	.3984	.4305	39
22	.69925	.71488	.97813	.0223	.3988	.4301	38
23	.69945	.71468	.97870	.0218	.3992	.4297	37
24	.69966	.71447	.97927	.0212	.3996	.4292	36
25	.69987	.71427	.97984	1.0206	.4000	1.4288	35
26	.70008	.71406	.98041	.0200	.4004	.4284	34
27	.70029	.71386	.98098	.0194	.4008	.4280	33
28	.70049	.71366	.98155	.0188	.4012	.4276	32
29	.70070	.71345	.98212	.0182	.4016	.4271	31
30	.70091	.71325	.98270	1.0176	.4020	1.4267	30
31	.70112	.71305	.98327	.0170	.4024	.4263	29
32	.70132	.71284	.98384	.0164	.4028	.4259	28
33	.70153	.71264	.98441	.0158	.4032	.4254	27
34	.70174	.71243	.98499	.0152	.4036	.4250	26
35	.70194	.71223	.98556	1.0146	.4040	1.4246	25
36	.70215	.71203	.98613	.0141	.4044	.4242	24
37	.70236	.71182	.98671	.0135	.4048	.4238	23
38	.70257	.71162	.98728	.0129	.4052	.4233	22
39	.70277	.71141	.98786	.0123	.4056	.4229	21
40	.70298	.71121	.98843	1.0117	.4060	1.4225	20
41	.70319	.71100	.98901	.0111	.4065	.4221	19
42	.70339	.71080	.98958	.0105	.4069	.4217	18
43	.70360	.71059	.99016	.0099	.4073	.4212	17
44	.70381	.71039	.99073	.0093	.4077	.4208	16
45	.70401	.71018	.99131	1.0088	.4081	1.4204	15
46	.70422	.70998	.99189	.0082	.4085	.4200	14
47	.70443	.70977	.99246	.0076	.4089	.4196	13
48	.70463	.70957	.99304	.0070	.4093	.4192	12
49	.70484	.70936	.99362	.0064	.4097	.4188	11
50	.70505	.70916	.99420	1.0058	.4101	1.4183	10
51	.70525	.70895	.99478	.0052	.4105	.4179	9
52	.70546	.70875	.99536	.0047	.4109	.4175	8
53	.70566	.70854	.99593	.0041	.4113	.4171	7
54	.70587	.70834	.99651	.0035	.4117	.4167	6
55	.70608	.70813	.99709	1.0029	.4122	1.4163	5
56	.70628	.70793	.99767	.0023	.4126	.4159	4
57	.70649	.70772	.99826	.0017	.4130	.4154	3
58	.70669	.70752	.99884	.0012	.4134	.4150	2
59	.70690	.70731	.99942	.0006	.4138	.4146	1
60	.70711	.70711	1.00000	1.0000	.4142	1.4142	0
′	cos	sin	cot	tan	cosec	sec	′

45°